Electrochemical Processes in Biological Systems

Electrochemical Processes in Biological Systems

Editor:

Valeria Severino, Ph.D.

ArclerPress

www.arclerpress.com

Electrochemical Processes in Biological Systems

Editor: Valeria Severino, Ph.D.

708 3rd Avenue, 6th Floor
New York
NY 10017
United States of America
www.arclerpress.com

ISBN: 978-1-68094-416-7

Library of Congress Control Number: 2016951081

About the Editor

Valeria Severino, Ph.D.

Valeria Severino obtained her PhD in Biological Processes and Biomolecules from the Second University of Naples in 2009. Her scientific interests concern the structural and functional characterization of proteins and peptides by using biochemical approaches. Fields of expertise comprises food chemistry, cellular biology, natural products and related applications. She published more than 40 papers on international peer-reviewed journals.

Contents

List of Contributors

Gerda Strifler
Institute of Surgical Research, University of Szeged, Szeged, Hungary

Eszter Tuboly
Institute of Surgical Research, University of Szeged, Szeged, Hungary

Edit Szél
Department of Dermatology and Allergology, University of Szeged, Szeged, Hungary

Enikő Kaszonyi
Institute of Surgical Research, University of Szeged, Szeged, Hungary

Chun Cao
Institute of Surgical Research, University of Szeged, Szeged, Hungary

József Kaszaki
Institute of Surgical Research, University of Szeged, Szeged, Hungary

András Mészáros
Institute of Surgical Research, University of Szeged, Szeged, Hungary

Mihály Boros
Institute of Surgical Research, University of Szeged, Szeged, Hungary

Petra Hartmann
Institute of Surgical Research, University of Szeged, Szeged, Hungary

Judit Doczi
Department of Medical Biochemistry, Semmelweis University MTA-SE Laboratory for Neurobiochemistry, Budapest, 1094, Hungary.
MTA-SE Lendület Neurobiochemistry Research Group, Budapest, Hungary.

BeataTorocsik
Department of Medical Biochemistry, Semmelweis University MTA-SE Laboratory for Neurobiochemistry, Budapest, 1094, Hungary.

Andoni Echaniz-Laguna
Department of Neurology, University Hospital, Hautepierre Hospital, 67098 Strasbourg cedex, France.

Bénédicte Mousson de Camaret
Hereditary Diseases Service Metabolism, Biology Center and East Pathology, CHU Lyon, 69677 Bron Cedex, France.

Anatoly Starkov
Feil Family Brain and Mind Research Institute, Weill Cornell Medical College, New York, NY 10065, USA

Natalia Starkova
Icahn School of Medicine at Mount Sinai, Department of Hematology and Medical Oncology, New York, NY 10029, USA

AnikoGál
Institute of Genomic Medicine and Rare Disorders, Semmelweis University, Budapest, 1083, Hungary.

Mária J Molnár
Institute of Genomic Medicine and Rare Disorders, Semmelweis University, Budapest, 1083, Hungary.

Hibiki Kawamata
Feil Family Brain and Mind Research Institute, Weill Cornell Medical College, New York, NY 10065, USA

Giovanni Manfredi
Feil Family Brain and Mind Research Institute, Weill Cornell Medical College, New York, NY 10065, USA

VeraAdam-Vizi
Department of Medical Biochemistry, Semmelweis University MTA-SE Laboratory for Neurobiochemistry, Budapest, 1094, Hungary

ChristosChinopoulos
Department of Medical Biochemistry, Semmelweis University MTA-SE Laboratory for Neurobiochemistry, Budapest, 1094, Hungary
MTA-SE Lendület Neurobiochemistry Research Group, Budapest, Hungary

Juliana S. Ruas
Department of Clinical Pathology, Faculty of Medical Sciences, State University of Campinas (Unicamp), Campinas, SP, Brazil

Edilene S. Siqueira-Santos
Department of Clinical Pathology, Faculty of Medical Sciences, State University of Campinas (Unicamp), Campinas, SP, Brazil

Ignacio Amigo
Department of Biochemistry, Institute of Chemistry, University of São Paulo (USP), São Paulo, SP, Brazil

Erika Rodrigues-Silva
Department of Clinical Pathology, Faculty of Medical Sciences, State University of Campinas (Unicamp), Campinas, SP, Brazil

Alicia J. Kowaltowski
Department of Biochemistry, Institute of Chemistry, University of São Paulo (USP), São Paulo, SP, Brazil

Roger F. Castilho
Department of Clinical Pathology, Faculty of Medical Sciences, State University of Campinas (Unicamp), Campinas, SP, Brazil

Shuqiang Yuan
Sun Yat-sen University Cancer Center; State Key Laboratory of Oncology in South China; Collaborative Innovation Center for Cancer Medicine, 651 E Dongfeng Road, Guangzhou, Guangdong 510060, China

Yunxin Lu
Sun Yat-sen University Cancer Center; State Key Laboratory of Oncology in South China; Collaborative Innovation Center for Cancer Medicine, 651 E Dongfeng Road, Guangzhou, Guangdong 510060, China

Jing Yang
Sun Yat-sen University Cancer Center; State Key Laboratory of Oncology in South China; Collaborative Innovation Center for Cancer Medicine, 651 E Dongfeng Road, Guangzhou, Guangdong 510060, China

Gang Chen
Department of Translational Molecular Pathology, The University of Texas MD Anderson Cancer Center, Houston, TX 77054, USA

Sangbae Kim
Department of Systems Biology, The University of Texas MD Anderson Cancer Center, Houston, TX 77054, USA

Li Feng
Department of Translational Molecular Pathology, The University of Texas MD Anderson Cancer Center, Houston, TX 77054, USA

Marcia Ogasawara
Department of Translational Molecular Pathology, The University of Texas MD Anderson Cancer Center, Houston, TX 77054, USA

Naima Hammoudi
Department of Translational Molecular Pathology, The University of Texas MD Anderson Cancer Center, Houston, TX 77054, USA

Weiqin Lu
Department of Translational Molecular Pathology, The University of Texas MD Anderson Cancer Center, Houston, TX 77054, USA

Hui Zhang
Department of Translational Molecular Pathology, The University of Texas MD Anderson Cancer Center, Houston, TX 77054, USA

Jinyun Liu
Department of Translational Molecular Pathology, The University of Texas MD Anderson Cancer Center, Houston, TX 77054, USA

Howard Colman
Department of Neuro-Oncology, University of Utah, Salt Lake City, UT, USA

Ju-Seog Lee
Department of Systems Biology, The University of Texas MD Anderson Cancer Center, Houston, TX 77054, USA

Xiao-Nan Li
Laboratory of Molecular Neuro-oncology, Texas Children's Cancer Center, Baylor College of Medicine, Houston, TX 77030, USA

Rui-hua Xu
Sun Yat-sen University Cancer Center; State Key Laboratory of Oncology in South China; Collaborative Innovation Center for Cancer Medicine, 651 E Dongfeng Road, Guangzhou, Guangdong 510060, China

Peng Huang
Sun Yat-sen University Cancer Center; State Key Laboratory of Oncology in South China; Collaborative Innovation Center for Cancer Medicine, 651 E Dongfeng Road, Guangzhou, Guangdong 510060, China
Department of Translational Molecular Pathology, The University of Texas MD Anderson Cancer Center, Houston, TX 77054, USA

Feng Wang
Sun Yat-sen University Cancer Center; State Key Laboratory of Oncology in South China; Collaborative Innovation Center for Cancer Medicine, 651 E Dongfeng Road, Guangzhou, Guangdong 510060, China

Maria Peiris-Pagès
The Breast Cancer Now Research Unit, Institute of Cancer Sciences, CRUK Manchester Institute, Paterson Building, University of Manchester, Manchester, UK
The Manchester Centre for Cellular Metabolism (MCCM), Institute of Cancer Sciences, CRUK Manchester Institute, Paterson Building, University of Manchester, Manchester, UK

Ubaldo E. Martinez-Outschoorn
The Sidney Kimmel Cancer Center, Philadelphia, PA, USA

Richard G. Pestell
The Sidney Kimmel Cancer Center, Philadelphia, PA, USA

Federica Sotgia
The Breast Cancer Now Research Unit, Institute of Cancer Sciences, CRUK Manchester Institute, Paterson Building, University of Manchester, Manchester, UK
The Manchester Centre for Cellular Metabolism (MCCM), Institute of Cancer Sciences, CRUK Manchester Institute, Paterson Building, University of Manchester, Manchester, UK

Michael P. Lisanti
The Breast Cancer Now Research Unit, Institute of Cancer Sciences, CRUK Manchester Institute, Paterson Building, University of Manchester, Manchester, UK
The Manchester Centre for Cellular Metabolism (MCCM), Institute of Cancer Sciences, CRUK Manchester Institute, Paterson Building, University of Manchester, Manchester, UK

Isaac G. Onyango
Gencia Biotechnology, 706 B Forest St, Charlottesville, VA 22903, USA

Jameel Dennis
Gencia Biotechnology, 706 B Forest St, Charlottesville, VA 22903, USA

Shaharyah M. Khan
Gencia Biotechnology, 706 B Forest St, Charlottesville, VA 22903, USA

Hye Jeong Yang
Food Functional Research Division, Korean Food Research Institutes, Sungnam, Korea

Min Jung Kim
Food Functional Research Division, Korean Food Research Institutes, Sungnam, Korea

Dae Young Kwon
Food Functional Research Division, Korean Food Research Institutes, Sungnam, Korea

Bo Reum Moon
Department of Food and Nutrition, Obesity/Diabetes Center, Hoseo University, Asan, Korea

A. Reum Kim
Department of Food and Nutrition, Obesity/Diabetes Center, Hoseo University, Asan, Korea

Suna Kang
Department of Food and Nutrition, Obesity/Diabetes Center, Hoseo University, Asan, Korea

Sunmin Park
Department of Food and Nutrition, Obesity/Diabetes Center, Hoseo University, Asan, Korea
Department of Food and Nutrition, Hoseo University, 165 Sechul-Ri, BaeBang-Yup Asan-Si, Asan, ChungNam-Do 336-795, South Korea.

Laura-Roxana Stingaciu
JCNS1, outstation at SNS, Oak Ridge National Laboratory, Oak Ridge, TN 37831, USA

HughO'Neill
NScD, Oak Ridge National Laboratory, Oak Ridge, TN 37831, USA

Michelle Liberton
Department of Biology, Washington University, St. Louis, MO 63130, USA

Volker S. Urban
NScD, Oak Ridge National Laboratory, Oak Ridge, TN 37831, USA

Himadri B. Pakrasi
Department of Biology, Washington University, St. Louis, MO 63130, USA

MichaelOhl
JCNS1, outstation at SNS, Oak Ridge National Laboratory, Oak Ridge, TN 37831, USA

Alvar R.Garrigues
Department of Physics, University of Central Florida, Orlando, Florida 32816 – USA

LiYuan
Department of Chemistry, National University of Singapore 3 Science Drive 3, 117543 Singapore

LejiaWang
Department of Chemistry, National University of Singapore 3 Science Drive 3, 117543 Singapore

Eduardo R. Mucciolo
Department of Physics, University of Central Florida, Orlando, Florida 32816 – USA

DamienThompon
Department of Physics and Energy, University of Limerick, Ireland
Materials and Surface Science Institute, University of Limerick, Ireland

Enrique del Barco
Department of Physics, University of Central Florida, Orlando, Florida 32816 – USA

ChristianA. Nijhuis
Department of Chemistry, National University of Singapore 3 Science Drive 3, 117543 Singapore
Centre for Advanced 2D Materials, National University of Singapore, 6 Science Drive 2, 117546 Singapore

Eda Mehmeti
Institute of Chemistry, Analytical Chemistry, Karl-Franzens University, 8010 Graz, Austria

Dalibor M. Stanković
Department of Analytical Chemistry, Innovation Center of the Faculty of Chemistry, University of Belgrade, Studentski trg 12-16, Belgrade 11000, Serbia

Sudkate Chaiyo
Electrochemistry and Optical Spectroscopy Research Unit, Department of Chemistry, Faculty of Science, Chulalongkorn University, 254 Phayathai Road, Pathumwan, Bangkok 10330, Thailand

Ľubomir Švorc
Institute of Analytical Chemistry, Faculty of Chemical and Food Technology, Slovak University of Technology in Bratislava, Radlinského 9, SK-812 37 Bratislava, Slovak Republic

Kurt Kalcher
Institute of Chemistry, Analytical Chemistry, Karl-Franzens University, 8010 Graz, Austria

I Ibrahim
Department of Chemistry, Faculty of Science, Universiti Putra Malaysia, 43400 UPM, Serdang, Selangor, Malaysia

H. N Lim
Department of Chemistry, Faculty of Science, Universiti Putra Malaysia, 43400 UPM, Serdang, Selangor, Malaysia
Functional Device Laboratory, Institute of Advanced Technology, Universiti Putra Malaysia, 43400 UPM, Serdang, Selangor, Malaysia

N. M Huang
Low Dimensional Materials Research Centre, Department of Physics, Faculty of Science, University of Malaya, Kuala Lumpur, 50603, Malaysia

A Pandikumar
Low Dimensional Materials Research Centre, Department of Physics, Faculty of Science, University of Malaya, Kuala Lumpur, 50603, Malaysia

Taek Lee
Department of Chemical and Biomolecular Engineering, Sogang University, 35 Baekbeom-ro (Sinsu-dong), Mapo-gu, Seoul 121-742, Korea
Research Center for Integrated Biotechnology, Sogang University, 35 Baekbeom-ro (Sinsu-dong), Mapo-gu, Seoul 121-742, Korea

Tae-Hyung Kim
School of Integrative Engineering, Chung-Ang University, Heukseok-dong, Dongjak-gu, Seoul 156-756, Korea

Jinho Yoon
Department of Chemical and Biomolecular Engineering, Sogang University, 35 Baekbeom-ro (Sinsu-dong), Mapo-gu, Seoul 121-742, Korea

Yong-Ho Chung
Department of Chemical Engineering, Hoseo University, 20, Hoseo-ro 79beon-gil, Baebang-Eup, Asan City, Chungnam 336-795, Korea

Ji Young Lee
Department of Chemical and Biomolecular Engineering, Sogang University, 35 Baekbeom-ro (Sinsu-dong), Mapo-gu, Seoul 121-742, Korea
Research Center for Integrated Biotechnology, Sogang University, 35 Baekbeom-ro (Sinsu-dong), Mapo-gu, Seoul 121-742, Korea

Jeong-Woo Choi
Department of Chemical and Biomolecular Engineering, Sogang University, 35 Baekbeom-ro (Sinsu-dong), Mapo-gu, Seoul 121-742, Korea
Research Center for Integrated Biotechnology, Sogang University, 35 Baekbeom-ro (Sinsu-dong), Mapo-gu, Seoul 121-742, Korea

Membranes Yuanjing Lin
Department of Electronic and Computer Engineering, The Hong Kong University of Science and Technology, Clear Water Bay, Kowloon, Hong Kong SAR, China

Qingfeng Lin
Department of Electronic and Computer Engineering, The Hong Kong University of Science and Technology, Clear Water Bay, Kowloon, Hong Kong SAR, China

Xue Liu
Department of Electronic and Computer Engineering, The Hong Kong University of Science and Technology, Clear Water Bay, Kowloon, Hong Kong SAR, China

Yuan Gao
Department of Electronic and Computer Engineering, The Hong Kong University of Science and Technology, Clear Water Bay, Kowloon, Hong Kong SAR, China

Jin He
Peking University Shenzhen SOC Key Laboratory, PKU-HKUST Shenzhen-Hong Kong Institution, Shenzhen 518051, China

Wenli Wang
College of Textile and Clothing Engineering, Soochow University, Suzhou 215021, China
National Engineering Laboratory for Modern Silk, Suzhou 215123, China.

Zhiyong Fan
Department of Electronic and Computer Engineering, The Hong Kong University of Science and Technology, Clear Water Bay, Kowloon, Hong Kong SAR, China

Phuc T. Ha
The Gene and Linda Voiland School of Chemical Engineering and Bioengineering, Washington State University, Pullman, WA, USA

Ryan S. Renslow
Environmental Molecular Sciences Laboratory, Pacific Northwest National Laboratory, Richland, WA, USA

Erhan Atci
The Gene and Linda Voiland School of Chemical Engineering and Bioengineering, Washington State University, Pullman, WA, USA

Patrick N. Reardon
Environmental Molecular Sciences Laboratory, Pacific Northwest National Laboratory, Richland, WA, USA

Stephen R. Lindemann
Biological Sciences Division, Pacific Northwest National Laboratory, Richland, WA, USA

James K. Fredrickson
Biological Sciences Division, Pacific Northwest National Laboratory, Richland, WA, USA

Douglas R. Call
Paul G. Allen School for Global Animal Health, Washington State University, Pullman, WA, USA

Haluk Beyenal
The Gene and Linda Voiland School of Chemical Engineering and Bioengineering, Washington State University, Pullman, WA, USA

Xiangmin Miao
Department of Chemistry, Hong Kong Baptist University, Kowloon Tong, Hong Kong, China

ChaoYang
State Key Laboratory of Quality Research in Chinese Medicine, Institute of Chinese Medical Sciences, University of Macau, Macao, China.

Chung-Hang Leung
State Key Laboratory of Quality Research in Chinese Medicine, Institute of Chinese Medical Sciences, University of Macau, Macao, China.

Dik-Lung Ma
Department of Chemistry, Hong Kong Baptist University, Kowloon Tong, Hong Kong, China

AlanaA. Zülke
Institute of Chemistry of São Carlos, University of São Paulo, 13560-970, São Carlos, SP, Brazil

HamiltonVarela
Institute of Chemistry of São Carlos, University of São Paulo, 13560-970, São Carlos, SP, Brazil

Peter Fromherz
Max-Planck-Institute for Biochemistry, Am Klopferspitz 18, Martinsried-München 82152, Germany

Preface

An electrochemical process is a chemical reaction that either causes or is caused by the movement of electrical current. These processes are a type of oxidation-reduction reaction in which one atom or molecule loses an electron to another atom or molecule. In electrochemical reactions, the atoms or molecules in the reaction are relatively far apart from each other compared to other reactions, forcing the electrons being transferred to travel a greater distance and thereby produce an electrical current. Many natural phenomena are based on electrochemical processes, such as the corrosion of metals, the ability of some sea creatures to generate electrical fields, and the workings of the nervous systems of humans and other animals. They also play an important role in modern technology, most prominently in the storage of electrical power in batteries, and the electrochemical process called electrolysis is important in modern industry. Electrochemical processes naturally occur every day, every moment. The energy necessary for all basic and vital functions is the result of an electrochemical process, as the energy flow in all living systems. Moreover, electrochemistry is currently a fast growing research field, being involved in a plethora of different fields of application, including medicine, environment, food science, and nanotechnology. In particular, electrochemical processes occurring in biological systems are continuously characterized in order to deeply understand mechanisms and regulation of such processes.

Electrolysis is an electrochemical process in which electrical current is used to trigger chemical reactions in a substance containing free ions, called an electrolyte. The electrolyte is either melted or dissolved in a solvent, and two electrodes, called the anode and cathode, are immersed in it. When an electrical potential is applied between the electrodes, electricity begins to flow between them, and each electrode begins attracting ions with the opposite of its own charge. The ions gain or lose electrons to the electrodes, causing oxidation of molecules near the anode and reduction of those near the cathode. Electrolysis is used in many areas of industrial processes, including metallurgy, the production of chemicals such as potassium chlorate and (KClO3) trifluoroacetic acid (C2HF3O2), and the extraction of highly reactive elements that are not found in their elemental form in nature, such as sodium and magnesium. The present book collects recent development both in terms of knowledge and applications. In particular, the first section comprises advanced studies on principal electrochemical processes occurring in cells, with a focus on the mitochondrion, the power station of the cell. The relation of mitochondrial electrochemical processes and some pathologies as cancer and Alzheimer is also discussed. Starting from the dynamics of thylakoid membranes in cyanobacterial, the second section of this book reports some interesting applications of electrochemical processes implemented in biological systems, such as the construction of specific single-molecule junctions, sensors, biosensors, and electrodes.

Editor
Valeria Severino, Ph.D.

INTRODUCTION

Electrochemistry is a branch within the field of chemistry which involves the intersection between chemical reactions and electrical currents. Some chemical reactions can be catalyzed by the presence of an electrical current, and conversely, it is possible to generate electricity through the process of a chemical reaction. While this pursuit may sound esoteric, chances are very high that you are benefiting from electrochemistry at this very moment, or that you will be at some point today, because it is the underlying process behind a wide range of things, from chemical signaling in your own body to the operation of a car battery.

Modern electrochemistry has diverged significantly from the 18th century roots of this field of study. In addition to inspiring a great deal of pure research, electrochemistry is used in a wide range of industrial processes, and in technology utilized in numerous settings. Earlier researchers were primarily interested in explaining phenomena they didn't understand, while modern researchers are interested in finding new applications for electrochemistry, and in understanding complex electrochemical reactions. They are also interested in understanding electrochemical reactions on a very small scale and basic level, now that the technology for precise observations of this type is available.

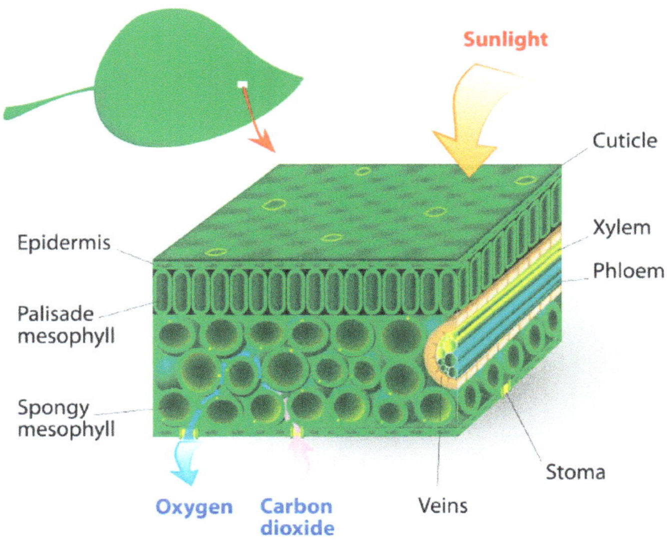

All living organisms use electrochemistry to one degree or another, from the electric eel to the humble houseplant. Living organisms have noticeable electric fields which are generated by chemical reactions in their bodies, and electrochemical reactions are involved in a number of biological processes. For this reason, some biologists are interested in electrochemistry, as are other people who work in the natural world or are interested in the natural environment.

One of the underlying concepts in electrochemistry is the reduction/oxidation or redox reaction, which describes a situation in which electrons are gained or lost. While a small scale reaction will not generate usable energy, it involves the very same electrons which move through the wiring in a home, and these reactions can be used to generate a usable electrical current. Processes like photosynthesis and respiration involve redox reactions, making them electrochemical in nature.

Electrochemistry is also used in scientific laboratories, for processing and analyzing a range of materials. It is also used in processes such as electroplating, in which the property of electrode position is harnessed, and in the operation of batteries, which utilize a chemical reaction to generate electrical energy. Another example of a natural electrochemical reaction is corrosion, especially iron oxidation, which is better known as "rust" among lay people.

Electrochemical Processes

An electrochemical process is a chemical reaction that either causes or is caused by the movement of electrical current. These processes are a type of oxidation-reduction reaction in which one atom or molecule loses an electron to another atom or molecule. In electrochemical reactions, the atoms or molecules in the reaction are relatively far apart from each other compared to other reactions, forcing the electrons being transferred to travel a greater distance and thereby produce an electrical current. Many natural phenomena are based on electrochemical processes, such as the corrosion of metals, the ability of some sea creatures to generate electrical fields, and the workings of the nervous systems of humans and other animals. They also play an important role in modern technology, most prominently in the storage of electrical power in batteries, and the electrochemical process called electrolysis is important in modern industry.

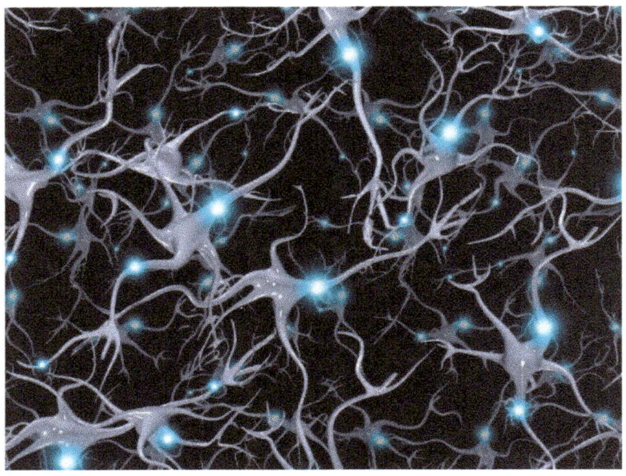

The activities of nervous systems, from the simple reactions and instinctive behaviors found in even primitive animals to the complex learning and reasoning abilities of human beings, depend on electrochemical processes. Neurons use electrochemical processes to transmit information through the nervous system, allowing the nervous system to communicate with itself and with the rest of the body. To send a signal, chemical processes in the neuron generate an electrical impulse that is sent through an elongated structure called an axon until it reaches the synapse, the point of contact between the neuron and neighboring cells. At the synapse, the electricity causes the release of chemicals called neurotransmitters, which cross the synapse to the cell being signaled. The neurotransmitters then chemically bond with structures called receptors on the target cell, setting off further biochemical processes within it.

The ability of fish such as electric eels, stargazers, and torpedo rays to produce electric fields is the result of an electrochemical process. Electric fish possess specialized cells called electrocytes. Transport proteins bind with positive potassium and sodium ions in the cell and carry them away, building up an electric charge in the cell. When this electricity is needed, a part of the nervous system called the medullary command nucleus sends an electric impulse to other nerve cells, which triggers the release of the neurotransmitter acetylcholine. The neurotransmitter bonds with the acetylcholine receptors of the electrocytes, which triggers the release of the electrocytes' charge.

Electric batteries use electrochemical processes to store and release electricity. Chemical reactions within the electric cells making up the battery create a difference in charge between the two halves of each cell, producing electrical current. Rechargeable batteries produce electricity with chemical reactions that are reversible, and so can be returned to their original chemical configuration if electricity is applied from an outside source. The reactions in nonrechargable batteries do not have this quality, though they usually produce more electric power than a rechargeable battery can provide in a single charge.

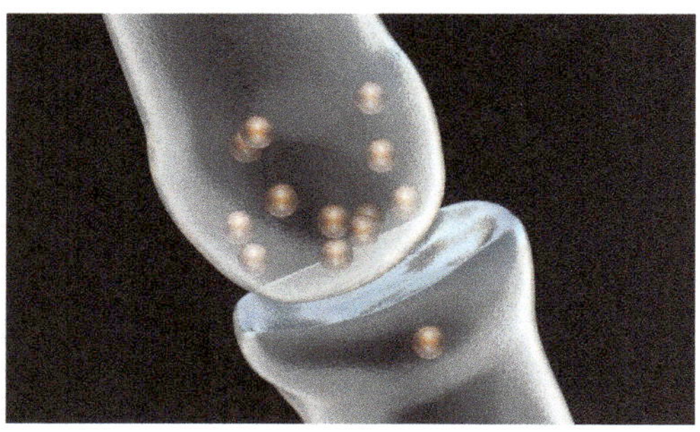

A variety of different chemical reactions are used in batteries. Nickel-cadmium batteries, which are commonly used in lights and household appliances, are based on separate reactions of cadmium and nickel with an alkaline, usually a solution of potassium hydroxide (KOH), and water. Nickel-metal hydride batteries are similar, but replace the cadmium with an intermetallic compound made from manganese, aluminum, or cobalt mixed with rare earth metals such as praseodymium, lanthanum, and cerium. Lithium batteries can use a variety of reactions involving lithium compounds, with the most common type using manganese dioxide (MnO_2) and a solution of lithium perchlorate ($LiClO_4$), dimethoxymethane ($C_4H_{10}O_2$), and propylene carbonate ($C_4H_6O_3$).

Electrolysis is an electrochemical process in which electrical current is used to trigger chemical reactions in a substance containing free ions, called an electrolyte. The electrolyte is either melted or dissolved in a solvent, and two electrodes, called the anode and cathode, are immersed in it. When an electrical potential is applied between the electrodes, electricity begins to flow between them, and each electrode begins attracting ions with the opposite of its own charge. The ions gain or lose electrons to the electrodes, causing oxidation of molecules near the anode and reduction of those near the cathode. Electrolysis is used in many areas of industrial processes, including metallurgy, the production of chemicals such as potassium chlorate and ($KClO_3$) trifluoroacetic acid ($C_2HF_3O_2$), and the extraction of highly reactive elements that are not found in their elemental form in nature, such as sodium and magnesium.

Chemical Process

A chemical process is any laboratory or industrial activity where products are chemically reacted with each other to form new products. The reactions change the structure of the product molecules, rather than adding them together in a mixing process. Mixtures can result from these activities, but a chemical change has occurred in one or more of the reacted materials.

Process designs often involve building larger molecules from smaller molecules found in raw materials. Some reactions can break larger molecules and rearrange them in different ways to produce a final product. Multiple reactions may be required to produce a desired molecule, and can include additional raw materials or changes in temperature or pressure to cause different products to form. A process may be a single reaction step, or an industrial operation creating finished chemical products on a large scale.

Chemicals are first produced in laboratory-scale equipment, both to confirm the necessary steps and to fine-tune the reaction temperatures and pressures often required for the reactions to occur. Many industrial chemical processes occur at high temperatures and/or pressures, but less often, the reactions can occur at very low pressures or in a vacuum, or at very low temperatures.

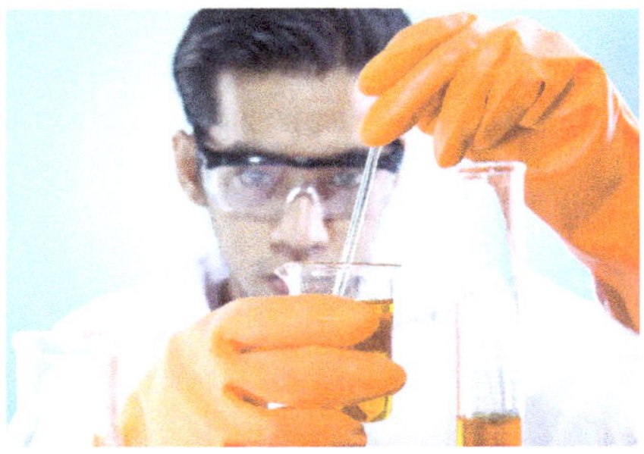

Catalysts may be used to aid the chemical reaction when designing a chemical process. These are products that speed up or accelerate the reaction, make more of the desired final product versus other byproducts, or in some cases allow the reactions to occur at lower temperatures or pressures to reduce costs. High-pressure or temperature reaction vessels can be expensive to manufacture, so a catalyst can be important when designing chemical plants.

Testing in pilot plant, or small scale, operations may be the next step in process design. Reaction times, temperatures and pressures must be reviewed and optimized to produce the correct chemicals and reduce waste materials. Products must be separated and purified in additional steps, and additional reactions may be needed before a final product is made. Waste materials need to be determined, and finding ways to neutralize or dispose them to minimize environmental issues. Pilot plant testing is important, because some laboratory tests do not operate correctly when run in larger equipment, and changes may be needed in the design before a full-scale plant is built.

A full-scale industrial chemical process can take years to build at great expense, so care is taken to verify all equipment and process control requirements before construction begins. Chemical plants require raw materials supply by train or truck, using tanks for liquids, cylinders or high-pressure vessels for gases, and warehouse space for bags or drums. Reaction tanks or towers are needed for producing the raw or unpurified finished products, and all piping, electrical and instrumentation must be designed to connect all parts of the process. Chemical process designs include methods to separate and recycle unused raw materials back to the front of the process to reduce waste and operating costs.

When the finished products have been purified and separated from the chemical process stream, storage may be needed until they are shipped to customers. This can include additional tanks for liquids, cylinders for gases, and warehousing for drums or bags, and some products may have specific temperatures or storage conditions to maintain quality and avoid safety issues. A manufacturer may have trucks or tank cars available to ship bulk quantities to customers, and these fleets may be owned or leased.

Each reaction in a chemical process is one step, and many may be combined in a complex chain of events called a chemical plant. Chemical process engineers and plant designers carefully consider the interaction of all steps to create an efficient operation. Product quality, energy efficiency and waste minimization are all carefully considered in process design, and are critical to produce a profitable chemical operation.

Electrolytic Cells

Voltaic cells are driven by a spontaneous chemical reaction that produces an electric current through an outside circuit. These cells are important because they are the basis for the batteries that fuel modern society. But they are not the only kind of electrochemical cell. The reverse reaction in each case is non-spontaneous and requires electrical energy to occur.

The general form of the reaction can be written as:

$$\text{Reactants} \underset{\longleftarrow \textit{Non spontaneous}}{\overset{\textit{Spontaneous} \longrightarrow}{\rightleftharpoons}} \text{Products} + \text{Electrical Energy}$$

It is possible to construct a cell that does work on a chemical system by driving an electric current through the system. These cells are called electrolytic cells, and operate through electrolysis. Electrolysis is used to drive an oxidation-reduction reaction in a direction in which it does not occur spontaneously by driving an electric current through the system while doing work on the chemical system itself, and therefore is non-spontaneous.

Electrolytic cells, like galvanic cells, are composed of two half-cells--one is a reduction half-cell, the other is an oxidation half-cell. The direction of electron flow in electrolytic cells, however, may be reversed from the direction of spontaneous electron flow in galvanic cells, but the definition of both cathode and anode remain the same, where reduction takes place at the cathode and oxidation occurs at the anode. Because the directions of both half-reactions have been reversed, the sign, but not the magnitude, of the cell potential has been reversed.

Differences between Galvanic and an Electrolytic cells

Electrolytic cells are very similar to voltaic (galvanic) cells in the sense that both require a salt bridge, both have a cathode and anode side, and both have a consistent flow of electrons from the anode to the cathode. However, there are also striking differences between the two cells. The main differences are outlined below:

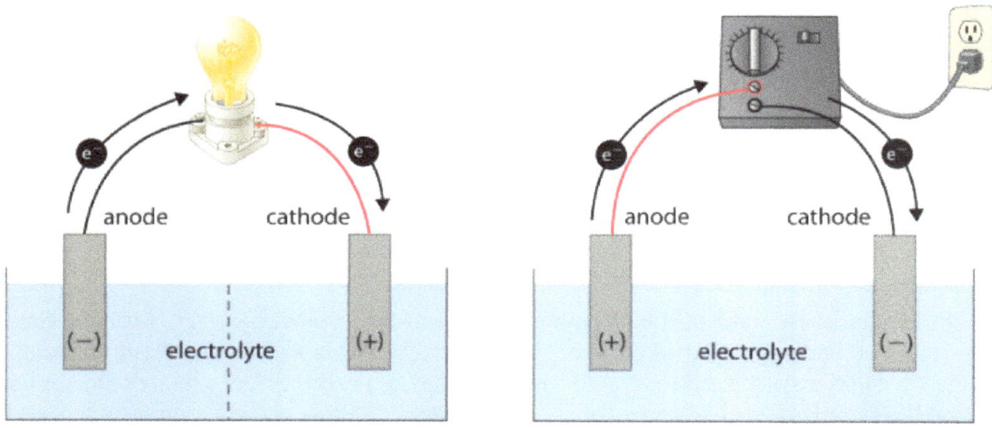

GALVANIC CELL

Energy released by spontaneous redox reaction is converted to electrical energy.

Oxidation half-reaction:
$Y \rightarrow Y^+ + e^-$

Reduction half-reaction:
$Z + e^- \rightarrow Z^-$

Overall cell reaction:
$Y + Z \rightarrow Y^+ + Z^-$ $(G < 0)$

ELECTROLYTIC CELL

Electrical energy is used to drive nonspontaneous redox reaction.

Oxidation half-reaction:
$Z^- \rightarrow Z + e^-$

Reduction half-reaction:
$Y^+ + e^- \rightarrow Y$

Overall cell reaction:
$Y^+ + Z^- \rightarrow Y + Z$ $(G > 0)$

Figure 1: Electrochemical Cells. A galvanic cell (left) transforms the energy released by a spontaneous redox reaction into electrical energy that can be used to perform work. The oxidative and reductive half-reactions usually occur in separate compartments that are connected by an external electrical circuit; in addition, a second connection that allows ions to flow between the compartments (shown here as a vertical dashed line to represent a porous barrier) is necessary to maintain electrical neutrality. The potential difference between the electrodes (voltage) causes electrons to flow from the reductant to the oxidant through the external circuit, generating an electric current. In an electrolytic cell (right), an external source of electrical energy is used to generate a potential difference between the electrodes that forces electrons to flow, driving a nonspontaneous redox reaction; only a single compartment is employed in most applications. In both kinds of electrochemical cells, the anode is the electrode at which the oxidation half-reaction occurs, and the cathode is the electrode at which the reduction half-reaction occurs.

Table 1: Properties of Galvanic and Electrochemical Cells

Electrochemical cell (Galvanic Cell)	Electrolytic cell
A Galvanic cell converts chemical energy into electrical energy.	An electrolytic cell converts electrical energy into chemical energy.
Here, the redox reaction is spontaneous and is responsible for the production of electrical energy.	The redox reaction is not spontaneous and electrical energy has to be supplied to initiate the reaction.
The two half-cells are set up in different containers, being connected through the salt bridge or porous partition.	Both the electrodes are placed in a same container in the solution of molten electrolyte.
Here the anode is negative and cathode is the positive electrode. The reaction at the anode is oxidation and that at the cathode is reduction.	Here, the anode is positive and cathode is the negative electrode. The reaction at the anode is oxidation and that at the cathode is reduction.
The electrons are supplied by the species getting oxidized. They move from anode to the cathode in the external circuit.	The external battery supplies the electrons. They enter through the cathode and come out through the anode.

To explain what happens in an electrolytic cell let us examine the decomposition of molten sodium chloride into sodium metal and chlorine gas. The reaction is written below.

---------> Non spontaneous (electrolytic cell)

$$\underset{(l)}{2\,NaCl} \rightleftharpoons 2\,Na\,(s) + Cl_2\,(g)$$

<--------- Spontaneous (electrochemical cell)

If molten $N_aCl_{(l)}$ is placed into the container and inert electrodes of $C_{(s)}$ are inserted, attached to the positive and negative terminals of a battery, an electrolytic reaction will occur.

1. Electrons from the negative terminal travel to the cathode and are used to reduce sodium ions into sodium atoms. The sodium will plate onto the cathode as it forms. The sodium ion are migrating towards the cathode.

$$Na^+ + e^- \rightleftharpoons Na_{(s)}$$

2. The negative Chlorine ions migrate towards the anode and release electrons as they oxidize to form chlorine atoms. The chlorine atoms will combine together to form chlorine gas which will bubble away.

$$2Cl^- \rightleftharpoons Cl_{2(g)} + 2e^-$$

3. Note that the site of oxidation is still the anode and the site of reduction is still the cathode, but the charge on these two electrodes are reversed. Anode is now positive charged and the cathode has a negative charged.
4. The conditions under which the electrolyte cell operates are very important. The substance that is the strongest reducing agent (the substance with the highest standard cell potential value in the table) will undergo oxidation. The substance that is the strongest oxidizing agent will be reduced. If an aqueous solution of sodium chloride were used in the above system, hydrogen would undergo reduction instead of sodium, because it is a stronger oxidizing agent that sodium.

Predicting Electrolysis Reaction

There are four primary factors that determine whether or not electrolysis will take place even if the external voltage exceeds the calculated amount:

1. An over potential or voltage excess is sometimes needed to overcome interactions at the electrode surface. This case happens more frequently with gases. E.g. H_2 (g) requires a 1.5 V over potential, while Pt (s) requires 0 V over potential
2. There might be more than one electrode reaction that occurs meaning that there may be more than one half-reaction leaving two or more possibilities for the cell reaction.

3. The reactants may be in nonstandard conditions which means that the voltage for the half cells may be less or more than the standard condition amount. For Example:

- Concentration of chloride ion = 5.5M not the unit activity of 1M. This means that the reduction of chloride = 1.31V not 1.36V
- The standard condition is to have a pH of 4 in the anode half-cell but sometimes during nonstandard states, the pH may be higher or lower changing the voltage.

4. An ***inert*** electrode's ability to electrolysis depend on the reactants in the electrolyte solution while an ***active*** electrode can run on its own to perform the oxidation or reduction half reaction.

If all four of these factors are accounted for, we can successfully predict electrode half reactions and overall reactions in electrolysis.

CHAPTER 1

INHALED METHANE LIMITS THE MITOCHONDRIAL ELECTRON TRANSPORT CHAIN DYSFUNCTION DURING EXPERIMENTAL LIVER ISCHEMIA-REPERFUSION INJURY

Gerda Strifler, Eszter Tuboly, Edit Szél, Enikő Kaszonyi , Chun Cao, József Kaszaki , András Mészáros, Mihály Boros, Petra Hartmann

INTRODUCTION

The mitochondria integrate the oxidation of substrates with the reduction of molecular oxygen (O_2) in the aerobic cell. A major threat to this equilibrium is hypoxia, when the lack of electron acceptor O_2 leads to less ATP generation, and the accumulation of metabolic by-products. Re-establishment of the O_2 flux is necessary but precarious, as the disturbed intracellular redox chemistry may lead to the formation of reactive oxygen species (ROS) with disturbances of the osmotic, ion and electric balances, structural membrane abnormalities and the activation of pro-death pathways.

In this system the availability of O_2 is a vital issue, but it has become clear that other gaseous components of the cellular atmosphere are also of importance to the mitochondrial biology. Methane (CH_4), a ubiquitous, small molecule, is a non-toxic, simple asphyxiant that can displace O_2 in a restricted area. There is good reason to assume that this feature can influence the biology of eukaryote cells, though the role of CH_4 in the mammalian physiology is largely unmapped and the effect of CH_4 on mitochondrial homeostasis has never been investigated.

Citation: Strifler G, Tuboly E, Szél E, Kaszonyi E, Cao C, Kaszaki J, et al. (2016) Inhaled Methane Limits the Mitochondrial Electron Transport Chain Dysfunction during Experimental Liver Ischemia-Reperfusion Injury. PLoS ONE 11(1): e0146363. doi:10.1371/journal.pone.0146363.

Mammalian methanogenesis is widely regarded as an indicator of the gastrointestinal (GI) carbohydrate fermentation by the anaerobic flora. Once generated by microbes or released by a nonbacterial process, CH_4 is generally considered to be biologically inactive. However, some data do hint at an association with the small bowel motility regulation, as CH_4 produced in the GI tract is usually associated with a decreased intestinal transit time, and other results suggest that CH_4 production (usually defined as a > 1 ppm elevation of exhaled CH_4 over the atmospheric level on breath testing) correlates with constipation in irritable bowel syndrome [1]. Information on the effects of exogenous CH_4 is sparse, but a previous study demonstrated that CH_4 supplementation can attenuate microcirculatory failure and the tissue accumulation of inflammatory cells in a large animal model of intestinal ischemia-reperfusion (IR) [2]. These data point to an anti-inflammatory potential for CH_4, but the identification of intracellular targets remains elusive [2].

Liver diseases are often accompanied by mitochondrial functional disorders, and diseases of the mitochondria appear to cause damage to liver cells. On this basis, we set out to investigate the effects of increased CH_4 inhalation on the function of the mitochondrial electron transport chain (ETC) in the liver of unstressed animals and after a standardized hypoxic insult. For this purpose, we employed a well-established IR model where the organ damage is mainly attributed to the enhanced activity of superoxide-generating enzymes and the failure of the mitochondrial ETC enzymes [3,4,5]. We postulated that, as they are critically involved in hypoxia-reoxygenation-induced intracellular respiratory damage, the mitochondria may be targets of CH_4 administration. In particular, we hypothesized that, if CH_4 is bioactive, it can exert its effect by influencing the respiratory activity and ROS production of the hepatic mitochondria.

MATERIALS AND METHODS

In vivo experiments

The experiments were carried out on male Sprague-Dawley rats (average weight 300±20 g) housed in an environmentally controlled room with a 12-h light-dark cycle, and kept on commercial rat chow and tap water ad libitum. The experimental protocol was in accordance with EU directive 2010/63 for the protection of animals used for scientific purposes and was approved by the Animal Welfare Committee of the University of Szeged. This study also complied with the criteria of the US National Institutes of Health Guidelines for the Care and Use of Laboratory Animals.

Surgical Procedures

The rats were anesthetized with sodium pentobarbital (45 mg/kg ip), and the trachea was cannulated to facilitate respiration. The right jugular vein and carotid artery were cannulated for fluid and drug administration, respectively. Further small supplementary doses of pentobarbital were given iv when necessary. The animals were placed in a supine position on a heating pad to maintain the body temperature between 36 and 37°C, and Ringer's lactate was infused at a rate of 10 ml/kg/h during the experiments. For the preparation of the liver, the fur covering the abdomen was shaved, and the skin was disinfected with povidone iodide. After midline laparotomy and bilateral subcostal incisions, the liver was carefully mobilized from all ligamentous attachments; complete ischemia of the median and left hepatic lobes was achieved by clamping the left lateral branches of the hepatic artery and the portal vein with a microsurgical clip for 60 min. After the period of ischemia, the clips were removed and measurements were performed during a 60-min reperfusion period [5]. The wound was temporarily covered with non-water-permeable foil during the reperfusion period. At the end of experiments the animals were over-anesthetized with a single overdose of pentobarbital.

Experimental Protocols

The animals were randomly assigned to one or other of the following groups. In the IR group (n = 6), the mitochondrial respiratory functions in response to 60-min complete ischemia and 60-min reperfusion in normoxic air were examined. Control tissue samples were taken to determine the baseline mitochondrial respiratory variables, and ischemia was then induced in the median and left hepatic lobes by clamping the left lateral branches of the hepatic artery and the portal vein. At 55 min of ischemia, liver samples were taken for analysis of the mitochondrial respiration in response to ischemia. Following release of the vascular occlusions, biopsies were obtained from the affected lobes at 5 min, 30 and 60 min of reperfusion. In the IR+CH_4 group (n = 6), the protocol was identical, except that inhalation with normoxic artificial air containing 2.2% CH_4 (Linde Gas, Budapest, Hungary) was started after 50 min of ischemia and continued throughout the reperfusion period. The sham-operated animals in the SH group (n = 6) underwent the same surgical procedure but liver ischemia was not induced and the animals inhaled normoxic air, while the sham-operated animals in the SH+CH_4 group (n = 6) were likewise not subjected to liver ischemia, but inhaled CH_4 for the same duration as in the IR+CH_4 (Fig 1).

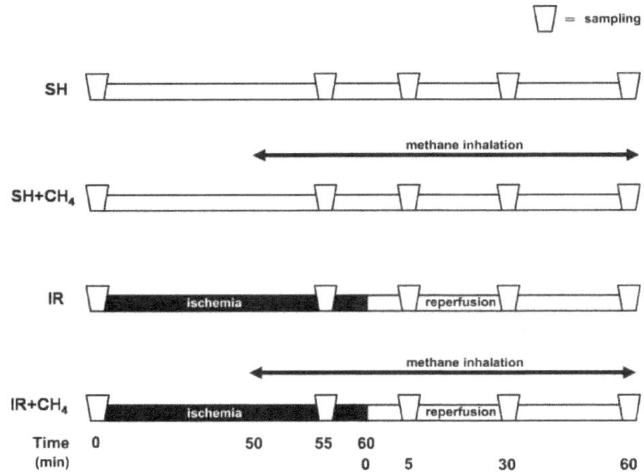

Figure 1. Experimental protocol. 60-min partial hepatic warm ischemia was induced with subsequent reperfusion (IR) in the presence or absence of 2.2% CH_4 inhalation (IR and IR+CH_4 groups). Sham-operated animals were subjected to the same surgical procedures, but without subsequent IR, and participated or not in 2.2% CH_4 inhalation (SH+CH_4 and SH groups). Liver samples for the analysis of the mitochondrial respiration were taken at baseline, at 55 min of ischemia, and at 5, 30 and 60 min of reperfusion.

Blood samples were taken for whole-blood superoxide, hydrogen peroxide (H_2O_2) and cytochrome c activity measurements at baseline and after 5, 15, 40 and 70 min of CH_4inhalation. Tissue biopsies for histological analysis (terminal deoxynucleotidyl transferase dUTP nick end labelling (TUNEL) and 4',6-diamidino-2-phenylindole (DAPI) staining) were taken at the end of the experiments and were stored at -80°C until assays.

Another series of animals (n = 3, each group) participated in identical treatment protocols and in vivo histology with laser-scanning confocal microscopy (LSCM) to determine the structural damage in the left liver lobe.

Analysis of Mitochondrial Respiratory Function

For determination of the respiratory activity in the liver mitochondria, tissue samples were homogenized in 3 ml MitOx2 (mitochondrial respiration medium) and then subjected to high-resolution respirometry with an Oxygraph-2k (Oroboros Instruments, Innsbruck, Austria). The steady-state basal O_2 consumption (respiratory flux) was observed, and the complex II-linked state II respiration rate was then determined with 10 mM succinate after the addition of 0.5 μM complex I inhibitor rotenone. For determination of the maximum respiratory capacity of complex II-linked (state III respiration), 2.5 mM ADP was added to the chamber. Finally, the intactness of the outer mitochondrial membrane was assessed after 10 μM cytochrome c addition. The respirometry data were normalized to wet weight [6].

Detection of Cytochrome C Oxidase Activity

Cytochrome c oxidase activity was calculated via the time-dependent oxidation of cytochrome c at 550 nm, as described previously [7]. Briefly, a cytochrome c stock solution was freshly prepared by dissolving 10.6 mg cytochrome c (Sigma-Aldrich, Budapest, Hungary) in 20 ml distilled water. The cytochrome c was then reduced by adding 50 µl 0.1 M sodium dithionite, and the absorbance of the solution was determined at 550 nm; the photometer was calibrated to this level. Liver samples were homogenized with a Potter grinder in 10x ice-cold MitOx2 medium and then centrifuged at 800g for 5 min at 4°C. 50 µl supernatant was added to 2.5 ml cytochrome c stock solution and the decrease in optical density at 550 nm was measured spectrophotometrically during 1-min intervals at 0, 30 and 60 min.

Malondialdehyde (MDA) Assay on Liver Tissue

The lipid peroxide MDA level was measured through the reaction with thiobarbituric acid, by the method of Placer et al.[8] and corrected for the tissue protein content.

Whole-Blood Superoxide and H_2O_2 Production

10 µl whole-blood and 50 µl zymosan were added to 1ml Hank's solution (PAA Cell Culture, Westborough, MA, USA) and the mixture was incubated at 37°C for 30 min, until assay [9]. The chemiluminometric response was measured with a Lumat LB9507 luminometer (Berthold Technologies, Wildbad, Germany) during a 30-min period after the addition of 100 µl of lucigenin and luminol reagent.

LSCM and Staining Protocol

After mobilization of the liver from its ligamentous attachments, the left and median lobes were exteriorized, placed on a specially designed pedestal, and turned on the left side, providing a suitable horizontal plane of the liver lobe for examinations with the LSCM device. This method provided free access to the appropriate vessels while ensuring an adequate blood supply of the investigated liver lobes without the twisting of vascular pedicles. Confocal imaging with the LSCM device (FIVE1 system, Optiscan, Victoria, Australia) was started at the end of the reperfusion period. The rigid confocal probe (excitation wavelength 488 nm; emission detected at 505–585 nm) was mounted on a specially designed metal frame and gently pressed onto the liver surface [10].

For the in vivo staining of liver cells, 10 mg/ml fluorescein isothiocyanate (FITC)-labeled dextran (150 kDa) (Sigma-Aldrich, Budapest, Hungary) and 0.5 ml 0.01% acriflavine (Sigma-Aldrich, Budapest, Hungary) was injected into the jugular vein at the beginning of the reperfusion period.

TUNEL and DAPI Staining

Apoptosis was detected by the TUNEL method. For apoptotic cell staining, samples (n = 4–6) were analyzed with In situ cell death detection kit, TMR red (Roche, Cat. No 12 156 792 910) according to the manufacturer's instructions. Briefly, tissue sections were fixed in 4% paraformaldehyde. For permeabilization, 0.1% Triton X-100 in 0.1% sodium citrate was used. The TUNEL reaction mixture comprised one part Enzyme Solution and nine parts Label Solution. Slides were incubated in a humidified atmosphere for 60 min at 37°C in the dark, followed by DAPI staining (Sigma-Aldrich®, 1:100). For each experimental series, one negative control (incubated with the Label Solution) and one positive control (digested with DNase I, grade I before application of the TUNEL reaction mixture) samples were used. Three pictures per sample were taken with a Zeiss AxioImager.Z1 microscope at 20x magnification. The number of apoptotic cells per field of view (524.19 µm x 524.19 µm) was determined by Image J 1.47 software.

Statistical Analysis

Data analysis was performed with SigmaStat statistical software (Jandel Corporation, San Rafael, CA, USA). Changes in variables within and between groups were analyzed by two-way repeated measures ANOVA, followed by the Bonferroni test in the cases of the mitochondrial respiratory function, cytochrome c activity from the mitochondria and whole-blood superoxide and H_2O_2 production; one-way ANOVA followed by the Holm-Sidak test was applied in the assay of MDA on liver tissue. Data were expressed as means ± SEM. For statistical analysis of TUNEL and DAPI staining, the Kruskal-Wallis and Dunn tests were applied. Histological data were expressed as median ± SD. Values of $P < 0.05$ were considered statistically significant.

In vitro Experiments

Pilot experiments were conducted to detect the changes in respiratory activity of different mitochondrial oxidative phosphorylation (OxPhos) system complexes in response to CH_4 in intact mitochondria. For this purpose, liver mitochondria were isolated by the method of Gnaiger et al.[11]. Briefly, mitochondria were isolated from the left liver lobe in isotonic sucrose medium (300 mM sucrose, 0.2 mM EDTA and 10 mM HEPES, adjusted to pH 7.4 with KOH at 4°C). After the last centrifugation, mitochondrial pellets were resuspended in sucrose medium.

For respirometric analysis, isolated mitochondria were suspended in 3 ml MitOx2 medium and weighed into the chambers while the gas phase contained the 2.2% CH_4-air mixture or room air (n = 8). The rate of respiration was determined after the addition of 2 mM malate and 10 mM glutamate for complex I-linked respiration, 2.5 mM ADP for complex I state III respiration, 10 mM succinate for complex I and complex II state III respiration, 0.5 µM rotenone for complex II-linked respiration and finally 2.5 µM antimycin A, 2 mM ascorbate and 20 µM N,N,N,N-tetramethyl-p-phenylenediamine (TMPD) for complex IV-linked respiration. DatLabTM software (Oroboros Instruments) was used for data analysis. The respirometry data were normalized to the mitochondrial biomass[6].

RESULTS

I. Mitochondrial respiration and integrity

In vivo experiments.

Respiratory activity in liver mitochondria.

The efficacy of the liver mitochondrial ETC was assessed from a liver homogenate through high-resolution respirometry (Fig 2). The basal respiratory flux values (complex II-linked state II respiration) were significantly lower than those of the SH animals at 55 min of ischemia and at 60 min of reperfusion. The IR-induced decreases in basal flux were reversed in response to CH_4 treatment. Interestingly, CH_4 treatment alone ($SH+CH_4$), elevated the basal O_2consumption throughout the observation period (Fig 2A).

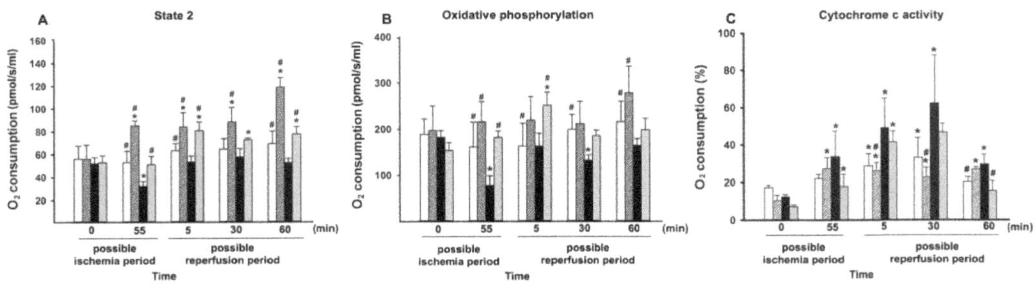

Figure 2. O_2 **consumption of liver homogenates measured by means of high-resolution** respirometry. Animals were subjected to 60 min of liver ischemia followed by 60 min of reperfusion (IR group, black column) or were sham-operated (SH group, white column). 2.2% CH_4inhalation was started 10 min before the end of ischemia and continued throughout the reperfusion ($IR+CH_4$ group, gray column), or during the identical interval in sham-operated animals ($SH+CH_4$ group, striped column). (A) Basal respiration (complex II-coupled state II respiration) (in pmol/s/ml). (B) Oxidative phosphorylation (complex II-coupled state III respiration) (in pmol/s/ml). (C) Cytochrome c activity (state III respiration augmented by adding cytochrome c to the medium) (as % of state III respiration). Data are presented as means ± SEM. * $P < 0.05$ vs baseline; # $P < 0.05$ vs IR group (two-way ANOVA, Bonferroni test).

In comparison with the SH group, IR resulted in a lower OxPhos capacity of the mitochondria (complex II-linked state III respiration) throughout the examination period. When CH_4 inhalation was applied, however, the respiratory capacity was preserved at 55 min of ischemia and at 30 min of reperfusion (Fig 2B).

Cytochrome c oxidase

The mitochondrial cytochrome c oxidase activity is an indicator of mitochondrial membrane damage. During our investigations, we determined the increase in mitochondrial respiration in response to exogenously administered cytochrome c by means of high-resolution respirometry (Fig 2C). The ability of exogenous cytochrome c to replace the enzyme in the inner mitochondrial membrane increased significantly as a result of the IR injury, while CH_4treatment restored the amount of exchanged enzyme to the baseline level. The cytochrome c oxidase activity was also determined with a spectrophotometric analysis (Fig 3A). In the SH animals, the cytochrome c level increased minimally as compared with the baseline during the experimental protocol. In the $SH+CH_4$ group, the enzyme activity decreased in response to CH_4 inhalation. In contrast, the IR group exhibited significantly higher cytochrome c oxidase activities during the reperfusion period, as an indication of functional damage. In the $IR+CH_4$group, the cytochrome c did not increase in response to the IR-induced damage (Fig 3A).

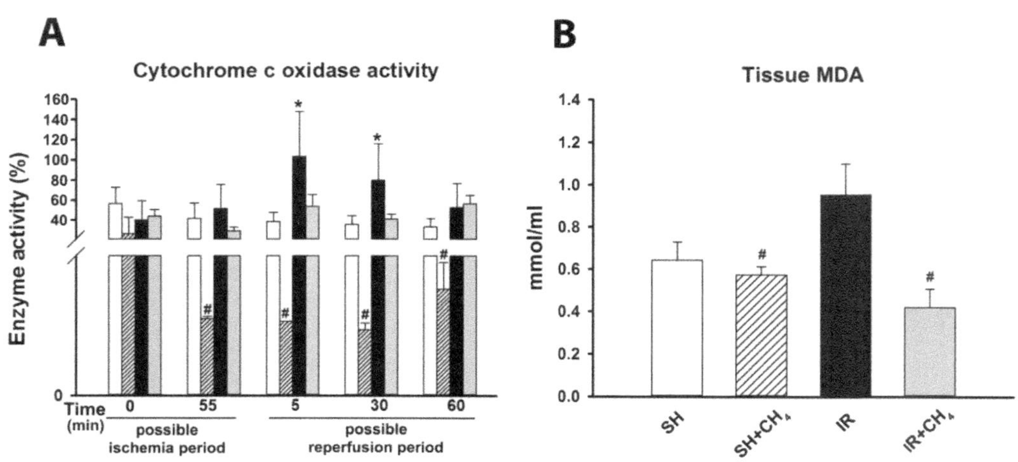

Figure 3. Cytochrome c activity and tissue MDA level in the rat liver. Animals were subjected to 60 min of liver ischemia followed by 60

min of reperfusion (IR group, black column) or were sham-operated (SH group, white column). 2.2% CH_4inhalation was started 10 min before the end of ischemia and continued throughout the reperfusion (IR+CH_4 group, gray column), or during the identical interval in sham-operated animals (SH+CH_4 group, striped column) (A) Cytochrome c activity (in %), (B) Tissue MDA level (in mmol/ml). Data are presented as means ± SEM. * $P < 0.05$ vs baseline; # $P < 0.05$ vs IR group (two-way ANOVA, Bonferroni test and one-way ANOVA, Holm-Sidak test).

In vitro experiments

Basal mitochondrial respiration in intact mitochondria was achieved by adding substrates of complex I (Table 1). The saturating concentration of ADP resulted in a 2-fold increase in complex I-linked respiration, which was not affected by CH_4 treatment. After a stable signal had been reached, complex II-dependent respiration was stimulated by adding succinate which caused a 6-fold increase in both groups. Complex I was then inhibited with rotenone to assess complex II-linked respiration. After complex III inhibition with antimycin-A, the residual O_2consumption was equal in the two groups. Finally, ascorbate and TMPD were added to the medium for the measurement of complex IV state III respiration; there was no significant difference in respiratory flux between the groups. Thus, incubation of the respiration medium with 2.2% CH_4 did not affect the activity of OxPhos complexes as compared with room air (Table 1).

Table 1. Effects of CH_4 incubation on O_2 consumption (pmol/s/ml) of isolated intact liver mitochondria.

	Glutamate+Malate	ADP	Succinate	Rotenone	Antimycin A	Ascorbate+TMPD
room air	16,4±0,9	34,5±4,5	200,1±15,9	207,1±15,6	9,2±0,6	322,9±37,8
2,2% CH_4	13,7±1,4	36,8±6,0	208,1±23,2	206,8±21,6	6,9±0,5	353,6±49,3

Data are presented as means ± SEM.

doi:10.1371/journal.pone.0146363.t001

II. Oxidative damage and lipid peroxidation products

Superoxide Production

The whole-blood superoxide-producing capacity was significantly higher in the IR group at 30 min of reperfusion in comparison with the SH animals. The CH_4 inhalation before the end of the ischemic period reduced the elevated superoxide production to the level in the SH animals (Fig 4A).

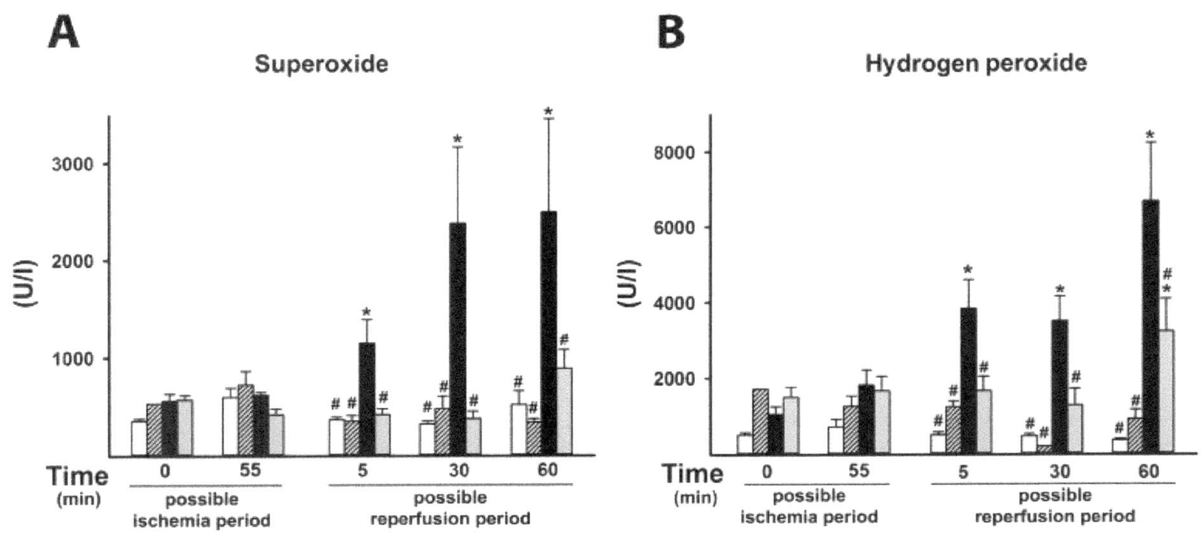

Figure 4. Superoxide and H_2O_2 production in the whole-blood.

Animals were subjected to 60 min of liver ischemia followed by 60 min of reperfusion (IR group, black column) or were sham-operated (SH group, white column). 2.2% CH_4inhalation was started 10 min before the end of ischemia and continued throughout the reperfusion (IR+CH_4 group, gray column), or during the identical interval in sham-operated animals (SH+CH_4 group, striped column) (A) Superoxide level (in U/L), (B) H_2O_2 level (in U/L). Data are presented as means ± SEM. * $P < 0.05$ vs baseline; # $P < 0.05$ vs IR group (two-way ANOVA, Bonferroni test).

H_2O_2 production

Significantly higher whole-blood H_2O_2 levels were measured at 5, 30 and 60 min of reperfusion in the IR group relative to the SH group. The CH_4 inhalation protocol effectively reversed H_2O_2production in the IR+ CH_4 group (Fig 4B).

MDA production.

A significantly higher MDA level was measured at the end of reperfusion in the IR group than in the SH group. The IR-induced elevation of the liver MDA level was effectively attenuated by CH_4 inhalation in the IR+CH_4 group (Fig 3B).

III. Structural damage

Apoptosis

As expected, few TUNEL-positive cells were observed in the liver specimens of the rats in the SH+CH_4 and SH groups. Conversely, liver IR was accompanied by an increased TUNEL positivity, which was diminished as a result of CH_4 inhalation (IR and IR+CH_4 groups) (Fig 5).

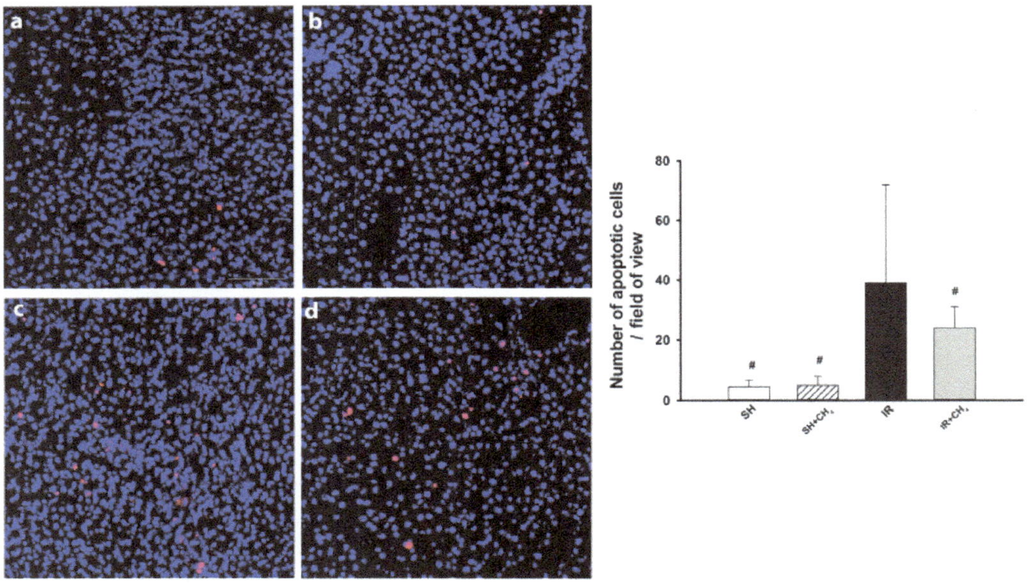

Figure 5. Liver cell apoptosis. Tissue sections are labeled with TUNEL/DAPI staining. (a) SH group, (b) SH+CH_4group, (c) IR group, (d) IR+CH_4 group. Nuclei are marked in blue, and apoptotic nuclei in red. Data are presented as median ± SD. * $P < 0.05$ vs baseline; # $P < 0.05$ vs IR group (Kruskal-Wallis and Dunn tests).

In vivo Morphological Changes

The morphological changes in the left liver lobe were evaluated by means of in vivo imaging, using confocal laser scanning endomicroscopy. The FITC-dextran and acriflavine staining demonstrated more dilated sinusoids in the IR group, and also histological signs of apoptosis: a loss of fluorescence intensity, changes in hexagonal cell shape and cytoplasm blebbing and vesicle formation relative to the SH group. CH_4 inhalation effectively attenuated these apoptosis-linked morphological changes (Fig 6).

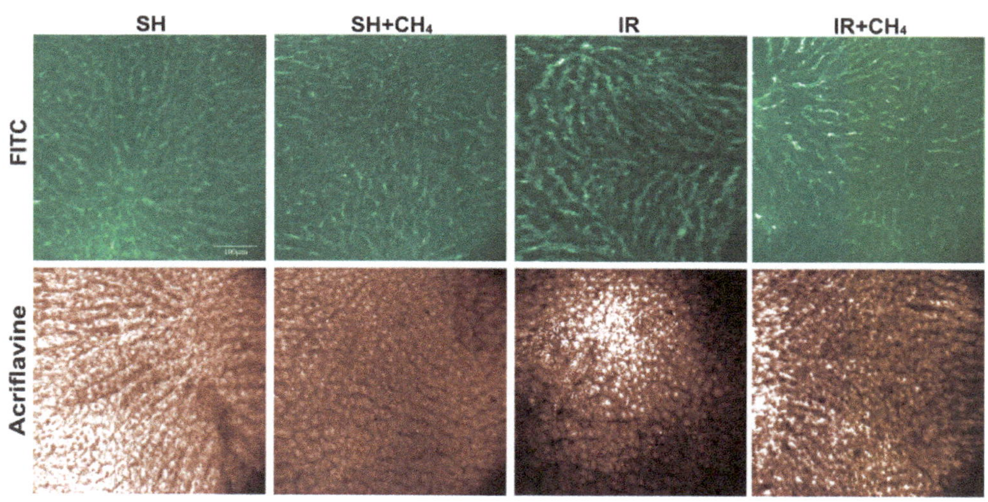

Figure 6. Histological changes in the rat liver. Tissue sections show the results of *in vivo* confocal laser scanning endomicroscopy with FITC dextran and acriflavine labeling. Apoptosis-related structural changes such as dilated sinusoids, loss of fluorescence intensity, changes in hexagonal cell shape, cytoplasmatic blebbing and vesicle formation can be observed in the IR group. The bar represents 100 μm.

DISCUSSION

A previous study has demonstrated the anti-inflammatory potential of CH_4, but the identification of intracellular targets remains elusive. Accordingly, we investigated the in vivo influence of an increased CH_4 intake on the respiratory activity of rat liver mitochondria by using controlled ventilation. This protocol did not influence the blood-gas chemistry in the anesthetized animals under the baseline conditions. More importantly, the inhalation of CH_4-containing normoxic artificial air preserved the OxPhos after a period of tissue ischemia, and significantly improved the basal mitochondrial respiration state after the onset of reperfusion. In line with this, the IR-induced ROS production, cytochrome c activity and hepatocyte apoptosis were also reduced significantly.

In our experiments the non-phosphorylating basal respiration (state II respiration) of complex II in the presence of the reducing substrate succinate, but in the absence of ADP. The state II respiration is the resting state of the physiologically uncoupled or pathologically dyscoupled respiration, when the O_2 flux is maintained in order to compensate for the proton leak at a high chemiosmotic potential, when ATP synthase is not active. ADP-dependent O_2 consumption is a test of mitochondrial functional integrity and the overall function which directly reflects mitochondrial OxPhos [12]. The ADP-stimulated respiration of coupled mitochondria (OxPhos or state III) was supported by high, saturating concentrations of ADP and was significantly increased, by ~ 60%, in the SH animals. In contrast, significantly decreased complex II-linked basal respiration and a lower OxPhos capacity with higher superoxide and H_2O_2 production were found during reperfusion. The ADP-stimulated respiration was also decreased at the end of ischemia and during reperfusion.

The ROS production measured in this study can be linked to the mitochondria or to other sources, such as an increased activity of cellular oxidases. It is known that ~ 0,1–4% of the electrons flowing through the mitochondrial ETC generate superoxide anions by the partial reduction of O_2 [13,14], i.e. involving precursors of ROS such as H_2O_2 and OH^-, and our data provide evidence of the influence of an increased CH_4 intake on mitochondrial ETC reactions. However, the exact pathway through which CH_4 influences the mitochondrial respiration remained unexplored and several mechanisms can be postulated to explain the in vivo effects of CH_4 in this system. Interestingly, CH_4 administration in sham animals elevated the state II O_2 consumption throughout the observation period, which refers to an increased leak-dependent respiration. Lower mitochondrial ROS formation can partially be originated from the increased leak respiration in response to CH_4 treatment.

In this respect, the data overall support a role for CH_4 membrane defence during ROS-induced damage. During ischemia, the mitochondrial NADH/NAD$^+$ and FADH/FAD$^+$ ratios remain elevated, leading to reductive stress, [15,16] while reperfusion of the previously ischemic tissue leads to oxidative stress with a burst of ROS generation following the start of reoxygenation. It was earlier suggested (without indicating the exact biochemistry, contributing compounds or enzymes) that ROS reactions could lead to a higher level of fixation of CH_4 in a lipid environment, such as the mitochondrium membrane [17,18]. Indeed, IR perturbs the heterogeneous lipid-bilayer membrane structure and changes the fluidity from fluid to gel. Disordered/fluidized bilayer states could therefore be analogous to physical damage to the ETC in these conditions. Inhaled exogenous CH_4 will move from the alveoli into the circulation, and diffuse into the plasma, throughout which it is distributed rapidly and evenly. If there are no barriers, the apolar CH_4 may enter the cytoplasm and mitochondrial matrix and dissolve in the hydrophobic nonpolar lipid tails of the phospholipid biomembranes [19]. Membrane rigidity relates to the degree of lipid peroxidation, when the level of oxidized lipids is increased and the fluidity of membranes is reduced. CH_4 dissolved in biological membranes may affect this process, thereby influencing the stereo structure of membrane proteins, which determines their accessibility and morphology [20]. The peroxidation of polyunsaturated fatty acids and direct triggering of cytochrome c activity from the mitochondria are well-known consequences of IR injury [21]. The degree of lipid peroxidation can be estimated via the amount of MDA, a marker of oxidative damage of lipid membranes. As a reactive aldehyde, MDA is one of the many reactive electrophilic compounds that cause further oxidative stress in cells and form covalent protein adducts referred to as advanced lipoxidation end-products. We have shown that the levels of both ROS and MDA were reduced by an increased CH_4 intake, indirectly demonstrating the decreased oxidative damage to the mitochondrial membranes.

Secondly, CH_4 may accumulate transiently at cell membrane interfaces, thereby transitorily changing the physico-chemical properties or the in situ functionality of proteins, ion channels and receptors embedded within this environment. In this case, an increased CH_4 intake may influence the function of membrane-bound structures. The pilot in vitro results demonstrated that the incubation of isolated mitochondria in a 2.2% CH_4 atmosphere has no effect on the respiratory activity of the ETC. As opposed to this, the in vivo CH_4 inhalation regimen unexpectedly increased the mitochondrial respiratory function, without influencing the baseline ROS production in the SH animals. These differing responses can also be explained by membrane injury, as compared to isolated mitochondria, homogenation can disrupt mitochondrial membranes [22]. In contrast, the protocol of mitochondria isolation permits rearrangement of the membranes and results in fully viable mitochondria whose function cannot be further ameliorated by CH_4 [11].

This study addressed the possibility of CH_4 bioactivity. This concept is supported by experimental data showing that gaseous CH_4 can delay the contraction of peristalsis and increase the amplitude of the peristaltic contractions in the guinea pig ileum [23]. A recent study provided in vitro evidence that CH_4 can inhibit the contractile activity of the smooth muscle by activating the voltage-dependent potassium channels and increasing the voltage-dependent potassium current [24].

Whereas the results indicate a bioactive role for higher concentrations of exogenous CH_4, this is not obvious for endogenous sources. It is widely recognized that large amounts of CH_4 can be produced by the anaerobic metabolism of methanogenic microorganisms in the GI tract [25,26], and CH_4 is present in measurable amounts in the breath of approximately one-third of humans [27]. Nevertheless, as opposed to the previous view, in vitro and in vivo studies have revealed the possibility of nonmicrobial CH_4 formation in both plants and animals [28,29]. CH_4generation has been demonstrated after site-specific inhibition of the ETC, and in association with a mitochondrial dysfunction, similarly to the effects of hypoxia [15,29].

Of interest, recent studies demonstrated the critical role of a ferryl species ($[Fe(IV) = O]^{2+}$) and CH_3 radicals, leading to the in vitro generation of CH_4 from methionine sulfoxide as substrate at ambient temperature[30]. In this chemical reaction, CH_4 is readily formed from the S-CH_3groups of organosulfur compounds in a model system containing iron(II/III), H_2O_2 and ascorbate that uses organic compounds with heterobonded CH_3 groups for CH_4 generation under ambient (1,000 mbar and 22°C) and aerobic (21% O_2) conditions. CH_3 radicals can be formed in the mitochondria through reaction between a reducing agent, a metal and a hydroperoxide. Methionine is known to be a key factor in many biochemical reactions in plants, fungi and animals, and methionine residues in the surface of proteins are highly susceptible to oxidation, the product generally being methionine sulfoxide. Importantly, the available data suggest that reversible methionine oxidation could be a novel mechanism in redox—regulation, which involves the repair mechanism of methionine sulfoxide reductases (MSRs) whose main function is to protect membranes from oxidative damage [20]. The MSRs were originally thought to be exclusively bacterial enzymes, but their presence was recently proven in mitochondria in mammals [31]. The scavenging action of these enzymes is based on the cyclic oxidation and reduction of the several methionine residues of the molecules that makes them the counterpart of the $NADH/NAD^+$ system[20]. In a continuous lack of oxygen, the methionine sulfoxide-CH_4 system may act as an alternative to $NADH/NAD^+$ and $FADH/FAD^+$as an electron donor, thereby mitigating ischemia-induced reductive stress. The capacity of mitochondria to reverse oxidant-induced changes upon reperfusion originates from the exposure of previously hidden epitopes of mitochondrial proteins, as proven in the case of mitochondrial MSRs [32].

IR injury is an antigen-independent stimulus that initiates intrinsic signaling pathways. Peroxidation, an immediate chain reaction, causes a breakdown of biomembranes, leading to decompartmentalization, and the loss of maintenance of a steady state. Cytochrome c is attached to the inner membrane, and is detached in response to a threshold disturbance in the membrane structure, which leads to activation of the apoptotic caspase cascade [33]. CH_4inhalation effectively attenuated the IR-induced elevation in MDA level and in parallel the activity of cytochrome c was diminished. Conventional and in vivo histology provided evidence of IR-related apoptosis and the observations suggest that CH_4 may influence the cell fate under stress conditions.

CONCLUSIONS

Liver IR injury is a progressive process starting from a depressed mitochondrial ETC, the abnormal formation of ROS leading to biomembrane damage and finally necrotic or apoptotic cell death. Our experiments permit the conclusion that the mitochondrial protection afforded by CH_4 inhalation involves different components under normal conditions, during ischemia and during reperfusion, similarly to the different pathomechanisms of damage during ischemia and after reperfusion. 2.2% CH_4 inhalation clearly influenced the IR-related disturbances of the mitochondrial ETC, and mitigated the severity of subsequent events. The protective potential of CH_4 was linked to reduced cytochrome c activity, and a reduced number of apoptotic hepatocytes. There is still no clear-cut evidence that CH_4 in the endogenously produced concentration range (1–10 ppm) has a role in the cellular physiology, and further studies should be performed to determine whether endogenously-generated or exogenously-administered CH_4participates in restricting membrane fluidity and preserving the optimal condition of membrane-bound structures. Much remains unknown about hypoxic reactions and the fate of intracellular CH_4 is an open question, but the results presented to date on the effects of CH_4 against oxido-reductive biomembrane damage indicate a bioactive role for CH_4.

Supporting Information

We have performed these additional *in vitro* experiments with isolated mitochondria in order to measure the maximal leak-dependent respiration.

Briefly, isolated mitochondria were suspended in 3 ml MitOx2 medium and placed into the chambers, while the gas phase contained the 2.2% CH_4-air mixture or room air (n=6). The complex II-linked state II respiration rate was then determined with 10 mM succinate after the addition of 0.5 µM complex I inhibitor rotenone. For determination of the complex II-linked state III respiration, 2.5 mM ADP was added to the chamber. Finally, 5 mM oligomycin was added to the medium. The respiratory control ratio was calculated by dividing LEAK by the state II respiration of the mitochondria. The results are summarized below. In brief, the leak respiration was 125 ± 4 and 127 ± 5 pmol/s/ml in the room air and the 2.2% CH_4-air mixture groups of isolated mitochondria; hence, these novel data do not indicate any effect of CH_4 on intact mitochondria under *in vitro* circumstances. (The explanation for this phenomenon is discussed in detail in para 2, on page 15. Nonetheless, these *in vitro* data cannot exclude a role of the increased leak respiration in *in vivo* experiments involving a lower ROS production.

S1 Table. Effects of CH_4 incubation on LEAK respiration (pmol/s/ml) in isolated intact liver mitochondria. Data are presented as means ± SEM

	Room air	**2.2% CH_4-air**
Rotenone / Succinate	157 ± 6	155 ± 6
OxPhos	684 ± 39	607 ± 28
LEAK oligomycin	125 ± 4	127 ± 5
Respiratory control ratio (RCR)	0.79	0.82

Acknowledgments

The authors are grateful to Nikolett Beretka, Ágnes Lilla Szilágyi, Csilla Mester and Gergely Kulinyi for their assistance and to Károly Tóth, Kálmán Vas and Péter Sárkány for their skillful work. The study was supported by Hungarian Science Research Fund OTKA-K75161.

REFERENCES

1. Pimentel M, Chang C, Chua KS, Mirocha J, DiBaise J, Rao S, et al. Antibiotic treatment of constipation-predominant irritable bowel syndrome. Dig Dis Sci. 2014;59: 1278–1285. doi: 10. 1007/S10620-014-3157-8. pmid:24788320
2. Boros M, Ghyczy M, Érces D, Varga G, Tőkés T, Kupai K, et al. The anti-inflammatory effects of methane. Crit Care Med. 2012;40: 1269–1278. doi: 10. 1186/2045-9912-2-25. pmid:22336723
3. Chance B, Seis H, Boveris A. Hydroperoxide metabolism in mammalian organs. Physiol Rev. 1979;59: 527–605. pmid:37532
4. Bedard K, Krause KH. The NOX family of ROS-generating NADPH oxidases: physiology and pathophysiology. Physiol Rev. 2007;87: 245–313. doi: 10. 1152/physrev. 00044. 2005. pmid:17237347
5. Hartmann P, Fet N, Garab D, Szabó A, Kaszaki J, Srinivasan PK, et al. L-α-glycerylphosphorylcholine reduces the microcirculatory dysfunction and nicotinamide adenine dinucleotide phosphate-oxidase type 4 induction after partial hepatic ischemia in rats. J Surg Res. 2014;189: 32–40. doi: 10. 1016/j. jss. 2013. 12. 025. pmid:24636100
6. Hutter E, Skovbro M, Lener B, Prats C, Rabol R, Dela F, et al. Oxidative stress and mitochondrial impairment can be separated from lipofuscin accumulation in aged human skeletal muscle. Aging Cell. 2007;6: 245–256. doi: 10. 1111/j. 1474-9726. 2007. 00282. x. pmid:17376148
7. Szarka A, Horemans N, Banhegyi G, Asard H. Facilitated glucose and dehydroascorbate transport in plant mitochondria. Arch Biochem Biophys. 2004;428: 73–80. doi: 10. 1016/j. abb. 2004. 05. 011. pmid:15234271
8. Placer ZA, Cusman L, Johnson B. C. Estimation of product of lipid peroxidation by malonyldialdehyde in biochemical systems. Anal Biochem. 1966;16: 359–364. doi: 10. 1016/0003-2697(66)90167-9. pmid:6007581
9. Ferdinandy P, Danial H, Ambrus I, Rothery RA, Schulz R. Peroxynitrite is a major contributor to cytokine-induced myocardial contractile failure. Circ Res. 2000;87: 241–247. doi: 10. 1161/01. RES. 87. 3. 241. pmid:10926876
10. Goetz M, Ansems JV, Galle PR, Schuchmann M, Kiesslich R. In vivo real-time imaging of the liver with confocal endomicroscopy permits visualization of the temporospatial patterns of hepatocyte apoptosis. Am J Physiol Gastrointest Liver Physiol. 2011;301: 764–772. doi: 10. 1152/ajpgi. 00175. 2011.
11. Gnaiger E, Kuznetsov AV, Schneeberger S, Seiler R, Brandacher G, Steurer W, Margreiter R. Mitochondria in the cold. In: Life in the Cold. New York: Springer; 2000.
12. Li Z, Graham BH. Measurement of mitochondrial oxygen consumption using a Clark electrode. Methods Mol Biol. 2012;837: 63–72. doi: 10. 1007/978-1-61779-504-6_5. pmid:22215541
13. Lesnefsky EJ, Moghaddas S, Tandler B, Kerner J, Hoppel CL. Mitochondrial dysfunction in cardiac disease: ischemia—reperfusion, aging, and heart failure. J Mol Cell Cardiol. 2001;33: 1065–1089. doi: 10. 1006/jmcc. 2001. 1378. pmid:11444914
14. Tahara EB, Navarete FD, Kowaltowski AJ. Tissue-, substrate-, and site-specific characteristics of mitochondrial reactive oxygen species generation. Free Radic Biol Med. 2009; 9: 1283–1297. doi: 10. 1016/j. freeradbiomed. 2009. 02. 008.
15. Ghyczy M, Torday C, Kaszaki J, Szabo A, Czobel M, Boros M. Hypoxia-induced generation of methane in mitochondria and eukaryotic cells: an alternative approach to methanogenesis. Cell Physiol Biochem. 2008;21: 251–258. doi: 10. 1159/000113766. pmid:18209491
16. Boros M, Wolfárd A, Ghyczy M. In vivo evidence of reductive stress-induced methane production. Shock. 1999;12: 56. doi: 10. 1097/00024382-199911001-00174
17. Carlisle SM, Burchart PA, McCauley C, Surette RA. Biokinetics of inhaled radioactive methane in rats: a pilot study. Appl Radiat Isot. 2005;62: 847–860. pmid:15799862 doi: 10. 1016/j. apradiso. 2005. 01. 010
18. Dougherty RW, O›Toole JJ, Allison MJ. Oxidation of intra-arterially administered carbon 14-labelled methane in sheep. Proc Soc Exp Biol Med. 1967;124: 1155–1157. pmid:6024824 doi: 10. 3181/00379727-124-31949
19. Meyer M, Tebbe U, Piiper J. Solubility of inert gases in dog blood and skeletal muscle. Pflugers Arch. 1980;384: 131–134. pmid:6247698 doi: 10. 1007/bf00584428
20. Levine RL, Mosoni L, Berlett BS, Stadtman ER. Methionine residues as endogenous antioxidants in proteins. Proc Natl Acad Sci USA. 1996;93: 15036–15040. pmid:8986759 doi: 10. 1073/pnas. 93. 26. 15036
21. McCord JM. Oxygen-derived free radicals in postischemic tissue injury. N Engl J Med. 1985;312: 159–163. doi: 10. 1056/NEJM198501173120305. pmid:2981404
22. Picard M, Taivassalo T, Ritchie D, Wright KJ, Thomas MM, Romestaing C, et al. Mitochondrial structure and function are disrupted by standard isolation methods. PloS One. 2011;6: 18317. doi: 10. 1371/journal. pone. 0018317.
23. Pimentel M, Lin HC, Enayati P, van den Burg B, Lee HR, Chen JH, et al. Methane, a gas produced by enteric bacteria, slows intestinal transit and augments small intestinal contractile activity. Am J Physiol Gastrointest Liver Physiol. 2006;290: G1089–1095. doi: 10. 1152/ajpgi. 00574. 2004. pmid:16293652
24. Liu Y, Luo HS, Liang CB, Tan W, Xia H, Xu WJ. Effects of methane on proximal colon motility of rats and ion channel mechanisms. Zhonghua Yi Xue Za Zhi. 2013;93: 459–463. pmid:23660269
25. Attaluri A, Jackson M, Valestin J, Rao SS. Methanogenic flora is associated with altered colonic transit but not stool characteristics in constipation without IBS. Am J Gastroenterol. 2010;105: 1407–1411. doi: 10. 1038/ajg. 2009. 655. pmid:19953090
26. Roccarina D, Lauritano EC, Gabrielli M, Franceschi F, Ojetti V, Gasbarrini A. The role of methane in intestinal diseases. Am J Gastroenterol. 2010;105: 1250–1256. doi: 10. 1038/ajg. 2009. 744. pmid:20216536

27. Levitt MD, Furne JK, Kuskowski M, Ruddy J. Stability of human methanogenic flora over 35 years and a review of insights obtained from breath methane measurements. Clin Gastroenterol Hepatol. 2006;4: 123–129. pmid:16469670 doi: 10. 1016/j. cgh. 2005. 11. 006
28. Keppler F, Boros M, Frankenberg C, Lelieveld J, McLeod A, Pirttilä AM, et al. Methane formation in aerobic environments. Environ Chem. 2009;6: 459–465. doi: 10. 1071/en09137
29. Tuboly E, Szabo A, Garab D, Bartha G, Janovszky A, Erős G, et al. Methane biogenesis during sodium azide-induced chemical hypoxia in rats. Am J Physiol Cell Physiol. 2013;304: C207–14. doi: 10. 1152/ajpcell. 00300. 2012. pmid:23174561
30. Althoff F, Jugold A, Keppler F. Methane formation by oxidation of ascorbic acid using iron minerals and hydrogen peroxide. Chemosphere. 2010;80: 286–292. doi: 10. 1016/j. chemosphere. 2010. 04. 004. pmid:20444486
31. Weissbach H, Resnick L, Brot N. Methionine sulfoxidereductases: history and cellular role in protecting against oxidative damage. Biochim Biophys Acta. 2005;1703: 203–212. doi: 10. 1016/j. bbapap. 2004. 10. 004. pmid:15680228
32. Cole NB, Daniels MP, Levine RL, Kim G. Oxidative stress causes reversible changes in mitochondrial permeability and structure. Exp Gerontol. 2010;45: 596–602. doi: 10. 1016/j. exger. 2010. 01. 016. pmid:20096768
33. Wang X. The expanding role of mitochondria in apoptosis. Genes Dev. 2001;15: 2922–2933. pmid:11711427

CHAPTER 2

ALTERATIONS IN VOLTAGE-SENSING OF THE MITOCHONDRIAL PERMEABILITY TRANSITION PORE IN ANT1-DEFICIENT CELLS

Judit Doczi, BeataTorocsik, Andoni Echaniz-Laguna, Bénédicte Mousson de Camaret, Anatoly Starkov, Natalia Starkova, AnikoGál, Mária J Molnár, Hibiki Kawamata, Giovanni Manfredi, VeraAdam-Vizi & Christos Chinopoulos

INTRODUCTION

Mitochondria that are subject to calcium overload exhibit a permeability transition mediated by a pore forming in the inner mitochondrial membrane1·2. The identity of the structural components of this pore has been until recently unknown; in the past three years though, the c-rings of the ATP synthase3·4, and the interface within ATP synthase dimers5 are being strongly favoured for filling this gap of knowledge6·7·2, but still with a number of questions unanswered, reviewed in8·9·10. Nonetheless, over the past three decades extensive amount of efforts have focused on the regulation and functional characteristics of this phenomenon11, and an inverse correlation of the probability of pore opening to the magnitude of the electrochemical gradient has been thoroughly characterized12·13·14·15. The molecular entity responsible for the voltage-dependence of the pore has not been identified, though the tuning of this sensor by the oxidation-reduction state of vicinal thiols as well as the contribution of critical arginines has been reported16·17·18·19. Relevant to this, glutathione depletion in cultured neurons by monochlorobimane was shown to initiate bioenergetic deficiency that was mediated by inhibition of ANT20. ANT exhibits a number of thiols that are amenable to oxidation by several agents21·22, some of which are unmasked in an energy-dependent manner23. Strong circumstantial evidence led to the formulation of the theory that the ANT is a structural part of the pore24, backed by hundreds of publications showing impacts of all known ANT ligands on pore opening probability25.

However in 2004, the Wallace's group showed that in the livers of genetically modified mice in which both isoforms of ANT were inactivated, mitochondria still exhibited Ca^{2+}-induced pore opening albeit requiring higher amounts of Ca^{2+} loading, and that the effects of ANT ligands were completely abolished26. From this report alone, the ANT was deemed as modulatory but not structural component of the mPT. On the other hand, mitochondria obtained from the crustacean *Artemia franciscana* do not exhibit a calcium-regulated pore in view of profound calcium storage27, an organism that was recently shown to expresses a unique ANT being refractory to bongkrekic acid28, a dual inhibitor for ANT and mPT29. Yet, ANTs expressed in related crustacean organisms that lack a Ca^{2+}-induced mPT, exhibited sensitivity of ANT-mediated adenine nucleotide exchange to bongkrekic acid30, while allogenic expression of Artemia ANT in yeasts also conferred sensitivity to this poison31.

Mindful of the above, one may conclude that the ANT is not a structural element of the pore, however, the involvement of this transporter in the regulation of the mitochondrial permeability transition is more than likely. Hereby we investigated the response of *in situ* mitochondria in intact cells

Citation: Doczi, J. et al. Alterations in voltage-sensing of the mitochondrial permeability transition pore in ANT1-deficient cells. Sci. Rep. 6, 26700; doi: 10.1038/srep26700 (2016).

lacking partially or completely ANT1 to Ca^{2+}-induced mPT opening conferred either by i) the Ca^{2+}-ionophore calcimycin or ii) H_2O_2, and as a function of the proton electrochemical gradient, also distinguishing between high- *vs* no electron flow in the respiratory chain. We also investigated the voltage thresholds of inducing mPT by submaximally loading permeabilized cells with calcium, and titrating mitochondrial membrane potential with cyanide or an uncoupler. Our results show that ANT1 is the voltage-sensor of the mPT, in line with earlier firm evidences proposing ANT as the voltage-sensor of the pore, as well as being the site of action of oxidative stress24·32·33.

RESULTS

Effect of Loss of ANT1gene Expression on *in situ* Mitochondrial Swelling Induced by Calcimycin in Human Fibroblasts at Different Bioenergetic States

The patient (AF) exhibited a complete loss of *ANT1* expression gene due to a homozygous G to A substitution (c.111+1G>A) abolishing the invariable consensus GT splice donor site of intron 134. The patient suffered from cardiomyopathy and myopathy, while her heterozygote mother (ST) was free from these clinical symptoms. In cultured fibroblasts from the patient, mitochondria appeared to be normal under ultrastructural examination and there were no alterations in the activities of respiratory chain complexes; mtDNA rearrangements or increased mtDNA copy number were absent. A thorough description of the clinical case, histopathological and biochemical findings in muscle and cultured fibroblasts appear in ref. 34. Regarding the fibroblasts from the individuals that were used in our study, ST exhibited a faint band of mRNA coding for ANT1, while AF fibroblasts exhibited no band of mRNA coding for this isoform, in northern blotting (panel Fig. 1A,B). mRNA coding for ANT2 appeared to be unaffected (see below under qPCR evaluation). Western blotting analysis (panel Fig. 1C) showed that patient AF had no detectable expression of ANT1, while the heterozygous mother exhibited moderate ANT1 protein expression. On the contrary, ANT2 protein expression was slightly elevated in AF cells compared to either ST cells or the three controls.

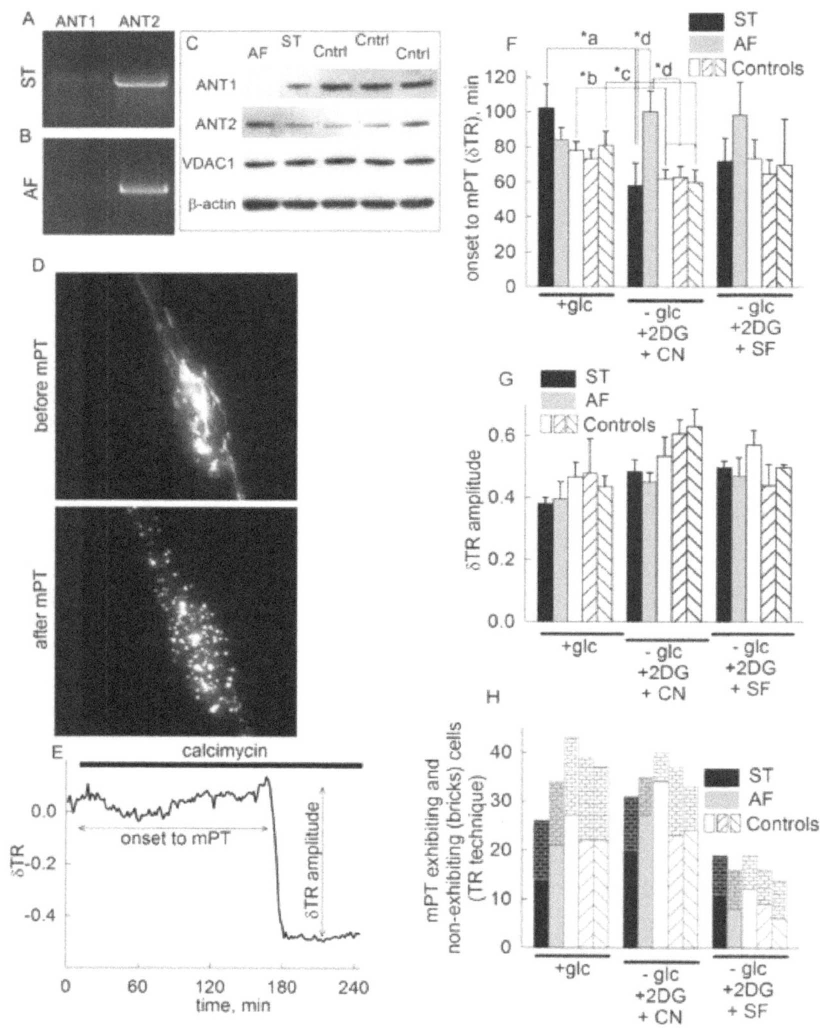

Figure 1: Effect of loss of *ANT1* gene expression on *in situ*mitochondrial swelling (visualized by DsRed2) induced by calcimycin in human fibroblasts during various metabolic conditions [+ glucose, no glucose+2-deoxyglucose (2 mM)+NaCN (5 mM), no glucose+2-deoxyglucose+SF 6847 (SF, 1 μM), detailed in the y-axis of the panels]. (**A**) Northern blot image of ST cells for ANT1 and ANT transcripts. (**B**) Northern blot image of AF cells for ANT1 and ANT2 transcripts. (**C**) Scanned images of Western blots of AF, ST and control cells

from three healthy donors (35 μg loading per lane) directed against ANT1, ANT2, VDAC1 and β-actin. (**D**) Epifluorescence image of a fibroblast transfected with DsRed2 before and after induction of mPT. (**E**) Representative trace of the effect of calcimycin (2 μM) on the thinness ratio (δTR) indicating mitochondrial volume as a function of time. (**F**) Time elapsed between calcimycin application and mitochondrial swelling (onset to mPT) detected by wide field imaging of mito-DsRed2 expressed in cultured fibroblasts from the heterozygous mother (ST, black bars) the patient (AF, grey bars), and three control subjects (white bars) during various metabolic conditions, detailed under the x-axis. The onset to mPT was determined by the sudden increase of mean mitochondrial diameter in the microscopic view field described as a sudden decrease of the thinness ratio (δTR). Bars indicate means ± S.E.M. of 14–43 cells from at least 8 different cultures as detailed in panel 1D (*$p < 0.05$ significance by Student's t-test for 'a' 'b' and 'c' or ANOVA 'd'; p-values: *a = 0.027, *b = 0.005, *c = 0.04, *d = 0.027. (**G**) Quantification of the δTR amplitude (as marked in panel F) for AF vs ST vs control cells during various metabolic conditions, detailed on the x-axis. (**H**) Comparison of mPT-exhibiting and non-exhibiting cells (bricks) in fibroblasts from ST (black bars), AF (grey bars) and control subjects (white bars) with the TR technique during various metabolic conditions, detailed on the x-axis, using Fisher exact test.

Swelling of AF, ST and three control subjects' mitochondria within fibroblasts were compared during Ca^{2+} overload, induced by addition of calcimycin (4Br-A23187, 2 μM). Although the impact of calcium ionophores on *in situ* mitochondria has not been attributed to the permeability transition in at least one study35, there, the effect of the ionophore was not studied for more than 10 minutes. On the other hand, induction of the mPT in *in situ* mitochondria has been succesfully employed in many other cell models, for more extended periods of time15·36·37·38·39·40, as in the present study. The bathing solution of the cells was deprived of Mg^{2+} because calcimycin conducts Ca^{2+} and Mg^{2+}almost equally well. Understandably, mitochondrial swelling in a Ca^{2+}-containing but Mg^{2+}-free medium may be due to opening of the inner membrane anion channel "IMAC"41 and/or K^{+}-uniport42, in addition to that attributed by the mPT; however, the first two mechanisms operate in all five cell lines, and thus would not serve as confounding factors in the results obtained due to differences in ANT1 expression. Mitochondria were visualized by wide field epifluorescence imaging of either i) mitochondrially targeted DsRed2, or ii) mitochondrially trapped calcein while cytoplasmic calcein fluorescence was quenched by cobalt (see below), a method devised by Petronilli *et al.*43. Swelling monitored by evaluating DsRed2-visualized mitochondrial morphology and changes in mean mitochondrial diameters were calculated using the thinness ratio technique, a methodology devised by Gerencser *et al.*44. In these assays the onset of swelling was defined by a sudden decrease in the thinness ratio, δTR. A representative epifluorescence image before and after induction of mPT are shown in panel Fig. 1D. A time-lapse series of such experiments are shown in the Supplemental Material, as a function of metabolic conditions (mPT_DsRed2_control_HF_calcimycin_plus_Glc.avi, mPT_DsRed2_control_HF_calcimycin_no_Glc_plus_CN.avi, and mPT_DsRed2_control_HF_calcimycin_no_Glc_plus_UNC.avi). A representative thinness ratio trace is shown in panel Fig. 1E, calculated from the recorded time-series images. In the experiments performed with AF, ST and control subjects' cells during various metabolic conditions described below, the following parameters were recorded: i) the time elapsed from the addition of calcimycin to the appearance of a large magnitude swelling (marked as 'onset to mPT', panel Fig. 1F) ii) the amplitude of swelling (marked as 'δTR amplitude', panel Fig. 1G), and iii) the number of cells exhibiting the large magnitude swelling considered as mPT within the experimental time frame (4 hours in the presence of 2 μM calcimycin), depicted in panel Fig. 1H.

As shown in panel Fig. 1F, the calcimycin-induced swelling of *in situ* ST (black bars) mitochondria was hastened by glucose deprivation and NaCN co-application (p-value $a^{*} = 0.027$, t-test); likewise, swelling of *in situ* mitochondria of two out of the three control subjects' cells (white bars) was hastened by glucose deprivation and NaCN co-application (p-value: $b^{*} = 0.005$, t-test and $c^{*} = 0.04$, t-test). This is consistent with the findings on mouse mitochondria reported previously by our group15, and in line with the established phenomenon that a diminished electrochemical gradient primes mitochondria to undergo Ca^{2+}-induced mPT. However, the calcimycin-induced swelling of *in situ* AF (grey bars) mitochondria was delayed, compared to other cell types during glucose deprivation and NaCN co-application (p-value: d* = 0.027, one-way ANOVA); furthermore, this parameter was not statistically significantly different from the other cell types during energized conditions. Although the above tendencies were the same for ST and AF cells during glucose deprivation and co-application of the uncoupler SF 6847 compared to the fully energized state, no statistically significant difference was reached.

Within the 4 hours of the experimental time frame the amplitude of swelling (panel Fig. 1G) was not significantly different between ST, AF or control cells under any metabolic condition. This is important, because different cell types exhibit different amplitudes of swelling, for unknown reasons44. The absence of a difference in these amplitudes among AF, ST and control cell mitochondria imply that the absence of ANT1 in AF cells did not confer a large, confounding alteration, thus supporting the notion that the mechanism of swelling must be the same for all cells, likely the opening of the permeability transition pore. Accordingly, as shown in panel Fig. 1H, the number of cells exhibiting large-magnitude swelling within the 4 hours of the experimental time frame was also not statistically significant among any cell type.

Visualization of DsRed2-expressing mitochondria offers superior imaging quality and the fluorescent protein does not exit mitochondria upon swelling. However, transfection efficiency is very low, typically less than 5% in fibroblasts. Therefore, we complemented this approach with the cobalt-calcein technique, a method that outlines the mitochondrial network of all cells. On the other hand, the amount of calcein loading has to be carefully titrated, as it may be toxic to the cells, plus mitochondrially-trapped calcein fluorescence is lost rapidly after opening of the mPT. In our hands, 50 nM of calcein-AM loading at 4 °C was optimal; below this concentration the signal-to-noise ratio was unsatisfactory, while above 100 nM mitochondria from all cell types fragmented a few minutes after loading. Representative epifluorescence images of cells before and after induction of mPT are shown in panelFig. 2A,B, respectively. A time-lapse series of such an experiment is shown in the Supplemental Material(mPT_COBCA_control_HF_calcimycin_plus_Glc.avi). The thinness ratio technique cannot be applied on calcein-loaded mitochondria, so therefore organellar volume cannot be quantified; using the cobalt-calcein method we only investigated the time elapsing from addition of calcimycin to transformation of mitochondria from a thread-to-ball configuration followed by immediate loss of calcein fluorescence, implying mPT opening. Furthermore, we only investigated the effect of glucose (fully energized state) versus no glucose plus 2-deoxyglucose plus NaCN (fully de-energized state, no electron flow), because the uncoupler SF 6847 was rapidly quenching mitochondrially-trapped calcein fluorescence.

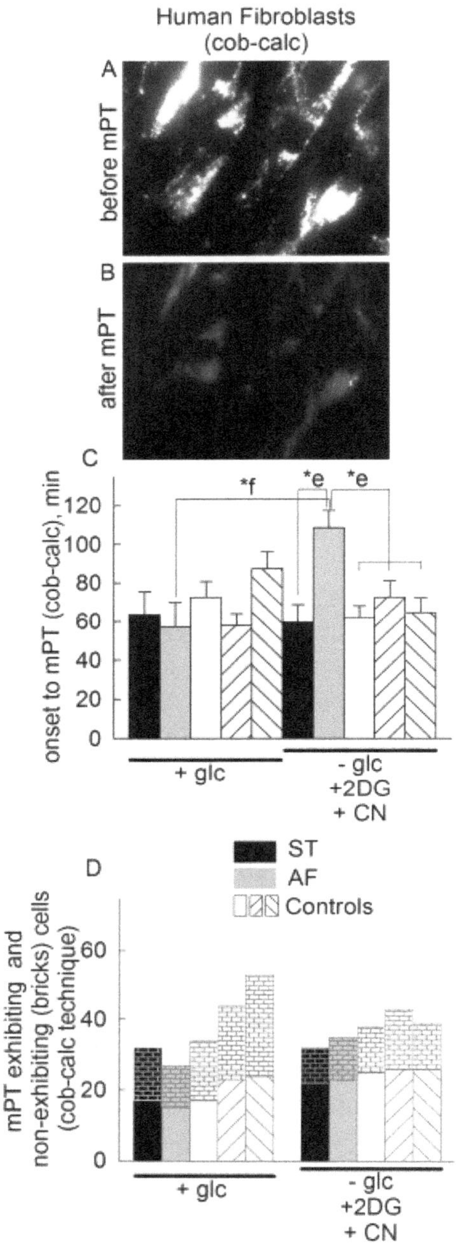

Figure 2: Effect of loss of *ANT1* gene expression on *in situ*mitochondrial swelling (visualized by the cobalt-calcein technique) induced by calcimycin in human fibroblasts during various metabolic conditions [+ glucose, no glucose+2-deoxyglucose (2 mM)+NaCN (5 mM), detailed in the y-axis of the panels]. (A) Epifluorescence image of calcein-loaded fibroblasts before induction of mPT. (**B**) Epifluorescence image of the same fibroblasts as shown in panel A, after induction of mPT. (**C**) Time elapsed between calcimycin application and mitochondrial swelling (onset to mPT) detected by wide field imaging of calcein-loaded cultured fibroblasts from the heterozygous mother (ST, black bars) the patient (AF, grey bars), and three control subjects (white bars) during various metabolic conditions, detailed under the x-axis, in the presence of $CoCl_2$. Bars indicate means ± S.E.M. of 29–53 cells from at least 8 different cultures (*$p<0.05$ significance by ANOVA; p: *e = 0.002, *f = 0.008; cob-calc: cobalt-calcein technique. (**D**) Comparison of mPT-exhibiting and non-exhibiting cells (bricks) in fibroblasts from ST (black bars), AF (grey bars) and control subjects (white bars) with the cobalt-calcein technique, using Fisher exact test.

The time elapsed until onset of mPT assessed by the cobalt-calcein technique (cob-calc) was less than that assessed by the thinness ratio technique (compare panel Fig. 1F with 2C), in the energized state; this likely reflects the toxicity of loading with calcein. Nevertheless, as shown in panel Fig. 2C, the calcimycin-induced swelling of *in situ* AF (grey bars) mitochondria was delayed, compared to other cell types during glucose deprivation and NaCN co-application (p-value: $e^{*}=0.002$, one-way ANOVA). The time elapsed until mPT onset of *in situ* AF (grey bars) mitochondria was also significantly increased in the de-energized compared to the energized state (p-value: $f^{*}=0.008$, t-test). The same tendency was observed in the experiments with DsRed2-transfected cells. Similar to the results obtained from DsRed2-expressing mitochondria, as shown in panel Fig. 2D, the number of cells exhibiting loss of mitochondrially-entrapped calcein fluorescence within the experimental time frame was also not statistically significant among any cell type. From the above data we concluded that the absence of ANT1 *increased* the elapsed time to mPT opening upon calcimycin addition only during the de-energized state.

Effect of Complete loss of ANT1 Gene Expression on Matrix Ca^{2+}Accumulation Upon Treatment with Calcimycin

To show that application of calcimycin to cells leads to matrix Ca^{2+}accumulation irrespective of the de-energization treatment and to address the possibility that the alterations seen in the elapsed time to mPT upon calcimycin addition between the cell lines (results obtained from the control cell lines were pooled) could be affected by alterations in the accumulation of Ca^{2+} in the mitochondrial matrix, we evaluated matrix Ca^{2+} levels by i) mitochondrially trapped X-rhod-1 (K*d* for $Ca^{2+}=0.7\,\mu M$) and ii) genetically encoded, mitochondrially-targeted, ratiometric calcium biosensor that contains enhanced CFP and circularly permuted Venus 173, 4mtD3cpv45, (K*d* for $Ca^{2+}=0.6\,\mu M$). This FRET-based ratiometric indicator exhibits a nearly exclusive preference for the mitochondrial compartment compared to X-rhod-1 distribution (not shown), though this methodology suffers from low transfection efficiency (typically less than 5%). On the other hand, with X-rhod-1 we could evaluate more cells, but the fluorescence signal could also be partially originating from the cytosolic compartment.

Mitochondrial matrix Ca^{2+} accumulation was induced in the various fibroblasts lines in the same manner as for the 'thinness ratio', while cells were either loaded with X-rhod-1 (panel Fig. 3A,C,E) or transfected with pcDNA-4mtD3cpv (panel Fig. 3B,D,F). As seen in panel Fig. 3A, representative traces (grey) of X-rhod-1 fluorescence signal representing matrix Ca^{2+} of *in situ* mitochondria are shown from a typical experiment; upon addition of calcimycin (2 µM), there is a gradual elevation in X-rhod-1 fluorescence that reaches a plateau within 30 minutes. The mean trace with S.E.M. bars is shown superimposed in black. In panel Fig. 3B, a representative trace of 4mtD3cpv FRET ratio fluorescence is shown, in response to addition of calcimycin. From such experiments we evaluated matrix Ca^{2+} levels in the various cell lines during the exact metabolic conditions as indicated for panel Fig. 1F, with the following two parameters: i) the rate of change in X-rhod-1 and 4mtD3cpv FRET ratio fluorescence within the first 10 minutes upon addition of calcimycin, and ii) the plateau of X-rhod-1 and 4mtD3cpv FRET ratio fluorescence reached within one hour upon addition of calcimycin. The results of the first parameter are shown in panel Fig. 3CD and for the second parameter in panel Fig. 3E,Ffor X-rhod-1 and 4mtD3cpv FRET ratio fluorescence, respectively. All of the results shown in Fig. 3 imply that there is a substantial elevation in matrix Ca^{2+} concentration induced by calcimycin even in the presence of cyanide or uncoupler. However, there was a statistically significant decrease in matrix Ca^{2+} accumulation recorded by 4mtD3cpv FRET ratio fluorescence (panel Fig. 3D) in AF cells compared to ST ($p=0.044$) and control fibroblasts ($p=0.042$) during de-energization by cyanide. No statistically significant difference was recorded for any other comparison.

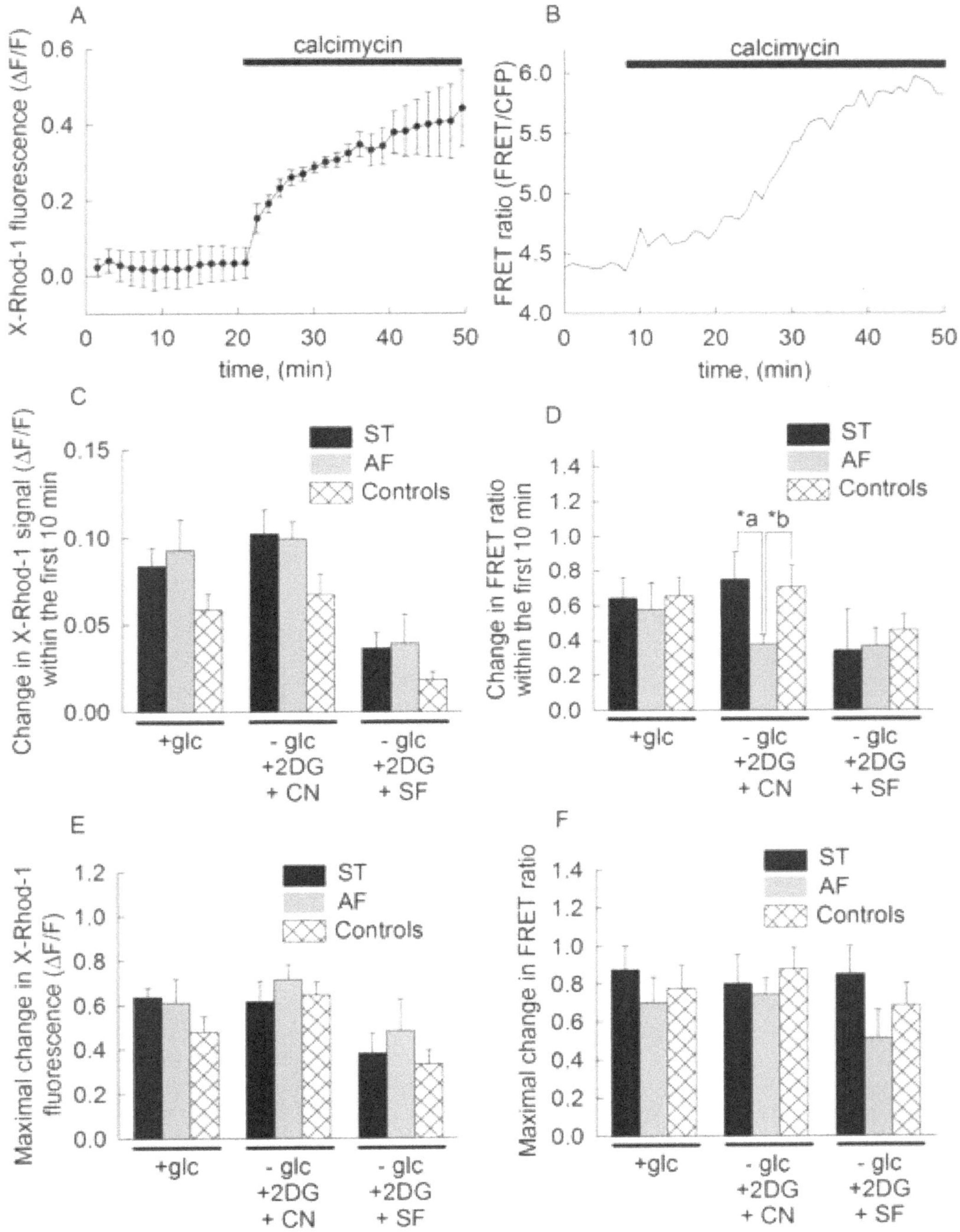

Figure 3: Effect of loss of ANT1 gene expression on matrix Ca^{2+}accumulation upon treatment with calcimycin.

(**A**) Representative traces (grey lines) and superimposed mean (black line with S.E.M.) of the effect of calcimycin (2 µM) on mitochondrially-trapped X-Rhod-1 fluorescence, reflecting matrix Ca^{2+} levels. (**B**) Representative trace of the effect of calcimycin (2 µM) on mitochondrially-targeted FRET-based 4mtD3cpv ratio fluorescence, reflecting matrix Ca^{2+}levels. (**C**) Quantification of changes in X-rhod-1 fluorescence within the first 10 min upon addition of calcimycin for various metabolic conditions indicated under the x-axis in cultured fibroblasts from the heterozygous mother (ST, black bars) the patient (AF, grey bars), and three control subjects (pooled cells shown in white hatched bars) during various metabolic conditions, detailed under the x-axis. Bars indicate means ± S.E.M. of 25–40 cells from at least 5 different cultures. (**D**) Quantification of changes in 4mtD3cpv ratio fluorescence within the first 10 min upon addition of calcimycin for various metabolic conditions indicated under the x-axis in cultured fibroblasts from the heterozygous mother (ST, black bars) the patient (AF, grey bars), and three control subjects (pooled cells shown in white hatched bars) during various metabolic conditions, detailed under the x-axis. Bars indicate means ± S.E.M. of 6–10 cells from at least 4 different cultures; p-values: *a = 0.042, *b = 0.044. (**E**) Quantification of maximal changes in X-rhod-1 fluorescence upon addition of calcimycin for various metabolic conditions indicated under the x-axis in cultured fibroblasts from the heterozygous mother (ST, black bars) the patient (AF, grey bars), and three control subjects (pooled cells shown in white hatched bars) during various metabolic conditions, detailed under the x-axis. Bars indicate means ± S.E.M. of 25–40 cells from at least 5 different cultures. (**F**) Quantification of maximal changes in 4mtD3cpv ratio fluorescence upon addition of calcimycin for various metabolic conditions indicated under the x-axis in cultured fibroblasts from the heterozygous mother (ST, black bars) the patient (AF, grey bars), and three control subjects (pooled cells shown in white hatched bars) during various metabolic conditions, detailed under the x-axis. Bars indicate means ± S.E.M. of 6–10 cells from at least 4 different cultures.

Effect of Complete Loss of ANT1 Gene Expression on *in situ* ΔΨm Upon Treatment with Calcimycin

As stated above, the threshold of mPT opening depends on the extent of ΔΨm; de-energization by cyanide in the abscence of glucose would lead to reversal of the mitochondrial ATP synthase in an attempt to maintain membrane potential, at the expense of hydrolysing ATP originating from either cytosol (glycolysis) or mitochondrial substrate-level phosphorylation[46,47,48,49,50]. Thus, during the de-energization protocol and in the presence of calcimycin, a complete collapse of ΔΨm in our cell lines is not warranted and the presence of a residual potential value is more than likely. To address the status of ΔΨm of the various fibroblast cell lines we measured mitochondrial membrane potential by TMRM and distinguishing it from a plasma membrane distribution signal by co-loading with DiBAC4(3), a negatively charged potentiometric dye. Our method is a variant of that described by Gerencser and colleagues[51] which used PMPI instead of DiBAC4(3). We used DiBAC4(3) because it exhibits less spectral overlap with TMRM. Representative epifluorescence images of TMRM and DiBAC4(3) fluorescence representing ΔΨm and plasma membrane potential (ΔΨp) respectively, are shown in Supplemental Fig. 1 during addition of the calibrants (see below); image processing flow of the spectral un-mixing performed as described in[52] is depicted in Supplemental Fig. 2, resulting in the images shown in Fig. 4A. Specifically, the calibration of plasma membrane potential was performed by establishing a K^+-equilibrium potential (K^+ equil) at the plasma membrane followed by stepwise increments of extracellular [K^+] (exact values indicated in panel Fig. 4B,C and Supplemental Fig. 1). The K^+-equilibrium potential was established by the application of 10 μM diazoxide. Finally, a cell membrane calibration cocktail (CDC) was applied. TMRM signals originating from the mitochondrial compartment was then recorded by spectrally un-mixing the DiBAC4(3) fluorescence signal as described in[51] using decomposition algorithms developed in[52].

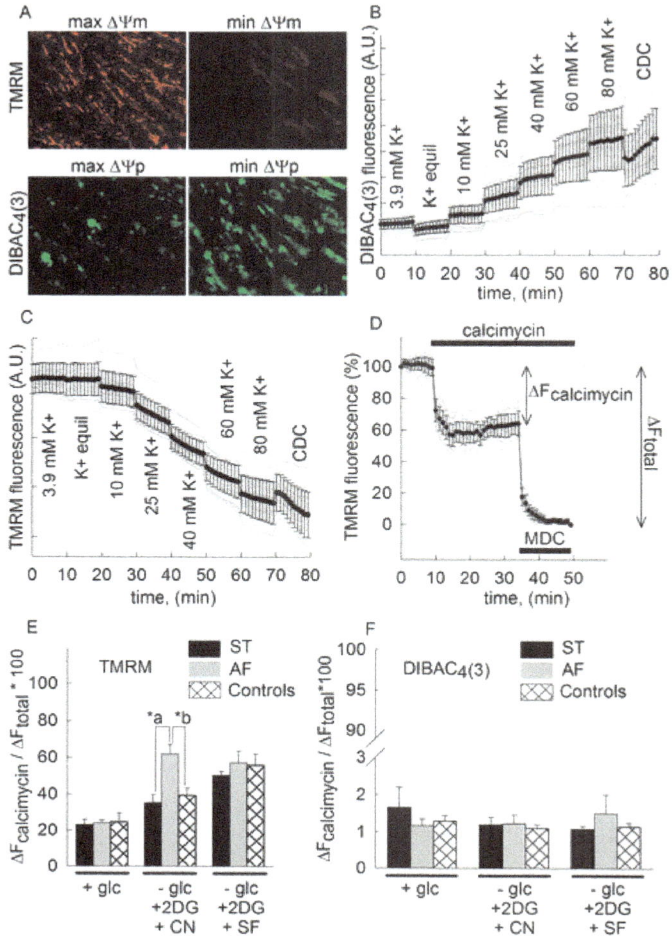

Figure 4: Effect of loss of ANT1 gene expression on in situ ΔΨm upon treatment with calcimycin. (**A**) Spectrally un-mixed epifluorescence images of TMRM and $DIBAC_4(3)$ fluorescence representing ΔΨm and ΔΨp respectively; the methodology of spectral un-mixing is outlined in Supplemental Figs 1 and 2. (**B**) $DIBAC_4(3)$ fluorescence traces (grey traces) after spectral un-mixing from TMRM fluorescence bleed-through. $DIBAC_4(3)$ fluorescence intensities (black scatters), indicating mean ± SEM of fluorescence measured over n = 20 fibroblasts (grey lines) in a representative experiment. (**C**) TMRM fluorescence traces after spectral un-mixing from $DIBAC_4(3)$ fluorescence bleed-through (corresponding $DIBAC_4(3)$ fluorescence shown in panel B). TMRM fluorescence intensities (black scatters), indicating mean ± SEM of fluorescence measured over n = 20 fibroblasts (grey lines) from the same experiment, shown in panel B. (**D**) Representative traces (grey lines) and superimposed mean (black line with S.E.M.) of the effect of calcimycin (2 μM) on TMRM fluorescence, spectrally un-mixed from $DIBAC_4(3)$ fluorescence, reflecting ΔΨm. (**E**) Quantification of changes in TMRM fluorescence, spectrally un-mixed from $DIBAC_4(3)$ fluorescence for the various metabolic conditions indicated under the x-axis, for cultured fibroblasts from the heterozygous mother (ST, black bars) the patient (AF, grey bars), and three control subjects (pooled cells shown in white cross-hatched bars). Bars indicate means ± S.E.M. of 168–227 cells from at least 5 different cultures; p-values: *a < 0.001, *b < 0.001. (**F**) Quantifica-

tion of changes in DIBAC$_4$(3) fluorescence for the various metabolic conditions indicated under the x-axis, for cultured fibroblasts from the heterozygous mother (ST, black bars) the patient (AF, grey bars), and three control subjects (pooled cells shown in white cross-hatched bars). Bars indicate means ± S.E.M. of 168–227 cells from at least 5 different cultures.

Representative traces (grey lines) of TMRM fluorescence (spectrally un-mixed from DiBAC4(3) fluorescence, see Supplemental Figs 1 and 2 and details under Materials and Methods) reflecting mitochondrial membrane potential from control fibroblasts under energized conditions are shown in panel Fig. 4D; the mean trace with S.E.M. bars is shown superimposed in black. As shown in panel Fig. 4D, addition of calcimycin decreased TMRM fluorescence by approximately 40%; residual ΔΨm was lost upon further addition of a mitochondrial membrane depolarization cocktail (MDC), the composition of which is detailed under Materials and Methods. From such experiments we evaluated ΔΨm levels in the various cell lines (results obtained from the control cell lines were pooled) during the exact metabolic conditions as indicated for panel Fig. 1F, by taking into account the ratio of calcimycin-induced ($\Delta F_{calcimycin}$) to MDC-induced (ΔF_{total}) change in TMRM fluorescence. The results are summarized in panel Fig. 4E; as shown in panel Fig. 4E, the mitochondrial de-energization induced by calcimycin -in addition to that by cyanide- is highly significantly ($p = 0.001$) greater in AF cells compared to either ST or control cells. Yet, the time lag until onset to mPT (shown in panel Fig. 1F) is greater under the same de-energization conditions for the AF cells. These two results are strongly supportive of the notion that lack of ANT1 renders mitochondria insensitive to modulation of mPT opening by the mitochondrial membrane potential. On the other hand, ΔΨp inferred by DiBAC4(3) fluorescence remained relatively unchanged (<2% changes), panel Fig. 4F. The fact that plasma membrane potential does not show large fluctuations, affords greater assurance that the TMRM signals are almost entirely due to fluctuations of the mitochondrial membrane potential, since TMRM is known to distribute across the plasma membrane as well53.

Effect of loss of ANT1 Gene Expression on *in situ* Mitochondrial Swelling Induced by H_2O_2 in Human Fibroblasts

To address the role of ANT1 on mPT induced by a different than Ca^{2+}-overload stimulus, we subjected the fibroblasts to an oxidant treatment by 1 mM H_2O_2. Reactive oxygen species are known to induce mPT54. Just like with calcimycin, treatment of DsRed2-transfected cells with H_2O_2(without omitting Mg^{2+} from the medium) led to an abrupt change in mitochondrial morphology which by the thinness ratio technique is quantified as swelling, thus implying mPT opening. Representative epifluorescence images of a cell before and after mPT induction are shown in panel Fig. 5A,B, respectively. A time-lapse series of such an experiment is shown in the Supplemental Material(mPT_DsRed2_control_HF_H2O2.avi). A representative thinness ratio trace is shown in panel Fig. 5C, calculated from the recorded time-series images. In the experiments using H_2O_2 as an mPT inducer, we did not variate the metabolic conditions by omitting glucose or including cyanide or uncoupler, because H_2O_2 is expected to exert a considerable metabolic challenge itself55. In AF, ST and pooled control subjects' cells, the following parameters were recorded: i) the time elapsed from the addition of H_2O_2 to the appearance of a large magnitude swelling (marked as 'onset to mPT', panel Fig. 5D), and ii) the number of cells exhibiting the large magnitude swelling considered as mPT within the experimental time frame (4 hours in the presence of 1 mM H_2O_2), depicted in panel Fig. 5E.

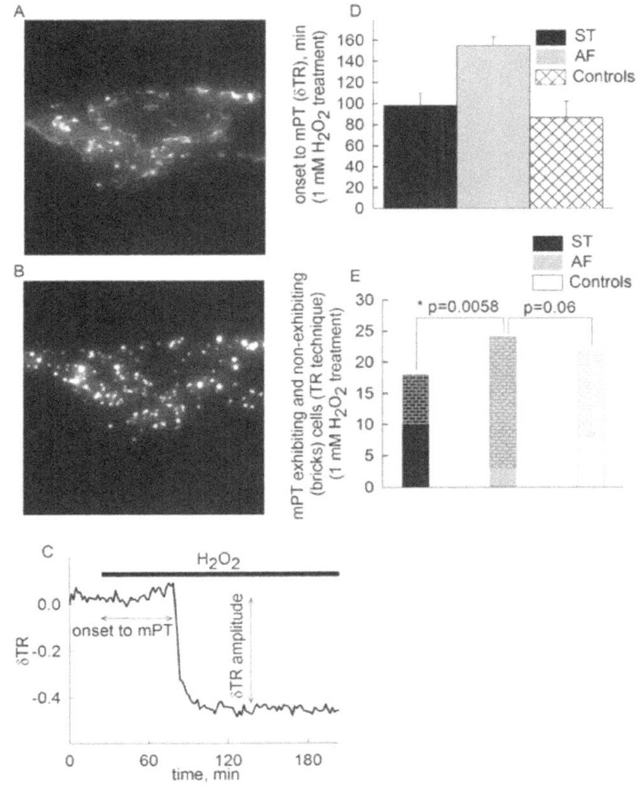

Figure 5: Effect of loss of *ANT1* gene expression on *in situ*mitochondrial swelling (visualized by DsRed2) induced by H_2O_2 in human fibroblasts. (**A**) Epifluorescence image of a fibroblast transfected with DsRed2 before induction of mPT. (**B**) Epifluorescence image of the same fibroblast as shown in panel 5A, after induction of mPT by H_2O_2. (**C**) Representative trace of the effect of H_2O_2 (1 mM) on the thinness ratio (δTR) indicating mitochondrial volume as a function of time. (**D**) Time elapsed between H_2O_2 application and mitochondrial swelling (onset to mPT) detected by wide field imaging of mito-DsRed2 expressed in cultured fibroblasts from the heterozygous mother (ST, black bars) the patient (AF, grey bars), and three control subjects (white cross-hatched bars) in the presence of glucose. The onset to mPT was determined by the sudden increase of mean mitochondrial diameter in the microscopic view field described as a sudden decrease of the thinness ratio (δTR). Bars indicate means ± S.E.M. of 18–24 cells from at least five different cultures. (**E**) Comparison of mPT-exhibiting and non-exhibiting cells (bricks) in fibroblasts from ST (black bars), AF (grey bars) and control subjects (white bars) with the TR technique, using Fisher exact test.

As shown in panel Fig. 5D, the H_2O_2-induced onset to mPT of *in situ* ST (black bar) mitochondria was similar to that from the control cells (cross-hatched white bar). However, the H_2O_2-induced onset to mPT of*in situ* AF (grey bar) mitochondria was delayed, compared to other cell types. Although this difference is large (~60% further delay in onset to mPT), it was not statistically significantly different from the other cell types but that is only because a very low number of cells (only three) underwent mPT, reducing the power of statistical analysis. Exactly because of this reason, as shown in panel Fig. 5E, the number of AF cells that underwent mPT was much smaller ($p = 0.0058$) that the other cell types, also leading to not significant differences between AF and control cells evaluated with the Fisher exact test ($p = 0.06$), but yielded significance using the less stringent Chi-square test ($p = 0.029$).

From the above data we concluded that the absence of ANT1 may increase the elapsed time to mPT upon H_2O_2 addition, but most importantly prevent cells from undergoing mPT. These results are in accordance to those by Kokoszka and colleagues26, where they showed that although tert-butyl hydroperoxide, a non-specific oxidant, and diamide, a specific –SH group oxidant, facilitated isolated mouse liver mitochondria to undergo mPT independent of the presence of ANT, ANT1−/− /ANT2−/− deficient mice exhibited more than three times higher maximum Ca^{2+} uptake capacity compared to wild-type mice26.

Effect of Knocking-down Ant1 Expression in C2C12 myotubes on *in situ* Mitochondrial Swelling Induced by calcimycin at Different Bioenergetic States

In order to evaluate the role of ANT1 on voltage-sensing properties of the mPT in a different cell context, we knocked-down its gene by stably expressing lentiviral shRNA targeted against mouse Ant1 as performed in56, and swelling of *in situ* mitochondria of these versus scramble RNA (scr) transfected cells were compared during Ca^{2+} overload induced by addition of calcimycin, as a function of various metabolic conditions. As in Kawamata *et al.*56, ANT1 expression in shRNA-treated cells was undetectable by Western blotting (see panel Fig. 6A), and qPCR for mRNA was reduced by over 90% (not shown). Expression of ANT2 was unaffected, (panel Fig. 6A). *In situ* mitochondria were visualized by wide field epifluorescence imaging of mitochondrially targeted DsRed2. Swelling was monitored by evaluating DsRed2-visualized mitochondrial morphology and changes in mean mitochondrial diameters were calculated using the thinness ratio technique. Representative epifluorescence images before and after induction of mPT are shown in panel Fig. 6B. Time-lapse series of such experiments from an shRNA-transfected and a scramble RNA-transfected C2C12 cell are shown in the Supplemental Material, (mPT_DsRed2_C2C12_shRNA_ANT1.avi, and mPT_DsRed2_C2C12_scramble_ANT1.avi). A representative thinness ratio trace is shown in panel Fig. 6C, calculated from the recorded time-series images. In the experiments performed with shRNA ANT1-transfected and a scramble RNA-transfected C2C12 cells during various metabolic conditions as described in the legend and panels, the following parameters were recorded: i) the time elapsed from the addition of calcimycin to the appearance of a large magnitude swelling (marked as 'onset to mPT', panel Fig. 6D) and ii) the number of cells exhibiting the large magnitude swelling considered as mPT within the experimental time frame (4 hours in the presence of 2 μM calcimycin), depicted in panel Fig. 6E. As shown in panel Fig. 6D, the calcimycin-induced onset to mPT in C2C12 cells with knocked-down ANT1 did not exhibit a statistically significant difference from scramble RNA transfected cells, under any metabolic conditions. However, as shown in panel Fig. 6E, the number of cells exhibiting large-magnitude swelling within the 4 hours of the experimental time frame was statistically significantly different among these cell types when comparing glucose deprivation and NaCN co-application versus no glucose deprivation. Nonetheless, the number of C2C12 cells with knocked-down Ant1 undergoing mPT increased less significantly ($p = 0.0089$ vs $p = 0.0237$) upon chemically-induced hypoxia than cells transfected with scramble RNA. This observation attests to the notion that ANT1 at least partially regulates mPT opening as a function of energization level of *in situ*mitochondria. However, in this experimental setting and just like in human fibroblasts, de-energization by uncoupler did not confer the same alterations in mPT exhibiting versus non-exhibiting cells.

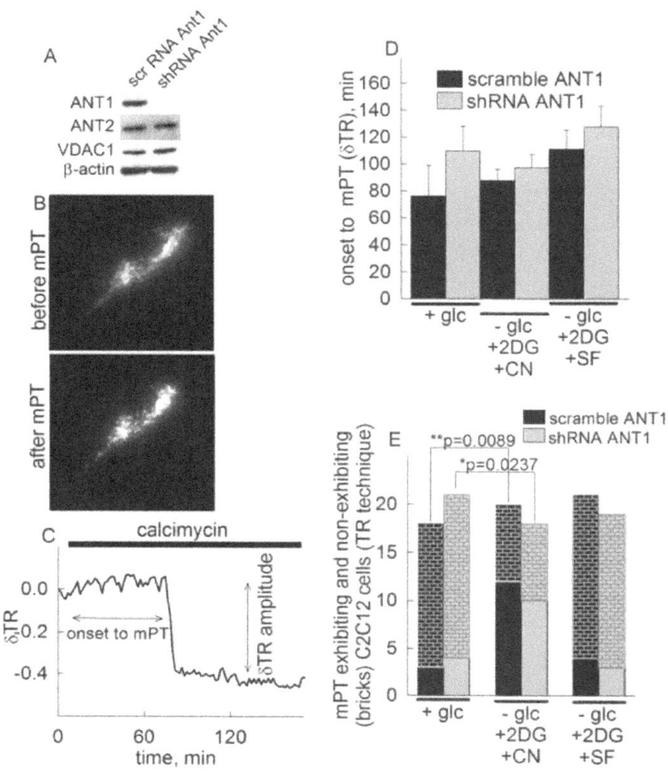

Figure 6: Effect of knocking-down Ant1 in C2C12 myotubes by shRNA using lentiviruses, versus scramble RNA-transfected cells on calcimycin-induced mitochondrial swelling or maximum calcium uptake capacity. (**A**) Scanned images of Western blots of C2C12 myotubes (10 μg loading per lane) transfected with shRNA directed against Ant1 or scramble (scr) RNA, for ANT1, ANT2, VDAC1 and β-actin. (**B**) Epifluorescence image of a C2C12 myotube transfected with DsRed2 before and after induction of mPT. (**C**) Representative trace of the effect of calcimycin (2 μM) on the thinness ratio (δTR) indicating mitochondrial volume as a function of time. (**D**) Time elapsed between calcimycin application and mitochondrial swelling (onset to mPT) detected by wide field imaging of mito-DsRed2 expressed in cultured C2C12 cells treated with shRNA for Ant1 or scramble RNA during various metabolic conditions, detailed under the x-axis. The onset to mPT was determined by the sudden increase of mean mitochondrial diameter in the microscopic view field described as a sudden decrease of the thinness ratio (δTR). Bars indicate means ± S.E.M. of 18–21 cells from at least five different cultures. (**E**) Comparison of mPT-exhibiting and non-exhibiting cells (bricks) detected with the TR technique during various metabolic conditions, using Fisher exact test.

Effect of Knocking-down Ant1 Expression in Permeabilized C2C12 Myotubes on the Voltage Threshold of Calcium-Induced mPT

We further evaluated the role of ANT1 on the voltage threshold of calcium-induced mPT under various metabolic conditions, in permeabilized C2C12 cells and measuring mitochondrial calcium uptake, ΔΨm, and cytochrome c release using the following protocols: i) maximum calcium uptake of *in situ* mitochondria supported by glutamate and malate (panels 7A and 7B, recording calcium uptake and ΔΨm, respectively), ii) challenge by submaximal $CaCl_2$ dose followed by titration with step-wise inhibition of respiration by NaCN (50 µM boluses) shown in panels 7C and 7D, recording calcium uptake and ΔΨm, respectively, iii) challenge by submaximal $CaCl_2$ dose followed by titration with step-wise dissipation of ΔΨm by SF6847 (10 nM boluses) shown in panels 7E and 7F, recording calcium uptake and ΔΨm, respectively, and iv) measuring cytochrome c released from the *in situ*mitochondria of the permeabilized cells during the following time points, indicated in panels 7C and 7E: 1 = before calcium additions, 2 = right after calcium additions, 3 = before loss of sequestered calcium and membrane potential, 4 = after loss of sequestered calcium in scramble RNA treated cells but not shRNA treated cells, 5 = after loss of sequestered calcium and membrane potential in both cells types, plus a time control for point 5 (cells were probed for cytochrome c release after 20 min of incubation but were not challenged by either $CaCl_2$ or NaCN or uncoupler) and an alamethicin (a pore-forming peptide) control. The amount of cytochrome c appearing in the supernatant upon treatment with alamethicin provides a measure of the maximum amount releasable57. Calcium uptake and ΔΨm measurements were performed in parallel but different samples (but from the same cultures) due to spectral overlap of calcium green and safranine O. The assessment of maximum calcium uptake capacity is a widely utilized protocol58,59, that can be performed in isolated60, or *in situ*mitochondria of permeabilized cells. In our hands, permeabilized human fibroblasts exhibited a very small maximum mitochondrial calcium uptake capacity, less than 10 nmol/mg protein, and therefore, experiments with these cells were not conducted. On the contrary, permeabilized C2C12 myotubes exhibited very large maximum calcium uptake capacities, providing the opportunity for a more thorough investigation of mPT opening by calcium loading as a function of various metabolic conditions. We have tried to isolate mitochondria from fibroblasts and C2C12 cells, but there were insurmountable obstacles with both cell types; with fibroblasts, an extremely small amount of isolated mitochondria was obtained, not suitable for functional experiments; with C2C12 cells only very harsh homogenization methods could yield sufficient amount of mitochondria, but they were damaged to the extent that they were too leaky for functional studies. For these reasons, we only evaluated C2C12 cells. shRNA-transfected and scramble RNA-transfected C2C12 cells were permeabilized as described under 'Materials and Methods', boluses of 100 µM $CaCl_2$ (panel Fig. 7A,B) were added where indicated, and extramitochondrial Ca^{2+} levels were followed by Calcium Green 5N hexapotassium salt fluorescence (Fig. 7A), or ΔΨm, using safranine O (Fig. 7B). As shown in panel Fig. 7A, C2C12 cells with knocked-down Ant1 exhibited larger mitochondrial calcium uptake capacity than scramble-transfected cells, in the presence of glutamate (5 mM) and malate (5 mM). It is to be noted, that this is not a condition in which mitochondria are continuously fully energized; it is well understood that upon progressive mitochondrial calcium sequestration, mitochondria depolarize accordingly61. Indeed, as shown in figure panel 7B, ΔΨm of both cell types gradually dissipates upon increasing calcium load.

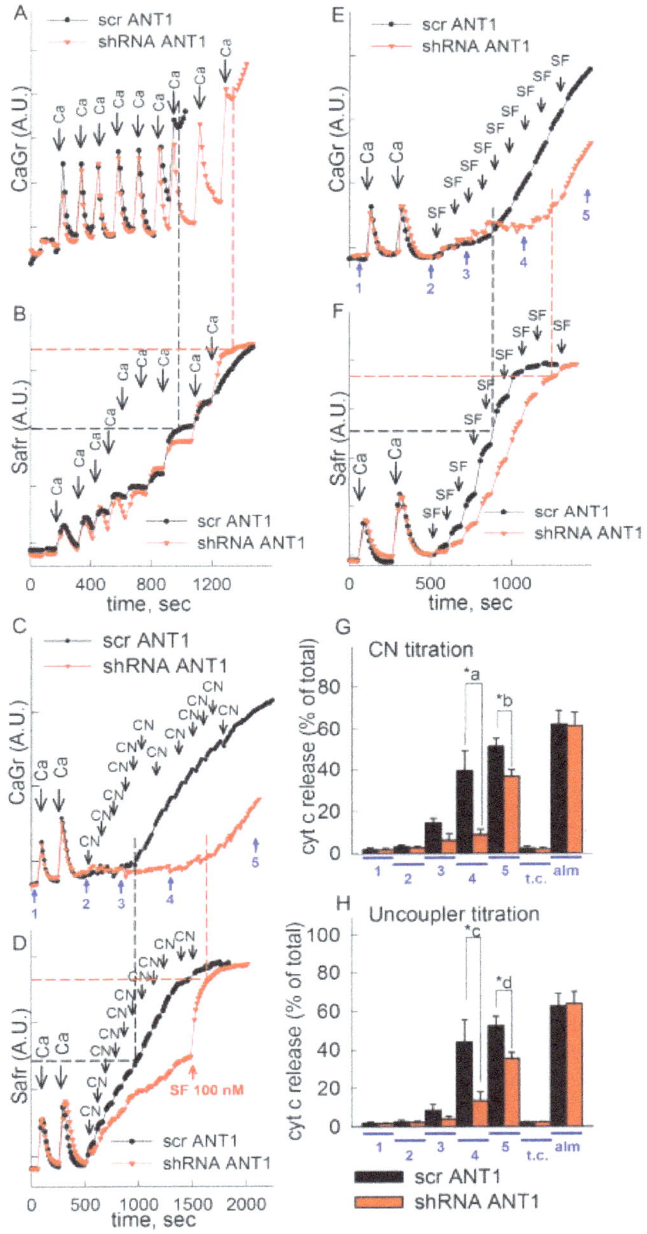

Figure 7: Determination of voltage thresholds of calcium-induced mPT in permeabilized C2C12 cells treated with shRNA for Ant1 (red inverted triangles) or scramble RNA (black closed circles), under various metabolic conditions. **(A)** Maximum calcium uptake capacity in permeabilized C2C12 cells treated with shRNA for Ant1 or scramble RNA. Each addition of $CaCl_2$ (Ca) is 100 µM. **(B)** Safranine O fluorescence traces of C2C12 cells treated exactly as in panel A. Mitochondrial substrates were 5 mM glutamate and 5 mM malate. **(C)** Cells were challenged by two additions of $CaCl_2$ (Ca, 100 µM each) followed by consecutive additions of 50 µM NaCN, and extramitochondrial calcium was measured by CaGr. Blue numbers and arrows signify time-points of probing for cytochrome c release (see panel **G**). **(D)** Safranine O fluo-

rescence traces of C2C12 cells treated exactly as in panel C, with the exception that shRNA-treated cells received a bolus of 100 nM SF6847 as shown by the red arrow, to indicate that prior to this addition a significant portion of ΔΨm remained. (**E**) Cells were challenged by two additions of $CaCl_2$ (Ca, 100 μM each) followed by consecutive additions of 10 nM SF6847, and extramitochondrial calcium was measured by CaGr. Blue numbers and arrows signify time-points of probing for cytochrome c release (see panel H). (**F**) Safranine O fluorescence traces of C2C12 cells treated exactly as in panel E. All traces shown in panels (A-F) are representative of 5 independent experiments. (**G**) Cytochrome c release (% of the total) estimated by enzyme-linked immunosorbent assay from permeabilized C2C12 cells probed at time points (1–5) indicated in panel C; *a=0.034, *b=0.049. (**H**) Cytochrome c release (% of the total) estimated by enzyme-linked immunosorbent assay from permeabilized C2C12 cells probed at time points (1–5) indicated in panel E; *c=0.047, *d=0.041. 'Time-controls (t.c.)' signify probing of C2C12 cells at time-point 5 for cytochrome c release that have not been treated with either $CaCl_2$ or NaCN nor SF6847. 'Alm' signifies probing of C2C12 cells for cytochrome c release that have been treated with 80 ngr of alamethicin. Bars indicate means±S.E.M. of ~270,000 cells from 3 different cultures.

However, as it is marked by black dashed line for scr RNA-treated cells and red dashed line for shRNA-treated cells, those cells with knocked-down ANT1 expression exhibited loss of sequestered calcium in a much more depolarized ΔΨm range, nearing that of complete depolarization. Next, to obtain a more accurate indication of how mPT opening relates to ΔΨm, we challenged *in situ* mitochondria of permeabilized C2C12 cells with an amount of $CaCl_2$ that is insufficient to trigger pore opening, and then dissipated ΔΨm gradually by inhibiting complex IV with cyanide (50 μM boluses, Fig. 7C,D) or using the uncoupler SF 6847 (10 nM boluses Fig. 7E,F), while recording extramitochondrial Ca^{2+} levels (Fig. 7C,E), or ΔΨm (Fig. 7D,F). In figure panel 7D the addition of 100 nM SF 6847 is also shown (red arrow), signifying that just prior to its addition there was a considerable extent of ΔΨm that could be dissipated. As shown in Fig. 7C–F and marked by black dashed line for scr RNA-treated cells and red dashed line for shRNA-treated cells, those cells with knocked-down ANT1 expression exhibited loss of sequestered calcium in a much more depolarized ΔΨm range, again nearing that of complete depolarization. Furthermore, as shown in figure panels 7G and 7H, a higher percentage of cytochrome c was released from scr RNA treated cells at time points 4 and 5 compared to that from shRNA treated cells, implying that the loss of ΔΨm and sequestered calcium during the same time frame is due to large amplitude swelling resulting in rupture of the outer mitochondrial membrane, implying mPT opening. From these experiments we concluded that ANT1 regulates the voltage threshold of calcium-induced mPT opening.

Effect of Complete Loss of ANT1 gene Expression on ADP-ATP Exchange Rates as a Function of ΔΨm

Human fibroblasts express three (1, 2 and 3) out of the four known isoforms of ANT. *ANT2* was not evaluated particularly for AF and ST cells in ref. 34, therefore we quantified this transcript by qPCR. As shown in panel Fig. 8A, AF cells did not exhibit a statistically significant increase of the *ANT2* gene expression compared to ST and control subject cells. This and the unchanged levels of *ANT3* expression reported in ref. 34 yielded ADP-ATP exchange rates as a function of ΔΨm that did not differ between AF, ST and control subjects' cells, shown in panel Fig. 8B. As reported in ref. 34, AF cells did not express*ANT1* (verified in results shown in panel Fig. 1B,C). From these results we concluded that the differences in onset of the calcimycin-induced mPT between AF, ST and control subjects' cells were unlikely to be attributed to alterations in the adenine nucleotide and/or ΔΨm levels. The determination of the amount of functional ANTs by means of titration with carboxyatractyloside while measuring respiration could not be reliably performed due to the low amount of mitochondria that could be accessed in the permeabilized cultures; this is because the amount of carboxyatractyloside required, was estimated to be in the low nanomolar range, *i.e.* near the K_i of the inhibitor for the ANT (not shown). Neither AF nor ST cells exhibited appreciable ANT reversal rates, probably due to the prominent expression of IF-1 in human fibroblasts62, also supported by our findings with the cells from three different control subjects, shown in panel figure bB.

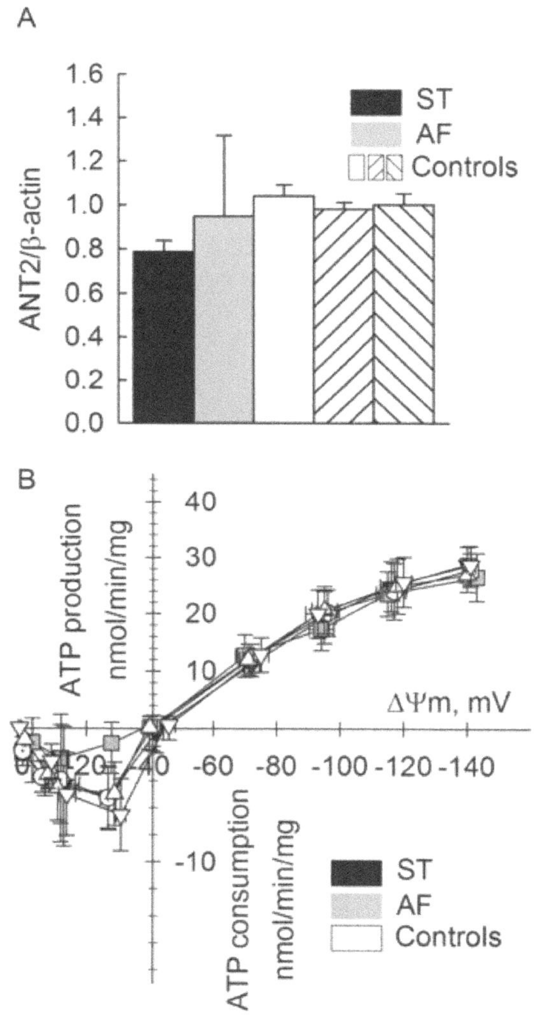

Figure 8: ***ANT2* expression and ADP/ATP exchange rates as a function of mitochondrial membrane potential.** (**A**) Reverse transcription PCR and amplification of cDNAs reflecting expression of *ANT2* gene in cultured fibroblasts of the patient (AF), the heterozygous mother (ST) and three control subjects, as a ratio to

β-actin expression. (**B**) Plot of ATP–ADP exchange rate mediated by ANT versus ΔΨm in *in situ* mitochondria of AF (grey squares), ST (black squares) and control subjects (white symbols) permeabilized cells depolarized to various voltages by increasing amounts of SF 6847; constructed from the data of 3 independent experiments.

DISCUSSION

mPT is implicated in many human diseases[2,63]. Extensive characterization of the phenomenon has assisted in the recognition of certain proteins as regulatory components, such as cyclophilin D and ANT1. VDAC was also initially considered as a structural component, however, tissues engineered to lack all known isoforms still exhibited mPT pore opening and regulation, reviewed in[64]. Furthermore, closure of VDAC hastened mPT pore opening by means of impeding the efflux of superoxide anions from the intermembrane space[65]. Likewise, the outer membrane translocator protein of 18 kDa, TSPO, was originally reported to participate in mPT regulation[66,67,68], a claim to be recently disproved[69]. Prior to the report that ANT-deficient mice still exhibit the mPT[26], the ANT was already dismissed as a structural element of the pore in dimer formation induced by pore-inducing oxidants[70]; however, formation of ANT dimers need not materialize for pore assembly. Oxidants and other chemical modifiers are known to modulate the mPT; there are at least three sites for oxidation tuning a sensor that translates the changes of both transmembrane voltage and surface potential into changes of the mPT pore open probability[63]. The critical arginines thought to tune the voltage sensor of the pore were considered not to reside in the ANT[18,19]. However, the experiments shown above support the notion that in human fibroblast mitochondria and in C2C12 myotubes the sensor -or at least a part of it- is likely to be isoform 1 of the ANT. Therefore, the sensing property of ANT1 is not mediated through its arginines. The possibility that other oxidation-prone sites tuning the pore sensor reside on ANT1 remains high; indeed, in the presence of oxidants, $ANT1^{-/-}$ /$ANT2^{-/-}$ deficient mice exhibited more than three times higher maximum Ca^{2+} uptake capacity compared to wild-type mice[26].

Most of the experiments by Bernardi's group that proved the inverse correlation of the proton electrochemical gradient to mPT pore opening probability were performed with rat liver mitochondria where ANT2 is the dominant isoform mediating ADP/ATP exchanges, however, in rats ANT1 also exhibits significant expression[71]. ANT1 (but not ANT2) is considered proapoptotic[72]. The proapoptotic property of ANT1 is extremely dominant: overexpression of ANT1 in tumors *in vivo* (where otherwise ANT2 is mostly expressed) induces apoptosis and tumor regression[73]. However, silencing of ANT1 expression in a human glioblastoma cell line induced a paraptotic-like process[74]. On the other hand, overexpression of ANT1 is associated with autosomal dominant facioscapulohumeral muscular dystrophy[75]. In addition, it was shown that mice deficient in ANT1 exhibit increased resistance to excitotoxic insults of the brain[76]. Brain mitochondria from these mice exhibited ~20% higher maximum Ca^{2+} uptake capacity; unfortunately they were not investigated for mPT under de-energized conditions[76]. In[26], WT and $ANT1^{-/-}$/$ANT2^{-/-}$ -deficient mouse liver mitochondria responded equally to uncoupler-induced swelling; we also found no statistically significant difference between ST, control and AF cells, even though there was a trend for AF mitochondria to exhibit smaller 'onset to mPT' values. As a word of caution though: the $ANT1^{-/-}$/$ANT2^{-/-}$ -deficient mice mitochondria exhibited stark differences in bioenergetic parameters and expression levels of other mitochondrial proteins[26], likely the result of compensatory mechanisms, while in this study fibroblasts and C2C12 myotubes were not confounded by other bioenergetic alterations, in line with the results obtained in[34]. However, it is interesting that in AF cells there seems to be a rebound upregulation in ANT2 expression. Since this was observed in Western blots but not Northern blots, it is likely to be due to regulation in the posttranslational modification of the ANT2 protein, and/or its degradation pathway. Relevant to this, different voltage-sensing properties of different ANT isoforms is a possibility. We do acknowledge though that to truly isolate ANT1's functions from possible compensatory effects of other ANT isoforms, re-introduction of expression of ANT isoform 1 in AF cells would prove that only ANT1 is sensitive to alterations of the electrochemical gradient. However, to achieve this, ANT1 level expression should be matched to those observed in the control cell lines, a task posing serious technological challenges.

There is an important difference between de-energization of cells by glucose deprivation and either chemical anoxia, or an uncoupler; in the former, there is no electron flow in the electron transport chain; in the latter, there is high electron flow, despite glucose deprivation due to the presence of endogenous substrates that can fuel the citric acid cycle and yield reducing equivalents. mPT is indeed tuned by a voltage sensor, but in addition to that, electron flow through complex I strongly dictates pore-opening probability[15,77,78] and is maybe part of the voltage-sensing mechanism. Furthermore, there is a potential additional confounding factor regarding the action of uncouplers: mitochondrial Ca^{2+} release by uncouplers is usually ascribed to matrix acidification[79,80], which induces dissociation of the Ca^{2+}phosphate complex. However, matrix acidification by nigericin in the presence of a complete collapse of ΔΨm by combined inhibition of the respiratory chain and the F_0-F_1 ATP synthase, lead to the release of only one-fifth of sequestered calcium[61,81]. Thus, uncouplers exert some other, yet to be identified, effect regarding matrix Ca^{2+} release, which may share a common denominator with induction of the permeability transition. In the present report we show that in *in situ* mitochondria of ANT1-deficient human fibroblast and C2C12 cells with knocked-down ANT1 by a lentiviral approach, the hastening of Ca^{2+}-induced mPT pore opening during glucose deprivation and chemical anoxia is reduced or abolished. A similar picture emerged from or H_2O_2- treated cells shown to induce mPT opening, with well-documented adverse bioenergetic effects. Finally, voltage-thresholds of calcium-induced mPT opening in permeabilized C2C12 cells were higher if ANT1 was knocked down, than scramble RNA treated cells. We therefore conclude that ANT1 is the module conferring sensitivity of the pore to voltage likely through the electrochemical proton gradient across the inner membrane and/or the electron flow though the electron transport chain. This is in line with previous postulations that ANT either exerts an indirect effect on the mPT by influencing surface potential[82,83], or through altered ANT conformation induced by adenine nucleotide binding affected by membrane potential[32].

MATERIALS AND METHODS

Cell Cultures and Transfections

Fibroblast cultures from skin biopsies from the patient with no ANT1 expression (AF) the heterozygous mother (ST) and three control subjects were prepared. All experiments were conducted in accordance with the guidelines of Ethical Rules for Using Human Tissues for Medical Research in Hungary (HM 34/1999) and the Code of Ethics of the World Medical Association (Declaration of Helsinki), and were approved by the Strasbourg University Hospital Ethics Committee. Informed consent was obtained from the patient, her mother, and three healthy donors. Cells were grown on poly-L-ornithine coated 8-well LabTek II chambered coverglasses (Nunc, Rochester, NY, USA) for 2–3 days, at a density of approximately $3*10^5$ cells/well in RPMI1640 medium (GIBCO, Life technologies, Carlsbad, CA, USA) supplemented with 10% fetal bovine serum and 2 mM glutamine and kept at 37 °C in 5% CO_2. The medium was also supplemented with penicillin, streptomycin and amphotericin (item A5955, Sigma-Aldrich St. Louis, MO, USA). Cultures were transfected on the 1st or 2nd DIV with mito-DsRed2 using Lipofectamine 2000 (Life technologies) in RPMI1640 medium at a 3:2 ratio of Lipofectamine (μl) to plasmid DNA (μg) or with mitochondria targeted FRET-based calcium indicator, 4mtD3cpv using flashFECTIN (Oxford Expression Technologies, Oxford, UK) according to the manufacturer's instructions. Experiments were carried out at day 1–2 post transfection. Typical transfection rates were <5% and therefore individual, non-overlapping cells were visualized.

Imaging of Cultured Cells and Estimation of *in situ* Mitochondrial Swelling with the TR Technique

Time lapse epifluorescence microscopy was carried out to image cells expressing mito-DsRed2 at 34 °C and swelling of *in situ* mitochondria was measured by the thinness ratio (TR) technique exactly as described in44. Basically, cells were imaged directly on the LabTek chamber without superfusion in a medium containing (in mM): 120 NaCl, 3.5 KCl, 1.3 $CaCl_2$, 20 HEPES, 15 glucose (where indicated) at pH 7.4. Experiments were performed on an Olympus IX81 inverted microscope equipped with a 60 × 1.4 NA oil immersion lens, a Bioprecision-2 *xy*-stage (Ludl Electronic Products Ltd., Hawthorne, NY) and a 75 W xenon arc lamp (Lambda LS, Sutter Instruments, Novato, CA). For mito-DsRed2 an 535/20 nm exciter, a 555LP dichroic mirror and an 570LP emitter (Omega Optical, Brattleboro, VT) were used. Time lapses of *z*-series of 16 planes of 512 × 512 pixels frames (digitized at 14bit with no binning, 250 ms exposure time, yielding 0.1 μm pixel size and 0.8 μm *z*-spacing) were acquired using an ORCA-ER2 cooled digital CCD camera (Hamamatsu Photonics, Hamamatsu, Japan) under control of MetaMorph 6.0 software (Molecular Devices; Sunnyvale, CA, USA). The TR technique measures changes of average diameters of thread-like or punctate structures in fluorescence images using a pair of (high and low frequency) bandpass spatial filters. A calibration image series of mito-DsRed2 fluorescence showing mitochondrial swelling by valinomycin (200 nM) was recorded and used to train a spatial bandpass filter set in Image Analyst MKII (Image Analyst Software, Novato, CA). To calculate TR, for each time point the z-stack was mean-intensity projected and the projection image was duplicated. Then, both images were spatially filtered and the absolute value of pixels was taken. The TR was calculated as the ratio of the average fluorescence intensity in the high frequency band pass filtered over the low frequency band pass filtered image. Mitochondrial swelling causes the loss of high spatial frequency image details, therefore a decrease in the TR value. Baseline normalized TR is given as $\delta TR = (TR - TR_0)/TR_0$.

Mitochondrial Ca^{2+} imaging

Two imaging approaches were used: X-Rhod-1 imaging or mitochondria targeted FRET-based calcium indicator, 4mtD3cpv imaging. In X-Rhod-1 experiments cultured cells were loaded with 4 micromole X-Rhod-1-AM for 15 min at 37 °C, in a medium containing (in mM): 120 NaCl, 3.5 KCl, 1.3 $CaCl_2$, 20 HEPES, 15 glucose (where indicated) at pH 7.4, then for 25 min cultures were kept in dark for 'de-esterification'. Experiments were performed at 34 °C on the setup mentioned above with a 60 × 1.4 NA oil immersion lens. For XRhod1 a 535/20 nm exciter, a 555LP dichroic mirror and a 570 LP emitter were used, all from Chroma Technology Corp., Bellows Falls, VT, USA. Time lapses of *z*-series of 5 planes of 1342 × 1024 pixels frames (digitized at 14bit with no binning, 250 ms exposure time, yielding 0.1 μm pixel size and 1.6 μm *z*-spacing) were acquired. For each time point the z-stack was mean-intensity projected. In 4mtD3cpv experiments cultured cells were imaged in a medium containing (in mM): 120 NaCl, 3.5 KCl, 1.3 $CaCl_2$, 20 HEPES, 15 glucose (where indicated) at pH 7.4. Experiments were performed at 34 °C on the setup mentioned above with a 60 × 1.4 NA oil immersion lens. For 4mtD3cpV the following filter sets were used: a 440/20 exciter, a dual bandpass (450–490 nm and 520–560 nm transmitting) dichroic mirror, a 480/10 emitter for CFP (CFP channel) and 535/20 emitter for YFP variant (FRET channel), all from Chroma Technology Corp., Bellows Falls, VT. Time lapses of 1344 × 1024 pixels frames for both channels (digitized at 14 bit with no binning, 400 msec exposure times) were acquired.

Estimation of Onset of mPT with the Cobalt-calcein Technique

Because transfection efficiency with mito-DsRed2 was low, in addition to the thinness ratio technique we assessed onset of mPT implying *in situ* mitochondrial swelling by the cobalt-calcein method, developed by Petronilli *et al.*43. Briefly, cells were loaded with 50 nM calcein-AM (above 100 nM, calcein was toxic for these cells) for 20 min at 4 °C, in a medium containing (in mM): 120 NaCl, 3.5 KCl, 1.3 $CaCl_2$, 20 HEPES, 15 glucose (where indicated) at pH 7.4. Subsequently, 1 mM $CoCl_2$ was added, and the LabTek chamber was mounted on the microscope stage. Experiments were performed at 34 °C on the setup mentioned above with a 60 × 1.4 NA oil immersion lens. For calcein an 488/6 nm exciter, a 505LP dichroic mirror and an 535/25 emitter (all from Chroma Technology Corp., Bellows Falls, VT) were used. Time lapses of *z*-series of 5 planes of 1342 × 1024 pixels frames (digitized at 14 bit with no binning, 250 ms exposure time, yielding 0.1 μm pixel size and 1.6 µm *z*-spacing) were acquired. For each time point the z-stack was mean-intensity projected.

Simultaneous Measurement of Mitochondrial (ΔΨm) and Plasma Membrane Potential (PMP)

Cultures were incubated at 37°C in imaging medium containing (in mM): 120 NaCl, 3.5 KCl, 1.3 $CaCl_2$, 20 HEPES, 15 glucose (where indicated) at pH 7.4 with TMRM (180 nM) plus the bis-oxonol type plasma membrane potential indicator, $DIBAC_4(3)$ (250 nM, Life Technologies Inc.) for 60 min before the experiment. Experiments were performed at 34 °C on the setup mentioned above with a UAPO 20× air 0.75NA lens. Time lapses of 1342 × 1024 pixels frames (digitized at 12 bit with 4 × 4 binning, 250 msec exposure time for TMRM and 100 msec exposure time for $DIBAC_4(3)$ were acquired. For the illumination of $DIBAC_4(3)$ a 490/10 nm exciter, a 505LP dichroic mirror and 535/25 emission filter were used, for TMRM a 535/20 nm exciter, a 555LP dichroic mirror and a 630/75 emitter were used, all from Chroma Technology Corp., Bellows Falls, VT. The cross-talk of TMRM and $DIBAC_4(3)$ emissions was eliminated by a linear spectral un-mixing algorithm implemented in Image Analyst MKII (Image Analyst Software, Novato, CA) as previously described51, using decomposition algorithms developed in52. The algorithm calculates un-mixed fluorescence intensities by solving a linear equation with the cross-talk coefficient matrix. Coefficient matrix was determined by loading the cultures either with TMRM or $DIBAC_4(3)$ while both emissions (TMRM and $DIBAC_4(3)$) were recorded, see Supplementary Fig. 1. At the end of every experiment, full mitochondrial depolarization and plasma membrane depolarization was achieved by the application of mitochondrial (MDC) and plasma membrane depolarization cocktails (CDC). MDC contained (in µM): 1 valinomycin, 1 SF 6847, 2 oligomycin; CDC contained (in µM): 1 valinomycin, 1 SF 6847, 2 oligomycin, 10 nigericin, 10 monensin.

Northern Blotting

Total RNA was isolated using Trizol reagent (Life technologies). RT-PCR was performed with the Phusion RT-PCR kit (Thermo Scientific, Waltham, MA, USA) according to the manufacturer's instructions. Primers (Sigma-Aldrich, St. Louis, MO, USA) were designed using Primer384, and were identical to those used by Le Bras *et al.*85. The primers were the following: hANT1-F: ATGGGTGATCACGCTTGGAGCTTCCTAAAG and hANT1-R: TTAGACATATTTTTTGATCTCATCATACAA, hANT2-F: cagcagtctgcctcctcttt and hANT2-R: aagctttgcctccttcatca.

qPCR

Total RNA from the fibroblast cultures was isolated using Trizol reagent (Life technologies). cDNA was synthesized with Oligo d(T)23 VN primer (New England Biolabs, Ipswich, MA, USA) and M-MuLV Reverse Transcriptase (New England Biolabs) as recommended by the manufacturer. Real-time PCR was performed using Power SYBR PCR Mastermix (Life Technologies) and ABI 7900HT Fast Real Time PCR cycler. Primer sequences were obtained from PrimerBank (http://pga.mgh.harvard.edu/primerbank/index.html): ATCACGCTTGGAGCTTCCTAA - ANT1-F, TGCTTCTCAGCACTGATCTGT – ANT1-R; TTATAGACTGCGTGGTCCGTA – ANT2-F; GGCGAAGTTAAGAGCCTGGG – ANT2-R. The results were normalized with the housekeeping human β-actin gene (primers: CTGGCACCCAGCACAATG– HsACTB-F; CCGATCCACACGGAGTACTTG– HsACTB-R86). The data were obtained from 3 independent experiments.

Ant1-silenced C2C12 stable cells

Mouse Ant1 in C2C12 cells was silenced with MISSION shRNA lentiviral transduction particles (Sigma, St Louis, MO), as described previously56. Briefly, stable cells were created based on the manufacturer's protocol in C2C12 myoblasts and selected with 1 mg/ml puromycin. Scrambled shRNA lentivirus was used to create the control cell line.

Harvesting and Permeabilization of Cells for ADP-ATP Exchange rate Measurements and Determination of Membrane Potential

Approximately 1 million fibroblasts (per experiment) cultured as above in poly-L-ornithine coated T-175 cm^2 flasks (Nunc, Rochester, NY, USA) were washed once in phosphate-buffered saline and harvested with 1 ml of 0.25% trypsin-EDTA, inactivated by 9 ml calf serum, followed by centrifugation at 1,100 g for 2 minutes. Next, cells were washed once in ice-cold buffer containing, in mM: KCl 8, K-gluconate 110, NaCl 10, Hepes 10, KH_2PO_4 10, EGTA 0.005, mannitol 10, $MgCl_2$ 1, and 0.5 mg/ml bovine serum albumin (fatty acid-free), pH 7.25 without disturbing the pellet. After the wash, cells were resuspended in 0.2 ml of the same buffer but pre-warmed to 37 °C and also including mitochondrial substrates (5 mM glutamate and 5 mM malate), 25 µM sodium orthovanadate, beryllium trifluoride (see below regarding final concentration) and 50 µM digitonin. Both ADP-ATP exchange rate- and ΔΨm measurements of *in situ* mitochondria in permeabilized cells occurred in this buffer. The rationale for using this particular buffer composition is elaborated in87·88. Glutamate and malate were chosen as mitochondrial substrates because they support mitochondrial substrate-level phosphorylation, and as such, they contribute to greater ATP efflux rates46·49·50.

$[Mg^{2+}]_{free}$ Determination From Magnesium Green Fluorescence in the Experimental Volume Containing Permeabilized Fibroblasts and Conversion to ADP-ATP Exchange rate

ADP-ATP exchange rate was estimated using the recently described fluorimetric method by our laboratory87, exploiting the differential affinity of ADP and ATP to Mg^{2+}, with the modifications required for digitonin-permeabilized cells, using sodium orthovanadate and beryllium trifluoride, see below and ref. 88. The rate of ATP appearing in the medium following addition of ADP (2 mM) to energized *in situ*mitochondria was calculated from the measured rate of change in free extramitochondrial $[Mg^{2+}]$ using standard binding equations. $[Mg^{2+}]$ in the extramitochondrial environment was measured using the fluorescent indicator, Magnesium Green $5K^+$ salt (1 µM). The assay is designed such that the ANT is the sole mediator of changes in $[Mg^{2+}]_{free}$in the extramitochondrial volume, as a result of ADP-ATP exchange. For the calculation of [ATP] or [ADP] from $[Mg^{2+}]_{free}$, the apparent K_d values are identical to those in (26) due to identical experimental conditions (K_{ADP}=0.906±0.023 mM, and K_{ATP}=0.114±0.005 mM). Magnesium Green fluorescence was recorded in Tecan Infinite® 200 PRO series plate reader (Tecan Deutschland GmbH, Crailsheim, Germany) at a 0.5 Hz acquisition rate, using 505 and 535 nm excitation and emission wavelengths, respectively. Experiments were performed at 37 °C.

Mitochondrial Membrane Potential (ΔΨm) Determination of *in Situ*mitochondria of Permeabilized Cells

ΔΨm was estimated using fluorescence quenching of the cationic dye safranin O due to its accumulation inside energized mitochondria46, also taking into account the considerations discussed in89 and90. Briefly, ~30,000 C2C12 cells (per experiment) were cultured in 35 mm wells or ~1 million fibroblasts (per experiment) cultured in poly-L-ornithine coated T-175 cm^2 flasks were washed once in phosphate-buffered saline and harvested with 0.25% trypsin-EDTA, inactivated by calf serum, followed by centrifugation at 1,100 g for 2 minutes. Next, cells were washed once in an ice-cold buffer containing, in mM: KCl 8, K-gluconate 110, NaCl 10, Hepes 10, KH_2PO_4 10, EGTA 0.005, mannitol 10, $MgCl_2$ 1, and 0.5 mg/ml bovine serum albumin (fatty acid-free), pH 7.25 without disturbing the pellet. After the wash, cells were resuspended in 0.2 ml of the same buffer but pre-warmed to 37 °C and also including mitochondrial substrates (5 mM glutamate and 5 mM malate), 5 µM safranin O and 50 µM digitonin. For results obtained from fibroblasts, traces obtained from safranin O fluorescence were calibrated to millivolts as described in ref. 50. For results obtained from C2C12 cells, traces are shown as arbitrary units of safranine O fluorescence. Fluorescence was recorded in a Tecan Infinite® 200 PRO series plate reader at a 0.08 Hz acquisition rate, using 495 and 585 nm excitation and emission wavelengths, respectively. After the completion of each experiment, cells were harvested from each well and protein content was determined. Only those data were considered in which protein content variability among wells was <5%. All experiments were performed at 37 °C and in triplicates.

Mitochondrial Calcium Uptake Capacity in Permeabilized C2C12 Cells

Maximum Ca^{2+} uptake capacity of *in situ mitochondria* of permeabilized cells was estimated as detailed previously57, with minor modifications: Briefly, approximately 30,000 C2C12 cells (per experiment) were cultured in 35 mm wells, washed once in phosphate-buffered saline and harvested with 0.2 ml of 0.25% trypsin-EDTA, inactivated by 1.8 ml calf serum, followed by centrifugation at 1,100 g for 2 minutes. Next, cells were washed once in an ice-cold buffer containing, in mM: KCl 8, K-gluconate 110, NaCl 10, Hepes 10, KH_2PO_4 10, EGTA 0.005, mannitol 10, $MgCl_2$ 1, and 0.5 mg/ml bovine serum albumin (fatty acid-free), pH 7.25 without disturbing the pellet. After the wash, cells were resuspended in 0.2 ml of the same buffer, but pre-warmed to 37 °C and also including mitochondrial substrates (5 mM glutamate and 5 mM malate), 1 µM Calcium Green 5N hexapotassium salt, (CaGr, Life Technologies) and 50 µM digitonin. Consecutive additions of $CaCl_2$ was added (100 µM boluses) or NaCN (50 µM boluses) or SF6847 (10 or 100 nM boluses) where indicated, until there was no further decrease in CaGr fluorescence, or upon reaching maximum depolarization inferred by parallel measurements of mitochondrial membrane potential measured by safranine O as detailed above. Fluorescence was recorded in a Tecan Infinite® 200 PRO series plate reader at a 0.08 Hz acquisition rate, using 506 and 531 nm excitation and emission wavelengths, respectively. All experiments were performed at 37 °C and in triplicates. After the completion of each experiment, cells were harvested from each well and protein content was determined. Only those data were considered in which protein content variability among wells was <5%.

Western blot Analysis

C2C12 cells or human fibroblasts were harvested by trypsinization, washed in phosphate-buffered saline, solubilised in RIPA buffer containing a cocktail of protease inhibitors (Protease Inhibitor Cocktail Set I, Merck Millipore, Billerica, MA, USA) and frozen at −80 °C for further analysis. Frozen pellets were thawed on ice, their protein concentration was determined using the bicinchoninic acid assay as detailed below and separated by sodium dodecyl sulfate – polyacrylamide gel electrophoresis (SDS-PAGE). Separated proteins were transferred to a methanol-activated polyvinylidene difluoride membrane. Immunoblotting was performed as recommended by the manufacturers of the antibodies. Rabbit polyclonal anti-ANT1 (Abcam, Cambridge, UK 1:500 dilution), rabbit polyclonal anti-ANT2 (Biogenesis, UK, 1:500 dilution), rabbit monoclonal anti-VDAC1 (Abcam, 1:5,000 dilution) and mouse monoclonal anti-β actin (Abcam, 1:5,000 dilution) primary antibodies were used. Immunoreactivity was detected using the appropriate peroxidase-linked secondary antibody (1:5,000, donkey anti-rabbit or donkey anti-mouse, Jackson Immunochemicals Europe Ltd, Cambridgeshire, UK) and enhanced chemiluminescence detection reagent (ECL system; Amersham Biosciences GE Healthcare Europe GmbH, Vienna, Austria).

Cytochrome c Release

Permeabilized C2C12 cells undergoing calcium uptake capacity determination using CaGr as detailed above were collected at specified time points, and centrifuged at 21,000 g for 3 min. The supernatants were carefully removed, and both the supernatant and pellet fractions were immediately frozen and stored at −20 °C. Cytochrome c immunoreactivity was quantified in both fractions using an enzyme-linked immunosorbent assay kit (Abcam). Before measurement, the supernatant and pellet samples were diluted 1:20 and 1:200, respectively. The release of cytochrome c from *in situ* mitochondria is expressed as the content of cytochrome c in the supernatant as a percentage of the total content of cytochrome c present in the supernatant plus pellet.

Preparation of Sodium Orthovanadate (Na_3VO_4) and Beryllium Trifluoride (BeF_3^-)

A 25 mM Na_3VO_4 solution was prepared in distilled water (>17 mOhm resistance). The pH of the solution was adjusted to 8.7 with HCl, upon which it turns yellow. This solution was boiled until it turns colorless and cooled to room temperature. The pH was reassessed, and readjusted to pH

8.7 with HCl, upon which the solution turned yellow again. This cycle of boiling until colorless and readjusting the pH was repeated until the solution remains colorless at pH 8.7. Finally, the solution was brought up to the initial volume with distilled water and stored in aliquots at −80 °C. This treatment removes all decavanadate ions present in the Na_3VO_4 solution, which induces mitochondrial membrane depolarization and inhibition of oxygen consumption91. The final concentration of orthovanadate used was 12.5 µM. Orthovanadate inhibits the oxidation of only disrupted mammalian mitochondria92. Likewise, fluoroberyllium nucleoside diphosphate complexes inhibit only the exposed F_1F_0-ATPase93. $BeSO_4$ and NaF are prepared as aqueous solutions of 0.2 M and 0.5 M, respectively, and kept at +4 °C for several years. 0.2 mM BeF_3^- is formed immediately in solution upon mixing of $BeSO_4$ (0.2 mM) and NaF (5 mM). Vanadate, beryllium and fluoride salts are highly toxic to tissues and to the environment, and thus require proper handling and disposal. The combination of orthovanadate and BeF_3^- inhibits kinases, mutases, phosphatases, and ATPases94·95. However, some kinases, such as pyruvate kinase, will remain uninhibited96. In this respect, upon permeabilization of the cells one has to totally separate pyruvate kinase from its substrate, phosphoenol pyruvate, i.e. there must be no glucose present in the medium prior to permeabilization, and a few minutes lag time must be allowed prior to ADP-ATP exchange rate measurements in order for the remaining reactions by kinases to 'die-out'.

Protein Determination

Protein content was measured by the bicinchoninic acid assay97 using bovine serum albumin protein as standards and calibrating by a 3 parameter power function, f=y0+a*xb, where y0 is background absorbance in the absence of protein, a and b are constants, and x is the amount of protein in the unknown samples.

Reagents

Standard laboratory chemicals, safranine O and digitonin were from Sigma. Calcium Green 5N 6K$^+$ salt, Magnesium Green 5K$^+$ salt and calcein-AM were from Life technologies. SF 6847 was from Biomol (BIOMOL GmbH, Hamburg, Germany). Fetal bovine serum was from Atlanta Biologicals (Lawrenceville, GA, USA) and all other tissue culture reagents were purchased from Life technologies. pDsRed2-mito (mito-DsRed2) was purchased from Clontech (Mountain View, CA). pcDNA-4mtD3cpv was a gift from Amy Palmer & Roger Tsien (Addgene plasmid # 36324).

Statistics

Data are presented as mean ± SEM; significant differences between two groups of data were evaluated by Student's t-test, with $p<0.05$ considered significant. Significant differences between three or more groups of data were evaluated by one-way analysis of variance followed by Tukey's *post-hoc* analysis, with $p<0.05$ considered statistically significant. If normality test failed, ANOVA on Ranks was performed.

Acknowledgements

We thank Katalin Zölde for excellent technical assistance. This work was supported by Országos Tudományos Kutatási Alapprogramok (OTKA) [112230] grant and Magyar Tudományos Akadémia (MTA) [02001] grant and the Hungarian Brain Research Program [KTIA_13_NAP-A-III/6] grant to V.A-V., grants TÁMOP [4.2.1./B-09/1/ KMR] and [NAP-A] to M.J.M. and A.G., grant NIH/NIA [AG014930] to A.A.S., and grants MTA-SE Lendület Neurobiochemistry Research Division [95003], OTKA NNF [78905], OTKA NNF2 [85658], OTKA K [100918] and Egészségügyi Tudományos Tanács [55160] to C. C.

REFERENCES

1. Azzolin, L. et al. The mitochondrial permeability transition from yeast to mammals. FEBS Lett. 584, 2504–2509 (2010).
2. Bonora, M. et al. Molecular mechanisms of cell death: central implication of ATP synthase in mitochondrial permeability transition. Oncogene 34, 1475–1486 (2015).
3. Bonora, M. et al. Role of the c subunit of the FO ATP synthase in mitochondrial permeability transition. Cell Cycle 12, 674–683 (2013).
4. Alavian, K. N. et al. An uncoupling channel within the c-subunit ring of the F1FO ATP synthase is the mitochondrial permeability transition pore. Proc. Natl. Acad. Sci. USA 111, 10580–10585 (2014).
5. Giorgio, V. et al. Dimers of mitochondrial ATP synthase form the permeability transition pore. Proc. Natl. Acad. Sci. USA 110, 5887–5892 (2013).
6. Jonas, E. A., Porter, G. A. Jr., Beutner, G., Mnatsakanyan, N. & Alavian, K. N. Cell death disguised: The mitochondrial permeability transition pore as the c-subunit of the F(1)F(O) ATP synthase. Pharmacol. Res. 99, 382–392 (2015).
7. Halestrap, A. P. The C Ring of the F1Fo ATP Synthase Forms the Mitochondrial Permeability Transition Pore: A Critical Appraisal. Front Oncol. 4, 234, doi: 10.3389/fonc.2014.00234 (2014).
8. Chinopoulos, C. & Szabadkai, G. What makes you can also break you: mitochondrial permeability transition pore formation by the c subunit of the F(1) F(0) ATP-synthase? Front Oncol. 3, 25, doi: 10.3389/fonc.2013.00025 (2013).
9. Szabadkai, G. & Chinopoulos, C. What Makes You Can Also Break You, Part II: Mitochondrial Permeability Transition Pore Formation by Dimers of the F1FO ATP-Synthase? Front Oncol. 3, 140, doi: 10.3389/fonc.2013.00140 (2013).
10. Chinopoulos, C. & Szabadkai, G. What Makes You Can also Break You, Part III: Mitochondrial Permeability Transition Pore Formation by an Uncoupling Channel within the C-Subunit Ring of the F1FO ATP Synthase? Front Oncol. 4, 235, doi: 10.3389/ fonc.2014.00235 (2014).
11. Bernardi, P. & von Stockum, S. The permeability transition pore as a Ca(2+) release channel: new answers to an old question. Cell Calcium 52, 22–27 (2012).
12. Bernardi, P. Modulation of the mitochondrial cyclosporin

A-sensitive permeability transition pore by the proton electrochemical gradient. Evidence that the pore can be opened by membrane depolarization. J. Biol. Chem. 267, 8834–8839 (1992).

13. Petronilli, V., Cola, C. & Bernardi, P. Modulation of the mitochondrial cyclosporin A-sensitive permeability transition pore. II. The minimal requirements for pore induction underscore a key role for transmembrane electrical potential, matrix pH, and matrix Ca2+. J. Biol. Chem. 268, 1011–1016 (1993).
14. Scorrano, L., Petronilli, V. & Bernardi, P. On the voltage dependence of the mitochondrial permeability transition pore. A critical appraisal. J. Biol. Chem. 272, 12295–12299 (1997).
15. Doczi, J. et al. Complex Contribution of Cyclophilin D to Ca2+-induced Permeability Transition in Brain Mitochondria, with Relation to the Bioenergetic State. J. Biol. Chem. 286, 6345–6353 (2011).
16. Petronilli, V. et al. The voltage sensor of the mitochondrial permeability transition pore is tuned by the oxidation-reduction state of vicinal thiols. Increase of the gating potential by oxidants and its reversal by reducing agents. J. Biol. Chem. 269, 16638–16642 (1994).
17. Palmeira, C. M. & Wallace, K. B. Benzoquinone inhibits the voltage-dependent induction of the mitochondrial permeability transition caused by redox-cycling naphthoquinones. Toxicol. Appl. Pharmacol. 143, 338–347 (1997).
18. Eriksson, O., Fontaine, E. & Bernardi, P. Chemical modification of arginines by 2,3-butanedione and phenylglyoxal causes closure of the mitochondrial permeability transition pore. J. Biol. Chem. 273, 12669–12674 (1998).
19. Linder, M. D., Morkunaite-Haimi, S., Kinnunen, P. K., Bernardi, P. & Eriksson, O. Ligand-selective modulation of the permeability transition pore by arginine modification. Opposing effects of p-hydroxyphenylglyoxal and phenylglyoxal. J. Biol. Chem. 277, 937–942 (2002).
20. Vesce, S., Jekabsons, M. B., Johnson-Cadwell, L. I. & Nicholls, D. G. Acute glutathione depletion restricts mitochondrial ATP export in cerebellar granule neurons. J. Biol. Chem. 280, 38720–38728 (2005).
21. Majima, E., Shinohara, Y., Yamaguchi, N., Hong, Y. M. & Terada, H. Importance of loops of mitochondrial ADP/ATP carrier for its transport activity deduced from reactivities of its cysteine residues with the sulfhydryl reagent eosin-5-maleimide. Biochemistry 33, 9530–9536 (1994).
22. Costantini, P. et al. Oxidation of a critical thiol residue of the adenine nucleotide translocator enforces Bcl-2-independent permeability transition pore opening and apoptosis. Oncogene 19, 307–314 (2000).
23. Michejda, J. & Vignais, P. V. The energy-dependent unmasking of -SH groups in the mitochondrial ADP/ATP carrier, and its prevention by nigericin. FEBS Lett. 132, 129–132 (1981).
24. Halestrap, A. P. & Brenner, C. The adenine nucleotide translocase: a central component of the mitochondrial permeability transition pore and key player in cell death. Curr. Med. Chem. 10, 1507–1525 (2003).
25. Zoratti, M. & Szabo, I. The mitochondrial permeability transition. Biochim. Biophys. Acta 1241, 139–176 (1995).
26. Kokoszka, J. E. et al. The ADP/ATP translocator is not essential for the mitochondrial permeability transition pore. Nature 427, 461–465 (2004).
27. Menze, M. A., Hutchinson, K., Laborde, S. M. & Hand, S. C. Mitochondrial permeability transition in the crustacean Artemia franciscana: absence of a calcium-regulated pore in the face of profound calcium storage. Am. J. Physiol Regul. Integr. Comp Physiol 289, R68–R76 (2005).
28. Konrad, C. et al. A distinct sequence in the adenine nucleotide translocase from Artemia franciscana embryos is associated with insensitivity to bongkrekate and atypical effects of adenine nucleotides on Ca(2+) uptake and sequestration. FEBS J. 278, 822–836 (2011).
29. Klingenberg, M. The ADP and ATP transport in mitochondria and its carrier. Biochim. Biophys. Acta 1778, 1978–2021 (2008).
30. Konrad, C., Kiss, G., Torocsik, B., Adam-Vizi, V. & Chinopoulos, C. Absence of Ca2+-induced mitochondrial permeability transition but presence of bongkrekate-sensitive nucleotide exchange in C. crangon and P. serratus. Plos. ONE. 7, e39839, doi: 10.1371/journal.pone.0039839 (2012).
31. Wysocka-Kapcinska, M. et al. The suppressor of AAC2 Lethality SAL1 modulates sensitivity of heterologously expressed artemia ADP/ATP carrier to bongkrekate in yeast. Plos. ONE 8, e74187, doi: 10.1371/journal.pone.0074187 (2013).
32. Halestrap, A. P., Woodfield, K. Y. & Connern, C. P. Oxidative stress, thiol reagents, and membrane potential modulate the mitochondrial permeability transition by affecting nucleotide binding to the adenine nucleotide translocase. J. Biol. Chem. 272, 3346–3354 (1997).
33. McStay, G. P., Clarke, S. J. & Halestrap, A. P. Role of critical thiol groups on the matrix surface of the adenine nucleotide translocase in the mechanism of the mitochondrial permeability transition pore. Biochem. J. 367, 541–548 (2002).
34. Echaniz-Laguna, A. et al. Complete loss of expression of the ANT1 gene causing cardiomyopathy and myopathy. J. Med. Genet. 49, 146–150 (2012).
35. Abramov, A. Y. & Duchen, M. R. Actions of ionomycin, 4-BrA23187 and a novel electrogenic Ca2+ ionophore on mitochondria in intact cells. Cell Calcium 33, 101–112 (2003).
36. Yun, B. et al. Serine hydrolase inhibitors block necrotic cell death by preventing calcium overload of the mitochondria and permeability transition pore formation. J. Biol. Chem 289, 1491–1504 (2014).
37. Wang, Z., Cai, F., Hu, L. & Lu, Y. The role of mitochondrial permeability transition pore in regulating the shedding of the platelet GPIbalpha ectodomain. Platelets. 25, 373–381 (2014).

38. Kahraman,S., Bambrick, L. L. & Fiskum, G. Effects of FK506 and cyclosporin a on calcium ionophore-induced mitochondrial depolarization and cytosolic calcium in astrocytes and neurons. J. Neurosci. Res. 89, 1973–1978 (2011).

39. Dumas, J. F. et al. Effect of transient and permanent permeability transition pore opening on NAD(P)H localization in intact cells. J. Biol. Chem 284, 15117–15125 (2009).

40. Oh, K. W., Qian, T., Brenner, D. A. & Lemasters, J. J. Salicylate enhances necrosis and apoptosis mediated by the mitochondrial permeability transition. Toxicol. Sci. 73, 44–52 (2003).

41. Beavis, A. D. Properties of the inner membrane anion channel in intact mitochondria. J. Bioenerg. Biomembr. 24, 77–90 (1992).

42. Beavis, A. D., Lu, Y. & Garlid, K. D. On the regulation of K+ uniport in intact mitochondria by adenine nucleotides and nucleotide analogs. J. Biol. Chem 268, 997–1004 (1993).

43. Petronilli,V. et al. Transient and long-lasting openings of the mitochondrial permeability transition pore can be monitored directly in intact cells by changes in mitochondrial calcein fluorescence. Biophys. J. 76, 725–734 (1999).

44. Gerencser, A. A., Doczi, J., Torocsik, B., Bossy-Wetzel,E. & Adam-Vizi, V. Mitochondrial swelling measurement in situ by optimized spatial filtering: astrocyte-neuron differences. Biophys. J. 95, 2583–2598 (2008).

45. Palmer, A. E. et al. Ca2+ indicators based on computationally redesigned calmodulin-peptide pairs. Chem Biol. 13, 521–530 (2006).

46. Chinopoulos, C. et al. Forward operation of adenine nucleotide translocase during F0F1-ATPase reversal: critical role of matrix substrate-level phosphorylation. FASEB J. 24, 2405–2416 (2010).

47. Chinopoulos, C. The "B Space" of mitochondrial phosphorylation. J. Neurosci. Res. 89, 1897–1904 (2011).

48. Chinopoulos, C. Mitochondrial consumption of cytosolic ATP: not so fast. FEBS Lett. 585, 1255–1259 (2011).

49. Kiss, G. et al. The negative impact of alpha-ketoglutarate dehydrogenase complex deficiency on matrix substrate-level phosphorylation. FASEB J. 27, 2392–2406 (2013).

50. Kiss, G. et al. Mitochondrial diaphorases as NAD(+) donors to segments of the citric acid cycle that support substrate-level phosphorylation yielding ATP during respiratory inhibition. FASEB J. 28, 1682–1697 (2014).

51. Gerencser, A. A. et al. Quantitative measurement of mitochondrial membrane potential in cultured cells: calcium-induced de- and hyperpolarization of neuronal mitochondria. J. Physiol 590, 2845–2871 (2012).

52. Neher, R. A. et al. Blind source separation techniques for the decomposition of multiply labeled fluorescence images. Biophys. J. 96, 3791–3800 (2009).

53. Ward, M. W., Rego, A. C., Frenguelli, B. G. & Nicholls, D. G. Mitochondrial membrane potential and glutamate excitotoxicity in cultured cerebellar granule cells. J. Neurosci. 20, 7208–7219 (2000).

54. Halestrap, A. P. & Richardson, A. P. The mitochondrial permeability transition: a current perspective on its identity and role in ischaemia/reperfusion injury. J. Mol. Cell Cardiol. 78, 129–141 (2015).

55. Chinopoulos, C., Tretter, L. & Adam-Vizi, V. Depolarization of in situ mitochondria due to hydrogen peroxide-induced oxidative stress in nerve terminals: inhibition of alpha-ketoglutarate dehydrogenase. J. Neurochem. 73, 220–228 (1999).

56. Kawamata, H., Tiranti, V., Magrane, J., Chinopoulos, C. & Manfredi,G. adPEO mutations in ANT1 impair ADP-ATP translocation in muscle mitochondria. Hum. Mol. Genet. 20, 2964–2974 (2011).

57. Chinopoulos, C., Starkov, A. A. & Fiskum, G. Cyclosporin A-insensitive permeability transition in brain mitochondria: inhibition by 2-aminoethoxydiphenyl borate. J. Biol. Chem. 278, 27382–27389 (2003).

58. Kristian, T., Weatherby, T. M., Bates, T. E. & Fiskum, G. Heterogeneity of the calcium-induced permeability transition in isolated non-synaptic brain mitochondria. J. Neurochem. 83, 1297–1308 (2002).

59. Murphy, A. N., Bredesen, D. E., Cortopassi, G., Wang, E. & Fiskum,G. Bcl-2 potentiates the maximal calcium uptake capacity of neural cell mitochondria. Proc. Natl. Acad. Sci. USA 93, 9893–9898 (1996).

60. Marcu, R., Neeley, C. K., Karamanlidis, G. & Hawkins, B. J. Multi-parameter measurement of the permeability transition pore opening in isolated mouse heart mitochondria. J. Vis. Exp. 67, 4131, doi: 10.3791/4131 (2012).

61. Chinopoulos, C. & Adam-Vizi, V. Mitochondrial Ca2+ sequestration and precipitation revisited. FEBS J. 277, 3637–3651 (2010).

62. Das, A. M., Byrd,D. J. & Brodehl, J. Regulation of the mitochondrial ATP-synthase in human fibroblasts. Clin. Chim. Acta 231, 61–68 (1994).

63. Rasola, A. & Bernardi, P. The mitochondrial permeability transition pore and its involvement in cell death and in disease pathogenesis. Apoptosis. 12, 815–833 (2007).

64. Bernardi, P. The mitochondrial permeability transition pore: a mystery solved? Front Physiol 4, 95, doi: 10.3389/fphys.2013.00095 (2013).

65. Tikunov, A. et al. Closure of VDAC causes oxidative stress and accelerates the Ca(2+)-induced mitochondrial permeability transition in rat liver mitochondria. Arch. Biochem. Biophys. 495, 174–181 (2010).

66. Roy, S. S. et al. Bad targets the permeability transition pore independent of Bax or Bak to switch between Ca2+-dependent cell survival and death. Mol. Cell 33, 377–388 (2009).

67. Sileikyte, J. et al. Regulation of the inner membrane mitochondrial permeability transition by the outer membrane translocator protein (peripheral benzodiazepine receptor). J. Biol. Chem. 286, 1046–1053 (2011).

68. Batarseh, A. & Papadopoulos, V. Regulation of translocator protein 18kDa (TSPO) expression in health

and disease states. Mol. Cell Endocrinol. 327, 1–12 (2010).

69. Sileikyte, J. et al. Regulation of the mitochondrial permeability transition pore by the outer membrane does not involve the peripheral benzodiazepine receptor (Translocator Protein of 18 kDa (TSPO)). J. Biol. Chem 289, 13769–13781 (2014).

70. Costantini, P., Colonna, R. & Bernardi, P. Induction of the mitochondrial permeability transition by N-ethylmaleimide depends on secondary oxidation of critical thiol groups. Potentiation by copper-ortho-phenanthroline without dimerization of the adenine nucleotide translocase. Biochim. Biophys. Acta 1365, 385–392 (1998).

71. Dorner, A., Olesch, M., Giessen, S., Pauschinger, M. & Schultheiss, H. P. Transcription of the adenine nucleotide translocase isoforms in various types of tissues in the rat. Biochim. Biophys. Acta 1417, 16–24 (1999).

72. Bauer, M. K., Schubert, A., Rocks, O. & Grimm, S. Adenine nucleotide translocase-1, a component of the permeability transition pore, can dominantly induce apoptosis. J. Cell Biol. 147, 1493–1502 (1999).

73. Jang, J. Y., Choi,Y., Jeon, Y. K., Aung, K. C. & Kim, C. W. Over-expression of adenine nucleotide translocase 1 (ANT1) induces apoptosis and tumor regression in vivo. BMC. Cancer 8, 160, doi: 10.1186/1471-2407-8-160 (2008).

74. Lena, A. et al. The silencing of adenine nucleotide translocase isoform 1 induces oxidative stress and programmed cell death in ADF human glioblastoma cells. FEBS J. 277, 2853–2867 (2010).

75. Laoudj-Chenivesse, D. et al. Increased levels of adenine nucleotide translocator 1 protein and response to oxidative stress are early events in facioscapulohumeral muscular dystrophy muscle. J. Mol. Med. (Berl) 83, 216–224 (2005).

76. Lee, J., Schriner, S. E. & Wallace, D. C. Adenine nucleotide translocator 1 deficiency increases resistance of mouse brain and neurons to excitotoxic insults. Biochim. Biophys. Acta 1787, 364–370 (2009).

77. Fontaine, E., Eriksson, O., Ichas, F. & Bernardi, P. Regulation of the permeability transition pore in skeletal muscle mitochondria. Modulation By electron flow through the respiratory chain complex i. J. Biol. Chem. 273, 12662–12668 (1998).

78. Fontaine, E. & Bernardi, P. Progress on the mitochondrial permeability transition pore: regulation by complex I and ubiquinone analogs. J. Bioenerg. Biomembr. 31, 335–345 (1999).

79. Chalmers, S. & Nicholls, D. G. The relationship between free and total calcium concentrations in the matrix of liver and brain mitochondria. J. Biol. Chem. 278, 19062–19070 (2003).

80. Nicholls, D. G. & Chalmers, S. The integration of mitochondrial calcium transport and storage. J. Bioenerg. Biomembr. 36, 277–281 (2004).

81. Vajda, S. et al. A re-evaluation of the role of matrix acidification in uncoupler-induced Ca2+ release from mitochondria. FEBS J. 276, 2713–2724 (2009).

82. Bernardi, P., Broekemeier, K. M. & Pfeiffer,D. R. Recent progress on regulation of the mitochondrial permeability transition pore; a cyclosporin-sensitive pore in the inner mitochondrial membrane. J. Bioenerg. Biomembr. 26, 509–517 (1994).

83. Broekemeier, K. M. & Pfeiffer, D. R. Inhibition of the mitochondrial permeability transition by cyclosporin A during long time frame experiments: relationship between pore opening and the activity of mitochondrial phospholipases. Biochemistry 34, 16440–16449 (1995).

84. Rozen, S. & Skaletsky, H. Primer3 on the WWW for general users and for biologist programmers. Methods Mol. Biol. 132, 365–386 (2000).

85. Le, B. M. et al. Chemosensitization by knockdown of adenine nucleotide translocase-2. Cancer Res. 66, 9143–9152 (2006).

86. Vilchez, D. et al. Increased proteasome activity in human embryonic stem cells is regulated by PSMD11. Nature 489, 304–308 (2012).

87. Chinopoulos, C. et al. A novel kinetic assay of mitochondrial ATP-ADP exchange rate mediated by the ANT. Biophys. J. 96, 2490–2504 (2009).

88. Kawamata, H., Starkov, A. A., Manfredi, G. & Chinopoulos, C. A kinetic assay of mitochondrial ADP-ATP exchange rate in permeabilized cells. Anal. Biochem. 407, 52–57 (2010).

89. Perevoshchikova, I. V., Sorochkina, A. I., Zorov, D. B. & Antonenko, Y. N. Safranine O as a fluorescent probe for mitochondrial membrane potential studied on the single particle level and in suspension. Biochemistry (Mosc.) 74, 663–671 (2009).

90. Figueira, T. R., Melo, D. R., Vercesi, A. E. & Castilho, R. F. Safranine as a fluorescent probe for the evaluation of mitochondrial membrane potential in isolated organelles and permeabilized cells. Methods Mol. Biol. 810, 103–117 (2012).

91. Aureliano, M. & Crans, D. C. Decavanadate (V10 O28 6-) and oxovanadates: oxometalates with many biological activities. J. Inorg. Biochem. 103, 536–546 (2009).

92. Byczkowski, J., Zychlinski, L. & Tluczkiewicz, J. Interaction of vanadate with respiratory chain of rat liver and wheat seedling mitochondria. Int. J. Biochem. 10, 1007–1011 (1979).

93. Issartel, J. P., Dupuis, A., Lunardi, J. & Vignais, P. V. Fluoroaluminum and fluoroberyllium nucleoside diphosphate complexes as probes of the enzymatic mechanism of the mitochondrial F1-ATPase. Biochemistry 30, 4726–4733 (1991).

94. Ray, B. D., Moore, J. M. & Rao, B. D. 31P NMR studies of enzyme-bound substrate complexes of yeast 3-phosphoglycerate kinase: III. Two ADP binding sites and their Mg(II) affinity; effects of vanadate and arsenate on enzymic complexes with ADP and 3-P-glycerate. J. Inorg. Biochem. 40, 47–57 (1990).

95. Climent, F., Bartrons, R., Pons, G. & Carreras, J. Effect of vanadate on phosphoryl transfer enzymes involved in

glucose metabolism. Biochem. Biophys. Res. Commun. 101, 570–576 (1981).

96. Lord, K. A. & Reed, G. H. Vanadyl(IV) complexes with pyruvate kinase: activation of the enzyme and electron paramagnetic resonance properties of ternary complexes with the protein. Arch. Biochem. Biophys. 281, 124–131 (1990).

97. Smith, P. K. et al. Measurement of protein using bicinchoninic acid. Anal. Biochem. 150, 76–85 (1985).

CHAPTER 3 MITOCHONDRIAL ELECTRON TRANSPORT SYSTEM IN OLIGOMYCIN-TREATED CELLS

Juliana S. Ruas, Edilene S. Siqueira-Santos, Ignacio Amigo, Erika Rodrigues-Silva, Alicia J. Kowaltowski, Roger F. Castilho

INTRODUCTION

Oxygen consumption rate (OCR) measurements are one of the preferred methods for mitochondrial function or dysfunction evaluation in cultured cells (for reviews see [1], [2]). When the plasma membrane is permeabilized or isolated mitochondria are studied, the respiratory control ratio can be assessed in a medium supplemented with respiratory substrates by measuring the increase in OCR after the addition of ADP. When using intact cells, specific respirometric protocols are usually employed to evaluate mitochondrial function and can provide valuable information such as that described below. The fraction of basal OCR (or ROUTINE respiration, as it is referred to in [2]) inhibited by addition of the ATP synthase inhibitor oligomycin gives an estimate of the respiration rate necessary to sustain cellular ATP turnover under basal conditions. The respiration remaining in the presence of oligomycin is linked to the proton leak rate across the mitochondrial membrane and to other processes such as reactive oxygen species formation and energy-driven ion/metabolite transport. Maximal capacity of the mitochondrial electron transport system (ETS) can be estimated by inducing maximal OCR via chemical dissipation of the mitochondrial membrane potential, generally by the addition of a potent protonophore such as CCCP or FCCP. However, assessment of this maximal OCR usually requires caution (to avoid underestimating the results and drawing incorrect conclusions) as well as titration of the protonophore [1, 2]. The presence of oligomycin during the estimation of maximal OCR is widespread in such assays and it seems to be important to prevent the reverse activity of ATP synthase with rapid intracellular ATP depletion, which may lead to cellular metabolic dysfunction and death. Spare respiratory capacity (SRC) is given by the difference between maximal OCR and basal respiration and is an estimative of the cell's ability to cope with large increases in ATP turnover. Finally, the addition of a potent respiratory chain inhibitor, such as antimycin A, allows non-mitochondrial OCR to be estimated.

Mitochondrial energy metabolism seems to play specific roles in the biology of tumor cells [3–6]. Although mutations in citric acid cycle enzymes are associated with tumor formation [7,8], most tumor cells present normal mitochondrial integrity and oxidative phosphorylation capacity [9–11]. Recently, two groups showed that mitochondrial respiration is essential for tumor cell proliferation since it promotes aspartate biosynthesis [12, 13]. Further assessment of mitochondrial function in tumor cells may contribute to a better understanding of the role of these organelles in tumorigenesis and to the development of effective new cancer therapies [11, 14, 15]. In this study we investigated the effect of ATP synthase inhibitors on maximum OCR measured in tumor cells in order to understand our experimental evidences of an undesirable inhibitory effect of the ATP synthase inhibitor oligomycin on maximal OCR obtained in studies evaluating mitochondrial function in glioma cell lines. The data presented here indicate that the presence of oligomycin significantly underestimates CCCP- or FCCP-induced maximal OCR and SRC. Moreover, alternative compounds and protocols to assess maximal OCR in cells with inhibited oxidative phosphorylation were evaluated.

Citation: Ruas JS, Siqueira-Santos ES, Amigo I, Rodrigues-Silva E, Kowaltowski AJ, Castilho RF (2016) Mitochondrial Electron Transport System in Oligomycin-Treated Cells. PLoS ONE 11(3): e0150967. doi:10.1371/journal.pone.0150967.

MATERIALS AND METHODS

Chemicals and Cell Lines

Most of the chemicals used, including adenosine diphosphate (ADP; A2754), bongkrekic acid (BKA; B6179), carbonyl cyanide 3-chlorophenylhydrazone (CCCP; C2759), carboxyatractyloside (CAT; C4992), carbonyl cyanide 4-(trifluoromethoxy)phenylhydrazone (FCCP; C2920), digitonin (D141), dimethyl sulfoxide (DMSO; D8418), 2,4-dinitrophenol (DNP; D198501), 4-(2-hydroxyethyl) piperazine-1-ethanesulfonic acid (HEPES; H3375), oligomycin (oligo; O4876), oligomycin A (oligo A; 75351) and sodium pyruvate (P4562) were obtained from Sigma-Aldrich (St Louis, MO, USA). Oligomycin (oligo*; item number 11341) was also obtained from Cayman Chemical Company (Ann Arbor, MI, USA), as was citreoviridin (citre; item number 11319). CCCP, citreoviridin, DNP, FCCP, oligomycin and oligomycin A stock solutions were prepared by dissolving the respective chemicals in DMSO. ADP and HEPES solutions were prepared in ultrapure (Type I) water, and their pH adjusted to 7.2 with NaOH. Carboxyatractyloside was dissolved in Type I water. The cocktail of NADH-linked respiratory substrates containing glutamate, α-ketoglutarate, malate and pyruvate was prepared by dissolving the respective acids in water and adjusting the pH to 7.2 with KOH.

Cell lines T98G and U-87MG were derived from human glioblastoma, while PC-3 was derived from a prostate cancer bone metastasis. All three cell lines were obtained from ATCC (Manassas, VA, USA). Antibiotics (penicillin and streptomycin); Dulbecco's modified Eagle's medium (DMEM), containing 2 g/L of glucose; fetal bovine serum (FBS); phosphate buffered saline (PBS); and trypsin/EDTA solution (0.25%) were purchased from Vitrocell (Campinas, SP, Brazil).

Cell Cultures

PC-3, T98G and U-87MG cells were maintained continuously at 37°C in a humidified 5% $CO_2$and 95% air atmosphere in 175 cm^2 tissue culture flasks and DMEM supplemented with 10% FBS and antibiotics (100 U/mL penicillin and 100 µg/mL streptomycin). Because the doubling times of these cells are between 24 and 36 h, the cells were seeded every two or three days (approximately 25,000 cells/cm^2) into new culture flasks and fresh medium. Briefly, the adhered cells were pre-washed with PBS and then incubated for 2–4 min with a trypsin/EDTA solution (0.25%) at 37°C to dissociate the cells from the flask surface. Trypsin was inactivated by adding supplemented DMEM to the flask. Passage of the cells into new flasks and fresh medium was performed after centrifuging the cell suspension at 400 *g* for 5 min and suspending the resulting pellet in supplemented DMEM. The experiments were conducted when the cells were between the 3rd and 16th passages.

For experiments with suspended cells, these were plated on 150 cm^2 culture dishes at an initial density of 15–20,000 cells/cm^2 and maintained for 3 or 4 days. On the day of the experiments, trypsin-dissociated cells were centrifuged at 400 *g* for 5 min and suspended in DMEM containing 20 mM HEPES-Na^+ at a density of 24–32 x 10^6/mL (>95% viability). Cell suspensions were kept at room temperature and used within 2.5 hours.

For the experiments with attached cells, T98G cells were seeded in Seahorse XF24 V7 cell culture microplates at a density of 40–50,000 cells/well in supplemented DMEM. After 24 h, the medium was replaced by supplemented DMEM containing 20 mM HEPES-Na^+ and left for 1 h in a CO_2 incubator to allow the cells to stabilize before the experiment.

Isolation of rat Forebrain Mitochondria

Male Wistar rats (*Rattus norvegicus albinus*) were obtained from the State University of Campinas (UNICAMP) Animal Breeding Center. Six rats were used for the present study and were euthanized by decapitation [16]. Rat brain mitochondria were isolated as described by Michelini et al. [17] using digitonin to permeabilize the synaptosomal plasma membrane [18]. The use of a similar procedure to isolate mitochondria from intact cultured glioma cells did not result in functional mitochondria in sufficient quantity to conduct experiments (data not shown). All procedures involving animal handling were approved by the Ethics Committee of Universidade Estadual de Campinas (protocol number CEUA 2534–1) and are in accordance with the Brazilian National Guide.

Oxygen Uptake Measurements in Suspended Tumor cells and Isolated rat Brain Mitochondria

The OCR in suspended intact cells and isolated rat brain mitochondria was measured using a high-resolution respirometer (OROBOROS Oxygraph-2k, Innsbruck, Austria) with DatLab 4 software for data acquisition and analysis. In intact tumor cells it was measured by incubating the cells (3–4 x 10^6) at 37°C in a 2 mL chamber containing supplemented DMEM and 20 mM HEPES-Na^+. The pH of the medium at the beginning of the experiments was between 7.30 and 7.45.

The OCR in permeabilized cells and isolated rat brain mitochondria was measured by incubating the cells (3–4 x 10^6) or mitochondria (0.6 mg) at 37°C in a 2 mL chamber containing 125 mM sucrose, 65 mM KCl, 10 mM HEPES-K^+ pH 7.2, 2 mM K_2HPO_4, 1 mM $MgCl_2$, 1 mM EGTA and a cocktail of respiratory substrates (α-ketoglutarate, malate, glutamate and pyruvate; 5 mM of each). Digitonin (30 µM) was also present in the reaction medium for the experiments with permeabilized cells.

Before each experiment, the oxygen concentration in the medium was equilibrated for 2–3 min with air in the respirometer chambers at 37°C until a stable signal was obtained at an oxygen concentration of approximately 195 µM. Cell suspensions or isolated rat brain mitochondria were then added to a final volume of 2 mL, and the chambers were closed by inserting the stoppers. The cell suspensions were constantly stirred (750 rpm) inside the chambers, and the chemicals were added through the titanium injection port of the stoppers using Hamilton syringes. All measurements were carried out when oxygen concentration was above 30 µM in the reaction medium. OCR was calculated as the negative time derivative of oxygen concentration measured in the closed respirometer chambers and expressed per million of cells or per mg of isolated mitochondria. Data were recorded at 2 s intervals, and

10 data points were used to calculate the slope of the OCR plot through a polynomial fit with DatLab 4 software [2].

Oxygen uptake Measurements in Attached T98G Cells

Oxygen consumption in attached T98G cells was measured using a Seahorse XF24 Analyzer. Initially, 300 μL of supplemented DMEM containing 20 mM HEPES-Na^+ was placed in each well, and each addition consisted of 75 μL, resulting in a final volume of 600 μL. The first injector was loaded with either medium plus oligomycin or medium containing the corresponding volume of DMSO. After the first addition, the oligomycin concentration in the wells was 1 μg/mL. The second injector was loaded with CCCP to achieve a concentration of 4 μM in the wells, while the other two injectors were loaded with CCCP to achieve concentrations of 5.4 and 6.8 μM. Four time points were measured before the first injection (basal respiration), three after the oligomycin/DMSO addition and two after each CCCP addition. At the end of each experiment, the cells were washed twice with PBS and their total protein content determined by the Bradford method in each well to normalize the OCR values. Data points for each condition were pooled and calculated as the percentage change from basal respiration.

Statistical Analysis

Results are presented as representative traces or as means ± standard errors (SEM). Each experimental protocol was conducted at least in duplicate, with a minimum of four different cell culture passages or mitochondrial preparations. Since the measurements were always made with matched samples, i.e., control and treated conditions, two-tailed paired Student's t-test was used to compare two different groups. Repeated-measures one-way ANOVA followed by the post hoc Bonferroni test was used when multiple comparisons were performed. Differences between the concentrations of CCCP and FCCP to reach maximal OCR were assessed by the Mann-Whitney test. The level of significance was set at $P<0.05$. The data were analyzed using GraphPad Prism 5 software.

RESULTS

The Inhibitory Effect of Oligomycin on CCCP-Induced Maximal OCR in Suspended Tumor Cells

Fig 1 shows the results of experiments conducted to determine the maximal OCR in suspended intact glioma cells. T98G and U-87MG cells were incubated in HEPES-buffered, FBS-supplemented (10%) DMEM. This medium contains 11 mM glucose, 4 mM glutamine and 1.25 mM pyruvate as major oxidative substrates. The purpose in choosing this incubation medium was to simulate similar conditions to those in the cell culture. First, maximal OCR was determined by titration with the protonophore CCCP. In most of the experiments, T98G and U-87MG cells reached maximal OCR in the presence of 8 and 4–5 μM CCCP, respectively (Fig 1A, 1B and 1F). Remarkably, when a standard experimental protocol to determine maximal OCR consisting of CCCP titration preceded by oligomycin-inhibited oxidative phosphorylation was used, a lower maximal OCR was observed for both glioma cells (Fig 1C and 1D). In most of the experiments with T98G and U-87MG cells previously treated with oligomycin, maximal OCR was obtained after the addition of 4 and 2 μM CCCP, respectively (Fig 1F). SRC, i.e., the difference between maximal and basal OCR, was underestimated by 42.8 ± 2.7% and 34.3 ± 2.8% in oligomycin-treated T98G and U-87MG cells, respectively, compared with cells treated only with the DMSO vehicle before CCCP titration (Fig 1E). As previously reported for isolated mitochondria and intact cells [2, 19–21], inhibition of OCR was observed in our experiments when supraoptimal concentrations of CCCP were present. This undesired effect of high concentrations of CCCP on OCR may be related to an excess of this compound in the mitochondrial membrane, leading to inhibition of respiratory activity [19], and/or due to a decline in the supply of respiratory substrates to mitochondria [1]. The concentrations of CCCP necessary to reach maximal OCR varied during our study. This may be related to the binding of CCCP to components of the different batches of FBS. In addition, we observed that this variation was minimized by avoiding repeated freezing and thawing of the aliquots of CCCP solution. Despite this variation, the concentration of CCCP required to reach maximal OCR in oligomycin-treated cells was always approximately half that required when only DMSO (the oligomycin solvent) was present.

To evaluate whether the inhibitory effect of oligomycin on CCCP-induced maximal OCR would also be observed with a different cell line, we conducted experiments with PC-3 cells, derived from prostatic adenocarcinoma. S1 File shows that compared with the value for PC-3 cells treated with DMSO, SRC was underestimated by 26.3 ± 3.6% in PC-3 cells treated with oligomycin.

Most of the next experiments used T98G cells, since they showed greater sensitivity to the inhibitory effect of oligomycin on CCCP-induced maximal OCR. Experiments were conducted to determine whether the inhibitory effect of oligomycin on CCCP-induced maximal OCR would be observed with a single addition of the optimal concentration of CCCP (S2 File). CCCP titration or a single addition of the optimal concentration of CCCP resulted in similar maximal OCRs for T98G cells (S2A and S2B File). However, when the optimal concentration of CCCP was added to T98G cells previously treated with oligomycin, a lower maximal OCR was observed and was followed by progressive inhibition of oxygen consumption (S2C File).

We also investigated whether the addition of oligomycin would have an inhibitory effect on the OCR of cells already stimulated by CCCP. A suboptimal concentration of CCCP was used to avoid any delayed inhibitory effect of this compound on OCR during the experiment (Fig 2A). The results indicated that the addition of oligomycin at a concentration high enough to inhibit oxygen consumption by oxidative phosphorylation, i.e., 0.1 μg/mL (Fig 2C), results in remarkable inhibition of CCCP-stimulated OCR for T98G cells (Fig 2B).

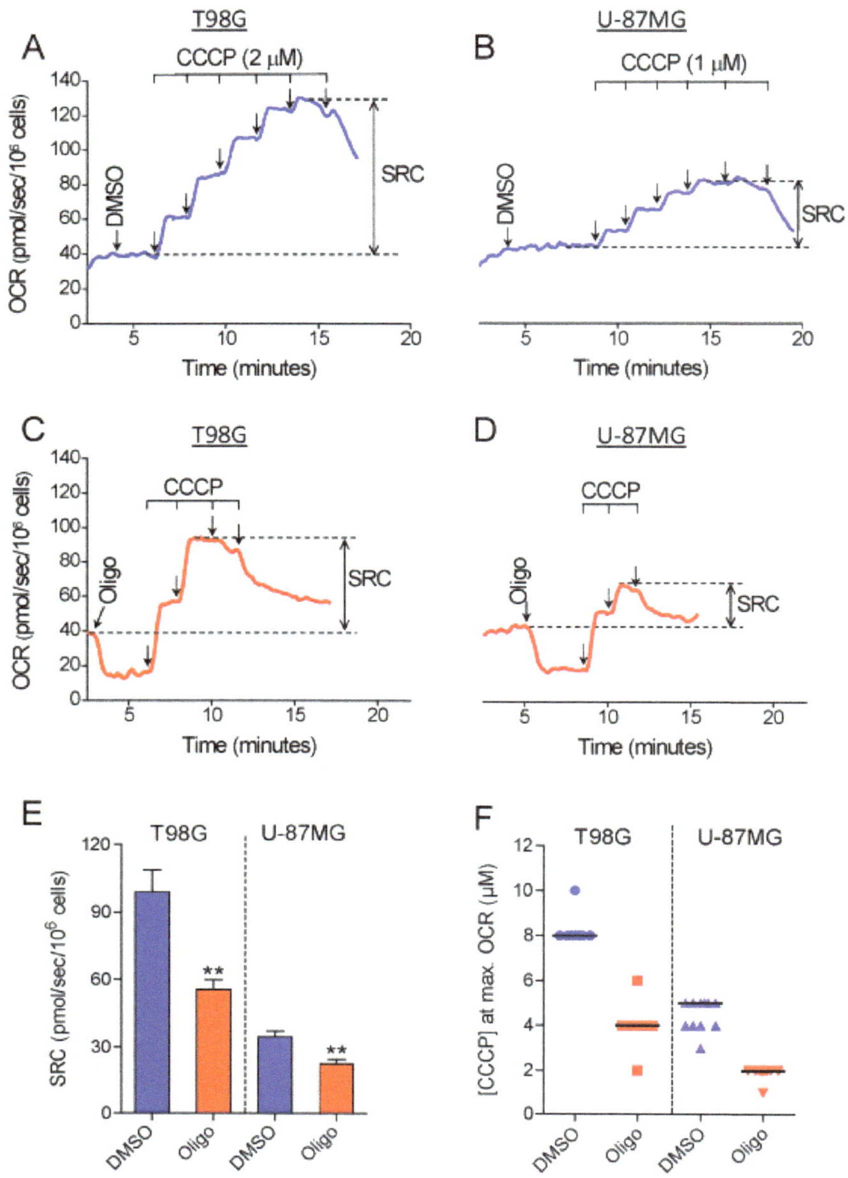

Figure 1. Inhibitory effect of oligomycin on CCCP-induced maximal oxygen consumption in intact human glioma cells. A and **C**: Representative oxygen consumption rate (OCR) traces in suspended T98G cells (1.5×10^6/mL). Where indicated by the arrows, 1 μg/mL oligomycin (Oligo) and DMSO (0.5 μL of each) were added, followed by sequential additions of CCCP (2 μM each). **B** and **D**: Representative traces of OCR in suspended U-87MG cells (2.0×10^6cells/mL). Where indicated by the arrows, 1 μg/mL oligomycin and 0.5 μL DMSO were added followed by sequential additions of CCCP (1 μM each). SRC was estimated as the difference between maximal and basal OCR, as shown in the representative traces. **E**: SRC values for T98G and U-87MG cells obtained in the presence and absence of oligomycin. **Statistically significant difference from the result for the respective control (DMSO), $P<0.01$. **F**: Values of CCCP concentrations required to achieve maximal OCR in T98G and U-87MG cells in the presence and absence of oligomycin.

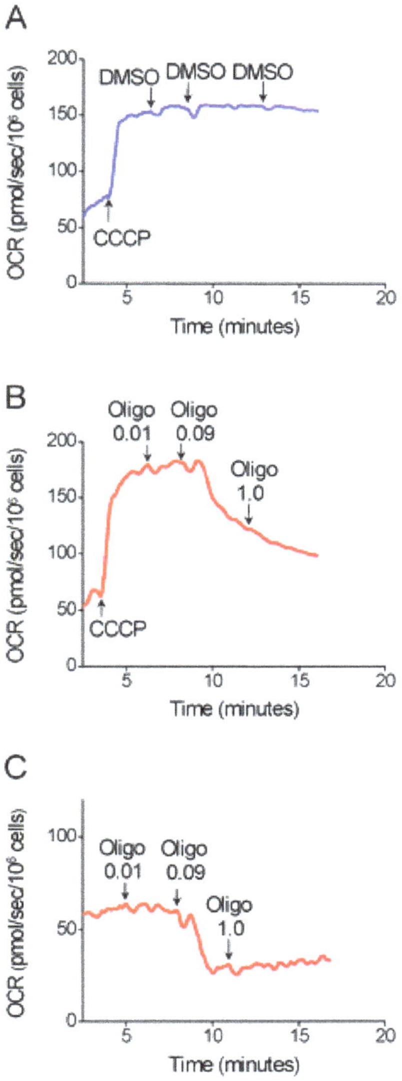

Figure 2. Inhibitory effect of oligomicyn addition on CCCP-induced maximal OCR in T98G cells. A–C: Representative OCR traces in suspended T98G cells (1.5×10^6 cells/mL). Where indicated by the arrows, 5 µM CCCP and 0.5 µL of oligomycin (Oligo; 0.01, 0.09 and 1.0 µg/mL) or DMSO were added.

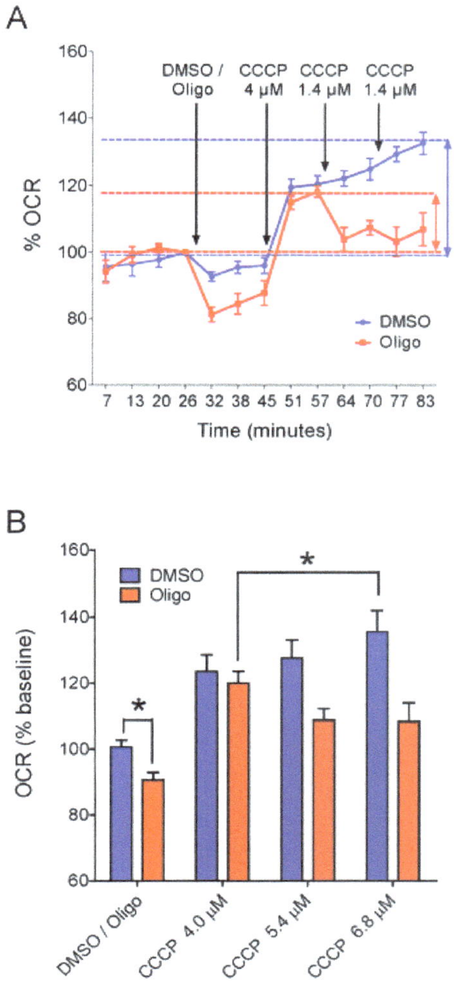

Figure 3. Inhibitory effect of oligomycin on CCCP-induced maximal OCR in attached T98G cells. A: Representative experiment to determine OCR in attached T98G cells. Arrows indicate additions of reagents and their concentrations. Results are shown as percentages in relation to the last point before the first addition. SRC (i.e., the difference between maximal respiratory rate and basal respiration) for each condition is indicated by vertical coloured arrows. **B**: Quantification of OCR with respect to basal respiration in cells treated with DMSO or oligomycin. Maximal respiratory rates with CCCP were smaller in oligomycin-treated cells and occurred at lower CCCP concentrations. *Statistically significant difference from the result for DMSO, $P < 0.05$.

The Inhibitory Effect of OLIGOMYCIN on CCCP-Induced maximal OCR in Attached T98G Cells

Fig 3 shows measurements of OCR in attached T98G cells. Maximal OCR was achieved by adding CCCP three consecutive times to give final protonophore concentrations of 4, 5.4 and 6.8 µM, respectively. Higher CCCP concentrations, e.g., 7.1 or 8.3 µM were not used because pilot experiments indicated that these did not result in further increases in OCR (results not shown). Like suspended T98G cells, attached T98G cells showed lower OCR and SRC when CCCP titration was preceded by addition of oligomycin (Fig 3A). In this case, maximal OCR was reached with 4 µM CCCP, while when only DMSO was added the corresponding figure was 6.8 µM CCCP (Fig 3B). Compared with the value for T98G cells treated with DMSO, SRC was underestimated by 43.7 ± 2.3% in T98G cells treated with oligomycin (Fig 3B).

Evaluation of alternative Protocols to Estimate Maximal OCR in Intact Tumor Cells

In view of the preceding results indicating that oligomycin has an important inhibitory effect on CCCP-induced maximal OCR in intact tumor cells, the next experiments were conducted to test different experimental protocols for maximal OCR assessment in T98G with inhibited oxidative phosphorylation. First, oligomycin from a different supplier (referred to here as oligo*) was tested. The effect of the A-isomer of oligomycin (oligo A) was also evaluated (Fig 4). The oligomycin complex is a mixture of oligomycins A, B and C, the first two being the most potent ATP synthase inhibitors [22]. The results in Fig 4 show that oligo* or oligo A had an inhibitory effect on SRC in T98G cells similar to that observed with oligomycin.

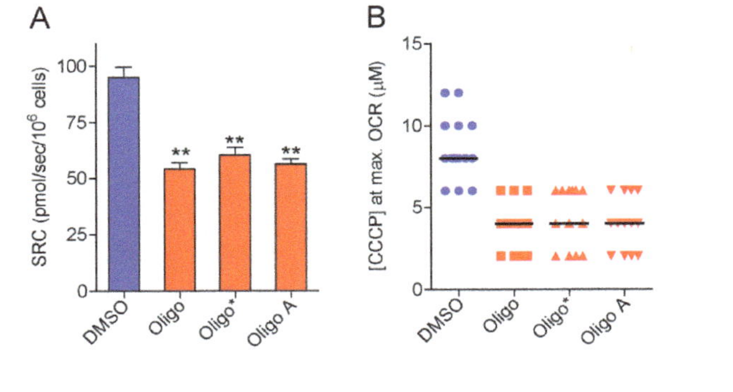

Figure 4. Inhibitory effect of oligomycin and its A-isomer on SRC for T98G cells. After incubation of T98G cells (1.5×10^6 cells/mL) for 3–5 min, 1 µg/mL oligomycin from Sigma-Aldrich (Oligo), 1 µg/mL oligomycin from Cayman Chemical (Oligo*), 1 µg/mL oligomycin A (Oligo A) or 0.5 µL DMSO were added, followed by sequential additions of CCCP (2 µM each). **A**: SRC values for suspended T98G cells in the absence and presence of oligomycin or its A-isomer. **Statistically significant difference from the result for DMSO, $P<0.01$. **B**: Values of CCCP concentrations required to achieve maximal OCR in T98G cells in the presence and absence of oligomycin or its A-isomer.

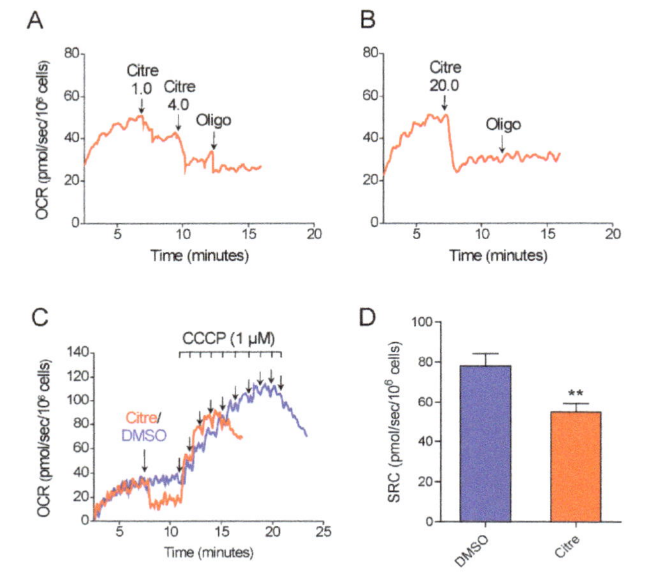

Figure 5. Inhibitory effect of citreoviridin on CCCP-induced maximal oxygen consumption in T98G cells. A and **B**: Representative OCR traces in suspended T98G cells (1.5×10^6 cells/mL). Where indicated by the arrows, citreoviridin (Citre; 1 µM, 4 µM and 20 µM) and 1 µg/mL oligomycin (Oligo) were added. **C**: Representative traces of OCR in T98G cells. Where indicated by the arrows, 1 µL DMSO or 20 µM citreoviridin were added followed by sequential additions of CCCP (1 µM each). **D**: SRC values for T98G cells in the presence and absence of 20 µM citreoviridin. **Statistically significant difference from the results for DMSO, $P<0.01$.

The effect of the ATP synthase inhibitor citreoviridin was also tested on CCCP-induced maximal OCR in T98G cells. Unlike oligomycin, which binds to the F_O subunit of ATP synthase [23, 24], citreoviridin targets the F_1 subunit [24, 25]. Fig 5A and 5B) shows that the addition of 5 µM citreoviridin partially inhibits oxygen consumption under basal conditions, indicating inhibition of oxidative phosphorylation. Nevertheless, sequential addition of oligomycin still resulted in further inhibition of oxygen consumption (Fig 5A). When a higher concentration of citreoviridin was used (20 µM), no further inhibitory effect was observed for oligomycin (Fig 5B). This 20 µM concentration of citreoviridin was used for the next experiments. The results in Fig 5C and 5D) show that, as with oligomycin (but to a lesser extent), when oxidative phosphorylation was fully inhibited by citreoviridin, CCCP-induced maximal OCR was underestimated. In the presence of citreoviridin, SRC was underestimated by 26.4 ± 6.5% compared with DMSO-treated T98G cells.

A different approach for the inhibition of oxidative phosphorylation in intact T98G cells was tested by using the adenine nucleotide translocator (ANT) inhibitors bongkrekic acid (BKA) and carboxyatractyloside (CAT) [26, 27]. Fig 6 shows the results with intact and permeabilized T98G cells. The additions of BKA or CAT (25 µM of each, **A, C**) to intact cells had no effect on basal oxygen consumption, indicating that these compounds do not inhibit oxidative phosphorylation in intact glioma cells. As expected, the sequential addition of oligomycin resulted in significant inhibition of oxygen consumption. Experiments with digitonin-permeabilized T98G cells indicated that BKA and CAT (2.5 µM of each, **B, D**) were effective in fully inhibiting ADP-stimulated oxygen consumption.

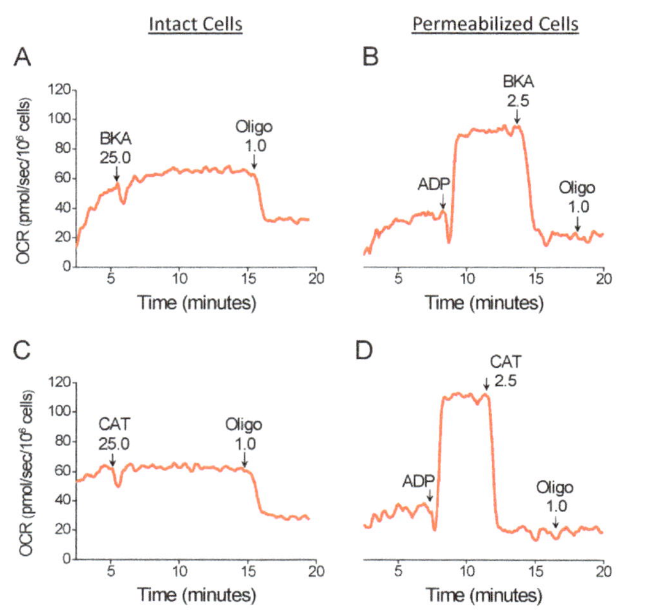

Figure 6. Effect of bongkrekic acid (BKA) and carboxyatractyloside (CAT) on oxygen consumption due to oxidative phosphorylation in intact and digitonin-permeabilized T98G cells. A and **C**: Representative OCR traces in intact T98G cells. Suspended T98G cells (1.5×10^6 cells/mL) were incubated in the medium for intact cells and, where indicated by the arrows, 25 µM BKA, 25 µM CAT and 1 µg/mL oligomycin (Oligo) were added. **Band D**: Representative traces of OCR in digitonin-permeabilized T98G cells. A 125-µL aliquot of T98G (3×10^6 cells) cell suspension was added to the final volume of 2 mL of reaction medium (125 mM sucrose, 65 mM KCl, 10 mM HEPES-K^+ pH 7.2, 2 mM K_2HPO_4, 1 mM $MgCl_2$, 1 mM EGTA and a cocktail of the respiratory substrates α-ketoglutarate, malate, glutamate and pyruvate, 5 mM of each) containing 30 µM digitonin. Where indicated by the arrows, 800 µM ADP, 2.5 µM BKA, 2.5 µM CAT and 1 µg/mL oligomycin (Oligo) were added.

Next, we tested whether similar results would also be observed when CCCP is replaced by another potent protonophore or by a weaker protonophore, respectively FCCP or 2,4-dinitrophenol (DNP); the latter is less potent than CCCP and FCCP by two orders of magnitude [28]. The results when CCCP was replaced by FCCP were similar to those when CCCP was used (S3 File). Compared with the values for T98G cells treated with the vehicle DMSO before FCCP titration, SRC was underestimated by 36.7 ± 1.8% in T98G cells treated with oligomycin (S3C File). As expected, the concentrations of FCCP (median concentration of 5 μM) required to achieve maximal OCR in T98G cells were lower ($P<0.01$) (S3D File) than those required when CCCP was used (median concentration of 8 μM).

Interestingly, when CCCP was replaced by DNP, lower maximal OCRs were observed for T98G and U-87MG cells (Fig 7, 7B, 7C and 7E), reaching similar values to those obtained with CCCP titration when oligomycin is present. The median DNP concentrations required for maximal OCR were 350 μM and 200 μM for T98G cells in the absence and presence of oligomycin and 250 μM and 150 μM for U-87MG cells in the absence and presence of oligomycin, respectively. The presence of oligomycin did not significantly decrease the SRC estimated by DNP titration, but the SRC values for both cells were similar to those obtained by CCCP titration in oligomycin-treated cells.

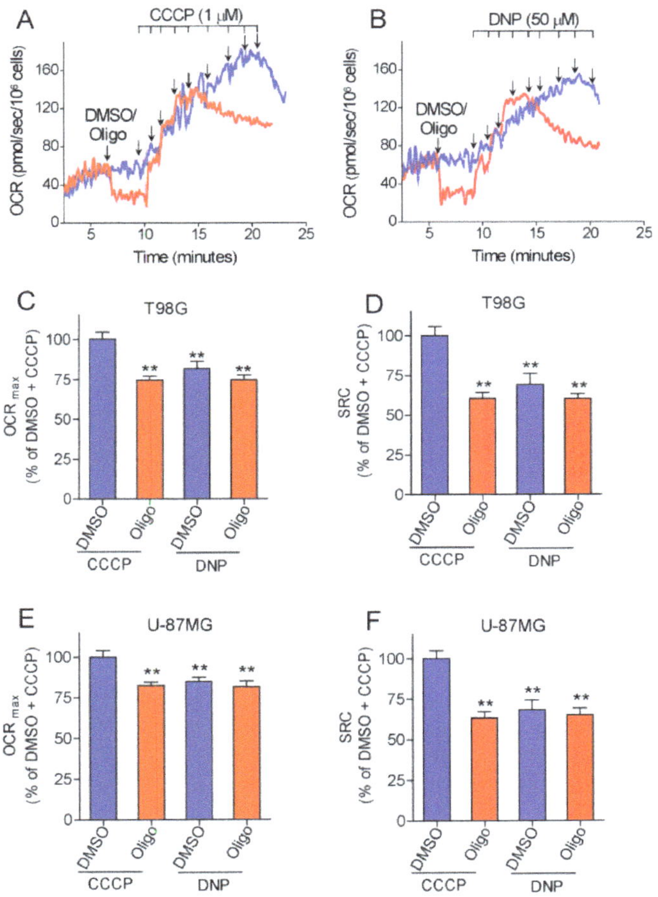

Figure 7. Effect of oligomycin on 2,4-dinitrophenol (DNP)-induced maximal oxygen consumption in intact human glioma cells. A and **B**: Representative oxygen consumption rate (OCR) traces of suspended T98G cells (1.5×10^6/mL). Where indicated by the arrows, 1 μg/mL oligomycin (Oligo) and DMSO (0.5 μL of each) were added, followed by sequential additions of CCCP (1 μM each) (**A**) or DNP (50 μM each) (**B**). **C** and **E**: Maximal OCR (OCR_{max}) values for T98G (**C**) and U-87MG cells (**E**) determined by sequential additions of CCCP or DNP, in the presence and absence of oligomycin. Data are expressed as percentage of OCR_{max} obtained with CCCP titration in DMSO-treated cells (% of DMSO + CCCP). **Statistically significant difference from the results for the "DMSO + CCCP" group, $P<0.01$. **D** and **F**: SRC values for T98G (**D**) and U-87MG cells (**F**) determined by sequential additions of CCCP or DNP, in the presence and absence of oligomycin. Data are expressed as percentage of SRC obtained with CCCP titration in DMSO-treated cells (% of DMSO + CCCP). **Statistically significant difference from the results for the "DMSO + CCCP" group, $P<0.01$.

The absence of pyruvate in the incubation medium can underestimate the maximal OCR induced by FCCP in some cell lines and synaptosomes [29, 30]. We thus tested if pyruvate supplementation could eliminate the effects of oligomycin inhibiting maximal OCR obtained with protonophore (CCCP). Our results (Fig 8) indicate that pyruvate supplementation (10 mM) does not prevent the inhibitory effect of oligomycin on SRC determined for T98G cells. In addition, no significant difference was observed in basal and maximal OCR with pyruvate supplementation (results not shown). The lack of effect of pyruvate under our conditions may be due to the weak mitochondrial oxidation of this substrate in high glycolytic tumor cells, which can use glutamine as a preferential respiratory substrate [31, 32].

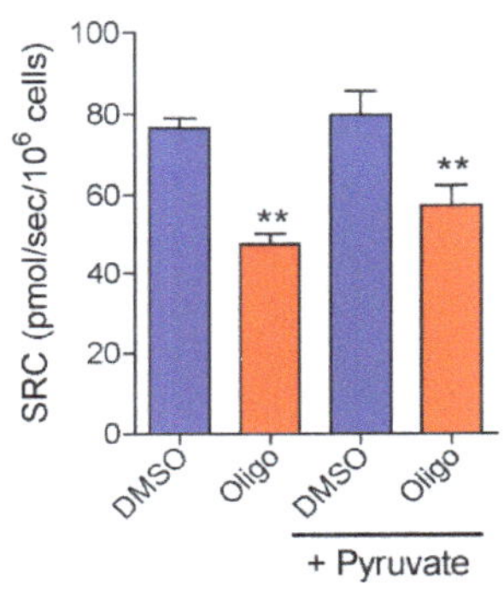

Figure 8. Pyruvate supplementation does not prevent the inhibitory effect of oligomycin on SRC determined for T98G cells. T98G cells (1.5×10^6 cells/mL) were incubated in the presence or absence of oligomycin (Oligo; 1 μg/mL). Where indicated, the incubation medium was supplemented with 10 mM sodium pyruvate (+ Pyruvate). Maximal OCR for suspenced T98G cells was determined by CCCP titration and SRC was calculated. **Statistically significant difference from the results for DMSO, $P<0.01$.

Oligomycin Has Little or no Inhibitory Effect on CCCP-Induced Maximal OCR in Permeabilized Tumor Cells or Isolated Brain Mitochondria

Fig 9 shows the results of CCCP-induced maximal OCR estimated in digitonin-permeabilized cells incubated in a reaction medium supplemented with exogenous respiratory substrates. Basal OCR was lower in permeabilized cells than in intact cells. This is due to the dilution of endogenous ADP after plasma membrane permeabilization with the consequent cessation of oxidative phosphorylation. In fact, the addition of oligomycin to respiring cells under basal conditions had little or no effect on OCR (Fig 9C and 9D). Unlike the results for intact tumor cells, the results in this case showed only a small inhibitory effect (11.5 ± 3.5%) of oligomycin on CCCP-induced maximal OCR estimated for U-87MG cells and no significant inhibitory effect when this parameter was estimated for T98G cells (Fig 9E). Maximal OCR in digitonin-permeabilized cells was about 75% of that observed with intact cells. Median concentrations of CCCP required to reach maximal OCR did not differ for glioma cells treated with oligomycin or with DMSO (Fig 9F). The CCCP concentration required to reach maximal OCR in permeabilized cells (0.5 μM for T98G and 0.25 μM U-87MG) was approximately 10 times lower than that for intact cells, mainly because of the absence of FBS in the incubation medium for the former. No significant increase in maximal OCR in permeabilized cells was observed when the medium was supplemented with 5 μM cytochrome *c* (results not shown), indicating that the cells did not lose cytochrome *c* following treatment with digitonin.

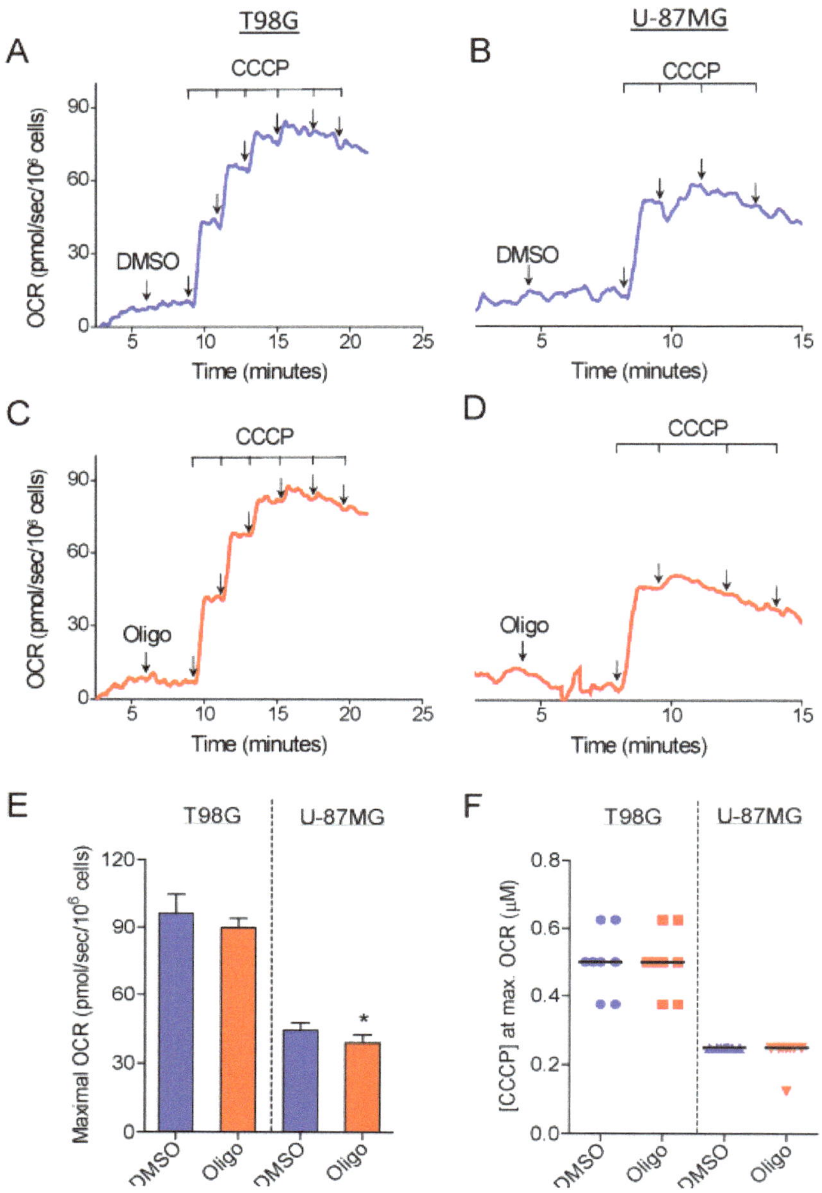

Figure 9. Effect of oligomycin on CCCP-induced maximal oxygen consumption in digitonin-permeabilized human glioma cells. A 125-μL aliquot of either T98G (3×10^6 cells) or U-87MG (4×10^6 cells) cell suspension was added to the final volume of 2 mL of reaction medium (125 mM sucrose, 65 mM KCl, 10 mM HEPES-K^+ pH 7.2, 2 mM K_2HPO_4, 1 mM $MgCl_2$, 1 mM EGTA and a cocktail of the respiratory substrates α-ketoglutarate, malate, glutamate and pyruvate, 5 mM of each) containing 30 μM digitonin. **A** and **C**: Representative OCR traces in suspended T98G cells. **B** and **D**: Representative OCR traces in suspended U-87MG cells. Where indicated by the arrows, 1 μg/mL oligomycin (Oligo) and 0.5 μL DMSO were added followed by sequential additions of CCCP (0.125 μM each). At the beginning of the measurements, it took 3–5 min to obtain a stable basal OCR. **E**: Values of CCCP-induced maximal OCR for T98G and U-87MG cells in the presence and absence of oligomycin. *Statistically significant difference from the results for the respective control (DMSO), $P<0.05$. **F**: Values of CCCP concentrations required for maximal OCR in T98G and U-87MG cells in the presence and absence of oligomycin.

As with permeabilized cells, CCCP-induced maximal OCR in isolated rat brain mitochondria was only slightly inhibited (10.0 ± 1.1%) when oligomycin was added before titration with CCCP (Fig 10).

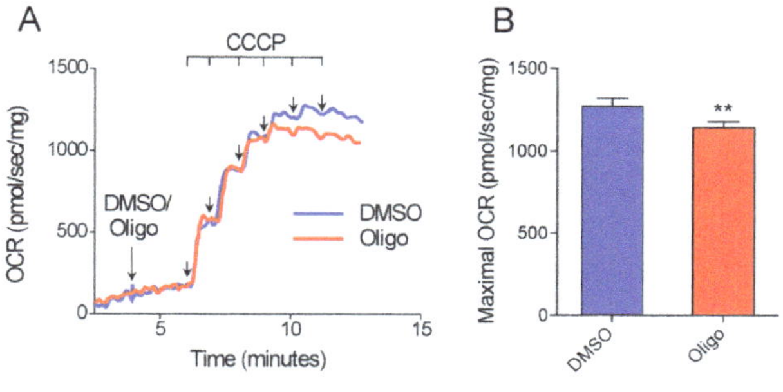

Figure 10. Oligomycin exerts only a minor inhibitory effect on CCCP-induced maximal oxygen consumption in isolated brain mitochondria. Isolated rat brain mitochondria (0.3 mg/mL) were incubated at 37°C in a 2 mL chamber containing 125 mM sucrose, 65 mM KCl, 10 mM HEPES-K^+ pH 7.2, 2 mM K_2HPO_4, 1 mM $MgCl_2$, and 1 mM EGTA. **A**: Representative traces of OCR in isolated brain mitochondria. Where indicated by the arrows, 1 μg/mL oligomycin (Oligo) or 0.5 μL DMSO were added, followed by sequential additions of CCCP (0.05 μM each). At the beginning of the measurements, it took 3–5 min to obtain a stable basal OCR. **B**: Values of CCCP-induced maximal OCR for brain mitochondria in the presence and absence of oligomycin. **Statistically significant difference from the results for DMSO,P<0.01.

DISCUSSION

The data reported here show that the presence of the ATP synthase inhibitors oligomycin or citreoviridin results in underestimation of CCCP- or FCCP-induced maximal OCR in intact cultured cells. This effect was observed in both attached (Fig 3) and suspended cells (Figs 1,4 and 5) (S1–S3 Files), but was nearly abolished in digitonin-permeabilized cells (Fig 9). The lower maximal OCR in cells when ATP synthase inhibitors were present results in underestimation of the SRC. Under our experimental conditions, the SRC was underestimated by 25 to 45% when the tumor cell lines PC-3, T98G and U-87MG were assayed in the presence of oligomycin or citreoviridin. Oligomycin did not result in underestimation of SRC when maximal OCR was induced by the weaker protonophore DNP, but the lack of effect of oligomycin under these conditions is most likely related to the lower maximal OCR induced by DNP (Fig 7) as compared to more potent protonophores (i.e. CCCP or FCCP).

In the absence of ATP synthase inhibitors, higher concentrations of the protonophores CCCP and FCCP were required to achieve maximal OCR for intact cells (Figs 1 and 3–5) (S1 and S3Files). This can be explained by the fact that when CCCP and FCCP promote a significant decrease in mitochondrial proton-motive force, reverse activity of ATP synthase occurs, with ATP hydrolysis driving proton pumping across the inner mitochondrial membrane [24, 33–35]. Under these conditions, the ATPase activity of ATP synthase occurs together with respiration-linked proton translocation through respiratory complexes I, III and IV into the intermembrane space. Thus, higher protonophore concentrations are required to dissipate the proton-motive force and promote maximal OCR. The ATP synthase inhibitors oligomycin [36] or citreoviridin (at a high concentration) [37] inhibit the forward and reverse activities of the ATP synthase. As discussed below, decreased respiratory chain activity may also contribute to the lower protonophore concentration required to achieve maximal OCR in the presence of ATP synthase inhibitors.

The inhibitory effect of oligomycin or citreoviridin on protonophore-induced maximal OCR seems to be associated with ATP synthase inhibition as the concentrations of these compounds required to achieve the reported undesired effect were similar to those that inhibit oxidative phosphorylation completely (Figs 2 and 5). ATP synthase inhibition may limit maximal OCR in cells by promoting changes in mitochondrial and cytosol adenylate energy charge, leading to decreased respiratory activity. Cells with high glycolytic activity, such as most tumor and proliferating non-tumor cells [3–5, 38], can partially maintain the intracellular ATP/ADP ratio even in the presence of respiratory chain inhibitors or protonophores as long as oligomycin is present to prevent consumption of ATP through reverse activity of ATP synthase [35, 38, 39]. A high intracellular ATP/ADP ratio may indirectly result in lower protonophore-induced maximal OCR because of a restraint in the availability or use of respiratory substrates by mitochondria. A high ATP/ADP ratio slows glycolysis [6, 40, 41], which generates the respiratory substrates glycerol-3-phosphate and pyruvate and limits the maximum activity of pyruvate dehydrogenase, glutamate dehydrogenase [42, 43] and the citric acid cycle enzymes citrate synthase and isocitrate dehydrogenase [44–46]. When cells are permeabilized or isolated mitochondria are used, a high concentration of exogenous respiratory substrate is provided and glycolysis is not present to promote ATP resynthesis. Under these conditions the inhibitory effect of oligomycin on CCCP-induced maximal OCR is minimized (Figs 9 and 10). In addition, ATP synthase inhibitors can also interact with non-specific targets, such as plasma membrane Na^+/K^+-ATPase and metabolic enzymes [47–49]. This may indirectly alter cellular energy metabolism and/or mitochondrial membrane conformation, limiting the maximal OCR induced by protonophores.

Interestingly, our results indicate that the ANT inhibitors BKA and CAT had no effect on oxidative phosphorylation in intact T98G cells (Fig 6A and 6C) and therefore preclude the use of these compounds to substitute ATP synthase inhibitors in the protocol for estimation of maximal OCR in cells with previously inhibited oxidative phosphorylation. Nonetheless, BKA and CAT were effective inhibitors of oxidative phosphorylation in digitonin-permeabilized cells (Fig 6B and 6D). The fact that BKA and CAT had no effect on intact tumor cells may be due to the poor permeability of these compounds through the plasma membrane of tumor cells, the effective removal of these compounds from tumor cells via multidrug resistant ATP-binding cassette (ABC) transporters [50], or, as recently proposed by Maldonado and Lemasters [6], a different mechanism of mitochondrial ADP/ATP transport in tumor cells, possibly involving the ATP-Mg/P_i carrier. In fact, the ATP-Mg/P_i carrier is overexpressed in many tumor cells [51]. In the case of this last hypothesis, why the mitochondrial ATP-Mg/P_i carrier does not transport ADP/ATP in digitonin-permeabilized T98G cells (Fig 6B and 6D) remains to be clarified.

CONCLUSION

Maximal OCR and SRC are valuable parameters that are widely used to assess cellular respiratory capacity in normal and tumor cells. Therefore, the effects of ATP synthase inhibitors on CCCP- or FCCP-induced maximal OCR and, consequently, on the estimation of SRC, should be carefully considered to avoid flawed interpretations. We conclude that unless a previously validated protocol is used, maximal OCR in intact cells should be determined in the absence of ATP synthase inhibitors.

SUPPORTING INFORMATION

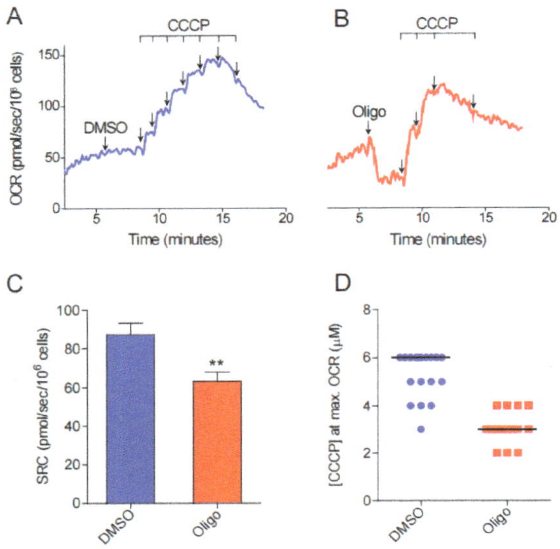

S1 File. Inhibitory effect of oligomycin on CCCP-induced maximal oxygen consumption in PC3 cells.

A and **B**: Representative OCR traces in suspended PC3 cells (1.5×10^6 cells/mL). Where indicated by the arrows, 1 μg/mL oligomycin (Oligo) or 0.5 μL DMSO were added, followed by sequential additions of CCCP (1 μM each). **C**: SRC values for PC3 cells in the presence and absence of oligomycin. **Statistically significant difference from the results for DMSO,$P<0.01$. **D**: Values of CCCP concentrations required to achieve maximal oxygen consumption rate in PC3 cells in the presence and absence of oligomycin.

doi:10.1371/journal.pone.0150967.s001

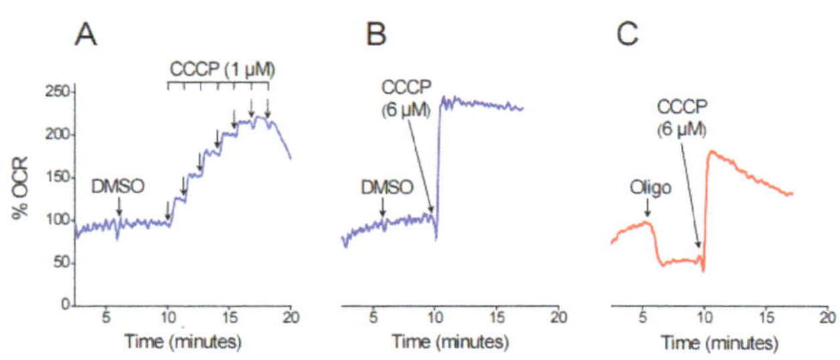

S2 File. Inhibitory effect of oligomycin on CCCP-stimulated oxygen consumption by T98G cells using a single addition of CCCP.

T98G cells (1.5×10^6 cells/mL) were incubated and where indicated by the arrows, 0.5 μL DMSO or 1 μg/mL oligomycin (Oligo) followed by CCCP were added (**A**: sequential additions of 1 μM CCCP; **B** and **C**: only one addition of 6 μM CCCP). Results are shown as percentages of OCR determined just before the addition of DMSO or oligomycin.

doi:10.1371/journal.pone.0150967.s002

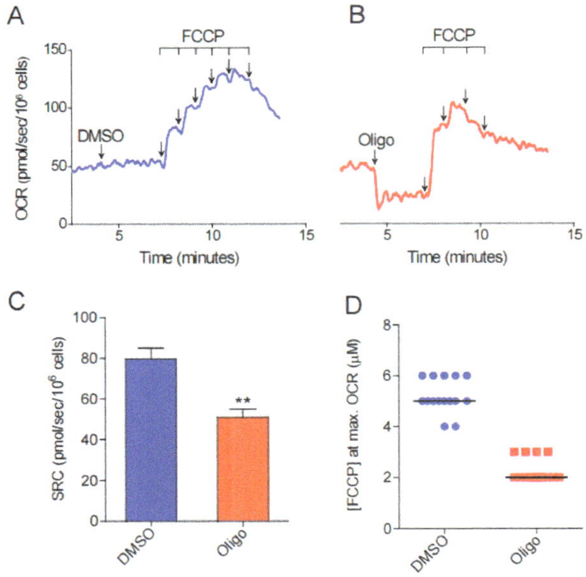

S3 File. Inhibitory effect of oligomycin on FCCP-induced maximal oxygen consumption by T98G cells.

A and **B**: Representative OCR traces by suspended T98G cells (1.5×10^6 cells/mL). Where indicated by the arrows, 1 μg/mL oligomycin (Oligo) or 0.5 μL DMSO were added followed by sequential additions of FCCP (1 μM each). **C**: SRC values for T98G cells in the presence and absence of oligomycin. **Statistically significant difference from the results for DMSO,$P<0.01$. **D**: FCCP concentrations required for maximal OCR in T98G cells in the presence and absence of oligomycin.

doi:10.1371/journal.pone.0150967.s003

ACKNOWLEDGMENTS

We thank Dr. Tiago R. Figueira for critical reading of the manuscript.

Author Contributions

Conceived and designed the experiments: JSR IA AJK RFC. Performed the experiments: JSR ESSS IA ERS. Analyzed the data: JSR IA AJK RFC. Contributed reagents/materials/analysis tools: AJK RFC. Wrote the paper: JSR IA AJK RFC.

REFERENCES

1. Brand MD, Nicholls DG. Assessing mitochondrial dysfunction in cells. *Biochem J.* 2011; 435: 297–312. doi: 10.1042/BJ20110162. pmid:21726199

2. Pesta D, Gnaiger E. High-resolution respirometry: OXPHOS protocols for human cells and permeabilized fibers from small biopsies of human muscle. *Methods Mol Biol.* 2012; 810: 25–58. doi: 10.1007/978-1-61779-382-0_3. pmid:22057559

3. DeBerardinis RJ, Lum JJ, Hatzivassiliou G, Thompson CB. The biology of cancer: metabolic reprogramming fuels cell growth and proliferation. *Cell Metab*. 2008; 7: 11–20. doi: 10.1016/j.cmet.2007.10.002. pmid:18177721

4. Vander Heiden MG, Cantley LC, Thompson CB. Understanding the Warburg effect: the metabolic requirements of cell proliferation. *Science*. 2009; 324: 1029–1033. doi: 10.1126/science.1160809. pmid:19460998

5. Ward PS, Thompson CB. Metabolic reprogramming: a cancer hallmark even Warburg did not anticipate. *Cancer Cell*. 2012; 21: 297–308. doi: 10.1016/j.ccr.2012.02.014. pmid:22439925

6. Maldonado EN, Lemasters JJ. ATP/ADP ratio, the missed connection between mitochondria and the Warburg effect. *Mitochondrion*. 2014; 19: 78–84. doi: 10.1016/j.mito.2014.09.002. pmid:25229666

7. Raimundo N, Baysal BE, Shadel GS. Revisiting the TCA cycle: signaling to tumor formation. *Trends Mol Med*. 2011; 17: 641–649. doi: 10.1016/j.molmed.2011.06.001. pmid:21764377

8. Senyilmaz D, Teleman AA. Chicken or the egg: Warburg effect and mitochondrial dysfunction. *F1000Prime Rep*. 2015; 7: 41. doi: 10.12703/P7-41. pmid:26097714

9. Koppenol WH, Bounds PL, Dang CV. Otto Warburg's contributions to current concepts of cancer metabolism. *Nat Rev Cancer*. 2011; 11: 325–337. doi: 10.1038/nrc3038. pmid:21508971

10. Moreno-Sánchez R, Rodríguez-Enríquez S, Marín-Hernández A, Saavedra E. Energy metabolism in tumor cells. *FEBS J*. 2007; 274: 1393–1418. pmid:17302740 doi: 10.1111/j.1742-4658.2007.05686.x

11. Viale A, Corti D, Draetta GF. Tumors and Mitochondrial Respiration: A Neglected Connection. *Cancer Res*. 2015; 75: 3687–3691. doi: 10.1158/0008-5472.can-15-0491

12. Birsoy K, Wang T, Chen WW, Freinkman E, Abu-Remaileh M, Sabatini DM. An essential role of the mitochondrial electron transport chain in cell proliferation is to enable aspartate synthesis. *Cell*. 2015; 162: 540–551. doi: 10.1016/j.cell.2015.07.016. pmid:26232224

13. Sullivan LB, Gui DY, Hosios AM, Bush LN, Freinkman E, Vander Heiden MG. Supporting aspartate biosynthesis is an essential function of respiration in proliferating cells. *Cell*. 2015; 162: 552–563. doi: 10.1016/j.cell.2015.07.017. pmid:26232225

14. Fulda S, Galluzzi L, Kroemer G. Targeting mitochondria for cancer therapy. *Nat Rev Drug Discov*. 2010; 9: 447–464. doi: 10.1038/nrd3137. pmid:20467424

15. Obre E, Rossignol R. Emerging concepts in bioenergetics and cancer research: metabolic flexibility, coupling, symbiosis, switch, oxidative tumors, metabolic remodeling, signaling and bioenergetic therapy. *Int J Biochem Cell Biol*. 2015; 59: 167–181. doi: 10.1016/j.biocel.2014.12.008. pmid:25542180

16. Chweih H, Castilho RF, Figueira TR. Tissue and sex specificities in Ca^{2+} handling by isolated mitochondria in conditions avoiding the permeability transition. *Exp Physiol*. 2015; 100: 1073–1092. doi: 10.1113/EP085248. pmid:26096641

17. Michelini LG, Figueira TR, Siqueira-Santos ES, Castilho RF. Rotenone exerts similar stimulatory effects on H_2O_2 production by isolated brain mitochondria from young-adult and old rats. *Neurosci Lett*. 2015; 589: 25–30. doi: 10.1016/j.neulet.2015.01.030. pmid:25596437

18. Rosenthal RE, Hamud F, Fiskum G, Varghese PJ, Sharpe S. Cerebral ischemia and reperfusion: prevention of brain mitochondrial injury by lidoflazine. *J Cereb Blood Flow Metab*. 1987; 7: 752–758. pmid:3693430 doi: 10.1038/jcbfm.1987.130

19. Chance B, Williams GR, Hollunger G. Inhibition of electron and energy transfer in mitochondria. III. Spectroscopic and respiratory effects of uncoupling agents. *J Biol Chem*. 1963; 238: 439–444. pmid:14020002

20. Steinlechner-Maran R, Eberl T, Kunc M, Margreiter R, Gnaiger E. Oxygen dependence of respiration in coupled and uncoupled endothelial cells. *Am J Physiol*. 1996; 271: C2053–C2061. pmid:8997208

21. Kenwood BM, Weaver JL, Bajwa A, Poon IK, Byrne FL, Murrow BA, et al. Identification of a novel mitochondrial uncoupler that does not depolarize the plasma membrane. *Mol Metab*. 2013; 3: 114–123. doi: 10.1016/j.molmet.2013.11.005. pmid:24634817

22. Lardy HA, Witonsky P, Johnson D. Antibiotics as tools for metabolic studies. IV. Comparative effectiveness of oligomycins A, B, C, and rutamycin as inhibitors of phosphoryl transfer reactions in mitochondria. *Biochemistry*. 1965; 4: 552–554. pmid:14311628 doi: 10.1021/bi00879a027

23. Devenish RJ, Prescott M, Boyle GM, Nagley P. The oligomycin axis of mitochondrial ATP synthase: OSCP and the proton channel. *J Bioenerg Biomembr*. 2000; 32: 507–515. pmid:15254386

24. Hong S, Pedersen PL. ATP synthase and the actions of inhibitors utilized to study its roles in human health, disease, and other scientific areas. *Microbiol Mol Biol Rev*. 2008; 72: 590–641. doi: 10.1128/MMBR.00016-08. pmid:19052322

25. Gause EM, Buck MA, Douglas MG. Binding of citreoviridin to the beta subunit of the yeast F_1-ATPase. *J Biol Chem*. 1981; 256: 557–559. pmid:6450205

26. Henderson PJ, Lardy HA. Bongkrekic acid. An inhibitor of the adenine nucleotide translocase of mitochondria. *J Biol Chem*. 1970; 245: 1319–1326. pmid:4245638

27. Vignais PV, Vignais PM, Defaye G. Adenosine diphosphate translocation in mitochondria. Nature of the receptor site for carboxyatractyloside (gummiferin). *Biochemistry*. 1973; 12: 1508–1519. pmid:4699983 doi: 10.1021/bi00732a007

28. Lou PH, Hansen BS, Olsen PH, Tullin S, Murphy MP, Brand MD. Mitochondrial uncouplers with an extraordinary dynamic range. *Biochem J*. 2007; 407: 129–140. pmid:17608618 doi: 10.1042/bj20070606

29. Choi SW, Gerencser AA, Nicholls DG. Bioenergetic analysis of isolated cerebrocortical nerve terminals on a microgram scale: spare respiratory capacity and

stochastic mitochondrial failure. *J Neurochem*. 2009; 109: 1179–1191. doi: 10.1111/j.1471-4159.2009.06055.x. pmid:19519782

30. Nicholls DG, Darley-Usmar VM, Wu M, Jensen PB, Rogers GW, Ferrick DA. Bioenergetic profile experiment using C2C12 myoblast cells. *J Vis Exp*. 2010; 46: pii: 2511. doi: 10.3791/2511. pmid:21189469
31. Wise DR, Thompson CB. Glutamine addiction: a new therapeutic target in cancer. *Trends Biochem Sci*. 2010; 35: 427–433. doi: 10.1016/j.tibs.2010.05.003. pmid:20570523
32. Israël M, Schwartz L. The metabolic advantage of tumor cells. *Mol Cancer*. 2011; 10: 70. doi: 10.1186/1476-4598-10-70. pmid:21649891
33. Feniouk BA, Yoshida M. Regulatory mechanisms of proton-translocating F_0F_1-ATP synthase. *Results Probl Cell Differ*. 2008; 45: 279–308. pmid:18026702 doi: 10.1007/400_2007_043
34. Chinopoulos C. Mitochondrial consumption of cytosolic ATP: not so fast. *FEBS Lett*. 2011; 585: 1255–1259. doi: 10.1016/j.febslet.2011.04.004. pmid:21486564
35. Ivanes F, Faccenda D, Gatliff J, Ahmed AA, Cocco S, Cheng CH, et al. The compound BTB06584 is an IF1 -dependent selective inhibitor of the mitochondrial F_1F_o-ATPase. *Br J Pharmacol*. 2014; 171: 4193–4206. doi: 10.1111/bph.12638. pmid:24641180
36. View Article
37. Roberton AM, Holloway CT, Knight IG, Beechey RB. A comparison of the effects of NN'-dicyclohexylcarbodiimide, oligomycin A and aurovertin on energy-linked reactions in mitochondria and submitochondrial particles. *Biochem J*. 1968; 108: 445–456. pmid:4299126 doi: 10.1042/bj1080445
38. Linnett PE, Mitchell AD, Osselton MD, Mulheirn LJ, Beechey RB. Citreoviridin, a specific inhibitor of the mitochondiral adenosine triphosphatase. *Biochem J*. 1978; 170: 503–510. pmid:148274 doi: 10.1042/bj1700503
39. Hao W, Chang CP, Tsao CC, Xu J. Oligomycin-induced bioenergetic adaptation in cancer cells with heterogeneous bioenergetic organization. *J Biol Chem*. 2010; 285: 12647–12654. doi: 10.1074/jbc.M109.084194. pmid:20110356
40. Castilho RF, Hansson O, Ward MW, Budd SL, Nicholls DG. Mitochondrial control of acute glutamate excitotoxicity in cultured cerebellar granule cells. *J Neurosci*. 1998; 18: 10277–10286. pmid:9852565
41. Hers HG, Van Schaftingen E. Fructose 2,6-bisphosphate 2 years after its discovery. *Biochem J*. 1982; 206: 1–12. pmid:6289809 doi: 10.1042/bj2060001
42. Mor I, Cheung EC, Vousden KH. Control of glycolysis through regulation of PFK1: old friends and recent additions. *Cold Spring Harb Symp Quant Biol*. 2011; 76: 211–216. doi: 10.1101/sqb.2011.76.010868. pmid:22096029
43. Patel MS, Roche TE. Molecular biology and biochemistry of pyruvate dehydrogenase complexes. *FASEB J*. 1990; 4: 3224–3233. pmid:2227213
44. Li M, Li C, Allen A, Stanley CA, Smith TJ. The structure and allosteric regulation of mammalian glutamate dehydrogenase. *Arch Biochem Biophys*. 2012; 519: 69–80. doi: 10.1016/j.abb.2011.10.015. pmid:22079166
45. Williamson JR, Cooper RH. Regulation of the citric acid cycle in mammalian systems. *FEBS Lett*. 1980; 117 Suppl: K73–85. pmid:6998729 doi: 10.1016/0014-5793(80)80572-2
46. Gabriel JL, Zervos PR, Plaut GW. Activity of purified NAD-specific isocitrate dehydrogenase at modulator and substrate concentrations approximating conditions in mitochondria. *Metabolism*. 1986; 35: 661–667. pmid:3724458 doi: 10.1016/0026-0495(86)90175-7
47. Wiegand G, Remington SJ. Citrate synthase: structure, control, and mechanism. *Annu Rev Biophys Biophys Chem*. 1986; 15: 97–117. pmid:3013232 doi: 10.1146/annurev.bb.15.060186.000525
48. Datta SC, Ghosh JJ. Effect of citreoviridin, a toxin from *Penicillium citreoviride* NRRL 2579, on glycogen metabolism of rat brain. *Toxicon*. 1981A; 19: 217–222. doi: 10.1016/0041-0101(81)90024-6
49. Datta SC, Ghosh JJ. Production and purification of *Penicillium citreoviride* toxin and its effect on TPP-dependent liver transketolase. *Folia Microbiol (Praha)*. 1981B; 26: 408–412. doi: 10.1007/bf02927336
50. Arato-Oshima T, Matsui H, Wakizaka A, Homareda H. Mechanism responsible for oligomycin-induced occlusion of Na^+ within Na/K-ATPase. *J Biol Chem*. 1996; 271: 25604–25610. pmid:8810335 doi: 10.1074/jbc.271.41.25604
51. Gottesman MM, Fojo T, Bates SE. Multidrug resistance in cancer: role of ATP-dependent transporters. *Nat Rev Cancer*. 2002; 2: 48–58. pmid:11902585 doi: 10.1038/nrc706
52. Traba J, Del Arco A, Duchen MR, Szabadkai G, Satrústegui J. SCaMC-1 promotes cancer cell survival by desensitizing mitochondrial permeability transition via ATP/ADP-mediated matrix Ca^{2+} buffering. *Cell Death Differ*. 2012; 19: 650–660. doi: 10.1038/cdd.2011.139. pmid:22015608

CHAPTER 4

METABOLIC ACTIVATION OF MITOCHONDRIA IN GLIOMA STEM CELLS PROMOTES CANCER DEVELOPMENT THROUGH A REACTIVE OXYGEN SPECIES-MEDIATED MECHANISM

Shuqiang Yuan, Yunxin Lu, Jing Yang, Gang Chen, Sangbae Kim, Li Feng, Marcia Ogasawara, Naima Hammoudi, Weiqin Lu, Hui Zhang , Jinyun Liu, Howard Colman, Ju-Seog Lee, Xiao-Nan Li, Rui-hua Xu1, Peng Huang and Feng Wang

INTRODUCTION

Recent studies indicate the existence of cancer stem cells (CSCs) in various types of cancers, including leukemia and solid tumors [1, 2]. Similar to normal stem cells, CSCs are able to self-renew and to generate the downstream progeny. Although CSCs constitute a very small fraction of the total cancer cells in the tumor bulk, this special subpopulation of malignant cells is thought to play a major role in cancer initiation and development and may be a key cause of resistance to chemotherapy and radiotherapy, leading to persistence of residual disease and cancer recurrence [3]. This phenomenon is due in part to the unique biological properties of CSCs, including high capacity of DNA repair, high expression of certain ATP-dependent drug exporting pumps, high levels of glutathione synthesis, and high expression of cell survival factors [4–6]. A detailed understanding of factors that affect the ability of CSCs to maintain their self-renewal and promote disease progression is important for developing new strategies to effectively kill CSCs. Mounting evidence suggests that the tissue microenvironment may profoundly affect the biological properties and the fates of stem cells and CSCs [7]. *In vivo*, normal stem cells or CSCs reside in special tissue locations known as stem cell niches, which are thought to provide the microenvironment important for the maintenance of their stemness [8]. Although the exact nature of the stem cell niches remains to be defined, it is known that low oxygen and proper levels of certain growth factors such as epidermal growth factor (EGF) and basic fibroblast growth factor (bFGF) are important to maintain the stemness of the cells [9]. Brain CSCs have been found in perivascular niches [8, 10]. Increasing the endothelial cells or blood vessels in orthotopic brain tumor xenografts enhances self-renewal of CSCs and accelerates the initiation and growth of tumors [10]. However, exposure of CSCs to serum *in vitro* usually induces differentiation and presumably may compromise their self-renewal ability [11, 12]. CSCs cultured in serum-free media seem to closely mimic the genotype and gene expression profiles of their primary tumors *in vivo* than do CSCs cultured in standard serum-containing medium [9]. Although the ability of serum to induce apparent differentiation of CSCs has been known for a long time, the underlying mechanisms remain largely unknown. It is also unclear whether exposure of CSCs to serum negatively or positively affects their ability to form tumor *in vivo*.

Reactive oxygen species (ROS) are known to play a role in affecting the fates of normal stem cells [13, 14]. Elevated ROS has been observed to induce differentiation of embryonic stem cells into cardiovascular and mesendodermal cells [7, 15]. The neural stem cells and hematopoietic stem cells contain lower levels of ROS than their mature progeny, whereas increased

Citation: Yuan S, Lu Y, Yang J, et al. Metabolic activation of mitochondria in glioma stem cells promotes cancer development through a reactive oxygen species-mediated mechanism. Stem Cell Research & Therapy. 2015;6:198. doi:10.1186/s13287-015-0174-2.

ROS levels are associated with lowered self-renewal capacity, increased cell cycling, and reduced viability [16–18]. Previous study showed that breast CSCs might have high ROS-scavenging capacity and contain lower cellular ROS compared with the corresponding non-tumorigenic cells [6]. A recent study suggests that ROS might affect the differentiation state of CSC by activation of p38 MAPK [19]. However, the role of ROS in serum-induce differentiation of CSCs and their physiological relevance in tumor development *in vivo* remain largely unclear. The present study was designed to investigate these important questions. We showed that serum could activate mitochondrial respiration and promote generation of mitochondrial ROS, leading to apparent loss of certain stem cell markers and lower ability to form neurospheres. However, despite these seemingly differentiation phenotypes *in vitro*, the serum-induced glioma stem cells exhibited greater capacity to form tumor *in vivo*. Our study revealed a novel role of mitochondrial ROS in serum activation of CSCs to produce the downstream progeny and promote tumor development *in vivo*. The regulation of this redox signaling mechanism has potential implications in developing new strategies to target CSCs.

METHODS

Cell lines and Cell Culture

GSC11, GSC23, and GBM3752 cell lines were originally established from fresh surgical specimens of glioblastoma multiforme at the University of Texas MD Anderson Cancer Center [20, 21]. GSC11 and GSC23 were maintained in Dulbecco's modified Eagle's medium with nutrient mixture F-12 (DMEM/F12) (Mediatech Inc., Manassas, VA, USA) supplemented with B27 (Invitrogen, Carlsbad, CA, USA), 20 ng/ml epidermal growth factor (Miltenyi Biotec, Auburn, CA, USA), 20 ng/ml of basic fibroblast growth factor (Miltenyi Biotec), and 2 mM L-glutamine (Mediatech Inc.) without serum (designated as "stem cell medium"). Cells were cultured in a humidified incubator maintained at 37 °C with 5 % CO_2. GBM3752 cells were obtained from GBM patients undergoing surgery at Texas Children's Hospital and maintained in severe combined immunodeficiency (SCID) mice orthotopically [22]. The cells were freshly isolated from the tumors and cultured in stem cell medium for *in vitro* study within the first five passages. For serum treatment, cells were cultured in the stem cell medium with 5 % fetal bovine serum (FBS) with or without various concentrations of N-acetyl-cysteine (NAC) (Sigma-Aldrich, St. Louis, MO, USA).

RNA Isolation, RNA Microarray Analyses, and Reverse Transcription-Polymerase Chain Reaction

GSC11 and GSC23 cells were cultured in stem cell medium with or without serum for 1, 3, or 7 days in triplicate. Total RNA was isolated from the cells by using an RNeasy Mini kit (Qiagen Inc., Valencia, CA, USA). Sample labeling was performed with an RNA amplification kit in accordance with the conditions recommended by the manufacturer (Applied Biosystems, Foster City, CA, USA). Total RNA was reverse-transcribed by using a complementary DNA (cDNA) synthesis kit (Fermentas Inc., Glen Burnie, MD, USA). The quantitative polymerase chain reaction analyses were carried out in a 25-µl reaction mixture that contained 1 µl cDNA, 0.1 µg oligonucleotide primer pairs, 12.5 µl SYBR Green Mix (Invitrogen), and diethylpyrocarbonate-treated water. Human HT-12v3 expression beadchips containing 48,000 probes of 25,000 annotated genes were obtained from Illumina Inc. (San Diego, CA, USA). The gene expression microarray analysis was performed at the System Biology Department of the UT MD Anderson Cancer Center. Total RNA was extracted from GSC11 cells and used for labeling and hybridization to human expression beadchips in accordance with the protocols of the manufacturer. All experiments were performed in triplicate. Primary microarray data in this study are available in the National Cancer for Biotechnology Information Gene Expression Omnibus (GEO) database (GSE28220). The following primer sets were used for quantitative reverse transcription-polymerase chain reaction (RT-PCR) analysis: SOX2-sense, 5′-GCCTGGGCGCCGAGTGGA-3′; SOX2-antisense, 5′-GGGCGAGCCGTTCATGTAGGTCTG-3′); Olig2-sense, 5′-TGCGCAAGCTTTCCAAGA-3′; Olig2-antisense, 5′-CAGCGAGTTGGTGAGCATGA-3′.

Flow Cytometric Analyses

Cells were dissociated into single-cell suspension by using accutase reagents (Sigma-Aldrich), stained with allophycocyanin (APC)-conjugated CD133 antibody (clone AC133 from MACS) or the control APC-IgG2b antibody (MACS) by using the conditions recommended by the manufacturer. APC fluorescence was quantitated by flow cytometry analysis. To measure intracellular ROS, cells were collected and dissociated into single-cell suspension by accutase, washed with phosphate-buffered saline (PBS) once, and resuspended in pre-warmed PBS containing freshly prepared CM-H2DCFDA (1 µM) or MitoSOX-Red (5 µM; Molecular Probes, Eugene, OR, USA). After incubation at 37 °C for 30 min (H2DCFDA) or 15 min (MitoSOX-Red), the cells were washed with PBS twice and then subjected to flow cytometric analyses.

Immunoblots

Cultured cells were washed with cold PBS before homogenization in lysate buffer. Whole cell lysate (20 µg protein/sample) was used in Western blot analysis. Cell lysates were separated by electrophoresis on 10–12 % sodium dodecyl sulfate polyacrylamide gel electrophoresis and transferred to nitrocellulose membranes. After blocking with 5 % non-fat milk/PBS with Tween 20 for 1 h, the membranes were incubated at 4 °C overnight with primary antibodies, including mouse anti-human CD133 (Miltenyi Biotec), rabbit anti-human SOX2 (Cell Signaling Technology Inc., Danvers, MA, USA), rabbit anti-human Olig2 (Abcam, Cambridge, MA, USA), rabbit anti-human Catalase (EMD Chemicals, Gibbstown, NJ, USA), sheep anti-human SOD1 (EMD Chemicals), rabbit anti-human SOD2 (Santa Cruz Biotechnology Inc., Santa Cruz, CA, USA), and anti-mouse total OXPHOS (Abcam). The Western blot signals were detected with horseradish peroxidase-conjugated secondary antibodies. The membranes were developed by using a Pierce Supersignal West Pico Chemiluminescent Substrate (Fisher Scientific Inc., Pittsburgh, PA, USA).

Immunofluorescence staining

Cells were fixed in 4 % formaldehyde, washed in PBS, and permeabilized for the analysis of intracellular markers (20 min, 0.25 % Triton X-100; Sigma-Aldrich). The monolayers were then incubated with a blocking solution (PBS with 5 % FBS) (45 min, room temperature), followed by incubation (overnight at 4 °C) with the primary antibodies: anti-glial fibrillary acidic protein (anti-GFAP) (Miltenyi Biotec), anti-β-III tubulin (Abcam), anti-Nestin (Abcam), and anti-O4 (Miltenyi Biotec). After extensive washing in PBS, a second incubation (1 h; 37 °C) with Alexa Fluor-488- or Alexa Fluor-547-specific anti-mouse or anti-rabbit secondary antibodies (all from Invitrogen) was performed. Cell nuclei were stained with 4′,6-diamidino-2-phenylindole (DAPI) (Sigma-Aldrich). Florescence labeling was observed by using a fluorescent microscope (Olympus, Tokyo, Japan).

Oxygen Consumption Assay

Samples were dissociated into singles cells, washed with PBS once, and suspended at 4 to approximately 10 million cells per milliliter in stem cell medium. Oxygen consumption was measured in 1-ml medium by using Oxytherm equipped with a Clark-type electrode (Hansatech Instruments Ltd, Norfolk, UK) as described previously [23].

Mouse Xenografts

Subcutaneous xenografts: GSC11 cells cultured under various conditions (stem cell medium without FBS, with 5 % FBS, or with FBS and 20 mM NAC for 7 days) were collected, treated with accutase to make single-cell suspension, and inoculated into the right flank of nude mice (2×10^6 cells per mouse). The mice were euthanized when the tumor diameter was greater than 1.5 cm. For orthotopic xenograft inoculation, GBM3752 cells were first cultured in stem cell medium with or without serum (5 % FBS). The cells were maintained *in vitro* under these two conditions for 60 passages. Cells were collected and inoculated intracranially into the brains of SCID mice (1×10^4 cells per mouse). The mice were euthanized when they developed signs of neurological deficit and became moribund. All experiments of the present study were performed in accordance with human protocols approved by the Institutional Review Board at UT MD Anderson Cancer Center and Baylor College of Medicine as well as animal protocols (ACUF 11-98-08136, AN-4548) approved by the Institutional Animal Care and Use Committee of Baylor College of Medicine. Signed informed consent was obtained from all patients or their legal guardians prior to sample acquisition.

RESULTS AND DISCUSSION

Induction of Apparent differentiation of GSCs by Serum and Association with ROS Stress Responses

Both established glioma stem cell lines and primary glioma cells isolated from fresh tumor tissues were used in our study. GSC11 and GSC23 are two glioblastoma stem cell lines originally derived from glioblastoma multiforme (GBM) surgical specimens and exhibit the *in vitro* stem cell characteristics of extensive self-renewal, the ability to differentiate to neurons and astrocytes, and the ability to initiate tumor *in vivo*. GSC11 and GSC23 cells were maintained in serum-free stem cell culture medium as described previously (He, 2010 #274) [24]. An orthotopic xenograft model (GBM3752) that preserves glioblastoma stem cells was originally established by directly inoculating primary tumor cells from fresh GBM specimen into the right cerebellum of SCID mice brain [22]. The xenograft tumor cells preserve tumorigenicity, multi-lineage differentiation, and $CD133^+$ expression after being subtransplanted in mice brain. GBM3752 cells were prepared freshly from tumors (maintained in SCID) mice for *in vitro* study. As shown in Fig. 1, GSC11 and GBM3752 cells grew well in stem cell medium containing EGF and bFGF without serum and exhibited the morphology of stem-like neurospheres (Fig. 1a) with high expression of CD133 (Fig. 1b, c). The addition of serum (% FBS) to the culture medium caused a significant change in cell morphology, manifested by a loss of neurosphere formation and the appearance of differentiated cells attaching to the culture dish (Fig. 1a). This was accompanied by a substantial decrease of CD133 expression in a time-dependent manner (Fig. 1b, c) and a decrease of Nestin (Additional file 1: Figure S1). Quantitative RT-PCR and Western blot analyses revealed a significant decrease in expression of Sox2 and Olig2, two transcription factors known to regulate neural stem cells and neural progenitor cells (Fig. 1d, e). The expression of Notch1, a molecule important for promoting neural stem cell function [25], was also downregulated (Fig. 1d). In contrast, the expression of differentiation markers, including GFAP, β-III tubulin, O4 (Additional file 1: Figure S1), and ANXA1, were increased after serum exposure (Additional file 1: Figure S1 and Fig. 1e). Similar results were observed in the third cell line GSC23 (Additional file 1: Figure S2). Surprisingly, this apparent differentiation induced by serum did not result in a decrease in tumorigenesis, and as will be described below, the glioma stem cells were activated by serum exposure (see *in vivo* study below).

To investigate the molecular events and potential alterations in the signaling pathways of GSCs in response to serum induction, we treated GSC11 cells with 5 % FBS for 1, 3, and 7 days in triplicate cultures, and RNA was isolated from each sample for determination of gene expression profiles using microarray analyses. As shown in Additional file 1: Figure S3A, clustering analysis of gene expression profiles revealed that the serum-treated GSC11 cells exhibited gene expression profiles clearly distinct from that of the GSC11 cells cultured in serum-free medium. There was a further shift of gene expression profiles as the time of serum exposure prolonged. The fact that the three separated samples of the same time point (biological triplicate) displayed similar gene expression patterns and clustered in the same group demonstrated the high reproducibility of this experimental system. Using the Ingenuity pathway analysis, we found that the oxidative stress response pathway was induced by serum most significantly ($P = 0.0005$) at all time points tested. Additional file 1: Figure S3B shows the genes involved in oxidative stress response identified by this analysis in GSC11 cells. Among these genes, SOD2, catalase, NQO1, peroxiredoxin 1, thioredoxin reductase 1, and glutamate-cysteine ligase are involved in ROS scavenging. These results suggested that the homeostasis of reduction/oxidation (redox) balance might have been disrupted in the serum-induced GSCs.

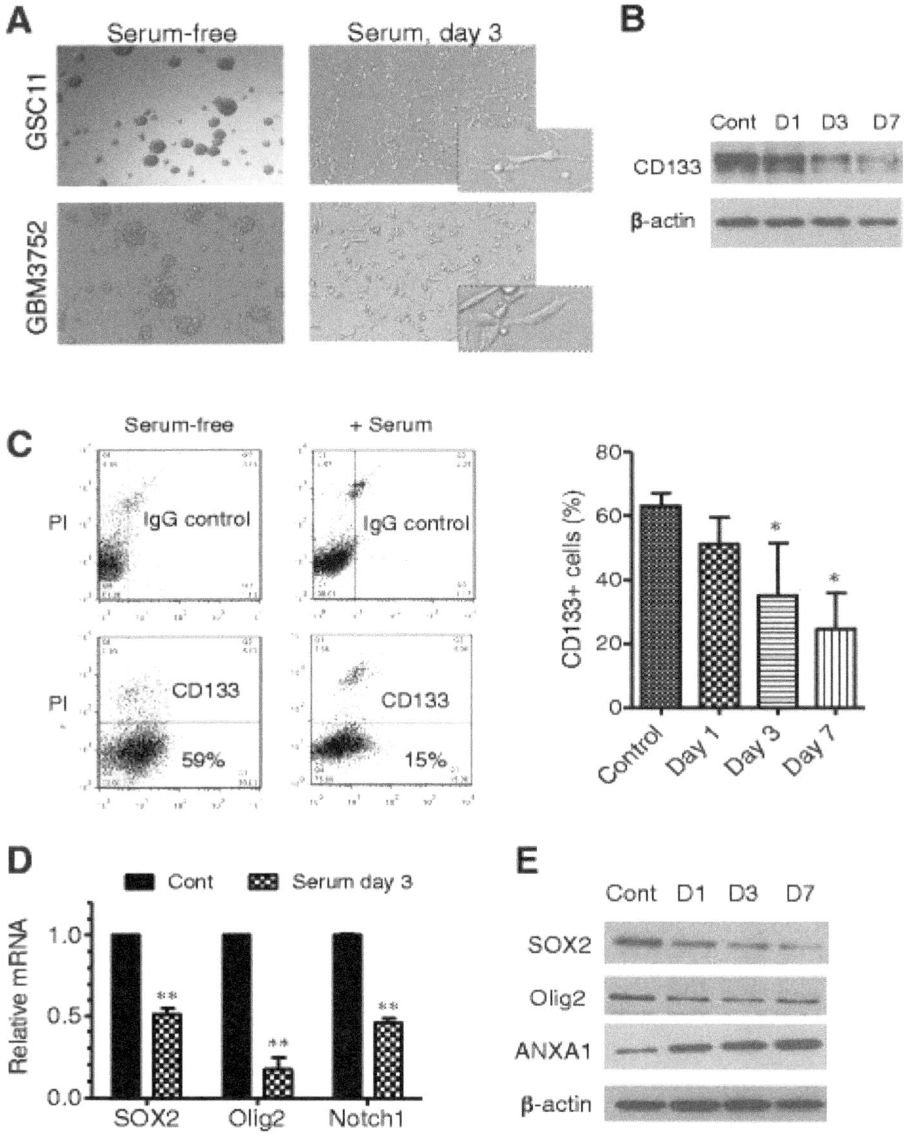

Figure 1: Effect of serum on neurosphere formation and the expression of stem cell markers in glioblastoma stem cells. a Glioblastoma stem cells (GSC11 and GBM3752) formed neurospheres in serum-free medium supplemented with epidermal growth factor and basic fibroblast growth factor. Exposure of the cells to serum (5 % FBS) for 3 days led to a loss of neurosphere formation in both clones. b Western blot analysis of CD133 in GSC11 cells before and after exposure to serum for 1, 3, and 7 days. c Flow cytometry analysis of CD133 expression in GSC11 cells before and after exposure to serum for 7 days. The right panel shows quantitation of the percentage of CD133+ cells before and after GSC11 cells were exposed to serum for 1, 3, and 7 days; $*P < 0.05$. d Expression of SOX2, Olig2, and Notch1 mRNA in GSC11 cells before and after exposure to serum for 3 days. Expression of mRNA was measured by quantitative reverse transcription-polymerase chain reaction. $**P < 0.001$. e Effect of serum on protein expression of stem cell markers SOX2 and Olig2 and differentiation marker ANXA1. GSC11 cells were exposed to 5 % FBS for 1, 3, and 7 days as indicated. SOX2, Olig2, ANXA1 were detected by Western blot analysis. Cont control, D day, FBS fetal bovine serum, GBM glioblastoma multiforme, GSC glioma stem cell, PI

Induction of Mitochondrial ROS Generation in Glioma Stem Cells by Serum Through Activation of Electron Transport Chain

The observations that exposure of GSCs to serum caused consistent oxidative stress response in all tested time points prompted us to explore possible changes in cellular redox status. Since mitochondria are major sites of ROS production, we used MitoSOX-Red to detect mitochondrial superoxide (O_2^-) and 5-(and-6)-chloromethyl-2,7-dichlorodihydrofluorescein diacetate acetyl ester (DCF-DA) to measure total cellular hydrogen peroxide (H_2O_2) and other ROS. The results showed that that serum induced a substantial increase of mitochondrial O_2^- in a time-dependent manner, with an increase of the median value from 46 units in control cells (serum-free) to 67, 188, and 268 units on days 1, 3, and 7 after serum exposure, respectively (Fig. 2a). Interestingly, total cellular ROS (as measured by DCF-DA) also showed a moderate increase, from 84 units in the control to 112, 126, and 159 units on days 1, 3, and 7, respectively (Fig. 2a). Similar results were observed in another glioblastoma stem cell line GSC23 (Additional file 1: Figure S4). Exposure of GSC23 cells to serum led to a 5- and

11-fold increase of mitochondrial O_2^- on days 3 and 7, respectively, and the total cellular ROS detected by DCF-DA also moderately increased (Additional file 1: Figure S4A).

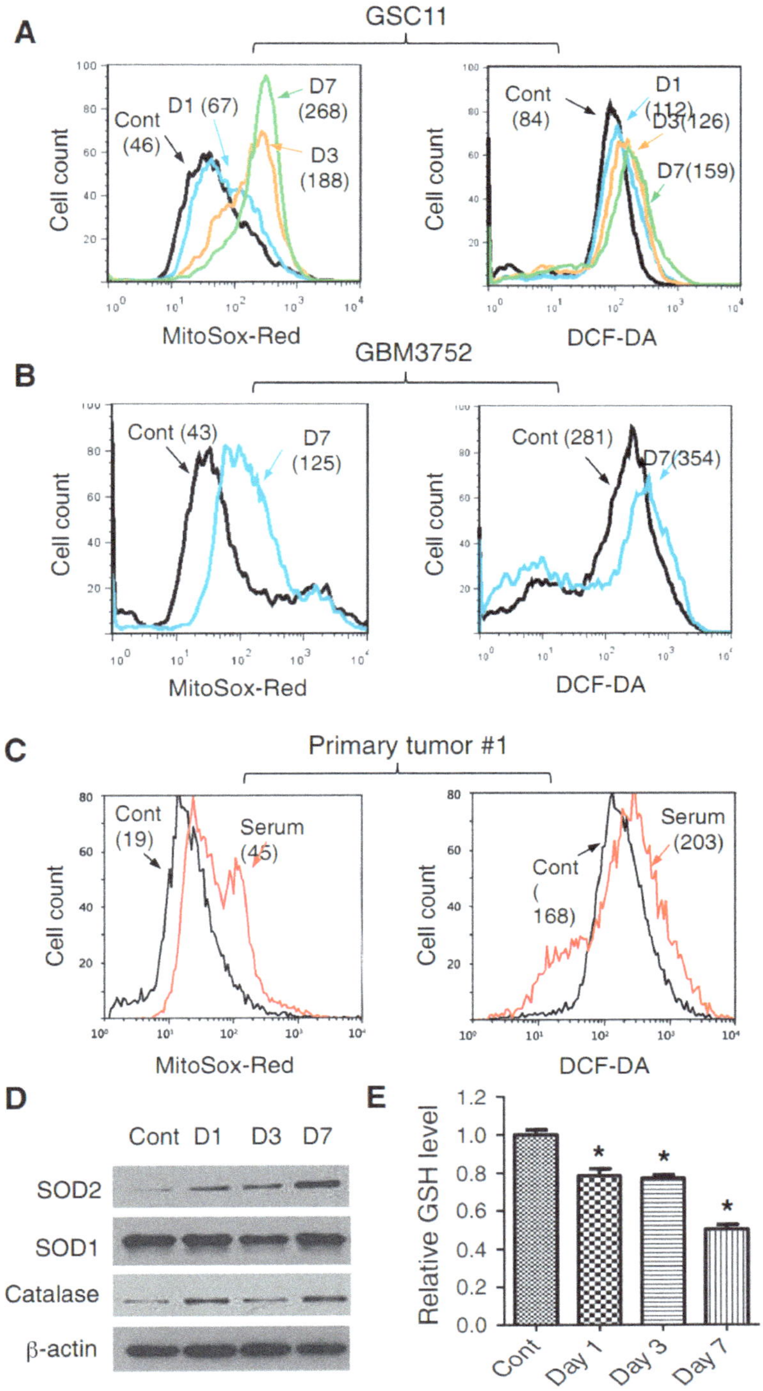

Figure 2: Induction of mitochondrial O_2^- generation and oxidative stress response in glioblastoma stem cells. a Effect of serum on mitochondrial O_2^- and total cellular ROS in GSC11 cells. Cells were incubated without or with 5 % FBS for 1, 3, or 7 days (D1, D3, D7). Mitochondrial O_2^- was measured by flow cytometry analysis after cells were stained with MitoSOX-Red, and total cellular ROS were detected by using CM-H2-DCFDA staining followed by flow cytometry analysis. The numbers in parentheses indicate the median fluorescent intensity. b Effect of serum exposure (1–7 days) on mitochondrial O_2^- and total cellular ROS in GBM3752 cells. c Effect of serum on mitochondrial O_2^- and total cellular ROS in primary tumor cells isolated from fresh GBM tumor tissue. The freshly isolated tumor cells were divided into two portions for incubation in stem cell medium without or with serum (5 % FBS) for 7 days, and mitochondrial O_2^- and total cellular ROS were then measured. The shaded curve shows the background (Bkg) fluorescent without dye. d Western blot analysis of SOD1, SOD2, and catalase in GSC11 cells before and after exposure to serum for 1, 3, and 7 days. e Cellular glutathione in GSC11 cells cultured in stem cell medium without or with

serum for 1, 3, and 7 days. *P < 0.05. Cont control, DCF-DA 5-(and-6)-chloromethyl-2,7-dichlorodihydrofluorescein diacetate acetyl ester, FBS fetal bovine serum, GBM glioblastoma multiforme, GSH glutathione, GSC glioma stem cell, O_2^- superoxide, ROS reactive oxygen species, SOD cytosolic superoxide dismutase

We then used two types of fresh glioma cells to further confirm the above observations. First, the stem-like GBM3752 cells [22] were obtained freshly from tumor xenografts and divided into two portions; one portion was cultured in serum-free stem cell medium and the other portion was cultured in serum-containing medium. After 7 days, the mitochondrial ROS and total cellular ROS in each culture condition were measured by MitoSOX and DC-FDA. As shown in Fig. 2b, a 3-fold increase in mitochondrial O_2^- and a moderate increase (26 %) in total cellular ROS were observed, consistent with that seen in GSC11 and GSC23 cells. Furthermore, this pattern of redox alterations was consistently observed in primary glioma cells isolated from fresh GBM tumor tissues (Fig. 2c and Additional file 1: Figure S4B), suggesting that the induction of mitochondrial O_2^- generation might be a highly consistent event in serum-induced changes in GSCs.

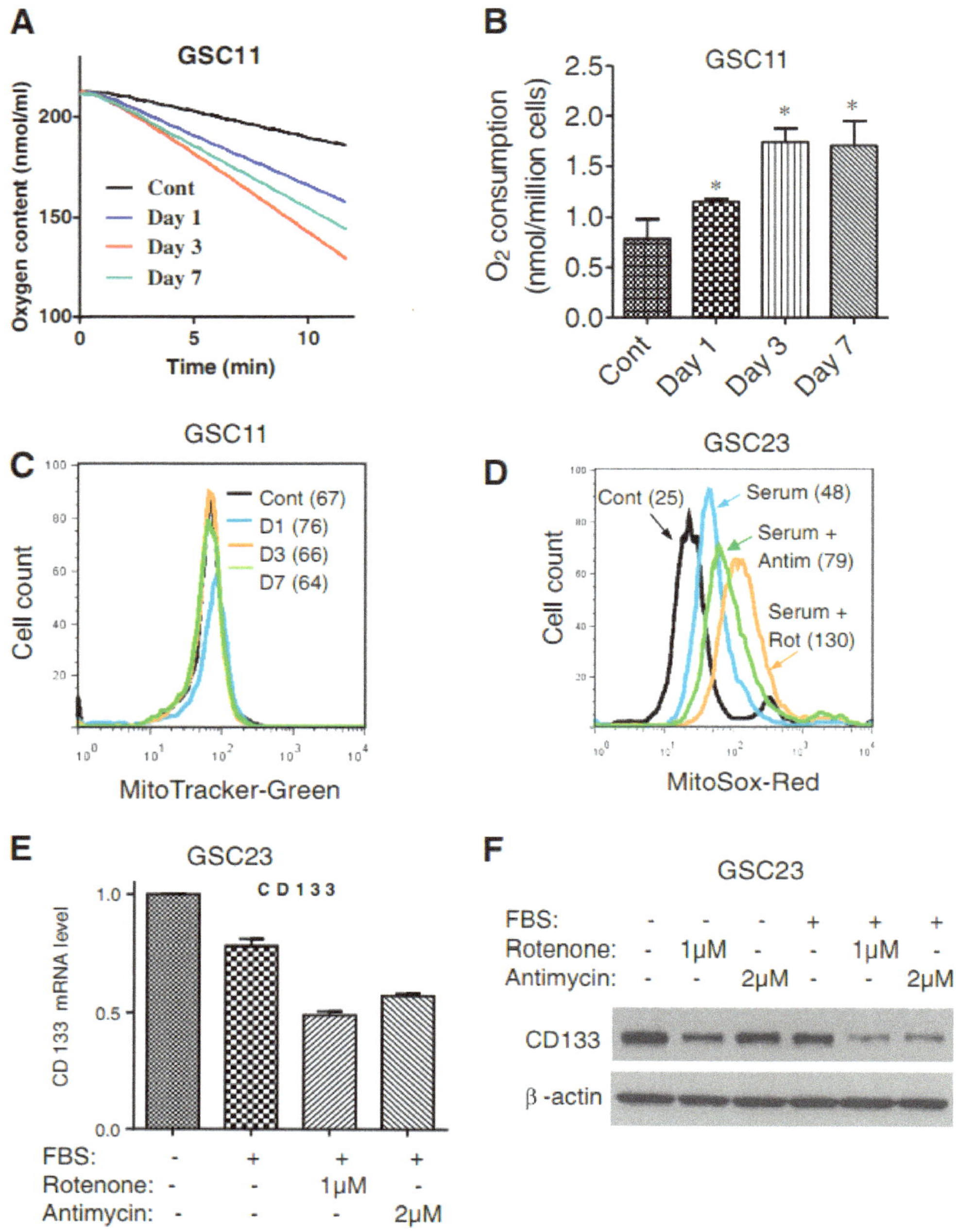

Figure 3: Novel role of mitochondrial activation and ROS generation in activation of glioblastoma stem cells. a Comparison of oxygen consumption in GSC11 cells before and after exposure to serum for 1, 3, and 7 days. Oxygen consumption was measured by using an Oxytherm system as described in Methods. b Quantitative analysis of oxygen consumption in GSC11 cells exposed to serum for 1, 3, and 7 days. *P < 0.01. c Serum did not induce increase of mitochondrial mass in GSC11 cells, measured by MitoTracker-Green. d GSC23 cells were cultured in stem cell medium without or with serum (5 % FBS) in the presence or absence of 1 μM rotenone (ETC. complex I inhibitor) or 2 μM antimycin (complex III inhibitor), and mitochondrial O_2^- was

measured by flow cytometric analysis after cells were stained with MitoSOX-Red. e Effect of serum and ETC. inhibitors on CD133 expression. GSC23 cells were cultured in stem cell medium in the presence or absence of serum (5 % FBS), rotenone (1 μM), or antimycin (2 μM). CD133 mRNA expression levels were measured by quantitative reverse transcription-polymerase chain reaction. f GSC23 cells were cultured in stem cell medium in the presence or absence of serum (5 % FBS), rotenone (1 μM), or antimycin (2 μM). CD133 protein expression was measured by Western blot. Cont control, ETC. electron transport chain, FBS fetal bovine serum, GSC glioma stem cell, O_2^- superoxide

To test whether the increase in mitochondrial O_2^- and cellular ROS induced by serum in GSCs might cause stress response, we used Western blotting to analyze the expression of antioxidant molecules before and after serum exposure. As shown in Fig. 2d, there was a time-dependent increase in expression of SOD2, a mitochondrial superoxide dismutase that converts O_2^- to H_2O_2. Interestingly, the cytosolic superoxide dismutase (SOD1) did not exhibit significant change after GSC11 and GSC23 were exposed to serum (Fig. 2d, Additional file 1: Figure S4C), suggesting that the main source of ROS stress might be mainly from mitochondria and was consistent with the increase in mitochondrial O_2^- shown in Fig. 2a-c. The expression of catalase, an enzyme that converts cellular H_2O_2 to water and oxygen, increased after serum incubation (Fig. 2d, Additional file 1: Figure S4C). Cellular glutathione (GSH), a major endogenous antioxidant, decreased after serum exposure in GSC11 cells (Fig. 2e) and G23 cells (Additional file 1: Figure S4D), reflecting a consumption of this antioxidant. These data together suggest that the increase in SOD2 expression might be a stress response to elevated mitochondrial O_2^- generation induced by serum. SOD2 converted O_2^- to H_2O_2, which was then able to pass the mitochondrial membranes to cytosol, where it was converted to O_2 and H_2O by catalase or neutralized by GSH, resulting in only a moderate increase of overall cellular ROS and a decrease in GSH.

Since mitochondrial O_2^- is generated mainly during respiration because of the release of electrons from complexes I and III of the electron transport chain, we speculated that the increased mitochondrial O_2^- might be a result of active mitochondrial respiration induced by serum and not a consequence of a slower O_2^-elimination since SOD2 expression was increased. To test this possibility, we measured oxygen consumption in GSCs as an indicator of mitochondrial respiration. As shown in Fig. 3a and b, exposure of GSC11 cells to serum led to a time-dependent increase of oxygen consumption, with approximately a 100 % increase by day 3. Interestingly, measurement of mitochondrial mass by using MitoTracker Green as well as mitochondria electron transport chain (ETC.) and the ATP synthase complex antibodies showed that serum did not cause any significant change in mitochondrial mass and complexes (Fig. 3c and Additional file 1: Figure S5), suggesting that the increase in respiration was mainly a functional activation of the pre-existing mitochondria. GBM3752 cells from freshly dissected orthotopic tumor xenografts were cultured in either serum-free medium or serum-containing medium for 7 days. A significantly higher oxygen consumption was observed in GBM3752 cells cultured with serum (Additional file 1: Figure S6A) without any significant changes in mitochondrial mass (Additional file 1: Figure S6B). The increase in respiration without increase of mitochondrial mass was also consistently observed in GSC23 cells (Additional file 1: Figure S6C). Despite the increase of mitochondrial respiration, cells at G_0/G_1 phase were decreased only on day 1 compared with cells cultured in serum-free medium (Additional file 1: Figure S7). Cells at G_2/M phase were not changed.

Important Role of Mitochondrial ROS in Mediating the Serum Effect on Glioma Stem Cells

To evaluate the role of activation of mitochondrial respiration and ROS generation in serum-induced apparent differentiation of GSCs, we first used ETC. complex I inhibitor rotenone and complex III inhibitor antimycin to inhibit mitochondrial respiration in glioma stem cells and then tested whether this affected the ability of serum to induce apparent differentiation in GSCs. The results showed that both ETC. inhibitors disrupted mitochondrial respiration and caused a further increase of mitochondrial O_2^- in the presence of serum (Fig. 3d), but neither of them prevented the serum-induced GSCs from attaching to the flasks and exhibiting apparent differentiation morphology (data not shown). In fact, adding rotenone or antimycin even caused a further decrease of CD133 at both mRNA (Fig. 3e) and protein levels (Fig. 3f).

Considering the observations that exposure of GSCs to serum caused an increase in mitochondrial respiration and O_2^- generation but inhibition of mitochondria respiration by ETC. inhibitors (rotenone and antimycin) did not prevent serum to induce changes in GSCs, we speculate it was the increase in mitochondrial ROS generation, not the respiration *per se*, that plays a key role in mediating the serum effect on GSCs. To test this possibility, we used exogenous H_2O_2 to cause a level of increase in mitochondrial ROS comparable to that caused by serum in GSC11 and GSC23 cells (Fig. 4a, b). Interestingly, a short-term treatment of GSCs with such exogenous H_2O_2 for 6 h led to a significant decrease of SOX2, Olig2, and CD133 mRNA expression in GSC11 cells (Fig. 4c) and GSC23 cells (Fig. 4d), similar to those observed in serum-induced cells.

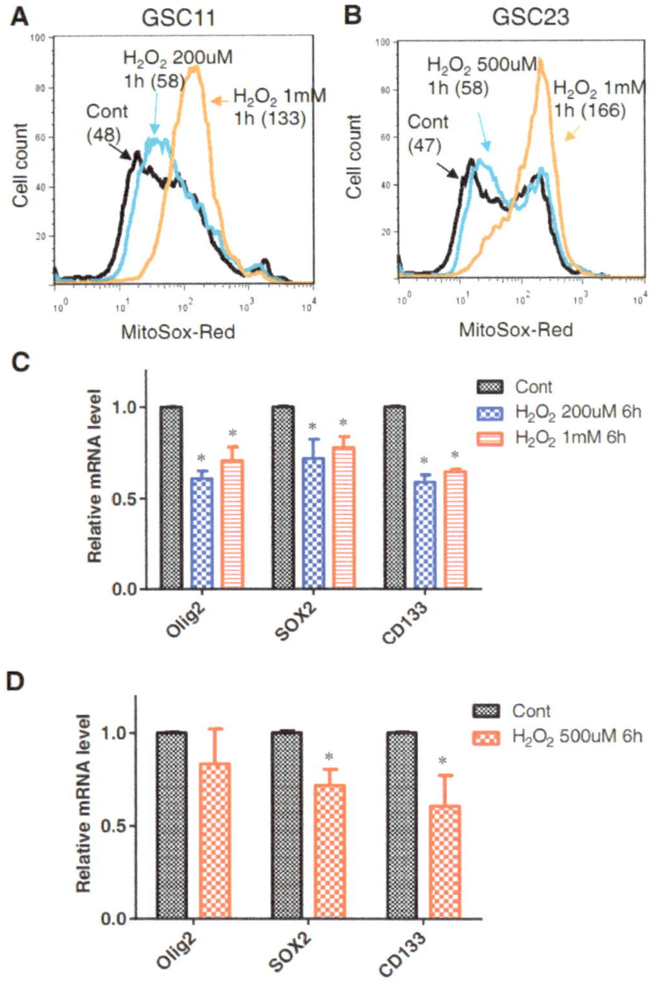

Figure 4: Exogenous H_2O_2 suppressed the expression of Olig2, SOX2, and CD133 in GSCs. a Exogenous H_2O_2 caused an increase in mitochondrial O_2^- in GSC11 cells. Cells were treated with 200 μM or 1 mM of H2O2 for 1 h, and mitochondrial O_2^- was measured by flow cytometric analysis. b Exogenous H2O2 caused an increase in mitochondrial O_2^- in GSC23 cells. Cells were treated with 500 μM or 1 mM of H2O2 for 1 h, and mitochondrial O_2^- was then measured by flow cytometric analysis. c GSC11 cells were treated with 200 μM or 1 mM H_2O_2 for 6 h with addition of H_2O_2 at 1-h intervals. Expression of SOX2, Olig2, and CD133 mRNA in GSC11 cells was measured by quantitative reverse transcription-polymerase chain reaction. *P < 0.05. d Effect of exogenous H_2O_2 on the expression of SOX2, Olig2, and CD133 in GSC23 cells under the same conditions as in (c). *P < 0.05. Cont control, GSC glioma stem cell, H_2O_2 hydrogen peroxide, O_2^- superoxide

To further validate this novel role of ROS, we used NAC, a precursor for glutathione synthesis with potent antioxidant property to reduce ROS stress, to test whether it could prevent the effect of serum on GSCs. As shown in Fig. 5a, incubation of GSC11 cells with serum for 7 days caused a significant increase of mitochondrial O_2^-, and the presence of NAC effectively suppressed such ROS increase. Importantly, NAC also prevented serum-induced loss of ability to form stem-like neurospheres (Fig. 5b) and partially preserved the expression of CD133 in serum-induced cells (Fig. 5c). These results suggest that the increase in mitochondrial ROS generation might play an important role in mediating the serum effect on GSCs. Since there was a significant decrease in expression of SOX2, Olig2, and Notch1 in serum-induced GSCs, we tested whether NAC might also suppress this serum effect. As shown in Fig. 5d, quantitative RT-PCR revealed that the addition of NAC to the serum-treated GSC11 cells largely blocked the decrease of SOX2 and Olig2 expression, suggesting that the expression of these two molecules might be redox-sensitive. Similar results were observed in GSC23 (Additional file 1: Figure S8). Furthermore, gene expression analysis of molecules involved in the Notch pathway revealed that serum caused a significant decrease in the expression of Notch1, MFNG, LFNG, HESs, DTX3, and DLL1 (Fig. 5e). Consistently, the presence of antioxidant NAC largely prevented the downregulation of the Notch-related genes (Fig. 5e), again suggesting the important role of ROS and redox signaling in regulation of GSCs.

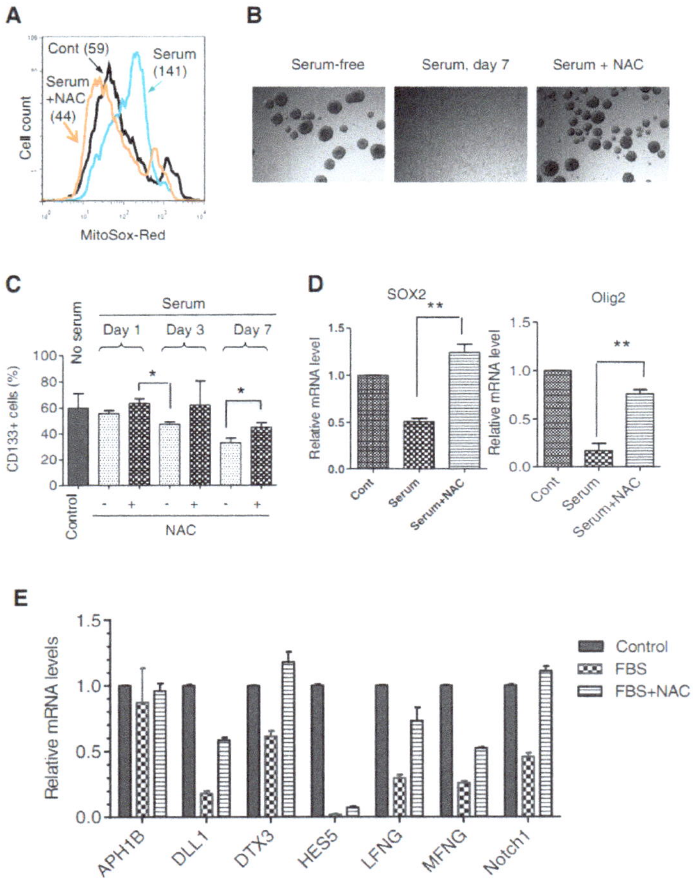

Figure 5: Effect of antioxidant NAC on serum-induced ROS generation and the expression of stem cell markers in GSCs. a Effect of NAC on mitochondrial O_2^- levels in GSC11 cells exposed to serum. Cells were cultured in stem cell medium with or without serum (5 % FBS) in the presence or absence of 20 mM NAC for 7 days. Mitochondrial O_2^- was measured by using MitoSOX-Red. b Comparison of neurosphere formation in GSC11 cells cultured in serum-free or serum-containing medium in the presence and absence of 20 mM NAC for 7 days. c NAC suppressed serum-induced loss of CD133 expression. GSC23 cells were exposed to serum for 1, 3, or 7 days in the presence or absence of 20 mM NAC as indicated, and CD133-positive cells were quantitated by flow cytometry analysis. *P < 0.05. d GSC11 cells were exposed to serum (5 % FBS) for 3 days in the presence and absence of 20 mM NAC. The expression of SOX2 and Olig2 mRNA was measured by quantitative RT-PCR. **P < 0.001. e NAC suppressed serum-induced downregulation of the Notch pathway. GSC11 cells were incubated without or with serum (5 % FBS) for 3 days in the presence or absence of 20 mM NAC. RNA was isolated from each sample, and the expression of molecules involved in Notch signaling was measure by quantitative RT-PCR. FBS fetal bovine serum, GSC glioma stem cell, NAC N-acetyl-cysteine, O_2^- superoxide, ROS reactive oxygen species, RT-PCR reverse transcription-polymerase chain reaction

Serum Induction of GSCs Promotes Tumorigenesis *in vivo*

Because exposure of GSCs to serum caused the cells to exhibit apparent differentiation morphology and a decrease in neurosphere formation, we used two *in vivo* models to test whether serum induction of such changes might alter the ability of GSCs to form tumors *in vivo*. First, GSC11 cells were incubated with or without serum for 7 days, and the same numbers (2×10^6) of the control cells or serum-induced cells were inoculated subcutaneously into the right flank of nude mice. Under these subcutaneous inoculation conditions, only two out of seven of the mice inoculated with the control GSC11 cells (serum-free) form tumors, and surprisingly seven out of seven mice inoculated with the serum-induced GSC11 cells developed tumor (Fig. 6a), suggesting that exposure of GSC11 cells to serum promotes their tumorigenesis. Interestingly, five out of seven mice inoculated with GSC11 cells treated with serum in the presence of the antioxidant NAC formed tumors (Fig. 6a). The overall survival of the mice inoculated with serum-induced GSC11 cells was significantly shorter than that of the control mice inoculated serum-free GSC11 cells (P=0.0019, Fig. 6b). No significant difference (P=0.064) in overall survival was found between the control group (serum-free) and the group of mice inoculated with GSC11 cells exposed to serum in the presence of NAC (20 mM).

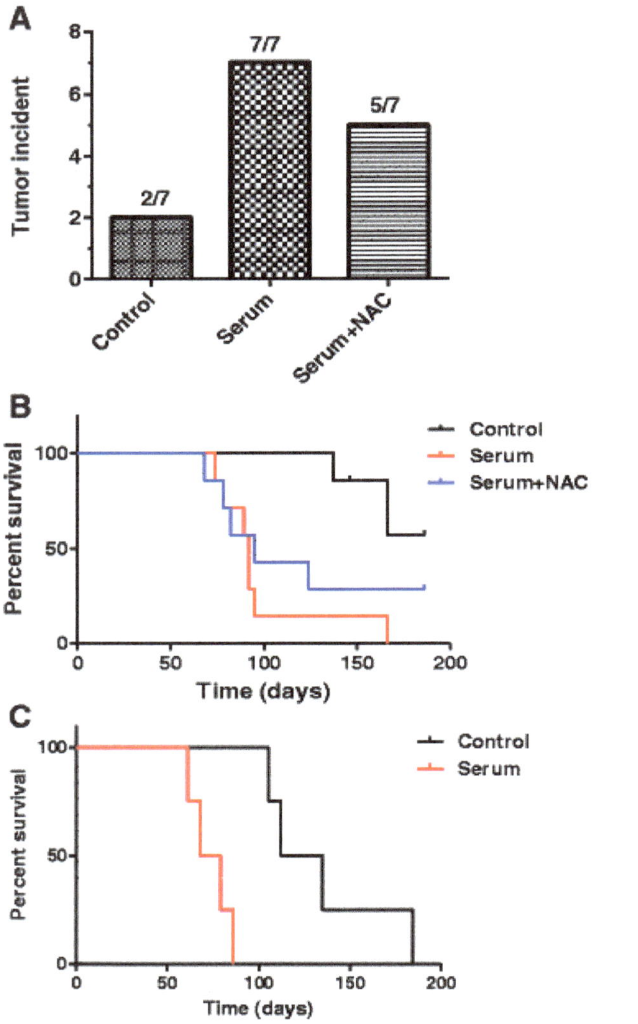

Figure 6: Serum promotes in vivo tumorigenesis of glioblastoma stem cells. a GSC11 cells were cultured in serum-free medium or serum-containing medium in the presence or absence of 20 mM NAC for 7 days. Equal numbers of cells (2×10^6) from each sample were inoculated subcutaneously into the right flank of nude mice (seven mice per group). The numbers of mice that developed tumors in each group are shown. b Survival curves for the mice in the three groups of mice described in (a). c GBM3752 cells were isolated from orthotopic tumor xenografts cultured in stem cell medium with or without serum as described in Methods. Equal numbers of cells (10,000 per injection) were inoculated into non-obese diabetic/severe combined immunodeficiency mice intracranially. The mice were observed for survival and euthanized when they developed signs of neurological deficit and became moribund. NAC N-acetyl-cysteine

We also used another mouse model, orthotopic inoculation of GBM3752 cells into the SCID mice (Shu et al. [22]), to further evaluate the role of serum exposure on tumorigenesis. The same numbers (1×10^4) of GBM3752 cells with or without serum exposure were inoculated into SCID mice intracranially, and the mice were observed for tumor development and survival. As shown in Fig. 6c, the overall survival of mice inoculated with serum-induced GBM3752 cells was significantly shorter than that of the mice bearing the control GBM3752 cells cultured in stem cell medium without serum (P=0.0067). These results were consistent with those of the subcutaneous tumor model and suggest that serum exposure activated glioma stem cells and promoted tumor formation.

Activation of the NF B Survival Pathway in Serum-Induced Glioma Stem Cells

To explore the possible mechanisms by which serum induction of mitochondrial ROS generation could lead to activation of GSCs and promote tumorigenesis, we further analyzed the gene expression microarray data from GSC11 cells exposed to serum for various times (Additional file 1: Figure S2) and found that the expression of multiple genes downstream of nuclear factor-kappa-B (NFκB) (including CD44, IL-8, IL-11, CCND1, TFPI2, and PLAUR) consistently increased after incubation with serum (Fig. 7a–g). Since oxidative stress is known to activate NFκB [26, 27], it is possible that *in vivo*. Indeed, incubation of GSC23 cells with serum caused a substantial increase in IκBα phosphorylation and p65 phosphorylation (Fig. 7h), two molecular events indicative of NFκB activation. Importantly, the addition of the antioxidant NAC partially suppressed the serum-induced phosphorylation of IκBα and p65, suggesting the role of ROS in mediating serum-induced NFκB activation. Similar results were observed in GSC11 cells (Fig. 7i).

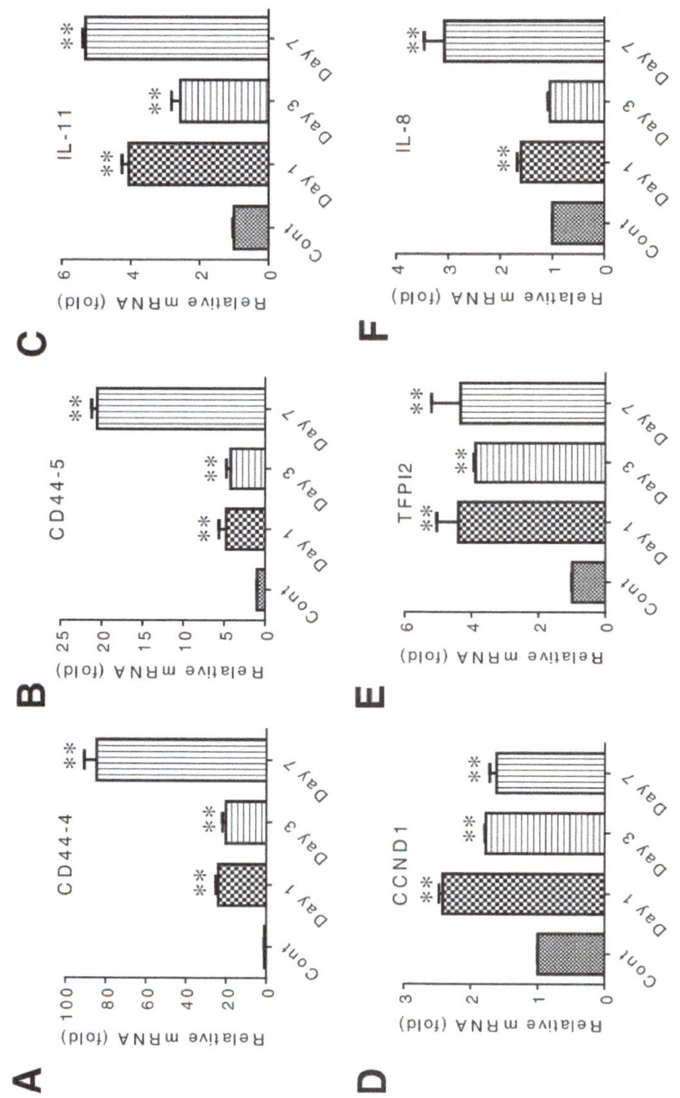

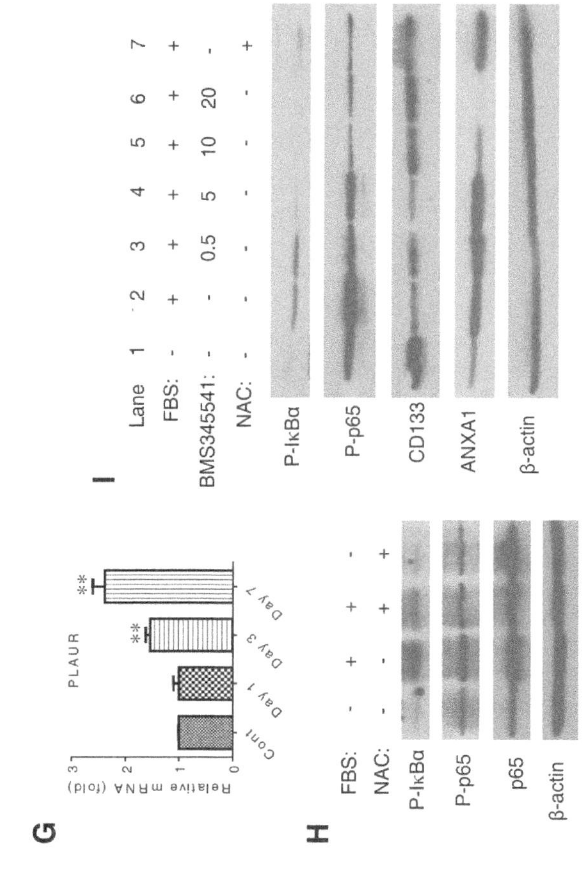

Figure 7: Activation of the NFκB pathway by serum in glioblastoma stem cells is associated with ROS stress. a–g Induction of expression of genes which are downstream of NFκB by serum exposure. GSC11 cells were cultured in serum-free medium or serum-containing medium for 1, 3, and 7 days. The relative mRNA expression levels of the target genes were analyzed by microarray assays in triplicate for each time point. h Antioxidant NAC suppressed serum-induced NFκB pathway activation in GSC23 cells. Protein samples were obtained from GSC23 cells cultured in serum-free or serum-containing medium in the presence or absence of 20 mM NAC for 3 days. The expression of phospho-IKBα, phosphor-p65, total p65, and β-actin was assayed by Western blot. i IKK inhibitor BMS-345541 and antioxidant NAC suppressed serum-induced NFκB activation in GSC11 cells. GSC11 cells were cultured in serum-free or serum-containing medium with the indicated concentrations of BMS345541 or 20 mM of NAC for 3 days. The expression of phospho-IKBα, phosphor-p65, CD133, ANXA1, and β-actin was measured by Western blot analysis. CCND1 cyclin D1, CD44-4 CD44 transcript variant 4, CD44-5 CD44 transcript variant 5, FBS fetal bovine serum, GSC glioma stem cell, IL-8 interleukin 8, IL-11 interleukin 11, NAC N-acetyl-cysteine, NFκB nuclear factor-kappa-B, PLAUR plasminogen activator urokinase receptor, ROS reactive oxygen species, TFP1-2 tissue factor pathway inhibitor 2

Serum exposure caused a significant increase in phosphorylated IκBα and p65 (Fig. 7i, lanes 1 and 2). The addition of NAC suppressed these phosphorylations (Fig. 7i, lane 7). These data together with the known function of IKK in phosphorylating IκBα and p65 suggest a possibility that serum might activate NFκB in GSCs through ROS-induced activation of IKK, a redox-sensitive molecule known to be activated by ROS [28]. To exam this possibility, we used a specific IKK inhibitor, BMS-345541, to test whether inhibition of IKK would prevent serum-induced phosphorylation of IκBα and p65. As shown in Fig. 7i, BMS-345541 at concentrations of 10–20 μM effectively suppressed serum-induced phosphorylation of IκBα and p65, associated with a preservation of CD133 expression of suppression of ANXA1. These data suggest that IKK might play an important role in mediating serum-induced activation of NFκB through a redox-sensitive mechanism.

It has been known for some time that CSCs, similar to normal stem cells, require a certain tissue microenvironment to maintain their stemness [29, 30] and that exposure of stem cells to serum *in vitro* usually induces differentiation phenotype [9]. However, the underlying mechanisms remain unclear. Although apparent differentiation phenotypes are also observed when CSCs are exposed to serum, it is unclear whether this would cause a decrease in tumorigenesis. Our study revealed a ROS-mediated mechanism by which serum induces apparent differentiation in glioma stem cells. We found that exposure of GSCs to serum resulted in activation of mitochondrial respiration, leading to an increase of oxygen consumption and high generation of mitochondrial O_2^-, which induced the expression of SOD2 to convert O_2^- to H_2O_2. Owing to its relatively long half-life and ability to cross biological membranes, H_2O_2 has been considered a second messenger that mediates redox-sensitive signaling in cellular response to growth factors [31, 32]. The ability of H_2O_2 to cause oxidation of protein thiol via catalytic cysteine can alter the function of the target proteins and thus provides a mechanism for redox signaling [32]. Since O_2^- has an extremely short half-life and cannot pass the mitochondrial membranes, the increased O_2^- within the mitochondria could not directly function as a second messenger to affect the nuclear gene expression. As such, conversion of the mitochondrial O_2^- to H_2O_2 by SOD2 seems important to relay the redox signal from mitochondria to cytosol and nucleus during serum-induced differentiation of GSCs. The upregulation of SOD2 in serum-treated GSCs would facilitate the conversion of O_2^- to H_2O_2.

Although the ability of ROS to induce normal stem cell differentiation has been noticed in some experimental systems [33, 34], the exact mechanisms remain unclear. Our study found that downregulation of SOX2, Olig2, and the Notch-related molecules by ROS might be a potential mechanism. The ability of exogenous antioxidant NAC to prevent the decreased expression of these genes and to block the serum-induced differentiation phenotype supports this notion. It is possible that the expression of SOX2, Olig2, and the Notch-related molecules is regulated via a redox-sensitive mechanism, with ROS being a negative regulator. Since SOX2, Olig2, and the Notch pathway are involved in regulation of neural stem cells [35, 36], downregulation of these genes could promote apparent differentiation of GSCs and drive them to enter a process in which GSCs progress to become the downstream progeny cancer cells.

Surprisingly, we found that although incubation of GSCs with serum induced apparent differentiation morphology and caused a downregulation of certain stem cell markers, including CD133, SOX2, and Olig2, this led to an increase of tumorigenicity and a reduction of survival in mice in two different mouse models (subcutaneous and orthotopic). These new findings seem to challenge the traditional view that CSCs are responsible for cancer development and their ability to form tumor would decrease once they are induced to undergo differentiation. Interestingly, a recent study showed that tumor cell self-renewal capacity did not predict tumor growth potential *in vivo* [37]. The study by Barrett et al. showed that glioma cells with low self-renewal capacity were more tumorigenic and generate tumor more rapidly than cells with high self-renewal capacity. These observations are consistent with our findings. Furthermore, our study revealed that activation of the nuclear factor-kappa-B (NFκB) survival pathway by ROS might be a mechanism that promotes GSC survival and enhances tumorigenesis. The role of the NFκB pathway in normal stem cell proliferation has been implicated previously. For instance, NFκB is activated during human embryonic stem cell (hESC) differentiation, and inhibition of NFκB leads to a reduction of hESC proliferation and suppression of their progression toward primitive extraembryonic and embryonic lineages [38]. As such, inhibition of NFκB may potentially prevent activation of CSCs and thus suppress tumor development. Interestingly, recent studies have shown that targeting the NFκB pathway may be effective against CSCs [39–41]. Their studies showed that inhibition of NFκB by using compounds such as niclosamide and disulfiram suppressed acute myeloid leukemia stem cells and breast CSCs.

The NFκB signaling pathway plays a crucial role in cancer development and progression [42]. It can either promote or inhibit carcinogenesis, depending on the cell types and experimental conditions. The NFκB signaling pathway is often activated by ROS through IKK phosphorylation of IκB [43] and p65 [44]. Our study found that serum could cause the activation of NFκB, which could be blocked by the antioxidant NAC. These results suggest that the activation of NFκB in the serum-induced cells is probably mediated by a redox-regulatory mechanism due to ROS generation in the serum-activated mitochondria. It is worth noting that deletion of NFκBIA which encodes the NFκB inhibitor IκBα seems to be an oncogenic event in GBM [45]. Interestingly, a recent study showed that the NFκB pathway was activated in glioblastoma-initiating cells (GICs) after induction of differentiation and that blockade of the NFκB pathway could drive differentiating GSCs into senescence [40], again suggesting that activation of NFκB is important in maintaining the proliferation of differentiating GSCs. This effect was partly mediated by reduced levels of the NFκB target gene cyclin D1. Furthermore, a novel small-molecule inhibitor of the NFκB pathway induced senescence of tumor cells in a mouse model bearing human GIC-derived tumors. These findings reveal that activation of NFκB may keep differentiating GICs from acquiring a mature postmitotic phenotype, thus allowing cell proliferation. Our study showed that, in the case of GSC exposure to serum, activation of NFκB is likely through ROS stimulation of IKK, which promotes phosphorylation of IκBα and p65. We found that antioxidant NAC or inhibition of IKK by BMS-345541 could effectively prevent the serum-induced NFκB activation and loss of CD133, suggesting a novel role of ROS in driving the progression of GSCs toward downstream progeny cells to promote tumor development.

CONCLUSIONS

In summary, our study showed that exposure of glioma stem cells to serum could stimulate mitochondrial respiration leading to increased generation of mitochondrial ROS, which activated NFκB to promote cancer cell survival and tumorigenesis. A key underlying mechanism is likely through a redox-mediated activation of IKK to phosphorylate IκBα and p65. Although the serum-induced elevation of ROS in GSCs also caused a decrease in neurosphere formation *in vitro* and a reduced expression of stem cell markers such as CD133, this apparent differentiation did not reduce the ability of the glioma stem cells to form tumor *in vivo*. Instead, the serum-induced GSCs exhibited greater tumorigenesis in both subcutaneous and orthotopic xenograft models. These new findings suggest that serum may activate glioma stem cells to progress toward the downstream cancer progenitor cells and promote tumor formation and that activation of mitochondrial respiration and ROS generation may play a key role in redox signaling during this tumorigenesis process. It is also important to note that the apparent differentiation phenotype such as neurosphere formation and CD133 expression *in vitro* observed might not necessarily predict tumorigenesis *in vivo*.

Accession numbers

The Gene Expression Omnibus (GEO) accession number for the RNA microarray data in this study isGSE28220.

Acknowledgments

The authors thank Zahid H. Siddik, Anthony Lucci, and Hector Martinez-Valdez for helpful discussion. This work was supported in part by a grant from the major science and technology project of the National Basic Research (973) Program of China (2012CB967004), grants from National Natural Science Foundation of China (81302194 and 81302144), a Specialized Research Fund for the Doctoral Program of Higher Education (20130171120047), a grant from the National Institutes of Health (CA100428), and the Fundamental Research Funds for the Central Universities (14ykpy40). FW is a recipient of the Rosalie B. Hite Fellowship from UT MD Anderson Cancer Center and the Outstanding Young Talents of Sun Yat-sen University Cancer Center.

Abbreviations

APC	Allophycocyanin
bFGF	Basic fibroblast growth factor
CSC	Cancer stem cell
DCF-DA	5-(and-6)-chloromethyl-2,7-dichlorodihydrofluorescein diacetate acetyl ester
EGF	Epidermal growth factor
ETC.	Electron transport chain
FBS	Fetal bovine serum
GBM	Glioblastoma multiforme

GIC	Glioblastoma-initiating cell
GSC	Glioma stem cell
GSH	Glutathione
H_2O_2	Hydrogen peroxide
hESC	Human embryonic stem cell
NAC	N-acetyl-cysteine
NFκB	Nuclear factor-kappa-B
O_2^-	Superoxide
PBS	Phosphate-buffered saline
ROS	Reactive oxygen species
RT-PCR	Reverse transcription-polymerase chain reaction
SCID	Severe combined immunodeficiency
SOD1	Cytosolic superoxide dismutase
UT	University of Texas

Additional file

Supplementary Figures

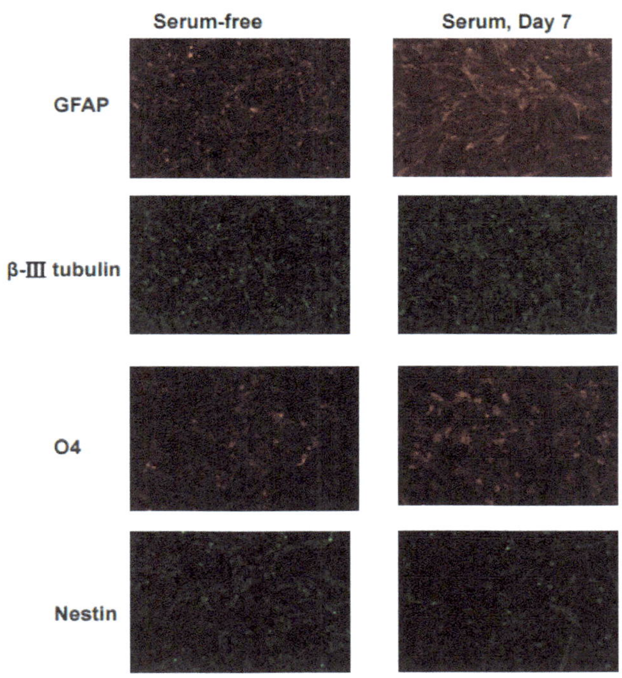

Fig S1. Effect of serum on the expression of stem cell marker and differentiation markers in GSC23 cells. Immunofluorescence staining of GFAP, β-III tubulin, O4 and Nestin in GSC11 cells before and after exposure to 5% FBS for 7 days.

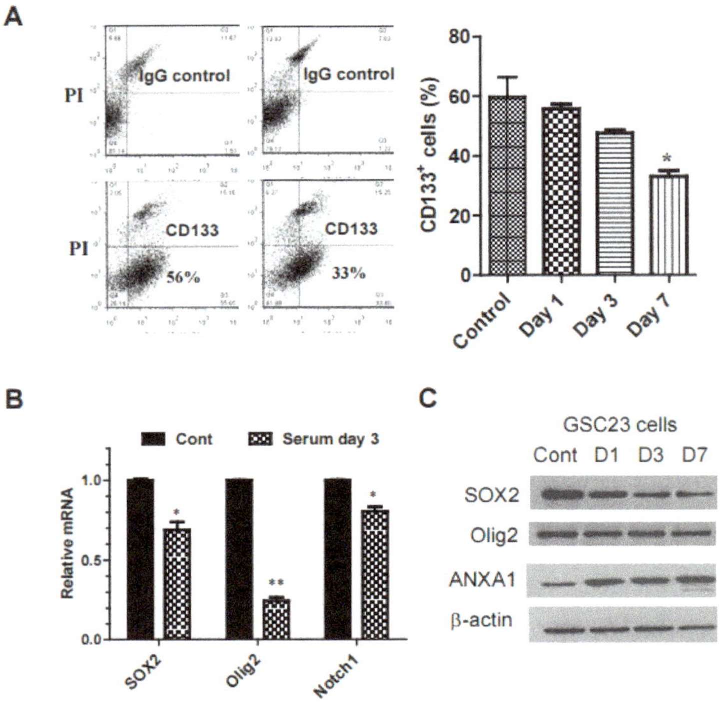

Fig S2. Effect of serum on the expression of stem cell markers in GSC23 cells. (A) Flow cytometry analysis of CD133 expression in GSC23 cells before and after exposure to serum for 7 days. Quantitation of the percentage of CD133+ cells before and after GSC23 cells were exposed to serum for various times (1, 3, and 7 days) is shown on the right panel. *. P

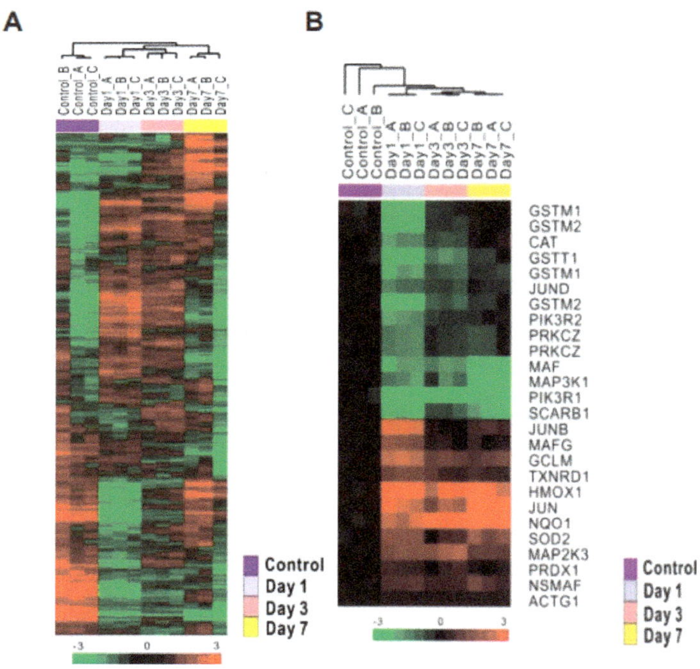

Fig S3. Analyses of gene expression profiles in glioblastoma stem cells before and after exposure to serum. (A) Comparison of gene expression patterns in GSC11 cells exposed to serum for various times. GSC11 cells maintained in serum-free stem cell medium were incubated with 5% FBS for 1, 3, and 7 days (each time point contained triplicate culture flasks A-C). After proper variance filtering, the gene expression data were used for unsupervised hierarchical clustering analysis with similarity metric of uncentered correlation. The intensity of red and green color represents the degree of high and low expression, respectively, as indicated in the log2-transformed scale bar. (B) List of genes associated with the Nrf2-mediated oxidative stress response pathway. Genes with expression showing significant differences between control (serum-free) and seruminduced conditions in GSC11 cells were identified by two-tail t-test ($p < 0.001$).

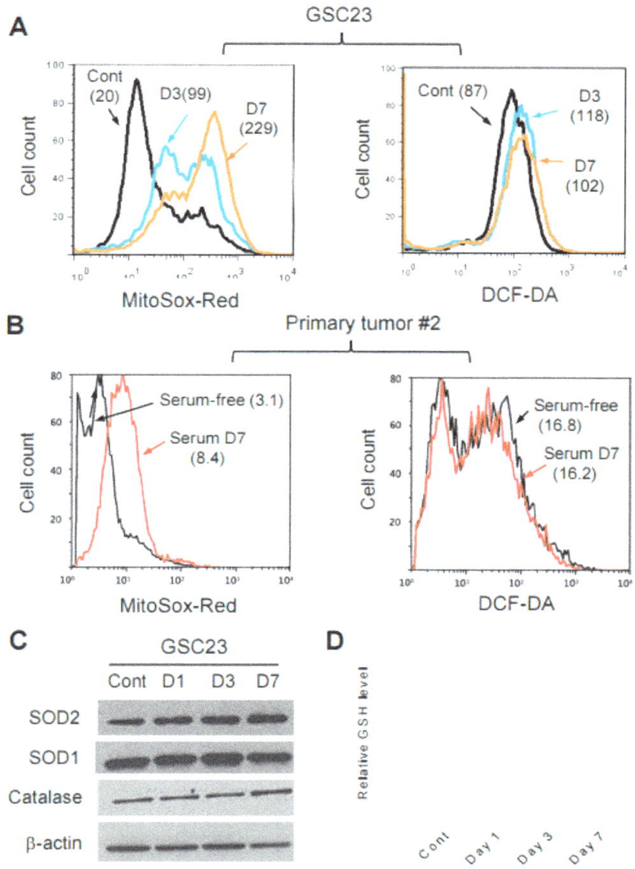

Fig S4. Effect of serum on mitochondrial superoxide, cellular ROS, and expression of antioxidant molecules in GSCs. (A) Effect of serum (5% FBS) on mitochondrial superoxide (measured by flow cytometry analysis with mitoSOX-Red staining) and total cellular ROS (measured by DC-FDA staining) in GSC23 cells. The numbers in parentheses indicate the median relative fluorescent intensity. (B) Effect of serum on mitochondrial superoxide and total cellular ROS in primary GBM cells (serum exposure for 7 days). (C) Western blot analysis of SOD1, SOD2, and Catalase in GSC23 cells before and after exposure to serum for 1-7 days. (D) Effect of serum on cellular glutathione (GSH) in GSC23 cells.

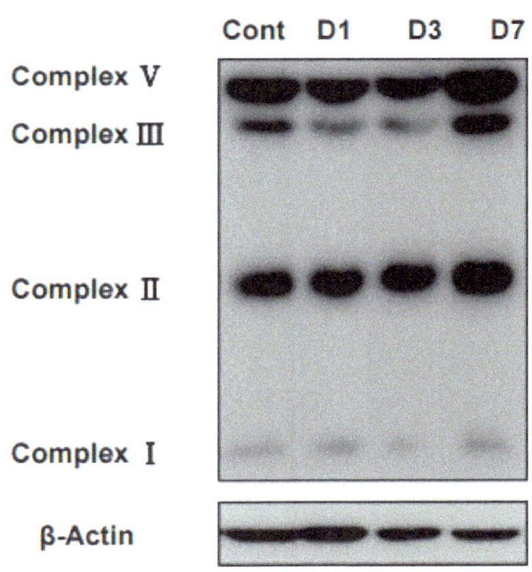

Fig S5. Effect of serum on the expression of mitochondria electron transport chain and the ATP synthase complexes. GSC11 cells were exposed to 5% FBS for 1, 3, and 7 days, and gene expression was analyzed by western blotting

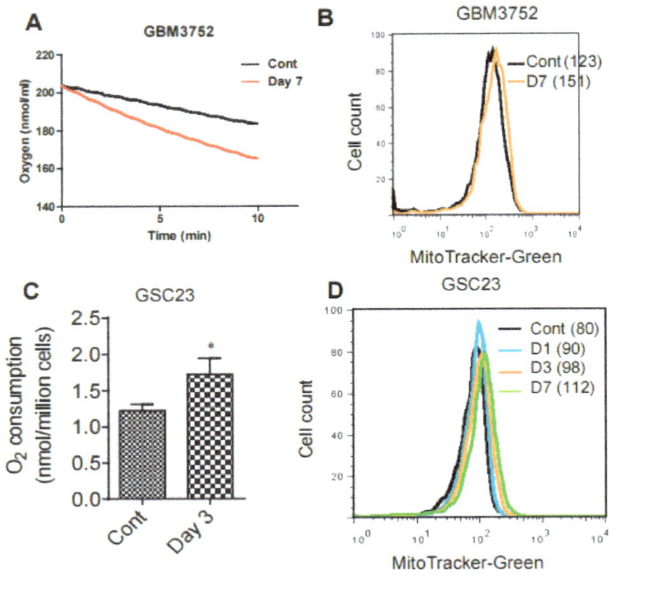

Fig S6. Activation of mitochondrial respiration by serum in GBM3752 and GSC23 cells. (A) Increase of oxygen consumption in GBM3752 cells after exposure to serum for 7 days. (B) No change in mitochondrial mass in GBM3752 cells exposed to serum for 7 days. The numbers in parentheses indicate the median fluorescent values. (C) Increase of oxygen consumption in GSC23 cells after exposure to serum for 3 days. *, P

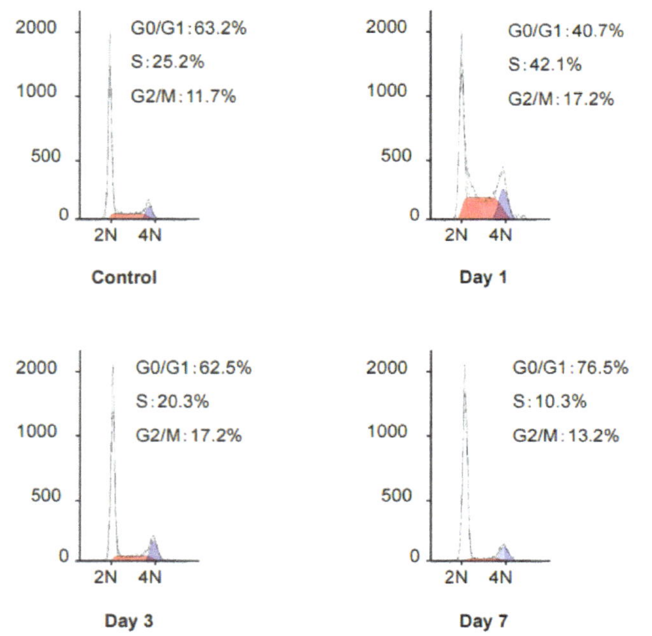

Fig S7. Effect of serum on cell cycle. GSC11 cells were exposed to 5% FBS for 1, 3, and 7 days, and cell cycle was analyzed by flow cytometry.

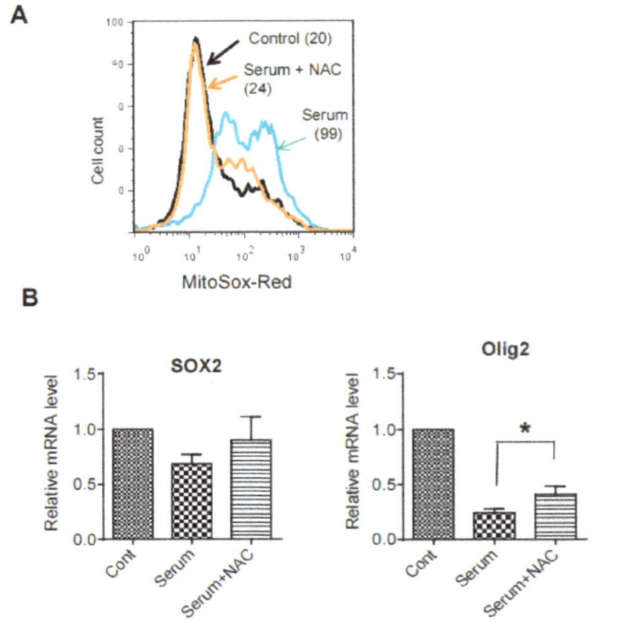

Fig S8. Effect of antioxidant N-acetylcysteine (NAC) on serum-induced changes in ROS and expression of SOX2 and Olig2. (A) Effect of NAC on mitochondrial superoxide levels in GSC23 cells exposed to serum. Cells were cultured with or without serum (5% FBS) in the presence or absence of 20 mM NAC for 3 or 7 days as indicated, and mitochondrial superoxide was measured using MitoSox Red. (B) NAC suppressed serum-induced loss of SOX2 and Olig2 expression. GSC23 cells were exposed to serum for 3 days in the presence or absence of 20 mM NAC as indicated, and the expression of SOX2 and Olig2 mRNA in each sample was measured by quantitative RT-PCR.

Footnotes

SY and YL are co-first authors.

Competing interests

The authors declare that they have no competing interests.

Authors' contributions

SY and YL contributed to conception and design, collection/assembly of data, data analysis/interpretation, manuscript writing, and final approval and contributed equally to this article. JY, MO, HZ, JL, HC, and JL contributed to provision of study materials, manuscript writing, and final approval. GC and SK contributed to collection/assembly of data, manuscript writing, and final approval. LF, NH, WL, and XL contributed to design of animal study, data analysis/interpretation, manuscript writing, and final approval. RX, PH, and FW contributed to conception and design, provision of study materials, data analysis/interpretation, manuscript writing, and final approval. All authors read and approved the manuscript.

REFERENCES

1. Pardal R, Clarke MF, Morrison SJ. Applying the principles of stem-cell biology to cancer. Nat Rev Cancer. 2003;3:895–902. doi: 10.1038/nrc1232.
2. Visvader JE, Lindeman GJ. Cancer stem cells in solid tumours: accumulating evidence and unresolved questions. Nat Rev Cancer. 2008;8:755–68. doi: 10.1038/nrc2499.
3. Eyler CE, Rich JN. Survival of the fittest: cancer stem cells in therapeutic resistance and angiogenesis. J Clin Oncol. 2008;26:2839–45. doi: 10.1200/JCO.2007.15.1829.
4. Morrison R, Schleicher SM, Sun Y, Niermann KJ, Kim S, Spratt DE, et al. Targeting the mechanisms of resistance to chemotherapy and radiotherapy with the cancer stem cell hypothesis. J Oncol.2011;2011:941876. doi: 10.1155/2011/941876.
5. Pajonk F, Vlashi E, McBride WH. Radiation resistance of cancer stem cells: the 4 R's of radiobiology revisited. Stem Cells. 2010;28:639–48. doi: 10.1002/stem.318.
6. Diehn M, Cho RW, Lobo NA, Kalisky T, Dorie MJ, Kulp AN, et al. Association of reactive oxygen species levels and radioresistance in cancer stem cells. Nature. 2009;458:780–3. doi: 10.1038/nature07733.
7. Ji AR, Ku SY, Cho MS, Kim YY, Kim YJ, Oh SK, et al. Reactive oxygen species enhance differentiation of human embryonic stem cells into mesendodermal lineage. Exp Mol Med. 2010;42:175–86. doi: 10.3858/emm.2010.42.3.018.
8. Borovski T, De Sousa EMF, Vermeulen L, Medema JP. Cancer stem cell niche: the place to be. Cancer Res. 2011;71:634–9. doi: 10.1158/0008-5472.CAN-10-3220.
9. Lee J, Kotliarova S, Kotliarov Y, Li A, Su Q, Donin NM, et al. Tumor stem cells derived from glioblastomas cultured in bFGF and EGF more closely mirror the phenotype and genotype of primary tumors than do serum-cultured cell lines. Cancer Cell. 2006;9:391–403. doi: 10.1016/j.ccr.2006.03.030.
10. Calabrese C, Poppleton H, Kocak M, Hogg TL, Fuller C, Hamner B, et al. A perivascular niche for brain tumor stem cells. Cancer Cell. 2007;11:69–82. doi: 10.1016/j.ccr.2006.11.020.
11. Gilbert CA, Ross AH. Cancer stem cells: cell culture, markers, and targets for new therapies. J Cell Biochem. 2009;108:1031–8. doi: 10.1002/jcb.22350.
12. Reynolds BA, Weiss S. Generation of neurons and astrocytes from isolated cells of the adult mammalian central nervous system. Science. 1992;255:1707–10. doi: 10.1126/science.1553558.
13. Pervaiz S, Taneja R, Ghaffari S. Oxidative stress regulation of stem and progenitor cells. Antioxid Redox Signal. 2009;11:2777–89. doi: 10.1089/ars.2009.2804.
14. Zhou D, Shao L, Spitz DR. Reactive oxygen species in normal and tumor stem cells. Adv Cancer Res.2014;122:1–67. doi: 10.1016/B978-0-12-420117-0.00001-3.
15. Sauer H, Wartenberg M. Reactive oxygen species as signaling molecules in cardiovascular differentiation of embryonic stem cells and tumor-induced angiogenesis. Antioxid Redox Signal. 2005;7:1423–34. doi: 10.1089/ars.2005.7.1423.
16. Smith J, Ladi E, Mayer-Proschel M, Noble M. Redox state is a central modulator of the balance between self-renewal and differentiation in a dividing glial precursor cell. Proc Natl Acad Sci U S A. 2000;97:10032–7. doi: 10.1073/pnas.170209797.
17. Tsatmali M, Walcott EC, Crossin KL. Newborn neurons acquire high levels of reactive oxygen species and increased mitochondrial proteins upon differentiation from progenitors. Brain Res. 2005;1040:137–50. doi: 10.1016/j.brainres.2005.01.087.
18. Ito K, Hirao A, Arai F, Matsuoka S, Takubo K, Hamaguchi I, et al. Regulation of oxidative stress by ATM is required for self-renewal of haematopoietic stem cells. Nature. 2004;431:997–1002. doi: 10.1038/nature02989.
19. Sato A, Okada M, Shibuya K, Watanabe E, Seino S, Narita Y, et al. Pivotal role for ROS activation of p38 MAPK in the control of differentiation and tumor-initiating capacity of glioma-initiating cells. Stem Cell Res. 2013;12:119–31. doi: 10.1016/j.scr.2013.09.012.
20. He H, Nilsson CL, Emmett MR, Marshall AG, Kroes RA, Moskal JR, et al. Glycomic and transcriptomic response of gsc11 glioblastoma stem cells to stat3 phosphorylation inhibition and serum-induced differentiation. J Proteome Res. 2010;9:2098–108. doi: 10.1021/pr900793a.
21. Jiang H, Gomez-Manzano C, Aoki H, Alonso MM, Kondo S, McCormick F, et al. Examination of the therapeutic potential of Delta-24-RGD in brain tumor stem cells: role of autophagic cell death. J Natl Cancer Inst. 2007;99:1410–4. doi: 10.1093/jnci/djm102.
22. Shu Q, Wong KK, Su JM, Adesina AM, Yu LT, Tsang YT, et al. Direct orthotopic transplantation of fresh surgical specimen preserves CD133+ tumor cells in clinically relevant mouse models of medulloblastoma and glioma. Stem Cells. 2008;26:1414–24. doi: 10.1634/stemcells.2007-1009.
23. Zhou Y, Shingu T, Feng L, Chen Z, Ogasawara M, Keating MJ, et al. Metabolic alterations in highly tumorigenic glioblastoma cells: preference for hypoxia and high dependency on glycolysis. J Biol Chem.2011;286:32843–53. doi: 10.1074/jbc.M111.260935.
24. Yuan S, Wang F, Chen G, Zhang H, Feng L, Wang L, et al. Effective elimination of cancer stem cells by a novel drug combination strategy. Stem Cells. 2013;31:23–34. doi: 10.1002/stem.1273.
25. Gaiano N, Fishell G. The role of notch in promoting glial and neural stem cell fates. Annu Rev Neurosci.2002;25:471–90. doi: 10.1146/annurev.neuro.25.030702.130823.
26. Li N, Karin M. Is NF-kappaB the sensor of oxidative stress? FASEB J. 1999;13:1137–43.
27. Klaunig JE, Kamendulis LM. The role of oxidative

stress in carcinogenesis. Annu Rev Pharmacol Toxicol. 2004;44:239–67. doi: 10.1146/annurev.pharmtox.44.101802.121851.

28. Gloire G, Legrand-Poels S, Piette J. NF-kappaB activation by reactive oxygen species: fifteen years later. Biochem Pharmacol. 2006;72:1493–505. doi: 10.1016/j.bcp.2006.04.011.

29. LaBarge MA. The difficulty of targeting cancer stem cell niches. Clin Cancer Res. 2010;16:3121–9. doi: 10.1158/1078-0432.CCR-09-2933.

30. Li L, Neaves WB. Normal stem cells and cancer stem cells: the niche matters. Cancer Res.2006;66:4553–7. doi: 10.1158/0008-5472.CAN-05-3986.

31. Cuddihy SL, Winterbourn CC, Hampton MB. Assessment of redox changes to hydrogen peroxide-sensitive proteins during EGF signaling. Antioxid Redox Signal. 2011;15:167–74. doi: 10.1089/ars.2010.3843.

32. Forman HJ, Maiorino M, Ursini F. Signaling functions of reactive oxygen species. Biochemistry.2010;49:835–42. doi: 10.1021/bi9020378.

33. Hernandez-Garcia D, Wood CD, Castro-Obregon S, Covarrubias L. Reactive oxygen species: a radical role in development? Free Radic Biol Med. 2010;49:130–43. doi: 10.1016/j.freeradbiomed.2010.03.020.

34. Bigarella CL, Liang R, Ghaffari S. Stem cells and the impact of ROS signaling. Development.2014;141:4206–18. doi: 10.1242/dev.107086.

35. Gangemi RM, Griffero F, Marubbi D, Perera M, Capra MC, Malatesta P, et al. SOX2 silencing in glioblastoma tumor-initiating cells causes stop of proliferation and loss of tumorigenicity. Stem Cells.2009;27:40–8. doi: 10.1634/stemcells.2008-0493.

36. Ligon KL, Huillard E, Mehta S, Kesari S, Liu H, Alberta JA, et al. Olig2-regulated lineage-restricted pathway controls replication competence in neural stem cells and malignant glioma. Neuron. 2007;53:503–17. doi: 10.1016/j.neuron.2007.01.009.

37. Barrett LE, Granot Z, Coker C, Iavarone A, Hambardzumyan D, Holland EC, et al. Self-renewal does not predict tumor growth potential in mouse models of high-grade glioma. Cancer Cell. 2012;21:11–24. doi: 10.1016/j.ccr.2011.11.025.

38. Yang C, Atkinson SP, Vilella F, Lloret M, Armstrong L, Mann DA, et al. Opposing putative roles for canonical and noncanonical NFkappaB signaling on the survival, proliferation, and differentiation potential of human embryonic stem cells. Stem Cells. 2010;28:1970–80. doi: 10.1002/stem.528.

39. Jin Y, Lu Z, Ding K, Li J, Du X, Chen C, et al. Antineoplastic mechanisms of niclosamide in acute myelogenous leukemia stem cells: inactivation of the NF-kappaB pathway and generation of reactive oxygen species. Cancer Res. 2010;70:2516–27. doi: 10.1158/0008-5472.CAN-09-3950.

40. Nogueira L, Ruiz-Ontanon P, Vazquez-Barquero A, Lafarga M, Berciano MT, Aldaz B, et al. Blockade of the NFkappaB pathway drives differentiating glioblastoma-initiating cells into senescence both in vitro and in vivo. Oncogene. 2011;30:3537–48. doi: 10.1038/onc.2011.74.

41. Yip NC, Fombon IS, Liu P, Brown S, Kannappan V, Armesilla AL, et al. Disulfiram modulated ROS-MAPK and NFkappaB pathways and targeted breast cancer cells with cancer stem cell-like properties. Br J Cancer. 2011;104:1564–74. doi: 10.1038/bjc.2011.126.

42. Karin M. NF-kappaB and cancer: mechanisms and targets. Mol Carcinog. 2006;45:355–61. doi: 10.1002/mc.20217.

43. Morgan MJ, Liu ZG. Crosstalk of reactive oxygen species and NF-kappaB signaling. Cell Res.2011;21:103–15. doi: 10.1038/cr.2010.178.

44. Sakurai H, Suzuki S, Kawasaki N, Nakano H, Okazaki T, Chino A, et al. Tumor necrosis factor-alpha-induced IKK phosphorylation of NF-kappaB p65 on serine 536 is mediated through the TRAF2, TRAF5, and TAK1 signaling pathway. J Biol Chem. 2003;278:36916–23. doi: 10.1074/jbc.M301598200.

45. Bredel M, Scholtens DM, Yadav AK, Alvarez AA, Renfrow JJ, Chandler JP, et al. NFKBIA deletion in glioblastomas. N Engl J Med. 2011;364:627–37. doi: 10.1056/NEJMoa1006312.

CHAPTER 5 CANCER STEM CELL METABOLISM

Maria Peiris-Pagès, Ubaldo E. Martinez-Outschoorn, Richard G. Pestell, Federica Sotgia, and Michael P. Lisanti

THE CANCER STEM CELL MODEL: *OMNIS CELLULA E CELLULA*

Adult stem cells, in contrast to most cells in our body, which are differentiated and have a specific role, are rare cells that harbour unique biological properties such as a lack of differentiation and indefinite self-renewal. Stem cell asymmetrical division into one new stem cell and a committed progenitor, which can give rise to a functionally mature progeny, helps maintain tissue homeostasis 1.

Cancer is characterized by an unrestrained proliferation of malignant cells that are morphologically and functionally different. Two models have been proposed in order to explain this cellular diversity within tumours. The traditional, stochastic way of explaining cancer initiation and development is through sequential accumulation of mutations, each of which promotes the loss of specific tissue traits until dedifferentiation and regression into a more primitive phenotype occurs. According to this clonal evolution model, each cancer cell has a similar potential to grow a tumour. A second model, the cancer stem cell (CSC) hypothesis, postulates that a reduced group of stem-like cells is responsible for the development of the disease. Accordingly, tumours are hierarchically organized and sustained by a distinct self-renewal subpopulation of cancer cells. These tumour-initiating cells (TICs) with stemness properties are located at the apex of a pyramid and are responsible for the generation of a varied progeny of highly proliferative cells forming the bulk of the tumour 1, 2. Both models are not mutually exclusive and can be viewed as integrated processes because CSCs can themselves undergo clonal evolution, through which a second more dominant population of CSCs may emerge. In addition, recent reports add more complexity to this scenario by demonstrating that cancer cells have a remarkable degree of plasticity. Indeed, it is thought that CSCs may arise from different cell types such as normal adult stem cells or differentiated cancer cells 2, 3.

CSCs share numerous properties with normal stem cells besides their ability to renew themselves by remaining in an undifferentiated state: the expression of surface markers, such as CD44, CD133 or the enzyme aldehyde dehydrogenase (ALDH), the activation of particular cell signalling pathways, such as Wnt, Notch or Hedgehog, a relative quiescence or an active DNA repair capacity 2. Given that CSCs are considered to be the source from which cancer cells arise, are therapy resistant and are responsible for metastatic dissemination, eliminating them could potentially achieve a permanent cure for the patient. In addition, if conventional therapy fails to kill CSCs, acting only against differentiated cancer cells, the tumour can eventually relapse 2. The specific elimination of CSCs may thus represent one of the most important challenges of current cancer research (Fig. 1). Because of their similarity, an accurate distinction between CSCs and normal stem cells is needed, and once these differences are identified new therapies can be developed to eliminate CSCs without damaging normal cells. In particular, the metabolic features of CSCs might represent a promising target. In this review we summarize the latest findings and most significant discoveries on CSC metabolism.

Citation: Peiris-Pagès M, Martinez-Outschoorn UE, Pestell RG, Sotgia F, Lisanti MP. Cancer stem cell metabolism. Breast Cancer Research : BCR. 2016;18:55. doi:10.1186/s13058-016-0712-6.

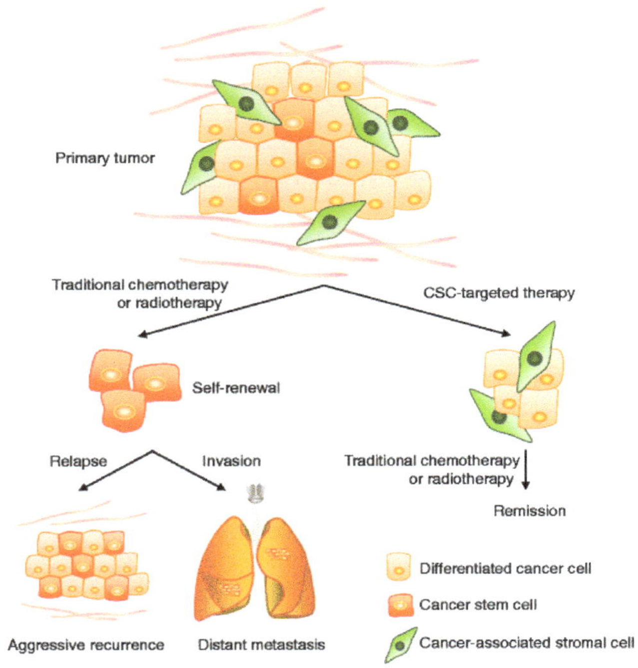

Figure 1: Potential impact of strategies that target cancer stem cells (CSCs) on the effectiveness of cancer treatment. Conventional cancer therapies result in a transient reduction of the tumour by killing non-stem cancer cells whilst failing to eliminate CSCs. Two major obstacles are limiting success in these cancer therapies: the ability of CSCs to survive cytotoxic treatments, and their potential to form metastases. The use of CSC specific inhibitors would reduce their therapy resistance and reduce relapse, and would prevent their spread, as the loss of stem cell properties reduces invasiveness and the capacity of disseminated cells to initiate distant secondary colonies

WHEN METABOLISM TAKES OVER

Metabolic adaptation is believed to be one of the hallmarks of cancer cells 4. The role of metabolism in cancer has become a dynamic field of research and a broad spectrum of novel strategies to target cancer metabolic pathways is emerging. However, the cellular heterogeneity present in tumours is not taken into account by most studies. It is important to highlight that different phenotypes such as hypoxic versus normoxic or quiescent versus proliferative will have substantially different metabolic requirements, which in turn may result in notably different responses to metabolic therapies. For example, CSCs – generally considered quiescent or slow-cycling compared with their differentiated cancer cell progeny – can re-enter into the cell cycle after exposure to radiotherapy, whereas most differentiated cells die or undergo cell cycle arrest 5.

Glycolysis is the enzymatic conversion of glucose into lactate, which concomitantly produces 2 molecules of ATP per molecule of glucose. In the presence of oxygen, cells generally adopt oxidative phosphorylation (OXPHOS) as the main pathway to produce energy, which is more efficient than glycolysis because it theoretically generates 36 molecules of ATP per molecule of glucose. Cancer cells can generate ATP via glycolysis even under normoxic concentrations in what is known as the Warburg effect. In fact, gycolysis can more rapidly produce ATP compared with OXPHOS in the presence of abundant levels of glucose 6. Stem cells also rely more on glycolysis when compared with their differentiated offspring, which preferentially metabolizes glucose via mitochondrial respiration. Of note, the metabolic reprogramming of normal somatic cells into induced pluripotent stem (iPC) cells actually requires a shift from mitochondrial respiration to a metabolism that is mainly glycolytic 7, a switch which precedes the acquisition of pluripotency markers, suggesting that changes in metabolism occur before changes in stemness 8. During differentiation, stem cells are also able to adjust their metabolic infrastructure, as they can rapidly shift from a preferentially glycolytic profile in undifferentiated cells to a more oxidative phenotype to generate the large amounts of energy needed for this process 9.

Studies of mitochondrial morphology and mitochondrial DNA levels indicate that stem cells have fewer mitochondria, which are less mature and relatively inactive compared with those of differentiated cells, resulting in reduced reactive oxygen species (ROS) levels 10. Low amounts of ROS are actually needed to maintain quiescence and the self-renewal potential 11. In sum, stem cells favour glycolysis and have less mitochondria, hence producing small amounts of ROS, which induce little mitochondrial DNA damage 9. Thus, a glycolytic phenotype seems to be a shared feature of normal stem cells and differentiated cancer cells in culture. However, very few studies have directly investigated the metabolism of CSCs.

THE METABOLIC PROFILE OF CANCER STEM CELLS

Are Cancer Stem Cells Mainly Glycolytic?

Glucose seems to be an essential nutrient for CSCs, as its presence in the microenvironment significantly increases the amount of stem-like cells in the overall cancer cell population. On the other hand, glucose deprivation induces the depletion of CSCs in vitro 12. However, the metabolic singularities of CSCs and the effects of different metabolites on CSC physiology remain largely unexplored, because the number of publications studying the metabolism of CSCs is small. Nevertheless, these studies indicate that CSCs have a distinctive metabolic phenotype compared with the bulk of the tumour and with normal stem cells, although there is so far no consensus on this 13. Several reports suggest that CSCs are more glycolytic than other differentiated cancer cells in vitro and in vivo. These studies were performed in many tumour types including osteosarcoma, glioblastoma, breast cancer, lung cancer, ovarian cancer and colon cancer 14–18. Glucose uptake, glycolytic enzyme expression, lactate production and ATP content are significantly increased in CSCs compared with their differentiated counterparts. This glycolytic phenotype seems to be linked to a decrease in mitochondrial oxidative metabolism 16–18. Likewise, mutations in mitochondrial DNA and low mitochondrial DNA copy number have been associated with increased metastasis and poor prognosis 19, 20. The mitochondrial DNA copy number not only affects the viability and functionality of the cell but also its differentiation potential. Cyclin D_1 regulates stemness of cancer cells and mitochondrial DNA copy number 21, 22. Also, during differentiation, the mitochondrial DNA copy number and the levels of mature cell gene expression patterns increase, whereas the expression of pluripotency genes such as *OCT4*, *TERT* and *MYC* decreases 23. Instead, partial depletion of mitochondrial DNA increases the levels of these pluripotency genes.

Or do they Rely More on Mitochondrial Respiration?

In clear contrast with these publications, growing evidence shows that CSCs have a preference for mitochondrial oxidative metabolism (Fig. 2). According to these other studies, CSCs are less glycolytic, consume less glucose, produce less lactate and maintain higher ATP levels than their differentiated progeny. Moreover, the mitochondria of CSCs have an increased mass and membrane potential, which is a reflection of mitochondrial function, higher mitochondrial ROS and enhanced oxygen consumption rates compared with the bulk of differentiated cancer cells, which generate their energy mainly via glycolysis 24–30. Mitochondrial mass confers stem-like traits and is associated with metastatic potential and resistance to DNA damage 31. Invasive migratory cancer cells also exhibit high mitochondrial metabolism via activation of a mitochondrial biogenesis mediator, the transcription co-activator peroxisome proliferator-activated receptor gamma co-activator 1 alpha (PGC1α) 32. PGC1α has also been found overexpressed in circulating tumour cells 33, and its expression in a subset of human melanomas produces an increase in OXPHOS that is necessary for survival 34. Moreover, PGC1α inhibition reduces the stemness properties of breast CSCs 24. Oncogene ablation-resistant pancreatic cancer cells with features of CSCs also rely more on mitochondrial function to survive, and depend less on glucose and glutamine and more on pyruvate and palmitate to fuel the tricarboxylic acid (TCA) cycle 10. Similarly, a population of CSCs isolated from ovarian cancer patients overexpressed genes associated with mitochondrial OXPHOS and fatty acid oxidation 28. This oxidative phenotype seems to be related to the capacity to resist apoptosis in CSCs 35. Despite mitochondrial ROS levels being high in these studies, total amounts of ROS are significantly lower in CSCs, which also show a more powerful antioxidant defence system compared with their progeny. A strong antioxidant response keeps ROS levels at bay, and helps in the maintenance of the stemness and tumourigenic capacities of CSCs, therefore contributing to therapy resistance 28, 36.

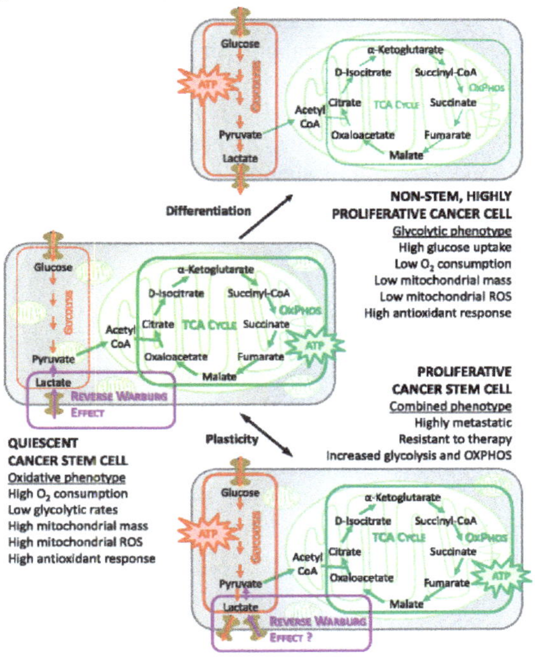

Figure 2: Bioenergetic pathways underlying CSC metabolism. In more differentiated cancer cells, the glycolytic phenotype might predominate over oxidative phosphorylation (OXPHOS). CSCs instead might rely more on an oxidative metabolism for their energy production. CSCs also appear to be metabolically plastic: when OXPHOS is blocked they can eventually develop resistance by acquiring an intermediate glycolytic/oxidative phenotype. ROS reactive oxygen species, TCA tricarboxylic acid

During differentiation under hypoxic conditions, CSCs from several tumour types are able to switch from an oxidative to a glycolytic metabolism in order to compensate for deficient mitochondrial machinery 37. Likewise, CSCs might be able to regulate their differentiation via subtle changes of the redox status, with transitory bursts of ROS production that stimulate differentiation of CSCs towards their non-stem cancer cell counterparts 38. Indeed, administration of antioxidants such as *N*-acetyl-cysteine (NAC) reduces ROS, suppressing the differentiation of CSCs and increasing metastatic burden 38, 39. Furthermore, a recent study shows that epithelial stem-like cells apportion aged mitochondria asymmetrically between the two daughter cells with different fates. Those daughter cells that receive fewer old mitochondria maintain stem cell traits, whereas cells with a higher content of aged mitochondria are more prone to differentiate. This asymmetrical division of mitochondria requires the pertinent functioning of the mitochondrial fission machinery that spatially restricts old mitochondria to the perinuclear region of the mother cell 40. Indeed, increased mitochondrial fission appears to be a characteristic of CSCs and its pharmacological or genetic inhibition leads to the loss of stemness traits and differentiation 40, 41. Hence, the control mechanisms involved in the asymmetrical sorting of aged and young mitochondria, such as mitochondrial fission, also play an important role in CSC maintenance.

Finally, CSCs seem to be vulnerable to mitochondria-targeted drugs and the inhibition of OXPHOS seems to inhibit tumour formation 29, 42–44. Treatment with metformin, an inhibitor of the OXPHOS complex I, induces the partial suppression of stemness traits, such as mammosphere formation, and in-vivo tumour growth delay, although the effects are not lasting and resistance to treatment is observed 29, 45, 46. CSCs treated with a mitochondrial ROS inducer such as menadione do not become resistant, suggesting that increasing mitochondrial ROS levels to non-viable levels might be a better approach to eliminate CSCs 29. Other studies indicate that the use of mitochondria-located antioxidants can prevent metastatic dissemination, suggesting that fine-tuning of oxidative stress to keep it below a critical threshold may be crucial for the maintenance of the CSC phenotype 47, 48. According to symbiogenesis, which states that the origin of eukaryotic mitochondria was the engulfment of aerobic bacteria 49, the use of antibiotics has been postulated as an effective treatment to target mitochondrial mass and metabolism. Indeed, several approved antibiotics such as salinomycin, erythromycins, tetracyclines or glycylcyclines have already shown effects on CSC survival in preclinical models and in clinical studies via reduction of stemness properties 50–55. Mitochondrial health is thus fundamental for the maintenance of CSCs and can be targeted for cancer therapy.

THE METABOLIC PLASTICITY OF CANCER STEM CELLS

One possible explanation for the discrepancies in CSC metabolism reported in the scientific literature may be found in the metabolic adaptability that CSCs show following microenvironmental fluctuations. For example, most in-vitro studies are carried out in non-physiological high glucose and oxygen concentrations, which favour a glycolytic phenotype. Ideally, the optimal experimental conditions to keep the metabolic traits of CSCs intact would be to isolate them directly from patients and analyse them immediately or within the first steps of in-vitro culture. In fact, when patient-derived, low-passage CSCs are used, OXPHOS seems to be the preferred metabolic pathway for the energy production of CSCs 56.

Breast CSCs in mouse and human tumours have a more glycolytic phenotype compared with their differentiated progeny 57. CSCs from low-passage, patient-derived glioblastoma specimens relied more on OXPHOS than their differentiated progeny 56. These glioma stem cells have high metabolic plasticity since they can switch to a glycolytic metabolism when OXPHOS is blocked 58. Leukaemia stem cells also rely primarily on OXPHOS 26.

This observed adaptive metabolic plasticity might allow CSCs to survive in changeable, sometimes hostile, environments or unfavourable circumstances encountered during tumour progression, such as at metastatic sites. In support of this metabolic malleability scenario, some publications show that CSCs are able to switch to a glycolytic metabolism when OXPHOS is blocked 56, 59.

CSCs have been shown to adapt to starvation and hypoxia by upregulating glucose transporters and switching to a more glycolytic phenotype to outcompete their differentiated progeny 58, 60. Hypoxia and changes in glucose concentration induce CSC enrichment, which is mediated by hypoxia-inducible factor 1 alpha (HIF1α) and the AKT/MTOR/beta-catenin (CTNNB1) stem cell regulatory pathway 61–63. Inhibiting HIF1α in combination with anti-angiogenic therapy reduces CSCs in mouse models of breast cancer and holds promise to be an effective therapy in breast cancer, which is currently being investigated in clinical trials 64.

K-Ras ablation-resistant pancreatic cells with stem-like characteristics are unable to increase compensatory fluxes such as glycolysis following OXPHOS inhibition, despite being more resistant to nutrient deprivation and other environmental stresses. This lack of plasticity may be attributed to the shutdown of the K-Ras-driven metabolic programme 65. Ont the contrary, in pancreatic CSCs the lack of plasticity seems to be independent of K-Ras. Instead, another oncogene, MYC, controls the observed limited metabolic adaptability of most metformin-treated CSCs, which undergo energetic crisis and die. However, MYC-dependent resistant clones eventually emerge 29. These data suggest that oncogene activation is sufficient for the induction of a particular metabolic pathway in CSCs, and the extent of its effects on metabolic reprogramming may depend on the context; for example, on the differentiation status of the cell. Determining the mechanisms behind this specificity will be critical to understanding tumour heterogeneity and complexity. Normal stem cells and iPS cells utilize glycolysis, while CSCs have the capacity to shift to OXPHOS and mitochondrial metabolism. One can speculate that differences in signalling pathways mediated by OCT3/4, SOX2, KLF4, MYC in iPS cells and by SHH, NOTCH, WNT, PTEN, MAPK, KRAS, HIF and TP53 in CSCs drives these divergent metabolic phenotypes. Understanding the metabolic differences between normal stem cells and CSCs and their regulators will be important for the process of developing novel therapeutics that metabolically target, but preserve, the key functions of normal stem cells.

Glycolysis or Mitochondrial Respiration: Is It Really One Or The Other?

The CSC phenotype may not necessarily be uniform between cancer subtypes or even between tumours of the same subtype. For instance, the preferred metabolic pathway to produce energy may depend on the metastatic site, indicating extensive metabolic variability 66. Moreover, different subpopulations of CSCs exhibit different metabolic patterns. Recent publications imply the existence of an epithelial-like (mesenchymal to epithelial transition (MET)) and a mesenchymal-like (epithelial to mesenchymal transition (EMT)) CSC phenotype, states that might be interconvertible 67, 68. In breast cancer, MET CSCs are characterized by high ALDH activity and enhanced proliferative capacity, whereas EMT CSCs are identified by the expression of the $CD44^{hi}/CS24^{-}$ surface markers and a slow-cycling, quiescent state 68. Mesenchymal-like EMT CSCs seem to favour glycolysis, and have a marked reduction in oxygen consumption, decreased mitochondrial mass and membrane potential, lower ROS production and higher antioxidant capacity compared with the epithelial-like fraction of CSCs 36, 67. Indeed, CD44 acts as a metabolic modulator, by activating glycolysis under hypoxia. CD44 ablation reduces glycolysis and the antioxidant response, and moves the energy production to the mitochondria, with an increase in ROS 69. Conversely, a proteomics study revealed that the ALDH-expressing population of CSCs express more glycolytic enzymes than the $CD44^{hi}/CS24^{-}$ CSCs 70. Finally, a recent study shows that highly metastatic murine cancer cells enhance both glycolysis and OXPHOS pathways compared with cells with the same genetic background that lack metastatic potential 66. Another report links metabolic plasticity to the acquisition of therapy resistance by showing that although most CSCs have limited metabolic malleability and predominantly rely on OXPHOS, a subpopulation of metformin-resistant CSCs is able to acquire a more adaptable intermediate glycolytic/respiratory phenotype. The metabolic phenotype of CSCs thus appears to be heterogeneous with distinct metabolic programmes activated in different subpopulations of cancer cells (Fig. 2).

These results suggest that the dual blockade of glycolysis and mitochondrial respiration may represent a better way to eradicate CSC heterogeneity than focusing exclusively on glycolysis inhibition or suppression of mitochondrial respiration. Indeed, combined inhibition of glycolysis and mitochondrial respiration has been shown to be effective in suppressing tumour growth and metastasis 71.

Genetic analysis of breast cancers has demonstrated different mutational profiles across the subtypes of breast cancer. For example, the most frequent genetic alteration found in luminal breast cancers is mutational activation of PI3K signalling 72. In contrast, triple negative breast cancer

(TNBC) almost always contain mutations in *TP53* and also frequently display deletions or epigenetic silencing of the *PTEN* tumour suppressor gene. In addition to *HER2* gene amplification, *HER2*-positive breast cancers frequently display deletions in *PTEN*, and indeed this is a likely cause of resistance to *HER2*-targeted therapies 73–75.*BRCA1* germline mutations or epigenetic silencing of the *BRCA1* locus are most frequently associated with TNBCs 76. All of these molecular alterations have been demonstrated to increase CSC frequency in pre-clinical models as well as in patient samples 3.

THE CONTRIBUTION OF THE MICROENVIRONMENT

The effects of the niche on CSC metabolism are also starting to be recognized. High catabolism in the microenvironment with NF-κB, HIF-1α and TGF-β activation coincides with glycolysis and ketogenesis, and promotes CSC features 77–80. A model of reverse Warburg metabolism in which non-glycolytic stem-like cells may be fed by more differentiated glycolytic cells in normoxic conditions has also been observed in breast cancer 60. Another study shows that EMT-induced cancer cells with CSC features have enhanced ability to utilize catabolites taken up from the extracellular microenvironment, such as the glycolytic end products pyruvate and lactate, the amino acids glutamine, glutamate and alanine, or ketone bodies, especially upon starvation, to support their mitochondrial energy production 81. Indeed, glutamine, glutamate and alanine have been identified as EMT-associated metabolites in another report, which demonstrates that this oncometabolite signature correlates with poor survival in breast cancer 82. Similarly, high lactate concentrations achieved by exogenous lactate administration increase the metastatic potential of breast cancer cells in vivo 83. Finally, recent studies show that mitochondrial DNA transfer from host cells of the tumour microenvironment to tumour cells with compromised respiratory function re-establishes not only their mitochondrial respiration but also their tumour-initiating capacity and resistance to therapy 84, 85. CSCs may thus enable the internalization of energy-rich nutrients or energy-producing mitochondrial components from the extracellular milieu to exploit in their own bioenergetic pathways. Although most studies focus on the interaction between cancer cells and host cells including immune cells, communication between heterogeneous populations of tumour cells might also be relevant.

Metabolic stresses also induce significant changes in non-malignant cells within tumours. T cells can preferentially develop into the immunosuppressive regulatory subtype (Tregs) following glucose restriction, which promotes tumour growth 86, 87. Hypoxia alters interactions between breast CSCs and macrophages with transformation of macrophages to an immunosuppressive phenotype with upregulation of HIF-1α and HIF-2α 88, 89.

Conversely, inflammatory cytokines generated by the tumour microenvironment (such as IL-6 and IL-8) with activation of NF-κB induce glycolysis with activation of PI3K and AKT and stimulate CSC self-renewal, which then may promote tumour growth and metastasis 62, 90–92. Drugs targeting IL-6 and IL-8 are being investigated as a therapeutic strategy specifically for CSCs 63. Future studies will need to determine the effects of metabolic modulating therapies on the phenotype of different tumour cell populations and the role of metabolism modulation in the anticancer effects of drugs targeting CSC-sensitive cytokines and signalling pathways.

ADDITIONAL METABOLIC SINGULARITIES OF CANCER STEM CELLS

The role of cell metabolism has evolved into an active area of research during the last decade, with a strong focus on glucose metabolism. Unfortunately, scarce attention has been directed to amino acid and lipid metabolism. A recent report shows that pancreatic CSCs are glutamine dependent. Inhibiting glutamine availability by targeting glutaminase or glutamine oxaloacetic transaminase (GOT), accountable for the conversion of glutamine into oxaloacetate, reduces the expression of stemness genes, inhibits self-renewal and sensitizes CSCs to radiation therapy via accumulation of ROS in vitro and in vivo 93. Colon adenocarcinoma circulating tumour cells are able to colonize hepatic tissue due to their elevated lysine catabolism, which reduces ROS levels, promotes self-renewal potential and activates the stemness-related Wnt signalling pathway 94. Other studies show the importance of fatty acid metabolism and in particular the mevalonate pathway in the generation of CSCs 28, 65, 95. Thus, very little is known regarding the role of protein and fatty acid metabolism in CSC biology, and further investigations will be required to elucidate their contribution to such CSC traits and their relationship with glucose metabolism. Likewise, a strong dependence of CSCs on other catabolic processes such as autophagy, which makes them more resistant to nutrient deprivation and other stresses 96, should be further investigated. An entire world of metabolic pathways might be awaiting future discovery.

FINAL REMARKS AND CHALLENGES AHEAD

Over the years, substantial evidence for the existence of CSCs has strengthened the view that these cells are accountable for cancer development. CSCs renew themselves and at the same time generate progenitors that lose their stemness, ultimately giving rise to the bulk of the tumour. In addition, the last decade of research highlights that metabolism is not a mere player subordinated to CSC physiology, but actually may orchestrate it. Given that changes in metabolism precede changes in stemness, a perturbation in the metabolic phenotype of CSCs could be essential for the acquisition of the CSC state. Conventional therapy targets rapidly proliferating cancer cells that make up the bulk of the tumour without necessarily having an effect on the CSC population. CSCs possess special metabolic traits that distinguish them from the bulk of the tumour and that may constitute the basis for the development of new therapeutic strategies to eradicate them. From a clinical point of view, targeting the particularities of CSC metabolism is more likely to translate into permanently curing cancer or at least providing long-term disease-free survival, because these cells are responsible for resistance to therapy and metastasis, the main cause of cancer-related deaths. The interest in exploiting CSC metabolism for drug targeting is therefore gaining ground. However, a well-defined portrait of the singularities of CSC metabolism still needs to be depicted and CSC metabolism remains a controversial issue, with studies supporting a glycolytic phenotype of CSC and others stating that CSC metabolism is mainly oxidative.

Many other unresolved issues need to be addressed. Elucidating the differences in metabolism between CSCs and non-stem cancer cells, and between CSCs and normal progenitor stem cells, will be crucial to develop new therapies and may reveal new ways to distinctively target these TICs. Whereas normal stem cells rely more on glycolysis, CSCs might depend more on mitochondrial oxidative metabolism. If this is the case, why would the stemness state of cancer cells require a different metabolic state than normal stem cells? In contrast with normal physiological development, tumourigenesis tends to be highly disorganized and cancer metabolic malleability could provide a niche more prone to CSC survival. Nevertheless, the stability or plasticity of the CSCs phenotype needs to be verified. Are CSCs really able to metabolically adapt depending on microenvironmental fluctuations? Is the CSC population metabolically heterogeneous or does it exhibit different degrees of stemness-related phenotypes? During tumourigenesis, characteristics of the CSCs might mutate, and distinct CSC populations could eventually emerge in what it would be a metabolically changeable or versatile target. In such a scenario, future therapies designed to eradicate CSCs via targeting their metabolism might need to simultaneously block glycolysis and mitochondrial respiration.

We therefore need to overcome multiple obstacles before we can effectively eliminate CSCs. First of all, to properly recognize CSCs and differentiate them from other cell types, greater efforts should be made towards the identification of specific CSC markers, because none of the markers so far defined is unique for CSCs. Likewise, a combination of markers could greatly improve the purity of CSCs for research purposes. Finally, it is important to note that CSCs are settled in a niche formed by multiple other cell types and cancer heterogeneity is clearly more complex than originally thought. Hence, studying the metabolism of CSCs in experimental settings that do not reflect the heterogenic architecture of tumours such as the absence of a pertinent microenvironment is not ideal. Better models that preserve the CSC physiological state and structure should be developed.

Despite the limited information we currently have on the role of metabolism in the ability of CSCs to self-renew, initiate tumours, metastasize and survive therapy, targeting CSCs by blocking their metabolic singularities holds great potential in improving current cancer treatments. In practice, combinational treatments involving both a standard cytotoxic therapy and a CSC-targeted therapy will probably be required to ablate all cancer cells (Fig. 1).

ACKNOWLEDGEMENTS

The Sotgia and Lisanti Laboratories in England were supported, in part, by funding from the European Union (ERC Advanced Grant), Breast Cancer Now, The Healthy Life Foundation and the Manchester Cancer Research Centre (MCRC). UEMO was supported by the National Cancer Institute (NCI) of the National Institutes of Health (NIH), under Award Number K08-CA175193. RGP's laboratory receives funding from the NIH and the NCI, as well as the Breast Cancer Research Foundation and the Ralph and Marian C. Falk Medical Research Trust.

Authors' contributions

MP-P drafted the manuscript. MPL, UEM-O, FS and RGP helped to write, edit and revise drafts of the manuscript. All authors read and approved the final manuscript.

Competing Interests

The authors declare that they have no competing interests.

Abbreviations

ALDH	aldehyde dehydrogenase
CSC	cancer stem cell
EMT	epithelial to mesenchymal transition
HIF	hypoxia-inducible factor
iPS	induced pluripotent stem
MET	mesenchymal to epithelial transition
NAC	*N*-acetyl-cysteine
OXPHOS	oxidative phosphorylation
PGC1α	peroxisome proliferator-activated receptor gamma co-activator 1 alpha
ROS	reactive oxygen species
TCA	tricarboxylic acid
TIC	tumour-initiating cell

REFERENCES

1. Aguilar-Gallardo C, Simon C. Cells, stem cells, and cancer stem cells. Semin Reprod Med. 2013;31(1):5–13. doi: 10.1055/s-0032-1331792.
2. Pattabiraman DR, Weinberg RA. Tackling the cancer stem cells—what challenges do they pose? Nat Rev Drug Discov. 2014;13(7):497–512. doi: 10.1038/nrd4253.
3. Brooks MD, Burness ML, Wicha MS. Therapeutic implications of cellular heterogeneity and plasticity in breast cancer. Cell Stem Cell. 2015;17(3):260–271. doi: 10.1016/j.stem.2015.08.014.
4. Hanahan D, Weinberg RA. Hallmarks of cancer: the next generation. Cell. 2011;144(5):646–674. doi: 10.1016/j.cell.2011.02.013.
5. Lagadec C, Meignan S, Adriaenssens E, Foveau B, Vanhecke E, Romon R, et al. TrkA overexpression enhances growth and metastasis of breast cancer cells. Oncogene. 2009;28(18):1960–1970. doi: 10.1038/onc.2009.61.
6. Guppy M, Greiner E, Brand K. The role of the Crabtree effect and an endogenous fuel in the energy metabolism of resting and proliferating thymocytes. Eur J Biochem. 1993;212(1):95–99. doi: 10.1111/j.1432-1033.1993.tb17637.x.
7. Zhou W, Choi M, Margineantu D, Margaretha L, Hesson J, Cavanaugh C, et al. HIF1alpha induced switch from bivalent to exclusively glycolytic metabolism during ESC-to-EpiSC/hESC transition. Embo J.2012;31(9):2103–2116. doi: 10.1038/emboj.2012.71.
8. Folmes CD, Nelson TJ, Martinez-Fernandez A, Arrell DK, Lindor JZ, Dzeja PP, et al. Somatic oxidative bioenergetics transitions into pluripotency-dependent glycolysis to facilitate nuclear reprogramming. Cell

Metab. 2011;14(2):264–271. doi: 10.1016/j.cmet.2011.06.011.

9. Jang H, Yang J, Lee E, Cheong JH. Metabolism in embryonic and cancer stemness. Arch Pharm Res.2015;38(3):381–388. doi: 10.1007/s12272-015-0558-y.
10. Prigione A, Fauler B, Lurz R, Lehrach H, Adjaye J. The senescence-related mitochondrial/oxidative stress pathway is repressed in human induced pluripotent stem cells. Stem Cells. 2010;28(4):721–733. doi: 10.1002/stem.404.
11. Jang YY, Sharkis SJ. A low level of reactive oxygen species selects for primitive hematopoietic stem cells that may reside in the low-oxygenic niche. Blood. 2007;110(8):3056–3063. doi: 10.1182/blood-2007-05-087759.
12. Liu PP, Liao J, Tang ZJ, Wu WJ, Yang J, Zeng ZL, et al. Metabolic regulation of cancer cell side population by glucose through activation of the Akt pathway. Cell Death Differ. 2012;21(1):124–135. doi: 10.1038/cdd.2013.131.
13. Dando I, Dalla Pozza E, Biondani G, Cordani M, Palmieri M, Donadelli M. The metabolic landscape of cancer stem cells. IUBMB Life. 2015;67(9):687–693. doi: 10.1002/iub.1426.
14. Ciavardelli D, Rossi C, Barcaroli D, Volpe S, Consalvo A, Zucchelli M, et al. Breast cancer stem cells rely on fermentative glycolysis and are sensitive to 2-deoxyglucose treatment. Cell Death Dis. 2014;5:e1336. doi: 10.1038/cddis.2014.285.
15. Emmink BL, Verheem A, Van Houdt WJ, Steller EJ, Govaert KM, Pham TV, et al. The secretome of colon cancer stem cells contains drug-metabolizing enzymes. J Proteomics. 2013;91:84–96. doi: 10.1016/j.jprot.2013.06.027.
16. Liao J, Qian F, Tchabo N, Mhawech-Fauceglia P, Beck A, Qian Z, et al. Ovarian cancer spheroid cells with stem cell-like properties contribute to tumor generation, metastasis and chemotherapy resistance through hypoxia-resistant metabolism. PLoS One. 2014;9(1):e84941. doi: 10.1371/journal.pone.0084941.
17. Palorini R, Votta G, Balestrieri C, Monestiroli A, Olivieri S, Vento R, et al. Energy metabolism characterization of a novel cancer stem cell-like line 3AB-OS. J Cell Biochem. 2014;115(2):368–379. doi: 10.1002/jcb.24671.
18. Zhou Y, Zhou Y, Shingu T, Feng L, Chen Z, Ogasawara M, et al. Metabolic alterations in highly tumorigenic glioblastoma cells: preference for hypoxia and high dependency on glycolysis. J Biol Chem.2011;286(37):32843–32853. doi: 10.1074/jbc.M111.260935.
19. Guha M, Srinivasan S, Ruthel G, Kashina AK, Carstens RP, Mendoza A, et al. Mitochondrial retrograde signaling induces epithelial-mesenchymal transition and generates breast cancer stem cells. Oncogene.2013;33(45):5238–5250. doi: 10.1038/onc.2013.467.
20. Petros JA, Baumann AK, Ruiz-Pesini E, Amin MB, Sun CQ, Hall J, et al. mtDNA mutations increase tumorigenicity in prostate cancer. Proc Natl Acad Sci U S A. 2005;102(3):719–724. doi: 10.1073/pnas.0408894102.
21. Wang C, Li Z, Lu Y, Du R, Katiyar S, Yang J, et al. Cyclin D1 repression of nuclear respiratory factor 1 integrates nuclear DNA synthesis and mitochondrial function. Proc Natl Acad Sci U S A.2006;103(31):11567–11572. doi: 10.1073/pnas.0603363103.
22. Sakamaki T, Casimiro MC, Ju X, Quong AA, Katiyar S, Liu M, et al. Cyclin D1 determines mitochondrial function in vivo. Mol Cell Biol. 2006;26(14):5449–5469. doi: 10.1128/MCB.02074-05.
23. Lee WT, St John J. The control of mitochondrial DNA replication during development and tumorigenesis. Ann N Y Acad Sci. 2015;1350:95–106. doi: 10.1111/nyas.12873.
24. De Luca A, Fiorillo M, Peiris-Pages M, Ozsvari B, Smith DL, Sanchez-Alvarez R, et al. Mitochondrial biogenesis is required for the anchorage-independent survival and propagation of stem-like cancer cells. Oncotarget. 2015;6(17):14777–14795. doi: 10.18632/oncotarget.4401.
25. Janiszewska M, Suva ML, Riggi N, Houtkooper RH, Auwerx J, Clement-Schatlo V, et al. Imp2 controls oxidative phosphorylation and is crucial for preserving glioblastoma cancer stem cells. Genes Dev.2012;26(17):1926–1944. doi: 10.1101/gad.188292.112.
26. Lagadinou ED, Sach A, Callahan K, Rossi RM, Neering SJ, Minhajuddin M, et al. BCL-2 inhibition targets oxidative phosphorylation and selectively eradicates quiescent human leukemia stem cells. Cell Stem Cell. 2013;12(3):329–341. doi: 10.1016/j.stem.2012.12.013.
27. Lamb R, Bonuccelli G, Ozsvari B, Peiris-Pages M, Fiorillo M, Smith DL, et al. Mitochondrial mass, a new metabolic biomarker for stem-like cancer cells: understanding WNT/FGF-driven anabolic signaling. Oncotarget. 2015;6(31):30453–30471.
28. Pasto A, Bellio C, Pilotto G, Ciminale V, Silic-Benussi M, Guzzo G, et al. Cancer stem cells from epithelial ovarian cancer patients privilege oxidative phosphorylation, and resist glucose deprivation.Oncotarget. 2014;5(12):4305–4319. doi: 10.18632/oncotarget.2010.
29. Sancho P, Burgos-Ramos E, Tavera A, Bou Kheir T, Jagust P, Schoenhals M, et al. MYC/PGC-1alpha balance determines the metabolic phenotype and plasticity of pancreatic cancer stem cells. Cell Metab.2015;22(4):590–605. doi: 10.1016/j.cmet.2015.08.015.
30. Vlashi E, Lagadec C, Vergnes L, Reue K, Frohnen P, Chan M, et al. Metabolic differences in breast cancer stem cells and differentiated progeny. Breast Cancer Res Treat. 2014;146(3):525–534. doi: 10.1007/s10549-014-3051-2.
31. Farnie G, Sotgia F, Lisanti MP. High mitochondrial mass identifies a sub-population of stem-like cancer cells that are chemo-resistant. Oncotarget. 2015;6(31):30472–

30486.

32. LeBleu VS, O'Connell JT, Gonzalez Herrera KN, Wikman H, Pantel K, Haigis MC, et al. PGC-1alpha mediates mitochondrial biogenesis and oxidative phosphorylation in cancer cells to promote metastasis. Nat Cell Biol. 2014;16(10):992–1003. doi: 10.1038/ncb3039.

33. Wey A, Knoepfler PS. c-myc and N-myc promote active stem cell metabolism and cycling as architects of the developing brain. Oncotarget. 2010;1(2):120–130. doi: 10.18632/oncotarget.116.

34. Vazquez F, Lim JH, Chim H, Bhalla K, Girnun G, Pierce K, et al. PGC1alpha expression defines a subset of human melanoma tumors with increased mitochondrial capacity and resistance to oxidative stress.Cancer Cell. 2013;23(3):287–301. doi: 10.1016/j.ccr.2012.11.020.

35. Yajima T, Ochiai H, Uchiyama T, Takano N, Shibahara T, Azuma T. Resistance to cytotoxic chemotherapy-induced apoptosis in side population cells of human oral squamous cell carcinoma cell line Ho-1-N-1. Int J Oncol. 2009;35(2):273–280.

36. Diehn M, Cho RW, Lobo NA, Kalisky T, Dorie MJ, Kulp AN, et al. Association of reactive oxygen species levels and radioresistance in cancer stem cells. Nature. 2009;458(7239):780–783. doi: 10.1038/nature07733.

37. Cipolleschi MG, Marzi I, Santini R, Fredducci D, Vinci MC, D'Amico M, et al. Hypoxia-resistant profile implies vulnerability of cancer stem cells to physiological agents, which suggests new therapeutic targets.Cell Cycle. 2014;13(2):268–278. doi: 10.4161/cc.27031.

38. Yuan S, Lu Y, Yang J, Chen G, Kim S, Feng L, et al. Metabolic activation of mitochondria in glioma stem cells promotes cancer development through a reactive oxygen species-mediated mechanism. Stem Cell Res Ther. 2015;6(1):198. doi: 10.1186/s13287-015-0174-2.

39. Piskounova E, Agathocleous M, Murphy MM, Hu Z, Huddlestun SE, Zhao Z, et al. Oxidative stress inhibits distant metastasis by human melanoma cells. Nature. 2015;527(7577):186–191. doi: 10.1038/nature15726.

40. Katajisto P, Dohla J, Chaffer CL, Pentinmikko N, Marjanovic N, Iqbal S, et al. Stem cells. Asymmetric apportioning of aged mitochondria between daughter cells is required for stemness. Science.2015;348(6232):340–343. doi: 10.1126/science.1260384.

41. Xie Q, Wu Q, Horbinski CM, Flavahan WA, Yang K, Zhou W, et al. Mitochondrial control by DRP1 in brain tumor initiating cells. Nat Neurosci. 2015;18(4):501–510. doi: 10.1038/nn.3960.

42. Biasutto L, Dong LF, Zoratti M, Neuzil J. Mitochondrially targeted anti-cancer agents. Mitochondrion.2010;10(6):670–681. doi: 10.1016/j.mito.2010.06.004.

43. Roesch A, Vultur A, Bogeski I, Wang H, Zimmermann KM, Speicher D, et al. Overcoming intrinsic multidrug resistance in melanoma by blocking the mitochondrial respiratory chain of slow-cycling JARID1B(high) cells. Cancer Cell. 2013;23(6):811–825. doi: 10.1016/j.ccr.2013.05.003.

44. Lamb R, Ozsvari B, Lisanti CL, Tanowitz HB, Howell A, Martinez-Outschoorn UE, et al. Antibiotics that target mitochondria effectively eradicate cancer stem cells, across multiple tumor types: treating cancer like an infectious disease. Oncotarget. 2015;6(7):4569–4584. doi: 10.18632/oncotarget.3174.

45. Jung JW, Park SB, Lee SJ, Seo MS, Trosko JE, Kang KS. Metformin represses self-renewal of the human breast carcinoma stem cells via inhibition of estrogen receptor-mediated OCT4 expression. PLoS One. 2011;6(11):e28068. doi: 10.1371/journal.pone.0028068.

46. Mayer MJ, Klotz LH, Venkateswaran V. Metformin and prostate cancer stem cells: a novel therapeutic target. Prostate Cancer Prostatic Dis. 2015;18(4):303–309. doi: 10.1038/pcan.2015.35.

47. Goh J, Enns L, Fatemie S, Hopkins H, Morton J, Pettan-Brewer C, et al. Mitochondrial targeted catalase suppresses invasive breast cancer in mice. BMC Cancer. 2011;11:191. doi: 10.1186/1471-2407-11-191.

48. Porporato PE, Payen VL, Perez-Escuredo J, De Saedeleer CJ, Danhier P, Copetti T, et al. A mitochondrial switch promotes tumor metastasis. Cell Rep. 2014;8(3):754–766. doi: 10.1016/j.celrep.2014.06.043.

49. Zimorski V, Ku C, Martin WF, Gould SB. Endosymbiotic theory for organelle origins. Curr Opin Microbiol. 2014;22:38–48. doi: 10.1016/j.mib.2014.09.008.

50. Lamb R, Harrison H, Hulit J, Smith DL, Lisanti MP, Sotgia F. Mitochondria as new therapeutic targets for eradicating cancer stem cells: quantitative proteomics and functional validation via MCT1/2 inhibition. Oncotarget. 2014;5(22):11029–11037. doi: 10.18632/oncotarget.2789.

51. An H, Kim JY, Oh E, Lee N, Cho Y, Seo JH. Salinomycin promotes anoikis and decreases the CD44+/CD24– stem-like population via inhibition of STAT3 activation in MDA-MB-231 cells. PLoS One.2015;10(11):e0141919. doi: 10.1371/journal.pone.0141919.

52. Ferreri AJ, Ponzoni M, Guidoboni M, Resti AG, Politi LS, Cortelazzo S, et al. Bacteria-eradicating therapy with doxycycline in ocular adnexal MALT lymphoma: a multicenter prospective trial. J Natl Cancer Inst. 2006;98(19):1375–1382. doi: 10.1093/jnci/djj373.

53. Chu DJ, Yao DE, Zhuang YF, Hong Y, Zhu XC, Fang ZR, et al. Azithromycin enhances the favorable results of paclitaxel and cisplatin in patients with advanced non-small cell lung cancer. Genet Mol Res.2014;13(2):2796–2805. doi: 10.4238/2014.April.14.8.

54. Zhou J, Li P, Xue X, He S, Kuang Y, Zhao H, et al. Salinomycin induces apoptosis in cisplatin-resistant colorectal cancer cells by accumulation of reactive oxygen species. Toxicol Lett. 2013;222(2):139–145. doi: 10.1016/j.toxlet.2013.07.022.

55. Gupta PB, Onder TT, Jiang G, Tao K, Kuperwasser C, Weinberg RA, et al. Identification of selective

inhibitors of cancer stem cells by high-throughput screening. Cell. 2009;138(4):645–659. doi: 10.1016/j.cell.2009.06.034.

56. Vlashi E, Lagadec C, Vergnes L, Matsutani T, Masui K, Poulou M, et al. Metabolic state of glioma stem cells and nontumorigenic cells. Proc Natl Acad Sci U S A. 2011;108(38):16062–16067. doi: 10.1073/pnas.1106704108.

57. Feng W, Gentles A, Nair RV, Huang M, Lin Y, Lee CY, et al. Targeting unique metabolic properties of breast tumor initiating cells. Stem Cells. 2014;32(7):1734–1745. doi: 10.1002/stem.1662.

58. Flavahan WA, Wu Q, Hitomi M, Rahim N, Kim Y, Sloan AE, et al. Brain tumor initiating cells adapt to restricted nutrition through preferential glucose uptake. Nat Neurosci. 2013;16(10):1373–1382. doi: 10.1038/nn.3510.

59. Dong C, Yuan T, Wu Y, Wang Y, Fan TW, Miriyala S, et al. Loss of FBP1 by Snail-mediated repression provides metabolic advantages in basal-like breast cancer. Cancer Cell. 2013;23(3):316–331. doi: 10.1016/j.ccr.2013.01.022.

60. Gordon N, Skinner AM, Pommier RF, Schillace RV, O'Neill S, Peckham JL, et al. Gene expression signatures of breast cancer stem and progenitor cells do not exhibit features of Warburg metabolism. Stem Cell Res Ther. 2015;6:157. doi: 10.1186/s13287-015-0153-7.

61. Conley SJ, Gheordunescu E, Kakarala P, Newman B, Korkaya H, Heath AN, et al. Antiangiogenic agents increase breast cancer stem cells via the generation of tumor hypoxia. Proc Natl Acad Sci U S A.2012;109(8):2784–2789. doi: 10.1073/pnas.1018866109.

62. Korkaya H, Liu S, Wicha MS. Breast cancer stem cells, cytokine networks, and the tumor microenvironment. J Clin Invest. 2011;121(10):3804–3809. doi: 10.1172/JCI57099.

63. Sehl ME, Shimada M, Landeros A, Lange K, Wicha MS. Modeling of cancer stem cell state transitions predicts therapeutic response. PLoS One. 2015;10(9):e0135797. doi: 10.1371/journal.pone.0135797.

64. Conley SJ, Baker TL, Burnett JP, Theisen RL, Lazarus D, Peters CG, et al. CRLX101, an investigational camptothecin-containing nanoparticle-drug conjugate, targets cancer stem cells and impedes resistance to antiangiogenic therapy in mouse models of breast cancer. Breast Cancer Res Treat.2015;150(3):559–567. doi: 10.1007/s10549-015-3349-8.

65. Viale A, Pettazzoni P, Lyssiotis CA, Ying H, Sanchez N, Marchesini M, et al. Oncogene ablation-resistant pancreatic cancer cells depend on mitochondrial function. Nature. 2014;514(7524):628–632. doi: 10.1038/nature13611.

66. Dupuy F, Tabaries S, Andrzejewski S, Dong Z, Blagih J, Annis MG, et al. PDK1-dependent metabolic reprogramming dictates metastatic potential in breast cancer. Cell Metab. 2015;22(4):577–589. doi: 10.1016/j.cmet.2015.08.007.

67. Gammon L, Biddle A, Heywood HK, Johannessen AC, Mackenzie IC. Sub-sets of cancer stem cells differ intrinsically in their patterns of oxygen metabolism. PLoS One. 2013;8(4):e62493. doi: 10.1371/journal.pone.0062493.

68. Liu S, Cong Y, Wang D, Sun Y, Deng L, Liu Y, et al. Breast cancer stem cells transition between epithelial and mesenchymal states reflective of their normal counterparts. Stem Cell Reports. 2013;2(1):78–91. doi: 10.1016/j.stemcr.2013.11.009.

69. Tamada M, Nagano O, Tateyama S, Ohmura M, Yae T, Ishimoto T, et al. Modulation of glucose metabolism by CD44 contributes to antioxidant status and drug resistance in cancer cells. Cancer Res.2012;72(6):1438–1448. doi: 10.1158/0008-5472.CAN-11-3024.

70. Nie S, McDermott SP, Deol Y, Tan Z, Wicha MS, Lubman DM. A quantitative proteomics analysis of MCF7 breast cancer stem and progenitor cell populations. Proteomics. 2015;15(22):3772–3783. doi: 10.1002/pmic.201500002.

71. Cheong JH, Park ES, Liang J, Dennison JB, Tsavachidou D, Nguyen-Charles C, et al. Dual inhibition of tumor energy pathway by 2-deoxyglucose and metformin is effective against a broad spectrum of preclinical cancer models. Mol Cancer Ther. 2011;10(12):2350–2362. doi: 10.1158/1535-7163.MCT-11-0497.

72. Creighton CJ, Fu X, Hennessy BT, Casa AJ, Zhang Y, Gonzalez-Angulo AM, Lluch A, Gray JW, Brown PH, Hilsenbeck SG, et al. Proteomic and transcriptomic profiling reveals a link between the PI3K pathway and lower estrogen-receptor (ER) levels and activity in ER+ breast cancer. Breast Cancer Res.2010;12(3):R40. doi: 10.1186/bcr2594.

73. Dave B, Migliaccio I, Gutierrez MC, Wu MF, Chamness GC, Wong H, Narasanna A, Chakrabarty A, Hilsenbeck SG, Huang J, et al. Loss of phosphatase and tensin homolog or phosphoinositol-3 kinase activation and response to trastuzumab or lapatinib in human epidermal growth factor receptor 2-overexpressing locally advanced breast cancers. J Clin Oncol. 2011;29(2):166–173. doi: 10.1200/JCO.2009.27.7814.

74. Fabi A, Metro G, Di Benedetto A, Nistico C, Vici P, Melucci E, Antoniani B, Perracchio L, Sperduti I, Milella M, et al. Clinical significance of PTEN and p-Akt co-expression in HER2-positive metastatic breast cancer patients treated with trastuzumab-based therapies. Oncology. 2010;78(2):141–149. doi: 10.1159/000312656.

75. Park YH, Jung HA, Choi MK, Chang W, Choi YL, Do IG, Ahn JS, Im YH. Role of HER3 expression and PTEN loss in patients with HER2-overexpressing metastatic breast cancer (MBC) who received taxane plus trastuzumab treatment. Br J Cancer. 2014;110(2):384–391. doi: 10.1038/bjc.2013.757.

76. Foulkes WD, Stefansson IM, Chappuis PO, Begin LR, Goffin JR, Wong N, Trudel M, Akslen LA. Germline BRCA1 mutations and a basal epithelial phenotype in breast cancer. J Natl Cancer Inst.2003;95(19):1482–1485. doi: 10.1093/jnci/djg050.

77. Chiavarina B, Whitaker-Menezes D, Migneco G, Martinez-Outschoorn UE, Pavlides S, Howell A, et al.

HIF1-alpha functions as a tumor promoter in cancer associated fibroblasts, and as a tumor suppressor in breast cancer cells: autophagy drives compartment-specific oncogenesis. Cell Cycle. 2010;9(17):3534–3551. doi: 10.4161/cc.9.17.12908.

78. Chiavarina B, Martinez-Outschoorn UE, Whitaker-Menezes D, Howell A, Tanowitz HB, Pestell RG, et al. Metabolic reprogramming and two-compartment tumor metabolism: opposing role(s) of HIF1alpha and HIF2alpha in tumor-associated fibroblasts and human breast cancer cells. Cell Cycle. 2012;11(17):3280–3289. doi: 10.4161/cc.21643.

79. Guido C, Whitaker-Menezes D, Capparelli C, Balliet R, Lin Z, Pestell RG, et al. Metabolic reprogramming of cancer-associated fibroblasts by TGF-beta drives tumor growth: connecting TGF-beta signaling with 'Warburg-like' cancer metabolism and L-lactate production. Cell Cycle. 2012;11(16):3019–3035. doi: 10.4161/cc.21384.

80. Martinez-Outschoorn UE, Prisco M, Ertel A, Tsirigos A, Lin Z, Pavlides S, et al. Ketones and lactate increase cancer cell 'stemness', driving recurrence, metastasis and poor clinical outcome in breast cancer: achieving personalized medicine via metabolo-genomics. Cell Cycle. 2011;10(8):1271–1286. doi: 10.4161/cc.10.8.15330.

81. Cuyas E, Corominas-Faja B, Menendez JA. The nutritional phenome of EMT-induced cancer stem-like cells. Oncotarget. 2014;5(12):3970–3982. doi: 10.18632/oncotarget.2147.

82. Bhowmik SK, Ramirez-Pena E, Arnold JM, Putluri V, Sphyris N, Michailidis G, et al. EMT-induced metabolite signature identifies poor clinical outcome. Oncotarget. 2015;6(40):42651–42660.

83. Bonuccelli G, Tsirigos A, Whitaker-Menezes D, Pavlides S, Pestell RG, Chiavarina B, et al. Ketones and lactate 'fuel' tumor growth and metastasis: evidence that epithelial cancer cells use oxidative mitochondrial metabolism. Cell Cycle. 2010;9(17):3506–3514. doi: 10.4161/cc.9.17.12731.

84. Pasquier J, Guerrouahen BS, Al Thawadi H, Ghiabi P, Maleki M, Abu-Kaoud N, et al. Preferential transfer of mitochondria from endothelial to cancer cells through tunneling nanotubes modulates chemoresistance. J Transl Med. 2013;11:94. doi: 10.1186/1479-5876-11-94.

85. Tan AS, Baty JW, Dong LF, Bezawork-Geleta A, Endaya B, Goodwin J, et al. Mitochondrial genome acquisition restores respiratory function and tumorigenic potential of cancer cells without mitochondrial DNA. Cell Metab. 2015;21(1):81–94. doi: 10.1016/j.cmet.2014.12.003.

86. Shi LZ, Wang R, Huang G, Vogel P, Neale G, Green DR, et al. HIF1alpha-dependent glycolytic pathway orchestrates a metabolic checkpoint for the differentiation of TH17 and Treg cells. J Exp Med.2011;208(7):1367–1376. doi: 10.1084/jem.20110278.

87. Wang T, Liu G, Wang R. The intercellular metabolic interplay between tumor and immune cells. Front Immunol. 2014;5:358.

88. Williams CB, Yeh ES, Soloff AC. Tumor-associated macrophages: unwitting accomplices in breast cancer malignancy. NPJ Breast Cancer 2016, 2. http://dx.doi.org/10.1038/npjbcancer.2015.25.

89. Ostuni R, Kratochvill F, Murray PJ, Natoli G. Macrophages and cancer: from mechanisms to therapeutic implications. Trends Immunol. 2015;36(4):229–239. doi: 10.1016/j.it.2015.02.004.

90. Al-Zoughbi W, Huang J, Paramasivan GS, Till H, Pichler M, Guertl-Lackner B, et al. Tumor macroenvironment and metabolism. Semin Oncol. 2014;41(2):281–295. doi: 10.1053/j.seminoncol.2014.02.005.

91. Mauer J, Denson JL, Bruning JC. Versatile functions for IL-6 in metabolism and cancer. Trends Immunol. 2015;36(2):92–101. doi: 10.1016/j.it.2014.12.008.

92. Martinez-Outschoorn UE, Whitaker-Menezes D, Lin Z, Flomenberg N, Howell A, Pestell RG, et al. Cytokine production and inflammation drive autophagy in the tumor microenvironment: role of stromal caveolin-1 as a key regulator. Cell Cycle. 2011;10(11):1784–1793. doi: 10.4161/cc.10.11.15674.

93. Li D, Fu Z, Chen R, Zhao X, Zhou Y, Zeng B, et al. Inhibition of glutamine metabolism counteracts pancreatic cancer stem cell features and sensitizes cells to radiotherapy. Oncotarget. 2015;6(31):31151–31163.

94. Wu Z, Wei D, Gao W, Xu Y, Hu Z, Ma Z, et al. TPO-induced metabolic reprogramming drives liver metastasis of colorectal cancer CD110+ tumor-initiating cells. Cell Stem Cell. 2015;17(1):47–59. doi: 10.1016/j.stem.2015.05.016.

95. Ginestier C, Monville F, Wicinski J, Cabaud O, Cervera N, Josselin E, et al. Mevalonate metabolism regulates basal breast cancer stem cells and is a potential therapeutic target. Stem Cells. 2012;30(7):1327–1337. doi: 10.1002/stem.1122.

96. Ojha R, Bhattacharyya S, Singh SK. Autophagy in cancer stem cells: a potential link between chemoresistance, recurrence, and metastasis. Biores Open Access. 2015;4(1):97–108. doi: 10.1089/biores.2014.0035.

CHAPTER 6

MITOCHONDRIAL DYSFUNCTION IN ALZHEIMER'S DISEASE AND THE RATIONALE FOR BIOENERGETICS BASED THERAPIES

Isaac G. Onyango, Jameel Dennis, Shaharyah M. Khan

Keywords: Alzheimer's disease, mitochondria, mitophagy, oxidative stress, neuroinflammation, mitochondrial biogenesis, neuroinflammation

Alzheimer's disease (AD) is the most common form of dementia and affects millions worldwide. It is characterized by severe memory loss, with episodic memory being particularly impaired during the initial phases. Most AD cases occur sporadically, although inheritance of certain susceptibility genes enhances risk. A role for dysfunctional mitochondria in AD pathogenesis has been postulated [1-3].

Cumulative evidence reveal that the regulation of mitochondrial turnover and function becomes impaired as a function of age in the brain and may contribute to neurodegeneration in AD [4]. Cerebral hypometabolism is evident in affected brain regions [5, 6] where mitochondrial structure is altered [7, 8]. The expression and activity of mitochondrial enzymes important for metabolism, including cytochrome *c* oxidase (COX), α-ketoglutarate dehydrogenase complex, and pyruvate dehydrogenase complex is reduced [9-11]. AD brain mitochondria have reduced membrane potential, increased permeability, and produce excess reactive oxygen species (ROS) which damages proteins, lipids, and nucleic acids, and are believed to contribute to the pathogenesis of neurodegeneration. Growing evidence suggest that elevated amyloid-β (Aβ) levels contribute to the mitochondrial abnormalities and although the mechanism is not clearly established, both amyloid precursor protein (APP) and Aβ are found in mitochondrial membranes and interact with mitochondrial proteins. Overproduction of the APP and Aβ may affect dynamics of mitochondrial fusion/fission [12-14], impair mitochondrial transport, disrupt the electron transfer chain, increase ROS production [15-17], and impair mitochondrial function [18-20]. These findings build a strong case for mitochondrial dysfunction in AD and effective treatment will likely include targets that address mitochondrial function [21-23].

Citation: Onyango IG, Dennis J, Khan SM. Mitochondrial Dysfunction in Alzheimer's Disease and the Rationale for Bioenergetics Based Therapies. Aging and Disease. 2016;7(2):201-214. doi:10.14336/AD.2015.1007.

MITOCHONDRIAL BIOGENESIS

Mitochondrial biogenesis plays an essential role in maintaining an adequate functional neuronal mitochondrial mass by compensating for damaged mitochondria that have been eliminated. It is highly regulated and requires coordination and crosstalk between the nuclear and mitochondrial genomes [24]. While mitochondrial biogenesis occurs on a regular basis in healthy cells where mitochondria constantly divide and fuse with each other [25-27]; it also occurs in response to oxidative stress, increased energy demand, exercise training and certain diseases. The status of mitochondrial biogenesis in AD neurons is unclear [28]. Biogenesis is induced by the peroxisome proliferator-activated receptor γ coactivator-1α (PGC-1α) which activates different transcription factors, including nuclear respiratory factors 1 and 2 proteins (NRF-1 and NRF-2) and the mitochondrial transcription factor A (TFAM) [4, 29, 30];[28]. NRF-1 and NRF-2 regulate transcription of nuclear and mitochondrial genes involved in OXPHOS, electron transport (complex I-V), mtDNA transcription/replication, heme biosynthesis, protein import/assembly, ion channels, shuttles, and translation [31].

NRF-1 or NRF-2 also contribute to expression of nuclear encoded genes involved in biogenesis including [32]factor A (TFAM), mitochondrial transcription factor B1 or B2 (TFB1M or TFB2M), and mitochondrial RNA polymerase (POLRMT), and mitochondrial transcription termination factor (MTERF), mitochondrial DNA helicase (TWINKLE), single-stranded DNA-binding protein (mtSSB), and POLγB [33] [34] but not POLγA and MTERF3 [34]. When newly formed daughter mitochondria have been incorporated into the mitochondrial network, mitochondria that have been damaged or that have lost membrane potential are specifically targeted for degradation via an autophagy-like process termed as mitophagy [35, 36]. Mitochondrial biogenesis is thought to be impaired in AD where the quantity of mitochondria as well as levels of NRF 1, NRF 2, and TFAM along with nuclear levels of PGC-1α are reduced in hippocampal tissues from AD brain compared to age matched control brain [37] [38] [39].

PGC-1α activity at both the transcriptional and post-translational level is modulated by the nutrient supply and energy balance within the cell [40] and mitochondrial damage [41] [42]. Furthermore, PPARs, mTOR (acting on YY1), and CREB (downstream of PKA signaling) transcriptionally activate PGC-1α to initiate mitochondrial biogenesis [40]. At the post-translational level, PGC-1α is regulated by both phosphorylation and acetylation events. AMP-dependent kinase (AMPK) can phosphorylate and activate PGC-1α while GCN5-mediated acetylation inhibits PGC-1α activity [40]. Deacetylation of PGC-1α by NAD+ dependent SIRT1 promotes mitochondrial biogenesis and ensures that the activity of PGC-1α is sensitive to both the energy and the redox balance in the cell [32]. PGC-1α co-activation of ERRα in turn promotes expression of mitochondrial SIRT3 that ensures effective scavenging of ROS at the mitochondria through activation of mitochondrial superoxide dismutase, amongst other mitochondrial sirtuin targets [43]. PARIS, a Parkin substrate, represses mitochondrial biogenesis by transcriptionally inhibiting PGC-1α expression [44].

MITOPHAGY

Mitophagy is the process by which damaged or dysfunctional mitochondria are selectively engulfed by autophagosomes and delivered to lysosomes to be degraded and recycled by the cell [45]. An excess of reactive oxygen species (ROS) may function as an autophagy trigger [46] and dysfunctional mitochondria that overproduce ROS, are indeed selectively targeted for mitophagy [46].

Central to mitochondrial and cellular homeostasis, mitophagy is modulated by the PTEN-induced putative kinase 1 (PINK1)/Parkin pathway [47] which primarily targets mitochondria devoid of membrane potential (ΔΨm). PINK1 accumulate on the outer membrane of dysfunctional mitochondria and recruit the E3 ubiquitin ligase Parkin [48] [49] [50] that ubiquitinate several OMM proteins that are consequently targeted by P62/SQSTM1 [51].

P62 recognizes ubiquitinated substrates and directly interacts with autophagosome-associated LC3 to recruit autophagosomal membranes to the mitochondria [52]. Damaged mitochondria can also, independently of Parkin, increase FUNDC1 and Nix expression to recruit autophagosomes to mitochondria via direct interaction with LC3 [53] [54]. Ubiquitin ligases, like Smurf1, target depolarized mitochondria for mitophagy [55-57].

The transcription factor nuclear factor erythroid 2-related factor 2 (Nrf2) partly regulates P62 expression due to the presence of an antioxidant response element (ARE) in its promoter region [58, 59]. Electrophilic natural products such as isothiocyanate compound, sulforaphane which upregulate Nrf2 by interfering with its regulator protein, the redox sensitive ubiquitination facilitator Keap1 (Kelch-like ECH-associated protein 1) can potentially induce P62 expression [60-62]. P62-mediated mitophagy inducer (PMI) (HB229), was recently developed to upregulate P62 via stabilization of Nrf2 and promote mitophagy. This compound bypasses the upstream steps of the mitophagic cascade and acts independently of the ΔΨm collapse, and does not mediate any apparent toxic effects on mouse embryonic fibroblast (MEF) cells at the concentrations used in the assays [63]. Parkin also modulates transport of mitochondria along microtubules to a perinuclear region where autophagosomes are concentrated [64] [48]. This is likely due to Parkin-mediated turnover of Miro, a protein required to tether kinesin motor protein complexes to the OMM [65]. HDAC6, a ubiquitin-binding protein deacetylase is also recruited to mitochondria by Parkin [66] along microtubules [67, 68]. Mitophagy is crucial for cellular homeostasis and its impairment is linked to several neurodegenerative diseases [69] [70]. However, selective pharmacologic modulators of mitophagy that would facilitate dissection of the molecular steps involved in the removal of mitochondria from the network via this pathway are not presently available.

MITOCHONDRIAL FISSION

Mitochondrial fission occurs during mitochondrial biogenesis when intramitochondrial components are sorted and split into daughter mitochondria [71, 72] but also precedes the selective targeting of mitochondria for mitophagy or cellular apoptosis [73-76].

Dynamin-related protein 1 (Drp1), a member of the dynamin family of GTPases, is the major protein involved in the division of membranes through translocation from the cytosol to the outer mitochondrial membrane where constricting rings are formed [77]. While fission occurs regardless of mitochondrial membrane potential, it is upregulated following mitochondrial depolarization, oxidation or nitrosylation, and ETC inhibition which trigger posttranslational modifications, including phophorylation, S-nitrosylation, ubiquitylation, and sumoylation on Drp1[66, 78] which result in mitochondrial fragmentation [79, 80].

Drp1 lacks a pleckstrin-homology domain and requires membrane receptor proteins such as Fis1 to facilitate its association and polymerization at membranes. It can cause Bax oligomerization independent of its GTPase activity [81] while the anti-apoptotic Bcl-XL promotes mitochondrial fission in neurons through interactions with Drp1 that promote its GTPase activity [82].

In AD Aβ overproduction is associated with increased number of fragmented mitochondria, increased oxidative stress and loss of $\Delta\psi_m$ and ATP production that is associated with increased expression of Drp1 [83]. Accumulated Aβ enhances Drp1 activity in neurons by increasing Drp1 S-nitrosylation at Cys644. Likewise, AD patients are characterized by having Aβ-Drp1 mediated mitochondrial fragmentation, mtDNA mutations [84] and decrease in oxphos [85].

Mdivi-1, a small molecule noncompetitive inhibitor of Drp1 GTPase activity that attenuates Drp1 mediated mitochondrial-fission in response to pro-apoptotic stimuli [86], has been developed and may potentially have therapeutic utility. Mdivi-1 application *in vivo* has been shown to for example protect cardiomyocytes against ischemia/reperfusion injury and attenuate retinal ganglion cell death after ischemic injury [87-89]. Mdivi-1 also partially rescues the mitochondrial damage due to inactivation of PINK1 [90]. Further research targeting therapeutics aimed at preserving mitochondrial function for the treatment of disease and injury may lead to improved clinical outlook for neurodegenerative diseases such as AD.

MITOCHONDRIAL MEMBRANE POTENTIAL

The mitochondrial membrane potential ($\Delta\psi_m$) is created when protons are pumped from the mitochondrial matrix to the intermembrane space as electrons pass through the ETC and as a prerequisite for oxidative phosphorylation. However, the higher (more polarized) $\Delta\psi_m$, the more mtROS is generated presumably due to the slowed electron transport [91] [92] [93]. Indeed, ROS generation is decreased when $\Delta\psi_m$ is dissipated by either expressing mitochondrial uncoupling proteins (UCPs) [94] or using chemical uncouplers (Reynolds and Hastings, 1995), such as carbonyl cyanide p-(tri-fluromethoxy)phenyl-hydrazone (FCCP) [95] [96]. Small decreases in membrane potential (mild uncoupling) can reduce ROS formation by limiting the life span of reduced electron transport chain (ETC) intermediates capable of generating ROS, in addition to decreasing local oxygen tensions [97-99] [100]] without seriously compromising cellular energetics [98, 100]. However, several AD animal models, and AD patient brains show evidence of reduced ATP levels, declined complex IV activity, enhanced oxidative stress compared to controls [101] [102, 103] and decreased $\Delta\psi_m$ has been shown in AD animal models and in human cortical neurons ex vivo [102, 104, 105]. A redox-optimized ROS balance hypothesis, which states that physiological ROS signaling occurs within an optimized mitochondrial membrane potential, and oxidative stress can happen at either the extreme of high $\Delta\psi_m$ or low $\Delta\psi_m$ ([106] has been proposed to reconcile this obvious discrepancy. It is based on the fact that the redox couples involved in substrate oxidation (NADH) are closely linked to the redox couples involved in antioxidant defenses (NADPH). It is therefore vital to balance an adequate level of $\Delta\psi_m$ to maintain matrix NADPH rather than $NADP^+$, which is necessary for mitochondrial antioxidant enzyme systems. This means that an increase in mitochondrial uncoupling of the ETC can increase ROS production primarily because the antioxidant system of the cell is compromised. It has now been shown that ROS can stimulate mitochondrial uncoupling [107, 108] and that the processes of uncoupling and ROS generation exist in a feedback loop [109] [108, 110].

Fatty acid (FA) cycling across the inner mitochondrial membrane is an important endogenous mild uncoupling pathway that prevents ROS release [111, 112]. In the proton-rich intermembrane space, FA anions are protonated, become uncharged and flip-flop across the inner membrane lipid bilayer. Once in the mitochondrial matrix, the proton is released and the FA anion transported back to the intermembrane space by anion carriers, which include mitochondrial uncoupling proteins [113] [114], and the adenine nucleotide translocator [112] [115, 116]. Post-ischemic tissue survival in the brain has been shown to closely correlate with uncoupling proteins expression [117].

MITOHORMESIS

While ROS can generate detrimental oxidative damage, they also play a crucial role in numerous signaling and stress responses [118, 119]. Mild oxidative stress may in fact promote longevity and metabolic health through the concept of mitochondrial hormesis (mitohormesis).

Mitohormesis occurs when low levels of oxidative stress induced by either caloric restriction, exercise [120], or other stimuli trigger an adaptive response that improves overall stress resistance. This is likely via increased endogenous antioxidant defense, which eventually reduces chronic oxidative damage [121] and extends lifespan. Inhibition of glycolysis, impairment of insulin-like signaling and certain mutations in mitochondrial ETC components, are also conditions that may promote longevity via ROS-dependent mitohormesis [122]. Glucose restriction induces mitochondrial respiration and increases oxidative stress and extends *C. elegans* lifespan via the AMPK- pathway in a manner that is sensitive to the antioxidant N-acetyl cysteine, suggesting that oxidative stress is required for lifespan extension by dietary restriction [123]. Also treatment of *C. elegans* with low doses of the superoxide generating compound paraquat extend their lifespan [124]. Mild inhibition of mitochondrial respiration extends the lifespan of organisms as diverse as yeast, worms, flies and mice [125] [126] [127] presumably through ROS stimulated HIF-1 activation of gene expression that promote longevity [128]. Mild mitochondrial insults may also communicate a stress response to induce the expression of mitochondrial chaperones such as HSP-6 and HSP-60. This mitochondrial unfolded protein response (UPR^{mt}) is thought to extend the life span of C. elegans by inhibiting the ETC [129, 130]. However, deletion of atfs-1, encoding for a transcription factor required for the induction of the UPR^{mt}, does not avoid lifespan extension after inhibition of the ETC [129, 130], and constitutive activation of the UPR^{mt} by gain of function mutations in atfs-1 does not extend lifespan [131]. These along with complementary evidence obtained from diverse model organisms, has led to the mitohormesis model [129].

Although the evidence of mitohormesis in lifespan regulation in mammalian models is still lacking its translational implications should be considered as an ideal antioxidant therapy that prevents oxidative damage induced under pathological conditions without interfering with ROS needed for hormesis and cellular signaling.

CALORIC RESTRICTION

Caloric restriction (CR) involves consuming 20-40% lower calories than normal has been suggested as a promising intervention to increase both median and maximal lifespan in humans (Peterson et al, 2012). It can prevent or delay several diseases including cancer, cardiovascular diseases, neurodegenerative disorders, diabetes and autoimmune diseases [132] and has been reported to protect against age-related mitochondrial dysfunction [133] and reduce mtDNA damage [134]. In animal models of neurodegenerative diseases it promotes neurogenesis and enhances synaptic plasticity [135], improves cognitive capability, anti-inflammatory mechanisms, reduce neural oxidative stress, induce various stress and neurotrophic/neuroprotective factors and prevents Aβ neuropathology in AD transgenic models [136]. At the cellular level, CR alters $\Delta\psi_m$ and respiratory activity, which results in lower ROS generation and oxidative damage. CR also increases mitochondrial biogenesis and bioenergetic efficiency through Akt, which directly phosphorylates and activates endothelial nitric oxide synthase (eNOS) leading to nitric oxide (NO) production [137, 138] [139] [140, 141] [142]. NO activates a NO/cGMP-dependent signaling pathway that induces PGC-1α, increasing mitochondrial biogenesis [143, 144]. This increase in mitochondrial biogenesis elicits the beneficial effects of CR [138, 139, 142, 143]. Mice on 3 months CR have higher levels of mitochondrial DNA, PGC-1α, NRF-1, Tfam, expression of cytochrome c oxidase, and cytochrome c when compared with mice fed ad libitum, indicating increased mitochondrial biogenesis [139]. 2,4-dinitrophenol, a CR mimetic induces similar results [142]. Long term exposure to elevated ROS levels impairs eNOS activity [145, 146]. As a result eNOS functions in a negative feedback loop preventing the generation of excessive ROS. In a CR trial CALERIE based on 25% CR, CR patients were shown to have less mtDNA damage, more mtDNA content, and increased expression of some antioxidant enzymes, suggesting that CR improves mitochondrial function and delays mitochondrial aging through reducing oxidative stress. The increase in expression of several proteins involved in mitochondrial biogenesis such as PGC-1 , Tfam, and SIRT1 was reported in CR patients compared to controls [147]. CR also attenuates the age-related decline of autophagy by activating Sirt1 which deacetylates autophagy proteins [148-150] and this is associated with increased longevity as inefficient mitochondria are replaced with new functional mitochondria [151, 152]. The net result is that CR reduces oxidative stress and enhances mitochondrial biogenesis in order to produce mitochondria that are more efficient in ATP production, have optimal oxidative capacity, and generate less ROS.

EXERCISE

Exercise training alone or in combination with CR may also represent an efficient strategy to delay mitochondrial aging and age-related dysfunction in humans through mechanisms stimulating mitochondrial biogenesis and oxidative capacity and improving protein quality control [153]. Skeletal muscle biopsies of humans performing high- intensity interval training showed an increase in Sirt1, nuclear PGC-1α and Tfam, which lead to an increase in skeletal muscle mitochondria and improved exercise performance [154-156]. Biopsies performed in older men showed that even with aging, exercise increases mitochondrial DNA and mitochondrial respiratory chain activity which is likely related to increases in mitochondria biogenesis [157,158]. While exercise training optimized mitochondrial function in elderly individuals [159] [160] when combined with a low carbohydrate (glycogen) diet increases the expression of PGC-1 to optimize the oxidative capacity of human skeletal muscle [161]. In the CALERIE trial, CR with exercise training resulted in a 38% reduction in the estimated risk of cardiovascular disease, an important age-associated pathology, compared to controls [162]. Indeed, increased physical activity or even simply adopting active style habits may clearly reduce the rate of mitochondrial decline and attenuate the age-related phenotype. This exercise-induced increase in mitochondrial biogenesis is mediated through ROS as demonstrated by oral administration of antioxidants to rats impairs the exercise-induced increase in mRNA and protein levels of PGC-1α, NRF-1 and Tfam and cytochrome c [163]. Similar observations are made in humans. The exercise induced increase in PGC-1α and PGC-1β ameliorate insulin resistance and initiate an adaptive response promoting endogenous antioxidant defense capacity. However, when the subjects diet was supplemented with antioxidants these effects were not observed [120]. Importantly, it was found that exercise training also increases brain mitochondrial biogenesis (mtDNA, and PGC-1α, SIRT1, and citrate synthase) and this may have important implications, not only with regard to fatigue, but also with respect to various central nervous system diseases and age-related dementia that are often characterized by mitochondrial dysfunction [153]. Therefore, exercise could be considered as a therapeutic option to reduce the negative effects of aging and decrease the risk of AD.

MITOCHONDRIAL STRESS RESPONSE SIGNALING

Mitochondria are the major source of cellular ROS and hence stress signaling that induces cellular senescence and apoptosis [118] [164] [165, 166] [167]. One of the major consequences of increased ROS and altered cellular redox state is the oxidation of thiol groups in cysteine residues in relevant proteins [118]. FoxO transcription factors are activated in response to elevated ROS levels to induce anti-oxidant responses (increased expression of catalase and SOD2), cell cycle arrest and/or cell death [168, 169]. Kinases that modulate cellular stress responses include mitochondrial Akt, GSK-3β, PKA, Abl, PKC, Src and Atm [170] [171-177]. Akt phosphorylates and inactivates GSK-3β, which can localize to the mitochondria. Mitochondrial GSK-3β phosphorylates MCL-1 and VDAC [174, 178] Sheldon et al., 2011) leading to MCL-1 degradation and inducing apoptosis [178]; while phosphorylation of VDAC by GSK-3β results in increased mitochondrial membrane permeability, again predisposing to apoptosis [174, 179]. GSK-3β is also known to phosphorylate and promote the proteasomal degradation of c-Myc, cyclin D1, and β-catenin [180,181] [182]. PKA can be translocated to mitochondria by hypoxia and other physiological stresses [183, 184] where it associates with the mitochondria through Rab32 and other A-kinase AKAPs [171] and phosphorylates VDAC [172], Drp1 [183], and other mitochondrial proteins. For example, hypoxia destabilizes AKAP121 through induction of SIAH2, a mitochondrial ubiquitin ligase, thereby limiting oxidative capacity under conditions of low oxygen. Interestingly, AKAP121 also appears to promote mitochondrial localization of Src-tyrosine kinase [185] where Src appears to regulate CO activity and respiratory activity [185] [186], and other mitochondrial substrates for Src family kinases are likely [187]. Increased ROS induces protein kinase C-delta (PKCδ) association with the mitochondria and this in turn recruits other signaling molecules, including the Abl tyrosine kinase that is associated with loss of membrane potential and non-apoptotic cell death [175]. Impaired oxidative metabolism and decreased ATP levels in neurons activate AMPK [188]. AMPK can also be activated by drugs such as metformin that inhibits complex I or resveratrol that inhibits the F0F1 ATPase [170]. AMPK modulates mitochondrial metabolism and targets Acetyl CoA carboxylase-2 (ACC2) to the OMM where it regulates lipid metabolism by controlling production of malonyl CoA [170]. AMPK therefore plays a key role in mitochondrial homeostasis by ensuring that only functionally viable mitochondria are retained. Upon its activation it induces not only mitochondrial biogenesis through activation of PGC-1α [189, 190] but also initiates mitophagy through ULK1 activation and mTOR inhibition [183, 191]. ATM kinase inhibition has been shown to cause CNS neurodegeneration in animal models [192]. ATM kinase, which is partly located at the mitochondria, is activated upon mitochondrial uncoupling [193] and while its mitochondrial substrates are not known, loss of ATM in genetically engineered mouse models leads to mitochondrial dysfunction.

IMMUNOAGING AND MITOCHONDRIAL FUNCTION

Immune function is compromised during the course of aging as well as in AD. Investigations of changes in adaptive immune function associated with aging indicate disturbances of T- and B-cell homeostasis and activation as well as that of marcophages [194, 195]. Large scale (BLSA and MESA) transcriptomic studies of CD4+ T-cells and CD14+ monocytes from aged individuals show mitochondrial pathways, particularly OxPhos, as the most down-regulated (FDR <0.001) [196, 197]. The loss in OxPhos expression is significant as mitochondrial oxidative metabolism plays a critical regulatory role in immune function. Mitochondria serve as the scaffold for NLRP3 inflammasome formation, where mitochondrial ROS and oxidative metabolism regulate caspase-1 activation, the critical step in maturation of Il-1beta and Il-18. Mitochondrial oxidative metabolism regulates macrophage polarization, T-cell activation, differentiation and memory cell formation (for review see Weinberg et al., 2015 [198]). Thus, mitochondria not only sustain immune cell phenotypes but also are necessary for establishing immune cell phenotype and function. In a pro-inflammatory state this is accomplished by mitochondria shifting from producing ATP via oxidative metabolism to producing building blocks for macromolecule synthesis via anapleurosis and glutaminolysis. The shift from catabolism to anabolism is critical to affect cell expansion, production of inflammatory mediators and immune cell fate commitments. This may explain why the increase in serum pro-inflammatory cytokines occurs with age, giving rise to a chronic state of inflammation, termed inflammaging [199-201].

In AD, immune dysfunction has been identified in T- and B-cells, macrophages and microglia [202]. AD is associated with increased T cell infiltration, changes in immune populations associated with disease progression, reduction in T- and B-cell numbers and reductions in CD4+CD25+ Tregs [203]. CD8+CD28- suppressor cells are also decreased in PBMCs from AD patients. These data suggest that the immunosuppressive capabilities in AD patients are diminished and could represent a deficit in the ability to control Teff responses. As such, increased activities of Th17, levels of IL-21, IL-6, and IL-23, and the Th17-associated transcription factor RORγ, were increased among lymphocytes in AD patients [204]. This suggests AD specific overactivity of Th17 T-cell function and underactivity of Teff function. Given that Th17 T-cells primarily mobilize glycolysis and suppress OxPhos whereas Tregs and memory T cells oxidize fatty acids via mitochondrial oxidation, supports the concept that mitochondrial dysfunction fuels AD immune dysfunction [205].

CONCLUSION

Mitochondrial function is deregulated in AD and there is growing interest in understanding how altered mitochondrial function may be targeted to inhibit neurodegeneration. Proper modulation of mitochondrial turnover overall to eliminate dysfunctional mitochondria while maintaining efficient functional mitochondrial mass in response to stresses, including hypoxia and nutrient starvation may be relevant in delaying or managing the degenerative process in aging and AD. By preventing the generation of excessive ROS and conserving valuable nutrients neuronal survival may be promoted under conditions of energetic stress (Figure 1).

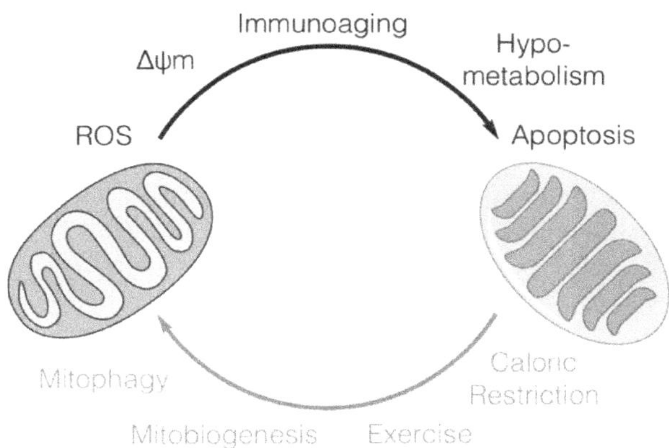

Figure 1. Factors regulating mitochondrial function in AD. In AD, neuronal injury, inflammation and aging may impair mitochondrial function by inducing fission, increasing Δψm and ROS production leading to decreased ATP production. Mitochondrial function may be improved by enhancing mitochondrial biogenesis through caloric restriction and exercise. Damaged and dysfunctional mitochondria can be selectively eliminated by mitophagy.

REFERENCES

1. Swerdlow RH,Burns JM,Khan SM (2014). The Alzheimer's disease mitochondrial cascade hypothesis: progress and perspectives. Biochim Biophys Acta, 1842:1219-1231 . .
2. Fukui H,Moraes CT (2008). The mitochondrial impairment, oxidative stress and neurodegeneration connection: reality or just an attractive hypothesis? Trends Neurosci, 31:251-256 ..
3. Picone P,Nuzzo D,Caruana L,Scafidi V,Di Carlo M

(2014). Mitochondrial Dysfunction: Different Routes to Alzheimer's Disease Therapy. Oxidative Medicine and Cellular Longevity. Oxid Med Cell Longev, 2014:780179. . .

4. Scarpulla RC (2008). Transcriptional paradigms in mammalian mitochondrial biogenesis and function. Physiol Rev, 88: 611-638. .
5. Mosconi L,Brys M,Switalski R,Mistur R,Glodzik L,Pirraglia E, et al. (2007). Maternal family history of Alzheimer's disease predisposes to reduced brain glucose metabolism. Proc Natl Acad Sci U S A, 104: 19067-19072. . .
6. Murray J,Tsui WH,Li Y,McHugh P,Williams S,Cummings M, et al. (2014). FDG and Amyloid PET in Cognitively Normal Individuals at Risk for Late-Onset Alzheimer's Disease. Adv J Mol Imaging, 4: 15-26.. .
7. Hirai K,Aliev G,Nunomura A,Fujioka H,Russell RL,Atwood CS, et al. (2001). Mitochondrial abnormalities in Alzheimer›s disease. J Neurosci, 21: 3017-3023 .
8. Moreira PI,Siedlak SL,Wang X,Santos MS,Oliveira CR,Tabaton M, et al. (2007). Increased autophagic degradation of mitochondria in Alzheimer disease. Autophagy, 3: 614-615. .
9. Parker WDJ,Parks JK (1995). Cytochrome c oxidase in Alzheimer›s disease brain: purification and characterization. Neurology, 45: 482-486 .
10. Gibson GE,Sheu KF,Blass JP (1998). Abnormalities of mitochondrial enzymes in Alzheimer disease. J NeuralTransm, 105: 855-870. .
11. Maurer I,Zierz S,Moller HJ (2000). A selective defect of cytochrome c oxidase is present in brain of Alzheimer disease patients. Neurobiol Aging, 21: 455-462 .
12. Wang X,Su B,Siedlak SL,Moreira PI,Fujioka H,Wang Y, et al. (2008). Amyloid-beta overproduction causes abnormal mitochondrial dynamics via differential modulation of mitochondrial fission/fusion proteins.Proc Natl Acad Sci USA, 105: 19318-19323. . .
13. Calkins MJ,Manczak M,Mao P,Shirendeb U,Reddy PH (2011). Impaired mitochondrial biogenesis, defective axonal transport of mitochondria, abnormal mitochondrial dynamics and synaptic degeneration in a mouse model of Alzheimer's disease. Hum Mol Genet, 20: 4515-4529. . .
14. Manczak M,Calkins MJ,Reddy PH (2011). Impaired mitochondrial dynamics and abnormal interaction of amyloid beta with mitochondrial protein Drp1 in neurons from patients with Alzheimer's disease: implications for neuronal damage. Hum Mol Genet, 20: 2495-2509. . .
15. Keller JN,Pang Z,Geddes JW,Begley JG,Germeyer A,Waeg G, et al. (1997). Impairment of glucose and glutamate transport and induction of mitochondrial oxidative stress and dysfunction in synaptosomes by amyloid beta-peptide: role of the lipid peroxidation product 4-hydroxynonenal. J Neurochem, 69: 273-284.
16. Abramov AY,Canevari L,Duchen MR (2004). Beta-amyloid peptides induce mitochondrial dysfunction and oxidative stress in astrocytes and death of neurons through activation of NADPH oxidase. J Neurosci,24: 565-575 .
17. Manczak M,Anekonda TS,Henson E,Park BS,Quinn J,Reddy PH (2006). Mitochondria are a direct site of A beta accumulation in Alzheimer's disease neurons: implications for free radical generation and oxidative damage in disease progression. Hum Mol Genet, 15: 1437-1449 .
18. Mattson MP,Partin J,Begley JG (1998). Amyloid beta-peptide induces apoptosis-related events in synapses and dendrites. Brain Res, 807: 167-176. .
19. Lustbader JW,Cirilli M,Lin C,Xu HW,Takuma K,Wang N, et al. (2004). ABAD directly links Abeta to mitochondrial toxicity in Alzheimer's disease. Science, 304: 448-452 .
20. Reddy PH (2008). Mitochondrial medicine for aging and neurodegenerative diseases. Neuromolecular Med, 10: 291-315 . .
21. Calkins MJ,Manczak M,Reddy PH (2012). Mitochondria-targeted antiox- idant SS31 prevents amyloid beta-induced mitochondrial abnormalities and synaptic degeneration in Alzheimer's disease. Pharmaceuticals,5:1103-1119. . .
22. Eckert GP,Renner K,Eckert SH,Eckmann J,Hagl S,Abdel-Kader RM, et al. (2012). Mitochondrial dysfunction a pharmacological target in Alzheimer's disease. Mol Neurobiol, 46: 136-150 .
23. Reddy PH,Tripathi R,Troung Q,Tirumala K,Reddy TP,Anekonda V, et al. (2012). Abnormal mitochondrial dynamics and synaptic degeneration as early events in Alzheimer's disease: implications to mitochondria-targeted antioxidant therapeutics. Biochim Biophys Acta, 1822: 639-649 ..
24. Ryan MT,Hoogenraad NJ (2007). Mitochondrial-nuclear communications. Annu Rev Biochem, 76: 701-722. .
25. Liu X,Hajnóczky G (2011). Altered fusion dynamics underlie unique morphological changes in mitochondria during hypoxia-reoxygenation stress. Cell Death Differ, 18: 1561-1572 ..
26. Chan DC (2007). Mitochondrial dynamics in disease. N Engl J Med, 356: 1707-1709 .
27. Medeiros DM (2008). Assessing mitochondria biogenesis. Methods, 46: 288-294 .
28. Onyango IG,Lu J,Rodova M,Lezi E,Crafter AB,Swerdlow RH (2010). Regulation of neuron mitochondrial biogenesis and relevance to brain health. Biochim Biophys Acta, 1802: 228-234. .
29. Yin W,Signore AP,Iwai M,Cao G,Gao Y,Chen J (2008). Rapidly increased neuronal mitochondrial biogenesis after hypoxic-ischemic brain injury. Stroke, 39:3057-3063 . .
30. Medeiros DM (2008). Assessing mitochondria biogenesis. Methods, 46: 288-294 .
31. Kelly DP,Scarpulla RC (2004). Transcriptional regulatory circuits controlling mitochondrial biogenesis and function. Genes Dev, 18: 357-368 .
32. Canto C,Auwerx J (2011). NAD+ as a signaling molecule modulating metabolism. Cold Spring Harb Symp Quant Biol, 76: 291-298. . .

33. Gleyzer N,Vercauteren K,Scarpulla RC (2005). Control of Mitochondrial Transcription Specificity Factors (TFB1M and TFB2M) by Nuclear Respiratory Factors (NRF-1 and NRF-2) and PGC-1 Family Coactivators. Molecular and Cellular Biology, 25: 1354-1366 . .

34. Bruni F,Polosa PL,Gadaleta MN,Cantatore P,Roberti M (2010). Nuclear Respiratory Factor 2 Induces the Expression of Many but Not All Human Proteins Acting in Mitochondrial DNA Transcription and Replication. J Bio Chem, 285: 3939-3948. . .

35. Twig G,Elorza A,Molina AJ,Mohamed H,Wikstrom JD,Walzer G, et al. (2008). Fission and selective fusion govern mitochondrial segregation and elimination by autophagy. Embo J, 27: 433-446. .

36. Youle RJ,Narendra DP (2011). Mechanisms of mitophagy. Nat Rev Mol Biol, 12: 9-14. .

37. Qin W,Haroutunian V,Katsel P,Cardozo CP,Ho L,Buxbaum JD, et al. (2009). PGC-1α expression decreases in the Alzheimer disease brain as a function of dementia. Arch Neurol, 66: 352-361.. .

38. Hirai K,Aliev G,Nunomura A,Fujioka H,Russell RL,Atwood CS, et al. (2001). Mitochondrial abnormalities in Alzheimer›s disease. J Neurosci, 21: 3017-3023 .

39. Sheng B,Wang X,Su B,Lee HG,Casadesus G,Perry G, et al. (2012). Impaired mitochondrial biogenesis contributes to mitochondrial dysfunction in Alzheimer›s disease. J Neurochem, 120: 419-429.. .

40. Dominy JE,Lee YS,Gerhart-Hines Z,Puigserver P (2010). Nutrient-dependent regulation of PGC-1a's acetylation state and metabolic function through the enzy- matic activities of Sirt1/GCN5. Biochim Biophys Acta, 1804: 1676-1683 . .

41. Rasbach KA,Schnellmann RG (2007). Signaling of mitochondrial biogenesis following oxidant injury. J Biol Chem, 282: 2355-2362. .

42. Rohas LM,St-Pierre J,Uldry M,Jager S,Handschin C,Spiegelman BM (2007). A fundamental system of cellular energy homeostasis regulated by PGC-1a. Proc Natl Acad Sci U S A, 104: 7933-7938.. .

43. Bell EL,Guarente L (2011). The SirT3 divining rod points to oxidative stress. Mol Cell, 42: 561-568. .

44. Shin JH,Ko HS,Kang HS,Lee YK,Lee YI,Pletinkova O (2011). PARIS (ZNF746) repression of PGC-1α contributes to neurodegeneration in Parkinson›s disease. Cell, 144: 689-702 . .

45. Kim I,Rodriguez-Enriquez S,Lemasters JJ (2007). Selective degradation of mitochondria by mitophagy. Arch Biochem. Biophys, 462: 245-253 . .

46. Scherz-Shouval R,Elazar Z (2011). Regulation of autophagy by ROS: physiology and pathology.Trends Biochem Sci, 36: 30-38 .

47. Narendra DP,Youle RJ (2011). Targeting mitochondrial dysfunction: role for PINK1 and Parkin in mitochondrial quality control. Antioxid. Redox Signal, 14: 1929-1938 . .

48. Narendra D,Kane LA,Hauser DN,Fearnley IM,Youle RJ (2010). p62/SQSTM1 is required for Parkin-induced mitochondrial clustering but not mitophagy; VDAC1 is dispensable for both. Autophagy, 6: 1090-1106 . .

49. Jin SM,Lazarou M,Wang C,Kane LA,Narendra DP,Youle RJ (2010). Mitochondrial membrane potential regulates PINK1 import and proteolytic destabilization by PARL. J Cell Biol, 191: 933-942. .

50. Valente EM,Abou-Sleiman PM,Caputo V,Muqit MM,Harvey K,Gispert S, et al. (2004). Hereditary early- onset Parkinson's disease caused by mutations in PINK1. Science, 304: 1158-1160 .

51. Geisler S,Holmstrom KM,Skujat D,Fiesel FC,Rothfuss OC,Kahle PJ, et al. (2010). PINK1/Parkin-mediated mitophagy is dependent on VDAC1 and p62/SQSTM1. Nat Cell Biol, 12: 119-131 .

52. Pankiv S,Clausen TH,Lamark T,Brech A,Bruun JA,Outzen H, et al. (2007). p62/SQSTM1 binds directly to Atg8/LC3 to facilitate degradation of ubiquitinated protein aggregates by autophagy. J Biol Chem,282: 24131-24145 .

53. Liu L,Feng D,Chen G,Chen M,Zheng Q,Song P, et al. (2012). Mitochondrial outer-membrane protein FUNDC1 mediates hypoxia-induced mitophagy in mammalian cells. Nat Cell Biol, 14: 177-185 .

54. Novak I,Kirkin V,McEwan DG,Zhang J,Wild P,Rozenknop A, et al. (2010). Nix is a selective autophagy receptor for mitochondrial clearance. EMBO Rep, 11: 45-51 . .

55. Ding WX,Yin XM (2012). Mitophagy, mechanisms, pathophysiological roles, and analysis. Biol Chem,393: 547-564 . .

56. Fu M,St-Pierre P,Shankar J,Wang PT,Joshi B,Nabi IR (2013). Regulation of mitophagy by the Gp78 E3 ubiquitin ligase. Mol Biol Cell, 24: 1153-1162. . .

57. Lokireddy S,Wijesoma IW,Teng S,Bonala S,Gluckman PD,McFarlane C, et al. (2012). The ubiquitin ligase Mul1 induces mitophagy in skeletal muscle in response to muscle-wasting stimuli. Cell Metab, 16: 613-624 .

58. Ishii T,Itoh K,Takahashi S,Sato H,Yanagawa T,Katoh Y, et al. (2000). Transcription factor Nrf2 coordinately regulates a group of oxidative stress-inducible genes in macrophages. J Biol Chem, 275: 16023-16029 .

59. Jain A,Lamark T,Sjøttem E,Larsen KB,Awuh JA,Øvervatn A, et al. (2010). p62/SQSTM1 is a target gene for transcription factor NRF2 and creates a positive feedback loop by inducing antioxidant response element-driven gene transcription. J Biol Chem, 285: 22576-22591 . .

60. Cheng X,Siow RC,Mann GE (2011). Impaired redox signaling and antioxidant gene expression in endothelial cells in diabetes: a role for mitochondria and the nuclear factor-E2-related factor 2-Kelch-like ECH-associated protein 1 defense pathway. Antioxid Redox Signal, 14: 469-487 .

61. Hayes JD,McMahon M,Chowdhry S,Dinkova-Kostova AT (2010). Cancer chemoprevention mechanisms mediated through the Keap1-Nrf2 pathway. Antioxid Redox Signal, 13: 1713-1748. .

62. Kensler TW,Wakabayashi N,Biswal S (2007). Cell

survival responses to environmental stresses via the Keap1-Nrf2-ARE pathway. Annu Rev Pharmacol. Toxicol, 47: 89-116 .

63. East DA,Fagiani F,Crosby J,Georgakopoulos ND,Bertrand H,M S, et al. (2014). PMI: A ΔΨm Independent Pharmacological Regulator of Mitophagy. Chemistry & Biology, 21: 1585-1596. .

64. Vives-Bauza C,Zhou C,Huang Y,Cui M,de Vries RL,Kim J (2010). PINK1- dependent recruitment of Parkin to mitochondria in mitophagy. Proc Natl Acad Sci U S A, 107: 378-383 . .

65. Wang X,Winter D,Ashrafi G,Schlehe J,Wong YL,Selkoe D (2011). PINK1 and Parkin target Miro for phosphorylation and degradation to arrest mitochondrial motility. Cell, 147: 893-906. ..

66. Cho DH,Nakamura T,Fang J,Cieplak P,Godzik A,Gu Z, et al. (2009). S-nitrosylation of Drp1 mediates beta-amyloid-related mitochondrial fission and neuronal injury. Science, 324: 102-105 ..

67. Lee JY,Koga H,Kawaguchi Y,Tang W,Wong E,Gao YS (2010). HDAC6 controls autophagosome maturation essential for ubiquitin-selective quality-control autophagy. EMBO J, 29: 969-980. .

68. Lee JY,Nagano Y,Taylor JP,Lim KL,Yao TP (2010). Disease-causing mutations in Parkin impair mitochondrial ubiquitination, aggregation, and HDAC6-dependent mitophagy. J Cell Biol, 189: 671-679.. .

69. de Castro IP,Martins LM,Tufi R (2010). Mitochondrial quality control and neurological disease: an emerging connection. Expert Rev Mol Med, 12: e12. . .

70. Karbowski M,Neutzner A (2012). Neurodegeneration as a consequence of failed mitochondrial maintenance. Acta Neuropathol, 123: 157-171 .

71. Horn SR,Thomenius MJ,Johnson ES,Freel CD,Wu JQ,Coloff JL (2011). Regulation of mitochondrial morphology by APC/CCdh1-mediated control of Drp1 stability. Mol Biol Cell, 22: 1207-1216. .

72. Kashatus DF,Lim KH,Brady DC,Pershing NL,Cox AD,Counter CM (2011). RALA and RALBP1 regulate mitochondrial fission at mitosis. Nat Cell Biol, 13: 1108-1115 . .

73. Twig G,Hyde B,Shirihai OS (2008). Mitochondrial fusion, fission and autophagy as a quality control axis: the bioenergetic view. Biochim Biophys Acta, 1777: 1092-1097 . .

74. Martinou JC,Youle RJ (2011). Mitochondria in apoptosis:Bcl-2 family members and mitochondrial dynamics. Dev Cell, 21: 92-101 . .

75. Autret A,Martin SJ (2009). Emerging role for members of the Bl-2 family in mito- chondrial morphogenesis. Mol Cell, 36: 355-363 .

76. Youle RJ,van der Bliek AM (2012). Mitochondrial fission, fusion and stress. Science, 337: 1062-1065.. .

77. Palmer CS,Osellame LD,Laine D,Koutsopoulos OS,Frazier AE,Ryan MT (2011). MiD49 and MiD51, new components of the mitochondrial fission machinery. EMBO reports, 12: 565-573 ..

78. Bossy B,Petrilli A,Klinglmayr E,Chen J,Lütz-Meindl U,Knott AB, et al. (2010). S-Nitrosylation of DRP1 does not affect enzymatic activity and is not specific to Alzheimer's disease. Journal of Alzheimer's disease, 20: S513-526 . .

79. Nakamura T,Lipton SA (2011). Redox modulation by S-nitrosylation contributes to protein misfolding, mitochondrial dynamics, and neuronal synaptic damage in neurodegenerative diseases. Cell Death Differ, 18: 1478-1486. . .

80. Sauvanet C,Duvezin-Caubet S,di Rago JP,Rojo M (2010). Energetic requirements and bioenergetic modulation of mitochondrial morphology and dynamics. Semin Cell Dev Biol, 21: 558-565 .

81. Montessuit S,Somasekharan SP,Terrones O,Lucken-Ardjomande S,Herzig S,Schwarzenbacher R (2010). Membrane remodeling induced by the dynamin-related protein Drp1 stimulates Bax oligomerization. Cell, 142: 889-901 . .

82. Li H,Chen Y,Jones AF,Sanger RH,Collis LP,Flannery R (2008). Bcl-xL induces Drp1-dependent synapse formation in cultured hippocampal neurons. Proc Natl Acad Sci U S A, 105: 2169-2174. .

83. Wang H,Lim PJ,Karbowski M,Monteiro MJ (2008). Effects of overexpression of Huntingtin proteins on mitochondrial integrity. Hum Mol Genet, 18: 737-752 . .

84. Hamblet NS,Ragland B,Ali M,Conyers B,Castora FJ (2006). Mutations in mitochondrial-encoded cytochrome c oxidase subunits I, II, and III genes detected in Alzheimer›s disease using single-strand conformation polymorphism. Electrophoresis, 27: 398-408 .

85. Chandrasekaran K,Hatanpää K,Brady DR,Rapoport SI (1996). Evidence for physiological down-regulation of brain oxidative phosphorylation in Alzheimer›s disease. Exp Neurol, 142: 80-88 .

86. Cassidy-Stone A,Chipuk JE,Ingerman E,Song C,Yoo C,Kuwana T, et al. (2008). Chemical inhibition of the mitochondrial division dynamin reveals its role in Bax/Bak-dependent mitochondrial outer membrane permeabilization. Dev Cell, 14: 193-204. . .

87. Brooks C,Wei Q,Cho SG,Dong Z (2009). Regulation of mitochondrial dynamics in acute kidney injury in cell culture and rodent models. J Clin Invest, 119: 1275-1285. . .

88. Ong SB,Subrayan S,Lim SY,Yellon DM,Davidson SM,Hausenloy DJ (2010). Inhibiting mitochondrial fission protects the heart against ischemia/reperfusion injury. Circulation, 121: 2012-2022 .

89. Park SW,Kim KY,Lindsey JD,Dai Y,Heo H,Nguyen DH, et al. (2011). Selective Inhibitor of Drp1, Mdivi-1, Increases Retinal Ganglion Cell Survival in Acute Ischemic Mouse Retina. Invest Ophthalmol Vis Sci, 52: 2837-2843. . .

90. Cui M,Tang X,Christian WV,Yoon Y,Tieu K (2010). Perturbations in mitochondrial dynamics induced by human mutant PINK1 can be rescued by the mitochondrial division inhibitor mdivi-1. J Biol Chem, 285: 11740-11752 . .

91. Madamanchi NR,Runge MS (2007). Mitochondrial dysfunction in atherosclerosis. Circ Res, 100: 460-473. .

92. Ohashi S,Nishio A,Nakamura H,Kido M,Ueno S,Uza N, et al. (2006). Protective roles of redox-active protein thioredoxin-1 for severe acute pancreatitis. Am J Physiol Gastrointest Liver Physiol, 290: G772-781.

93. Valko M,Leibfritz D,Moncol J,Cronin MT,Mazur M,Telser J (2007). Free radicals and antioxidants in normal physiological functions and human disease Int J Biochem Cell Biol, 39: 44-84. .

94. Ramsden DB,Ho PW-L,Ho JW-M, et al. (2012). Human neuronal uncoupling proteins 4 and 5 (UCP4 and UCP5): structural properties, regulation, and physiological role in protection against oxidative stress and mitochondrial dysfunction. Brain and Behavior, 2: 468-478. . .

95. (2001). Δψm dependent and independent production of reactive oxygen species by rat brain mitochondria. J Neurochem, 79: 266-277 .

96. Brennan JP,Southworth R,Medina RA,Davidson SM,Duchen MR,Shattock MJ (2006). "Mitochondrial uncoupling, with low concentration FCCP, induces ROS-dependent cardioprotection independent of KATP channel activation." Cardiovas Res, 72: 313-321. .

97. Skulachev VP (1998). Uncoupling: new approaches to an old problem of bioenergetics. Biochim Biophys Acta, 1363: 100-124 .

98. Korshunov SS,Skulachev VP,Starkov AA (1997). High protonic potential actuates a mechanism of production of reactive oxygen species in mitochondria. FEBS Lett, 416: 15-18. .

99. Brookes PS (2005). Mitochondrial H(+) leak and ROS generation: an odd couple. Free Radic Biol Med,38: 12-23 .

100. Brand MD,Buckingham JA,Esteves T (2004). Mitochondrial superoxide and aging: uncoupling-protein activity and superoxide production. Biochem Soc Symp, 203-213 .

101. Hauptmann S,Scherping I,Drose S,Brandt U,Schulz KL,Jendrach M, et al. (2009). Mitochondrial dysfunction: an early event in Alzheimer pathology accumulates with age in AD transgenic mice. Neurobiol aging, 30: 1574-1586 .

102. Reddy PH (2009). Amyloid beta, mitochondrial structural and functional dynamics in Alzheimer's disease. Exp Neurol, 2, 218: 286-292 . .

103. Xie H,Guan J,Borrelli LA,Xu J,Serrano-Pozo A,Bacskai BJ (2013). Mitochondrial Alterations near Amyloid Plaques in an Alzheimer's Disease Mouse Model. J Neurosci, 33: 17042-17051 ..

104. Deshpande A,Mina E,Glabe C,Busciglio J (2006). Different conformations of amyloid beta induce neurotoxicity by distinct mechanisms in human cortical neurons. J Neurosci, 26: 6011-6018 .

105. Cabezas-Opazo FA,Vergara-Pulgar K,Pérez MJ,Jara C,Osorio-Fuentealba C,Quintanilla RA (2015). Mitochondrial Dysfunction Contributes to the Pathogenesis of Alzheimer›s Disease. Oxid Med Cell Longev, 2015:509654. . .

106. Aon MA,Cortassa S,O'Rourke B (2010). Redox-optimized ROS balance: A unifying hypothesis.Biochim Biophys Acta, 1797: 865-877 . .

107. Echtay KS,Roussel D,St-Pierre J,Jekabsons MB,Cadenas S,Stuart J, et al. (2002). Superoxide activates mitochondrial uncoupling proteins. Nature, 415: 96-99 .

108. Murphy MP,Echtay KS,Blaikie FH,Asin-Cayuela J,Cocheme HM,Gree K, et al. (2003). Superoxide activates uncoupling proteins by generating carbon-centered radicals and initiating lipid peroxidation. J Biol Chem, 278: 48534-48545 .

109. Brookes PS,Land JM,Clark JB,Heales SJ (1998). Peroxynitrite and brain mitochondria: evidence for increased proton leak. J Neurochem, 70: 2195-2202. .

110. Miwa S,Brand MD (2003). Mitochondrial matrix reactive oxygen species production is very sensitive to mild uncoupling. Biochem Soc Trans, 311: 1300-1301. .

111. Jezek P,Engstova H,Zackova M (1998). Fatty acid cycling mechanism and mitochondrial uncoupling proteins. Biochim Biophys Acta, 1365: 319-327 .

112. Skulachev VP (1991). Fatty acid circuit as a physiological mechanism of uncoupling of oxidative phosphorylation. FEBS Lett, 294: 158-162 .

113. Pecqueur C,Alves-Guerra MC,Gelly C (2001). Uncoupling protein 2, in vivo distribution, induction upon oxidative stress, and evidence for translational regulation. J Biol Chem, 276: 8705-8712 .

114. Ricquier D,Bouillaud F (2000). The uncoupling protein homologues: UCP1, UCP2, UCP3, StUCP and AtUCP. Biochem J, 345: 161-179 . .

115. Andreyev A,Bondareva TO,Dedukhova VI (1988). Carboxyatractylate inhibits the uncoupling effect of free fatty acids. FEBS Lett, 226: 265-269 .

116. Hoerter J,Gonzalez-Barroso MD,Couplan E (2004). Mitochondrial uncoupling protein 1 expressed in the heart of transgenic mice protects against ischemic-reperfusion damage. Circulation, 110: 528-533.

117. McLeod CJ,Aziz A,Hoyt RFJ,McCoy JPJ,Sack MN (2005). Uncoupling proteins 2 and 3 function in concert to augment tolerance to cardiac ischemia. J Biol Chem, 280: 33470-33476 .

118. Hamanaka RB,Chandel NS (2010). Mitochondrial reactive oxygen species regulate cellular signaling and dictate biological outcomes. Trends Biochem Sci, 35: 505-513 . .

119. Mammucari C,Rizzuto R (2010). Signaling pathways in mitochondrial dysfunction and aging. Mech Ageing Dev, 131: 536-543 . .

120. Ristow M,Zarse K,Oberbach A, et al. (2009). Antioxidants prevent health-promoting effects of physical exercise in humans. Proc Natl Acad Sci U S A, 106: 8665-8670 . .

121. Ristow M,Zarse K (2010). How increased oxidative stress promotes longevity and metabolic health: the concept of mitochondrial hormesis (mitohormesis). Exp Gerontol, 45: 410-418 .

122. Kenyon CJ (2010). The genetics of ageing. Nature, 464: 504-512 .

123. Schulz TJ, et al. (2007). Glucose restriction extends Caenorhabditis elegans life span by inducing mitochondrial respiration and increasing oxidative stress. Cell Metab, 6: 280-293 .

124. Yun J,Finkel T (2014). Mitohormesis. Cell Metab, 19: 757-766 . .

125. Feng J,Bussière F,Hekimi S (2001). Mitochondrial electron transport is a key determinant of life span inCaenorhabditis elegans. Dev Cell, 1: 633-644 .

126. Copeland JM,Cho J,Lo TJ,Hur J,Bahadorani S,Arabyan T, et al. (2009). Extension of Drosophila life span by RNAi of the mitochondrial respiratory chain. Curr Biol, 19: 1591-1598 .

127. Dell'agnello C,Leo S,Agostino A,Szabadkai G,Tiveron C,Zulian A, et al. (2007). Increased longevity and refractoriness to Ca(2+)-dependent neurodegeneration inSurf1 knockout mice. Hum Mol Genet, 16: 431-444 .

128. Durieux J,Wolff S,Dillin A (2011). The cell-non-autonomous nature of electron transport chain-mediated longevity. Cell, 144: 79-91 . .

129. Bennett CF,Kaeberlein M (2014). The mitochondrial unfolded protein response and increased longevity: cause, consequence, or correlation? Exp Gerontol, 56: 42-46 . .

130. Bennett CF,Vander Wende H,Simko M,Klum S,Barfield S,Choi H, et al. (2014). Activation of the mitochondrial unfolded protein response does not predict longevity in Caenorhabditis elegans. Nat Commun,5: 3483. . .

131. Artal-Sanz M,Tavernarakis N (2009). Prohibitin and mitochondrial biology. Endocrinol Metab, 20: 394-401 .

132. Dirks AJ,Leeuwenburgh C (2006). Caloric restriction in humans: potential pitfalls and health concerns.Mech Ageing Dev, 127: 1-7 .

133. Sohal RS,Ku HH,Agarwal S,Forster MJ,Lal H (1994). Oxidative damage, mitochondrial oxidant generation and antioxidant defenses during aging and in response to food restriction in the mouse. Mech Ageing Dev, 74: 121-133 .

134. Melov S,Hinerfeld D,Esposito L,Wallace DC (1997). Multi-organ characterization of mitochondrial genomic rearrangements in ad libitum and caloric restricted mice show striking somatic mitochondrial DNA rearrangements with age. Nucleic Acids Research, 25: 974-982. . .

135. Mattson MP (2012). Energy Intake and Exercise as Determinants of Brain Health and Vulnerability to Injury and Disease. Cell metabolism, 16: 706-722. . .

136. Gillette-Guyonnet S,Secher M,Vellas B (2013). Nutrition and neurodegeneration: epidemiological evidence and challenges for future research. Br J Clin Pharmacol, 75: 738-755. . .

137. Fulton D,Gratton JP,McCabe TJ (1999). Regulation of endothelium-derived nitric oxide production by the protein kinase Akt. Nature, 399: 597-601 . .

138. Chen H,Montagnani M,Funahashi T,Shimomura I,Quon MJ (2003). Adiponectin stimulates production of nitric oxide in vascular endothelial cells. J Biol Chem, 278: 45021-45026. .

139. Nisoli E,Tonello C,Cardile A (2005). Calorie restriction promotes mitochondrial biogenesis by inducing the expression of eNOS. Science, 310: 314-317 .

140. Rashid G,Bernheim J,Green J,Benchetrit S (2007). Parathyroid hormone stimulates the endothelial nitric oxide synthase through protein kinase A and C pathways. Nephrol Dial Transplant., 22: 2831-2837..

141. Ritchie SA,Kohlhaas CF,Boyd AR,Yalla KC,Walsh K,Connell JM, et al. (2010). Insulin-stimulated phosphorylation of endothelial nitric oxide synthase at serine-615 contributes to nitric oxide synthesis.Biochem J, 426: 85-90 .

142. Cerqueiram FM,Laurindo FR,Kowaltowski AJ (2011). Mild mitochondrial uncoupling and calorie restriction increase fasting eNOS, akt and mitochondrial biogenesis. PLoS One, 6:e18433. ..

143. Nisoli E,Clementi E,Paolucci C (2003). Mitochondrial biogenesis in mammals: the role of endogenous nitric oxide. Science, 299: 896-899 .

144. Nisoli E,Falcone S,Tonello C (2004). Mitochondrial biogenesis by NO yields functionally active mitochondria in mammals. Proc Natl Acad Sci U S A, 101: 16507-16512 . .

145. Hu Z,Chen J,Wei Q,Xia Y (2008). Bidirectional actions of hydrogen peroxide on endothelial nitric- oxide synthase phosphorylation and function: co-commitment and interplay of Akt and AMPK. J Biol Chem,283: 25256-25263 . .

146. Zhou X,Bohlen HG,Unthank JL,SJ M (2009). Abnormal nitric oxide production in aged rat mesenteric arteries is mediated by NAD(P)H oxidase-derived peroxide. Am J Physiol Heart Circ Physiol, 297: H2227-2233. . .

147. Civitarese AE,Carling S,Heilbronn L (2007). Calorie restriction increases muscle mitochondrial biogenesis in healthy humans. PLoS Med, 4:e76. . .

148. Lee IH,Cao L,Mostoslavsky R (2008). A role for the NAD-dependent deacetylase Sirt1 in the regulation of autophagy. Proc Natl Acad Sci U S A, 105: 3374-3379. . .

149. Wohlgemuth SE,Julian D,Akin DE, et al. (2007). Autophagy in the heart and liver during normal aging and calorie restriction. Rejuvenation Res, 10: 281-292. .

150. Donati A,Cavallini G,Paradiso C (2001). Age-related changes in the autophagic proteolysis of rat isolated liver cells: effects of antiaging dietary restrictions. J Gerontol A Biol Sci Med Sci, 56: B375-383..

151. Bergamini E,Cavallini G,Donati A,Gori Z (2007). The role of autophagy in aging: its essential part in the anti-aging mechanism of caloric restriction. Ann N Y Acad Sci, 11145: 69-78 .

152. Simonsen A,Cumming RC,Brech A, et al. (2008). Promoting basal levels of autophagy in the nervous system enhances longevity and oxidant resistance in adult Drosophila. Autophagy, 4: 176-184. .

153. Steiner JL,Murphy EA,McClellan JL,Carmichael MD,Davis JM (2011). Exercise training increases mitochondrial biogenesis in the brain. J Appl

Physiol, 111: 1066-1071 .

154. Little JP,Safdar A,Wilkin GP,Tarnopolsky MA,Gibala MJ (2010). A practical model of low-volume high-intensity interval training induces mitochondrial biogenesis in human skeletal muscle: potential mechanisms. J Physiol, 588: 1011-1022 . .

155. Little JP,Safdar A,Cermak N,Tarnopolsky MA,Gibala MJ (2010). Acute endurance exercise increases the nuclear abundance of PGC-1alpha in trained human skeletal muscle. Am J Physiol Regul Integr Comp Physiol, 298: R912-917 .

156. Safdar A,Little JP,Stokl AJ, et al. (2011). Exercise Increases Mitochondrial PGC-1{alpha} Content and Promotes Nuclear-Mitochondrial Cross-talk to Coordinate Mitochondrial Biogenesis. J Biol Chem, 286: 10605-10617 . .

157. Menshikova EV,Ritov VB,Fairfull L (2006). Effects of exercise on mitochondrial content and function in aging human skeletal muscle. J Gerontol A Biol Sci Med Sci, 61: 534-540 . .

158. Vina J,Gomez-Cabrera MC,Borras C,Froio T,Sanchis-Gomar F,Martinez-Bello VE, et al. (2009).Mitochondrial biogenesis in exercise and in ageing. Adv Drug Deliv Rev, 61: 1369-1374 .

159. Short KR,Vittone JL,Bigelow ML,Proctor DN,Rizza RA,Coenen-Schimke JM, et al. (2003). Impact of aerobic exercise training on age-related changes in insulin sensitivity and muscle oxidative capacity.Diabetes, 52: 1888-1896 .

160. Ghosh S,Lertwattanarak R,Lefort N (2011). Reduction in reactive oxygen species production by mitochondria from elderly subjects with normal and impaired glucose tolerance. Diabetes, 60: 2051-2060.. .

161. Psilander N,Frank P,Flockhart M,Sahlin K (2013). Exercise with low glycogen increases PGC-1α gene expression in human skeletal muscle. European Journal of Applied Physiology, 113: 951-963 .

162. Lefevre M,Redman LM,Heilbronn LK,Smith JV,Martin CK,Rood JC, et al. (2009). Caloric restriction alone and with exercise improves CVD risk in healthy non-obese individuals. Atherosclerosis, 203: 206-213.. .

163. Gomez-Cabrera MC,Domenech E,M R (2008). Oral administration of vitamin C decreases muscle mitochondrial biogenesis and hampers training-induced adaptations in endurance performance. Am J Clin Nutr, 87: 142-149. .

164. Sena LA,Chandel NS (2012). Physiological roles of mitochondrial reactive oxygen species. Mol Cell,48: 158-167 . .

165. Havens CG,Ho H,Yoshioka N,Dowdy SF (2006). Regulation of late G1/S phase tran- sition and APCCdh1 by reactive oxygen. Mol Cell Biol, 26: 4701-4711 . .

166. Balaban RS,Nemoto S,Finkel T (2005). Mitochondria, oxidants and aging. Cell, 120: 483-495.

167. Tormos KV,Anso E,Hamanaka RB,Eisenbart J,Joseph J,Kalyanaraman B, et al. (2011). Mitochondrial complex III ROS regulate adipocyte differentiation. Cell Metab, 14: 537-544 . .

168. Fu Z,Tindall DJ (2008). FOXOs, cancer and regulation of apoptosis. Oncogene, 27: 2312-2319.. .

169. Kops GJ,Dansen TB,Polderman E,Saarloos I,Wirtz KWA, PJC (2002). Forkhead transcription factor FOXO3a protects quiescent cells from oxidative stress. Nature, 419: 316-321 .

170. Mihaylova MM,Shaw RJ (2011). The AMPK signaling pathway coordinates cell growth, autophagy and metabolism. Nat Rev Cell Biol, 13: 1016-1023 . .

171. Alto NM,Soderling J,Scott JD (2002). Rab32 is an A-kinase anchoring protein and participates in mitochondrial dynamics. J Cell Sci, 158: 659-6810 . .

172. Bera AK,Ghosh S,Das S (1995). Mitochondrial VDAC can be phosphorylated by cyclic AMP-dependent protein kinase. Biochem Biophys Res Commun, 209: 213-217 .

173. Das S,Wong R,Rajapakse N,Murphy E,Steenbergen C (2008). Glycogen synthase kinase 3 inhibition slows mitochondrial adenine nucleotide transport and regulates voltage-dependent anion channel phosphorylation. Circ Res, 103: 983-9110. . .

174. Juhaszova M,Zorov DB,Kim SH,Pepe S,Fu Q,Fishbein KW (2004). Glycogen synthase kinase-3beta mediates convergence of protection signaling to inhibit the mitochondrial permeability transition pore. J Clin Invest, 113: 1535-4910 . .

175. Kumar S,Bharti A,Mishra NC,Raina D,Kharbanda S,Saxena S (2001). Targeting of the c-Abl tyrosine kinase to mitochondria in the necrotic cell death response to oxidative stress. J Biol Chem, 276: 17281-17510.

176. Majumder PK,Mishra NC,Sun X,Bharti A,Kharbanda S,Saxena S (2001). Targeting of protein kinase C delta to mitochondria in the oxidative stress response. Cell Growth Differ, 12: 465-470 .

177. Robey RB,Hay N (2009). Is Akt the "Warburg kinase"? - Akt energy metabolism interactions and oncogenesis. Semin Cancer Biol, 19: 25-31 . .

178. Maurer U,Chavret C,Wagman AS,Dejardin E,Green DR (2006). Glycogen synthase kinase-3 regulates mitochondrial outer membrane permeabilization and apoptosis by destabilization of MCL-1. Mol Cell, 21: 749-6010 .

179. Martel C,Allouche M,Esposti DD,Fanelli E,Boursier C,Henry C (2013). Glycogen synthase kinase 3-mediated voltage-dependent anion channel phosphorylation controls outer mitochondrial membrane permeability during lipid accumulation. Hepatology, 57: 93-102 .

180. Diehl JA,Cheng M,Roussel MF,Sherr CJ (1998). Glycogen synthase kinase-3 beta regulates cyclin D1 proteolysis and subcellular localisation. Genes Dev, 12: 3499-3511 . .

181. Rubinfeld B,Albert I,Porfiri E,Fiol C,Munemitsu S,Polakis P (1996). Binding of GSK3B to the APC-b-catenin complex and regulation of complex assembly. Science, 272: 1023-1610. .

182. Yeh E,Cunningham M,Arnold H,Chasse D,Monteith T,Ivaldi G, et al. (2004). A signalling pathway controlling c-Myc degradation that impacts oncogenic

transformation of human cells. Nat Cell Biol, 6: 308-318 .

183. Kim J,Kundu M,Viollet B,Guan KL (2011). AMPK and mTOR regulate autophagy through direct phosphorylation of Ulk1. Nat Cell Biol, 13: 132-141 . .

184. Carlucci A,Adornetto A,Scorziello A,Viggiano D,Foca M,Cuomo O (2008). Proteolysis of AKAP121 regulates mitochondrial activity during cellular hypoxia and brain ischaemia. EMBO J, 27: 1073-8410.. .

185. Livigni A,Scorziello A,Agnese S,Adornetto A,Carlucci A,Garbi C (2006). Mitochondrial AKAP121 links cAMP and src signaling to oxidative metabolism. Mol Biol Cell, 17: 263-7110. ..

186. Miyazaki T,Neff L,Tanaka S,Horne WC,R B (2003). Regulation of cytochrome c oxidase activity by c-Src in osteoclasts. J Cell Biol, 160: 709-718 . .

187. Tibaldi E,Brunati AM,Massimino ML,Stringaro A,Colone M,Agostinelli E, et al. (2008). Src-Tyrosine kinases are major agents in mitochondrial tyrosine phosphorylation. J Cell Biochem, 104: 840-910 .

188. Hardie DG (2011). AMP-activated protein kinase - an energy sensor that regulates all aspects of cell function. Genes Dev, 25: 1895-1908 . .

189. Jäger S,Handschin C,St-Pierre J,BM S (2007). AMP-activated protein kinase (AMPK) action in skeletal muscle via direct phosphorylation of PGC- 1alpha. Proc Natl Acad Sci U S A,104: 12017-12022.. .

190. Zong H,Ren JM,Young LH,Pypaert M,Mu J,Birnbaum MJ, et al. (2002). AMP kinase is required for mitochondrial biogenesis in skeletal muscle in response to chronic energy deprivation. Proc Natl Acad Sci U S A, 99: 15983-15987 . .

191. Egan DF,Shackelford DB,Mihaylova MM,Gelino S,Kohnz RA,Mair W (2011). Phosphorylation of ULK1 (hATG1) by AMP-activated protein kinase connects energy sensing to mitophagy. Science, 331: 456-461 . .

192. Petersen AJ,Rimkus SA,Wassarman DA (2012). ATM kinase inhibition in glial cells activates the innate immune response and causes neurodegeneration in Drosophila. Proc Natl Acad Sci U S A, 109: E656-E664 . .

193. Valentin-Vega YA,MacLean KH,Tait-Mulder J,Milasta S,Steeves M,Dorsey FC, et al. (2012).Mitochondrial dysfunction in ataxia-telangiectasia Blood, 119: 1490-5001 . .

194. Fulop T,Le Page A,Fortin C,Witkowski JM,Dupuis G,Larbi A (2014). Cellular signaling in the aging immune system. Curr Opin Immunol, 29: 105-111. .

195. Fulop T,Witkowski JM,Pawelec G,Alan C,Larbi A (2014). On the immunological theory of aging.Interdiscip Top Gerontol, 39: 163-176. .

196. Bektas A,Zhang Y,Lehmann E,Wood WH,Becker KG,Madara K, et al. (2014). Age-associated changes in basal NF-κB function in human CD4+ T lymphocytes via dysregulation of PI3 kinase. Aging, 6: 957-969 . .

197. Reynolds LM,Taylor JR,Ding J,Lohman K,Johnson C,Siscovick D, et al. (2014). Age-related variations in the methylome associated with gene expression in human monocytes and T cells. Nat Commun,5:5366. . .

198. Weinberg SE,Chandel NS (2015). Targeting mitochondria metabolism for cancer therapy. Nat Chem Biol, 11: 9-15. . .

199. Mocchegiani E,Giacconi R,Cipriano C, et al. (2004). The variations during the circadian cycle of liver CD1d-unrestricted NK1.1+TCR gamma/delta+ cells lead to successful ageing. Role of metallothionein/ IL-6/gp130/PARP-1 interplay in very old mice. Exp Gerontol. 39:775-88. .

200. Franceschi C,Bonafè M,Valensin S,Olivieri F,De Luca M,Ottaviani E, et al. (2000). Inflamm-aging. An evolutionary perspective on immunosenescence. Ann N Y Acad Sci, 908: 244-254. .

201. Fulop T,Larbi A,Wikby A,Mocchegiani E,Hirokawa K,Pawelec G (2005). Dysregulation of T-cell function in the elderly:scientific basis and clinical implications. Drugs Aging, 22: 589-603 .

202. Heneka MT,Carson MJ,Khoury JE,Landreth GE,Brosseron F,Feinstein DL, et al. (2015). Neuroinflammation in Alzheimer›s disease. Lancet Neurol, 14: 388-405. .

203. Anderson KM,Olson KE,Estes KA,Flanagan K,Gendelman HE,Mosley RL (2014). Dual destructive and protective roles of adaptive immunity in neurodegenerative disorders. Transl Neurodegener, 3: 25.. .

204. Saresella M,Calabrese E,Marventano I,Piancone F,Gatti A,Alberoni M, et al. (2011). Increased activity of Th-17 and Th-9 lymphocytes and a skewing of the post-thymic differentiation pathway are seen in Alzheimer›s disease. Brain Behav Immun, 25: 539-547 .

205. Maciolek JA,Pasternak JA,Wilson HL (2014). Metabolism of activated T lymphocytes. Curr Opin Immunol, 27: 60-74 .

CHAPTER 7

THE COMBINATION OF ARTEMISIA PRINCEPS PAMP, LEONURUS JAPONICAS HOUTT, AND GARDENIA JASMINOIDES ELLIS FRUIT ATTENUATES THE EXACERBATION OF ENERGY, LIPID, AND GLUCOSE BY INCREASING HEPATIC PGC-1α EXPRESSION IN ESTROGEN-DEFICIENT RATS

Hye Jeong Yang, Min Jung Kim, Dae Young Kwon, Bo Reum Moon, A. Reum Kim, Suna Kang and Sunmin Park

Keywords: Estrogen deficiency, Glucose, Insulin, Lipid profiles, PGC-1α

BACKGROUND

Menopause is a transitional phase from a reproductive to a non-reproductive phase in a woman's life. Various symptoms are common during menopause and hot flashes, cognitive changes, anxiety and depression are included [1]. The symptoms are not deadly diseases but reduce the quality of a woman's life. In addition to menopausal symptoms, estrogen deficiency results in the decrease of energy, glucose, lipid and bone metabolism by reducing peroxisome proliferator-activated receptor-γ coactivator (PGC)-1α expression in various tissues [2]. This impairment may not influence daily life as much as menopausal symptoms but can eventually develop into metabolic diseases such as obesity, dyslipidemia, type 2 diabetes, osteoarthritis, and osteoporosis [3]. Therefore, the deterioration of metabolism needs to be prevented and/or delayed in post-menopausal women.

Menopause occurs due to the marked decrease of female hormones, especially estrogen. Hormone replacement therapy is effective for menopausal symptoms but is a limited option due to increased health risks for breast cancer, cardiovascular diseases, and dementia [4]. There is growing interest in alternative treatments for menopausal symptoms. Plant extracts such as lineseeds, red clover, St. John's wort, hop or black cohosh are most frequently used as phytochemical therapy [3, 5]. They contain some phytoestrogens, which have a similar chemical structure as estrogen and exhibit estrogenic or anti-estrogenic effects. There is growing evidence that herbs containing phytoestrogens have the potential to reduce menopausal symptoms without health risks, unlike hormone replacement therapy [5]. However, many herbal remedies do not have sufficient efficacy to reduce the symptoms of post-menopausal women [5]. It is necessary to explore alternatives

Citation: Yang HJ, Kim MJ, Kwon DY, et al. The combination of Artemisia princeps Pamp, Leonurus japonicas Houtt, and Gardenia jasminoides Ellis fruit attenuates the exacerbation of energy, lipid, and glucose by increasing hepatic PGC-1α expression in estrogen-deficient rats. BMC Complementary and Alternative Medicine. 2016;16:137. doi:10.1186/s12906-016-1109-x.

to attenuating menopausal symptoms without adverse effects to improve women's quality of life.

Various herbs were used traditionally to improve women's health as menopausal symptoms were not recognized in the past. Relatively recent studies have shown that these herbs attenuate menopausal symptoms [6, 7] and some are commercially available. However, they only improve some post-menopausal symptoms, not all. Thus, better herbal treatments to alleviate menopausal symptoms need to be explored. Many herbs contain phytoestrogen, but they have different functionalities and a combination of herbs may have better efficacy in alleviating menopausal symptoms. Among them, *Artemisia princeps Pamp* (APP; Ganghwayakssuk or mugwort), *Leonurus japonicas Houtt* (LJH; Chinese motherworth), and *Gardenia jasminoides Ellis* (GJE; Cape Jasmine) are traditionally used to improve women's health in East Asia including Korea. The major components of APP are eupatilin and jaceosidin, which are reported to reduce inflammation [8, 9]. LJH was mainly used for treating menoxenia, dysmenorrhea, amenorrhea, lochia, body edema, oliguresis, sores, ulcerations and other diseases in women in East Asia [10]. Pharmacological studies have demonstrated that the active components in LJH possess various functionalities such as cardioprotective, anti-oxidative, anti-cancer, analgesic, anti-inflammatory, neuroprotective and antibacterial actions [10]. Stachydrine is the main component of Chinese motherwort and is used as the official indicator to monitor its quality [11]. In addition, GJE has been reported to ameliorate hyperglycemia, hypertension, cerebral ischemia and dyslipidemia [12, 13]. It contains geniposide, ursolic acid, crocin and genipin. Geniposide and ursolic acid have the potential to inhibit glycogenolysis to increase glucose levels in the circulatory system, and improve lipid metabolism [14]. In addition, GJE is reported to protect liver function and neuronal cell death by activating anti-inflammatory activity through geniposide [15]. Although both mugwort and motherwort are known to be improve women's reproductive system to reduce primary dysmenorrhea and GJE has been shown to improve glucose and lipid metabolism, they have not been studied for the purpose of alleviating post-menopausal symptoms.

We were interested in APP for alleviating menopausal symptoms but it might not be sufficient to attenuate the deterioration of energy, glucose, lipid, and bone metabolism in estrogen deficient conditions. LJH and GJE were therefore combined with APP to increase the efficacy for anti-menopausal symptoms. We hypothesized that the mixture of APP, LJH and GJE would ameliorate the reduction in energy, glucose, lipid and bone metabolism caused by estrogen deficiency. We tested the hypothesis using ovariectomized rats and explored their mechanisms.

METHODS

Preparation of APP, LJH and GJE Water Extracts

APP, LJH and GJE were grown in Korea and APP leaves, LJH leaves and GJE fruit were purchased from Ganghwa Sajabal Ssook Inc. (Ganghwa, Korea) in 2013. They were identified by Dr. Byung Seob Ko (Korean Institutes of Oriental Medicine, Daejeon, Korea), and a voucher specimen (No. 2013–04, 2013–05 and 2013–06) deposited at the herbarium of Korean Institutes of Oriental Medicine. Dried and ground APP leaves, LJH leaves and GJE fruits (2 kg) were extracted three times by refluxing with water at 80 °C for 3 h, after which the filtered extracts were lyophilized. The yields of APP leaves, LJH leaves and GJE fruit were 14.8, 15.5 and 20.4 %, respectively. Each of the dried extract was dissolved in methanol. The total phenolic compound contents were then measured using Folin-Ciocalteu reagent [16] and expressed as mg gallic acid equivalents $\cdot g^{-1}$. The extracts were dissolved in ethanol and total flavonoid contents were measured using the modified methods described previously [17]. Rutin was used as a standard.

Analysis of Bioactive Compounds

The analyses were performed using an Acquity UPLC system (Waters, Miliford, MA, USA) with an Acquity UPLC BEH C18 column (2.1 mm × 100 mm, 1.7 μm). Mass spectrometric analyses were operated using a Waters Xevo TQ triple-quadrupole mass spectrometer in electrospray ionization (ESI) mode. Individual APP, LJH and GJE extracts were dissolved in methanol for quantifying the indicator compound. For eupatilin and jaceosidin analysis in APP, pyridine was added to methanol containing APP (1:10, v/v) and the mixture was injected into the UPLC. An isocratic mobile phase of 70 % methanol and 0.1 % TFA was used with a flow rate of 1.5 mL/min. The column temperature was 35 °C, the injection volume was 20 μL, and UV detection was performed at 285 nm.

For geniposide and ursolic acid analysis in GJE, there was an isocratic mobile phase of acetonitrile: methanol: water (45:45:10 v/v/v) with a flow rate of 0.4 mL/min. The injection volume was 5 μL and column temperature was maintained at 35 °C. The tandem MS was operated in negative ESI mode, and processed using MassLynx 4.1 (Waters) software. Quantification was performed using a single ion monitoring (SIM) mode of *m/z* 445.4 for ursolic acid. The detector was operated with a cone voltage of 35 V and a capillary voltage of 3.0 kV. The source temperature was set at 150 °C, while the desolvation flow rate and gas temperature were set at 800 L/h and 500 °C, respectively.

For stachydrine analysis in LJH, the mobile phase was composed of (A) 0.1 % formic acid aqueous solution and (B) 0.1 % formic acid in acetonitrile, at a flow rate of 0.6 mL/min. The conditions were as follows: initial condition of 99 % A, 0- 3 min at 99–70 % B, 3–5 min at 99 % A. The injection volume was 2 μL, column temperature was kept at 30 °C and the total run time was 5 min. The mass spectrometer was operated in positive ESI mode and scanned using the multiple reaction monitoring (MRM) mode. The MRM transitions were monitored at m/z 144.1 → 58.1, 84.1 for stachydrine. The voltage of capillary, cone and collision energy was set at 3.5 kV, 33 V and 22 V, respectively. The gas flow for desolvation and cone was set at 800 and 50 L/h.

Experimental Animals and Design

Ovariectomy (OVX) was performed on female Sprague–Dawley rats aged 8–10 weeks (219 ± 13 g) and they were housed individually in stainless steel cages in a controlled environment (23 °C with a 12-h light and dark cycle). All surgical and experimental procedures were approved by Hoseo University Animal Care and Use Review Committee (2013–04), which reviewed the procedures based on NIH Guidelines. The OVX rats were randomly separated into 5 groups. They freely consumed water and their assigned respective diets for 8 weeks. The high fat diet was a modified semi-purified AIN-93 formulation [18] consisting of 40 energy percent (En%) carbohydrates, 20 En% protein, and 40 En% fats. The major carbohydrate, protein and fat sources were starch plus sugar, casein (milk protein), and lard (CJ Co, Seoul).

Sixty OVX rats were randomly divided into five dietary groups: control, APP, APP+LJH, APP+LJH+GJE and 17β-estradiol (positive-control group). Their diets contained 2 % dextrose, 2 % APP, 2 % APP+LJH (15:5) or 2 % APP+LJH+GJE (10:5:5) in the high fat diet, respectively. The dosage of herb extracts used in the present study is equivalent to approximately 3–5 g/day for human usage. The diet for the positive-control group contained 30 μg/kg body weight of 17β-estradiol+2 % dextrose.

Tail Skin Temperature Measurement

Tail skin temperature was measured using an infrared thermometer (BIO-152-IRB, Bioseb, Chaville, France) designed for small rodents at the 1th and 8th weeks of the experimental periods during the sleep cycle. Three measurements were made 10 min apart and the average value for the animal was used as a single data point for each week [19].

Energy Expenditure by Indirect Calorimetry

After 7 weeks of the assigned diet, the rats were fasted for 6 h before the beginning of the dark phase and energy expenditure was measured. Energy expenditure was assessed by indirect calorimetry measuring average oxygen consumption (VO_2) and average carbon dioxide production (VCO_2): a rat was placed in the metabolic chambers (airflow=800 ml/min) with a computer-controlled O_2 and CO_2 measurement system (Biopac Systems Inc., Goleta, CA) for 30 min. The respiratory quotient (RQ) and resting energy expenditure were calculated using the equations described by Niwa et al. [19]. After the experiment, data were averaged over 1 min intervals and VO_2 and VCO_2 values were corrected for metabolic body size ($kg^{0.75}$) [20]. The amounts of carbohydrate and fat oxidation were calculated from non-protein oxygen consumption as were their relative oxidative proportions and the amount of oxygen consumed per gram of substrate oxidized [20].

Oral glucose Tolerance Test (OGTT) and Insulin Tolerance Test (ITT)

Two days after measuring energy expenditure, an OGTT was conducted in overnight-fasted animals by orally administering 2 g glucose/kg body weight. After glucose loading, blood samples were taken by tail bleeding at 0, 10, 20, 30, 40, 50, 60, 70, 80, 90, and 120 min to measure serum glucose levels with a Glucose Analyzer II (Beckman, Palo Alto, CA) and serum insulin levels were measured at 0, 20, 40, 60, 90 and 120 min with a radioimmunoassay kit (Linco Research, Billerica, MA). The average of the total areas under the curves of the serum glucose and insulin levels during the OGTT was calculated by the trapezoidal rule. In addition, since insulin is released in two phases after glucose, serum glucose and insulin levels were divided into two parts [21, 22]. In glucose tolerant condition, insulin release is peak at 15–30 min after glucose road (early phase) but in glucose intolerant condition, insulin release is delayed. In the present study, the early phase was defined as 0–40 min and the 2nd phase was 40–120 min.

Three days after OGTT, an ITT was conducted after the withdrawal of food for 6 h. Serum glucose levels were measured every 15 min for 90 min after intraperitoneal insulin injection (0.75 U/kg body weight). Serum glucose levels were measured by collecting blood through tail bleeding. Afterwards, food and water was freely provided for two days and then they were overnight-fasted to be scarified.

Body Composition Measurement

At the day scarifying the rats, body composition was measured by dual-energy X-ray absorptiometry (DEXA) using an absorptiometer (pDEXA Sabre; Norland Medical Systems Inc., Fort Atkinson, WI). Briefly, a densitometer was calibrated with a phantom supplied by the manufacturer on a daily basis [6]. The animals were laid in a prone position, with their hind legs maintained in external rotation with tape after the anesthetization with ketamine and xylazine (100 and 10 mg/kg body weight, respectively). Hip, knee and ankle articulations were in 90° flexion and body composition was measured. After the completion of scanning, bone mineral density (BMD) was determined in the right femur and lumbar spine. The pDEXA was equipped with the appropriate software for assessment of body composition in small animals. Similarly, abdominal fat mass and lean mass in abdomen, hip and leg were measured by DEXA.

After finishing DEXA analysis, blood samples were collected from the tail bleeding. After centrifugation of the blood, lipid profiles in circulation were determined by measuring serum levels of triglyceride, total cholesterol, and HDL cholesterol using the appropriate colorimetry kits (Asan Pharm., Seoul, Korea). In addition, liver function was measured by aspartate aminotransferase (AST) and alanine aminotransferase (ALT) in the circulation using colorimetry kits (Asan Pharm.). Serum leptin levels were determined using a radioimmunoassay kit (Linco Research). Insulin resistance was determined using the homeostasis model assessment estimate of insulin resistance (HOMA-IR) [HOMA-IR = fasting insulin (μIU/ml) × fasting glucose (mM) / 22.5].

After drawing blood, human regular insulin (unmodified insulin; 5 U/kg body weight) was injected through the inferior vena cava. Ten min later, the rats were killed by decapitation and tissues such as liver and gastrocnemius and quadriceps muscles were rapidly collected, frozen in liquid nitrogen, and stored at −70 °C for further experiments. Epididymal and retroperitoneal fat mass and uteruses were then excised and weighed. Uterus index was calculated as uterus weight divided by body weight.

Triglyceride Contents in the Liver and Skeletal Muscles

Triacylglycerol was extracted from the livers and gastrocnemius and quadriceps muscles with chloroform-methanol (2:1, vol/vol) and resuspended in pure chloroform [23]. After evaporating the chloroform, the residues were suspended with PBS with 0.1 % triton X-100 and the suspension was sonicated and boiled for 5 min. The triacylglycerol contents of the suspensions were assayed using a Trinder kit (Asan).

RNA Isolation and Reverse Transcription Polymerase Chain Reaction (RT-PCR)

The livers of four rats were randomly selected from each group. Total RNA was isolated from the liver using a monophasic solution of phenol and guanidine isothiocyanate (Trizol reagent, Gibco-BRL, Rockville, MD), followed by extraction and precipitation with isopropyl alcohol. The cDNA was synthesized from equal amounts of total RNA with superscript III reverse transcriptase, and the contents of cDNA was enlarged by PCR with high fidelity Taq DNA polymerase. Equal amounts of cDNA were mixed with sybergreen mix and analyzed using a realtime PCR machine (BioRad, Richmond, CA). The expression level of the gene of interest was corrected to that of the house keeping gene, β-actin. The primers used to detect rat peroxisome proliferator-activated receptor-gamma coactivator (PGC)-1α, carnitine palmitoyltransferase-1 (CPT-1), acetyl CoA carboxylase (ACC), sterol regulatory element-binding protein-1c (SREBP-1c), fatty acid synthase (FAS), and β-actin genes were described previously [7].

Immunoblot Analysis

The frozen livers of four rats were lysed with a 20 mM Tris buffer (pH 7.4) containing 2 mM EDTA, 137 mM NaCl, 1 % NP40, 10 % glycerol, 12 mM α-glycerol phosphate and protease inhibitors. Liver lysates containing equal amounts of protein (30–50 μg) were resolved by SDS-PAGE, and immunoblotting was performed with specific antibodies against phosphorylated Akt and glycogen synthase kinase-1β (GSK-1β) and Akt, GSK-1β, phosphoenolpyruvate carboxykinase (PEPCK) and β-actin. The intensity of protein expression was determined using Imagequant TL (Amersham Biosciences, Piscataway, NJ). Three sets of two samples per group were evaluated ($n=6$).

Statistical Analysis

All results are expressed as means±standard deviations. Statistical analysis was performed using the SAS software (SAS institutes, Cary, NC, USA). The variables that measured at different time points were analyzed with two-way repeated measures analysis of variance (ANOVA) with time and group as independent variables and interaction term between time and group. One-way ANOVA was used to determine the group (APP, APP+LJH, APP+LJH+GJE, positive-control and control groups) effect when the results were measured once at the end of experiment. Significant differences in the main effects among the groups were identified by Tukey's test at $p<0.05$.

RESULTS

The Contents of Total Phenolic Compounds, Flavonoids, and Bioactive Components

Total polyphenol and flavonoids contents were about 3 fold higher in GJE compared to APP and LJH (Table 1). The indicator compounds in each water extract were eupatilin and jaceosidin in APP, stachydrine in LJH and geniposide and ursolic acid in GJE. The amount of each of these compounds in the water extract was sufficient to use them as indicator compounds (Table 1).

Table 1: The contents of total polyphenols, total flavonoids and indicator compounds in *Artemisia princeps Pamp*, *Leonurus japonicas Houtt*, *Gardenia jasminoides Ellis* fruit water extracts

		Total polyphenols (mg/g)	Total flavonoids (mg/g)
Artemisia princeps Pamp (APP)		7.4 ± 0.6	3.3 ± 0.2
Leonurus japonicas Houtt (LJH)		6.8 ± 0.4	2.4 ± 0.1
Gardenia jasminoides Ellis fruit (GJE)		21.2 ± 0.7	18.4 ± 0.9
	Contents (mg/g)		Contents (mg/g)
Eupatilin in APP	1.34 ± 0.11	Geniposide in GJE	4.72 ± 0.33
Jaceosidin in APP	0.88 ± 0.09	Ursolic acid in GJE	0.02 ± 0.01
Stachydrine in LJH	1.50 ± 0.23		

Values are means ± SD ($n = 3$)

Tail Skin Temperature

Estrogen deficiency elevates skin temperature due to a vasomotor disorder and is known to increase tail skin temperature in OVX rats. None of the treatments changed the tail skin temperature in the first week, but rats in the control group exhibited higher tail skin temperature at week 8 than those in the positive-control group (Fig. 1). APP, APP+LJH and APP+LJH+GJE suppressed the increase in OVX rats as much as the positive-control (Fig. 1).

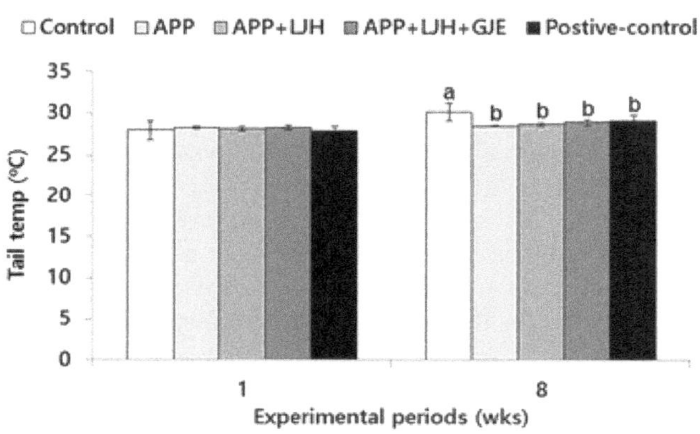

Figure 1: Tail skin temperature at 1st and 8th weeks of experimental period. Control, OVX rats fed a high-fat diet with 2 % dextrin; APP, OVX rats fed a high fat diet with 2 % Artemisia princeps Pamp water extract; LJH, OVX rats fed a high fat diet with 2 % Leonurus japonicas Houtt water extract; GJE, OVX rats fed a high fat diet with 2 % Gardenia jasminoides Ellis water extract; positive-control, OVX rats fed a high fat diet with 30 μg/kg body weight 17β-estradiol + 2 % dextrose. At 1st and 8th weeks, tail skin temperature was measured using infrared thermometer. Bars and error bars represent means ± SD (n = 12). [a,b] Significantly different among all groups at $p < 0.05$

Estrogen deficiency is also reported to reduce the size of the uterus. The uterine index was lower by about 3 folds in the control rats in comparison to the positive-control rats (Table 2). It indicated that APP, APP+LJH and APP+LJH+GJE have no uterine proliferation unlike estrogen treated rats (positive-control).

Table 2: Metabolic parameters related to energy metabolism at the end of experimental periods

	Control ($n = 12$)	APP ($n = 12$)	APP + LJH ($n = 12$)	APP + LJH + GJE ($n = 12$)	Positive-control ($n = 12$)
Body weight (g)	355 ± 18[a]	360 ± 25[a]	362 ± 19[a]	340 ± 19[b]	333 ± 18[b]
Body weight gain (g)	96.9 ± 7.7[a]	99.1 ± 9.4[a]	101.2 ± 7.3[a]	70.3 ± 7.3[b]	74.1 ± 7.9[c]
Peri-uterine fat (g)	11.2 ± 1.1[a]	8.9 ± 0.9[b]	9.2 ± 0.9[b]	6.6 ± 0.7[c]	6.1 ± 0.7[c]
Ratio of peri-uterine fat and body weight	0.032 ± 0.007[a]	0.025 ± 0.004[b]	0.025 ± 0.005[b]	0.019 ± 0.004[c]	0.018 ± 0.004[c]
Retroperitoneum fat (g)	13.3 ± 1.4[a]	12.2 ± 1.3[a]	13.0 ± 1.4[a]	8.7 ± 1.0[b]	8.2 ± 0.8[b]
Ratio of retroperitoneum fat and body weight	0.037 ± 0.008[a]	0.034 ± 0.005[a]	0.036 ± 0.007[a]	0.026 ± 0.005[b]	0.025 ± 0.005[b]
Uterine index	0.56 ± 0.06[c]	0.56 ± 0.06[c]	0.62 ± 0.07[c]	0.91 ± 0.09[b]	1.53 ± 0.14[a]
Overnight fasted leptin levels (ng/mL)	3.4 ± 0.6[a]	3.8 ± 0.6[ab]	3.9 ± 0.6[ab]	4.1 ± 0.6[b]	4.3 ± 0.7[b]

Control, OVX rats fed a high-fat diet (OVX-CON) with 2 % dextrin; APP, OVX rats fed a high fat diet with 2 % *Artemisia princeps Pamp* water extract; LJH, OVX rats fed a high fat diet with 2 % *Leonurus japonicas Houtt* water extract; GJE, OVX rats fed a high fat diet with 2 % *Gardenia jasminoides* Ellis water extract; positive-control, OVX rats fed a high fat diet with 30 µg/kg body weight 17β-estradiol + 2 % dextrose. Values are means ± SD
[a,b,c] Significantly different among all groups by Tukey test at $p < 0.05$

Energy Metabolism

OVX rats had greater weight gain than OVX rats administered with 17β-estrogen (positive-control). Peri-uterine and retroperitoneum fat pads were also higher in OVX rats than in the positive-control rats (Table 2). Despite higher visceral fat, OVX rats had lower serum leptin levels than OVX rats treated with 17β-estradiol (Table 2). Since 17β-estradiol has some adverse effects, alternative therapy needs to alleviate the deterioration of energy metabolism. APP+LJH+GJE, but not APP and APP+LJH treatments, suppressed the increase of body weight in OVX rats as much as the positive-control group (Table 2). Furthermore, peri-uterine and retroperitoneum fat pad weights were lower with APP, APP+LJH and APP+LJH+GJE treatments than the control group, with APP+LJH+GJE having a similar effect as the positive-control (Table 2). Unlike visceral fat amounts, APP+LJH+GJE had serum leptin to levels similar to the positive-control (Table 2). These results indicated that estrogen plays an important role in leptin secretion and that estrogen deficiency induced the impairment of leptin secretion.

Body weight and body fat are balanced by the sum of energy intake and energy expenditure. Food intake was not significantly different among the groups and the rats in all groups had about 20 g/day. Based on the food intake, the daily consumption of herbal extracts was calculated in rats. When the contents were adjusted to human equivalence using the conversion coefficient of 6.2 suggested by the US FDA [23], the daily amount for humans was approximately 3–5 g of the mixture of herbal water extracts. The increase of body weight and visceral fat in OVX rats was primarily due to lower energy expenditure without the modulation of energy intake (Table 3). Furthermore, in OVX rats, carbohydrate oxidation was higher but fat oxidation was lower than the positive-control (Table 3). None of the treatments altered the energy intake in OVX rats.

Table 3: Parameters of indirect calorimetry at the end of experiment

	Control ($n = 12$)	APP ($n = 12$)	APP + LJH ($n = 12$)	APP + LJH + GJE ($n = 12$)	Positive-control ($n = 12$)
Caloric intakes (Kcal/day)	95.4 ± 10	104 ± 11	103 ± 11	103 ± 10	92 ± 10
Energy expenditure (kcal/ $kg^{0.75}$/day)	106 ± 12[c]	111 ± 11[bc]	119 ± 12[b]	137 ± 13[a]	131 ± 13[a]
Respiratory quotient	0.84 ± 0.11	0.81 ± 0.10	0.82 ± 0.09	0.80 ± 0.09	0.79 ± 0.09
Carbohydrate oxidation (mg/ $kg^{0.75}$/min)	5.3 ± 0.7[a]	4.3 ± 0.6[b]	4.8 ± 0.6[ab]	4.5 ± 0.5[b]	4.1 ± 0.6[b]
Fat oxidation (mg/ $kg^{0.75}$/min)	6.2 ± 0.8[c]	7.6 ± 0.8[b]	8.0 ± 0.9[b]	10.3 ± 1.2[a]	9.7 ± 1.1[a]

Control, OVX rats fed a high-fat diet (OVX-CON) with 2 % dextrin; APP, OVX rats fed a high fat diet with 2 % *Artemisia princeps Pamp* water extract; LJH, OVX rats fed a high fat diet with 2 % *Leonurus japonicas Houtt* water extract; GJE, OVX rats fed a high fat diet with 2 % *Gardenia jasminoides* Ellis water extract; positive-control, OVX rats fed a high fat diet with 30 µg/kg body weight 17β-estradiol + 2 % dextrose. Values are mean ± SD
[a,b,c] Significantly different among all groups by Tukey test at $p < 0.05$

APP+LJH and APP+LJH+GJE had greater energy expenditure among OVX rats than the control and the daily energy expenditure of rats in the APP+LJH+GJE group was similar to that of the positive-control group (Table 3). In daily energy expenditure, carbohydrate oxidation was greater in the control rats than in the positive-control rats whereas fat oxidation contrasted with carbohydrate oxidation (Table 3). APP alone caused greater carbohydrate and fat oxidation in OVX rats whereas the APP+LJH+GJE group had the greatest carbohydrate and fat oxidation and reached the same amount as the positive-control (Table 3).

Body Composition

Consistent with the amounts of peri-uterine and retroperitoneum fat pads, the fat mass in the abdomen and leg measured by DEXA was significantly higher in the control rats than in the positive-control rats (Fig 2a). The fat mass in the abdomen was lowered in the descending order of the control < APP = APP + LJH < APP + LJH + GJE = positive-control. APP+LJH+GJE treatment was the only herbal treatment that reduced fat mass in the leg and the reduction was lower than in the positive-control (Fig. 2a).

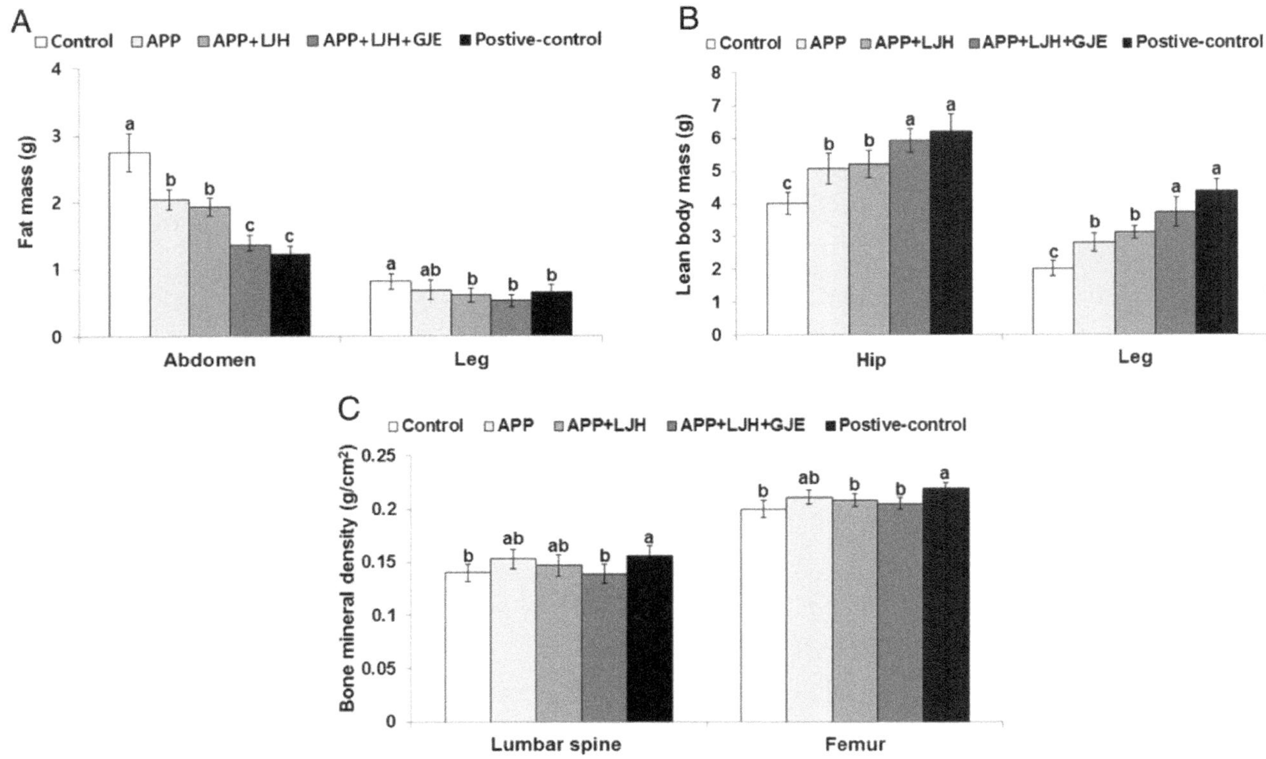

Figure 2: Body composition at 8th weeks of experimental period measured by DEXA. Control, OVX rats fed a high-fat diet with 2 % dextrin; APP, OVX rats fed a high fat diet with 2 % Artemisia princeps Pamp water extract; LJH, OVX rats fed a high fat diet with 2 % Leonurus japonicas Houtt water extract; GJE, OVX rats fed a high fat diet with 2 % Gardenia jasminoides Ellis water extract; positive-control, OVX rats fed a high fat diet with 30 μg/kg body weight 17β-estradiol + 2 % dextrose. At 8th weeks, lean body mass (a) and fat mass (b) were measured in the abdomen and leg by DEXA whereas bone mineral density (BMD) of the femur and lumbar spine (c) were also measured. Bars and error bars represent means ± SD (n = 12). a,b,c Significantly different among all groups by Tukey's test at $p < 0.05$

In contrast to the fat mass, lean body mass in the abdomen and leg was lower in the control rats than in the positive-control rats. The lower lean body mass was reduced by APP, APP+LJH, and APP+LJH+GJE in OVX rats and the lean body mass in the abdomens and legs of the APP+LJH+GJE group was similar to that of the positive-control group (Fig. 2b). APP suppressed the decrease of lean body mass in the leg as much as the APP+LJH+GJE and positive-control groups (Fig. 2b).

BMD in the lumbar spine and leg was lower in the control rats than in the positive-control rats (Fig. 2c). The decrease of BMD in OVX rats was attenuated by APP but the decrease was not significant. Other herbal extracts did not alter the BMD in OVX rats (Fig. 2c).

Lipid Profiles And liver Function Index

Serum levels of total and LDL cholesterol and triglyceride were higher in control rats than in 17β-estradiol-treated rats whereas HDL cholesterol levels in the circulatory system were lower in control rats (Table 4). APP and APP+LJH treatments suppressed the elevation of serum LDL cholesterol levels and only APP+LJH+GJE and normalized the lower of serum HDL cholesterol to levels similar to the positive-control group. APP, APP+LJH and APP+LJH+GJE all resulted in lower serum triglyceride levels than the control group (Table 4). Thus, all the treatments exhibited some beneficial effects on lipid metabolism in the circulatory system of OVX rats.

The major adverse effects of herbal treatments are generally liver damage. AST and ALT are found in various body tissues, and the elevation of serum ALT and AST levels are clinically used as a part of a diagnostic evaluation of hepatocellular injury. Serum AST levels were higher in the control rats than the positive-control rats and they decreased in the descending order of control, APP, APP+LJH, APP+LJH+GJE, and positive-control. APP+LJH+GJE lowered them as much as the positive-control group (Table 4). However, serum ALT levels were not significantly different between the control and positive-control groups and APP, APP+LJH, and APP+LJH+GJE did not modify serum ALT levels (Table 4).

Table 4: Lipid profiles, liver function index and serum glucose and insulin levels in overnight-fasted rats

	Control (*n* = 12)	APP (*n* = 12)	APP + LJH (*n* = 12)	APP + LJH + GJE (*n* = 12)	Positive-control (*n* = 12)
Total cholesterol (mg/dL)	109.9 ± 10.3[a]	101.3 ± 10.4[ab]	104.5 ± 10.4[ab]	108.5 ± 10.9[a]	99.4 ± 9.5[b]
LDL cholesterol (mg/dL)	56.3 ± 6.4[a]	50.7 ± 5.7[b]	51.6 ± 5.8[b]	53.4 ± 7.4[ab]	45.2 ± 5.6[b]
HDL cholesterol (mg/dL)	32.3 ± 3.3[b]	34.2 ± 3.7[ab]	35.7 ± 3.8[ab]	36.9 ± 3.6[a]	38.3 ± 3.7[a]
Triglyceride (mg/dL)	107.1 ± 10.4[a]	81.4 ± 8.8[b]	86.1 ± 9.8[b]	80.3 ± 8.1[b]	79.6 ± 8.5[b]
Aspartate aminotransferase (IU/L)	123 ± 9[a]	111 ± 12[ab]	109 ± 14[b]	90 ± 15[c]	87 ± 14[c]
Alanine aminotransferase (IU/L)	39.8 ± 4.6	41.5 ± 5.9	40.1 ± 6.8	35.6 ± 5.2	35.5 ± 5.5
Glucose (mg/dL)	100.3 ± 11.4[a]	87.5 ± 10.4[b]	92.8 ± 11.3[ab]	93.0 ± 10.8[ab]	94.9 ± 10.7[ab]
Insulin (ng/mL)	1.87 ± 0.28[a]	1.29 ± 0.25[b]	1.30 ± 0.24[b]	1.15 ± 0.22[c]	1.14 ± 0.25[c]
HOMA-IR	10.4 ± 1.8[a]	6.3 ± 0.8[bc]	6.7 ± 0.8[b]	5.9 ± 0.7[c]	6.0 ± 0.7[c]

Control, OVX rats fed a high-fat diet (OVX-CON) with 2 % dextrin; APP, OVX rats fed a high fat diet with 2 % *Artemisia princeps Pamp* water extract; LJH, OVX rats fed a high fat diet with 2 % *Leonurus japonicas Houtt* water extract; GJE, OVX rats fed a high fat diet with 2 % *Gardenia jasminoides* Ellis water extract; positive-control, OVX rats fed a high fat diet with 30 μg/kg body weight 17β-estradiol + 2 % dextrose. HOMA-IR, homeostasis model assessment estimate of insulin resistance. Values are means ± SD
[a,b,c] Significantly different among all groups by Tukey test at $p < 0.05$

Glucose Metabolism

Overnight-fasted serum glucose levels were higher in OVX rats in comparison to positive-control rats but not significantly (Table 4). However, the levels were significantly lower in APP rats than the control rats. Serum insulin levels in overnight-fasting states were much higher in OVX rats than in positive-control rats, but were lower in all OVX rats given herbal extract treatments than the control rats (Table 4). HOMA-IR, an index of insulin resistance, was higher in the OVX group than in the positive-control group. HOMA-IR was lower in the descending order of the control > APP + LJH > APP > APP + LJH + GJE > positive-control groups (Table 4). APP+LJH+GJE lowered the elevation of HOMA-IR similar to the positive-control group.

OGTT indicates the relative roles of insulin secretion and insulin resistance in the progression of glucose intolerance. Serum glucose levels were overall higher in OVX rats than in positive-control rats at all OGTT time points (Fig. 3a). Two-way repeated measures ANOVA revealed that serum glucose levels during OGTT has significant effects of time and treatment ($P<0.05$) but there were no interaction effects. APP+LJH and APP+LJH+GJE did not increase serum glucose levels at 10–40 min as much as the levels of the OVX-control group (Fig. 3a). Serum glucose levels at 10–40 min were higher in the ascending order of APP<APP+LJH<APP+LJH+GJE=positive control<control in OVX rats during OGTT (Fig. 3a). After peaking, serum glucose levels gradually decreased, with APP+LJH+GJE and the positive control exhibiting a faster decrease than the control (Fig. 3a). Serum glucose levels increased up to 40–50 min after glucose challenge whereas after that point serum glucose levels gradually decreased. Thus, area under the curves of glucose (AUCG) and insulin (AUCI) were separately calculated into the two parts in the 1st part (0–40 min) and 2nd parts (40-120 min). The AUCGs of the 1st and 2nd parts were much greater in the control group than in the positive-control group (Fig. 3b). The AUCGs of the 1st part were lowest in the APP group and was similar in both the APP+LJH+GJE and positive-control groups (Fig. 3b). Serum insulin levels during OGTT are shown in Fig. 3c. Serum insulin levels increased until 20 min and then they were lowered in herbal treatment groups and positive-control group but they were increased until 40 min and then slowly decreased in the control group (Fig. 3c). Two-way repeated measures ANOVA demonstrated that there was a significant effect of time and treatment effects and their interaction effect ($P<0.05$). Serum insulin levels increased until 40 mins and slightly decreased after 40 mins in OVX-control rats but in herbal treatment groups serum insulin levels peaked at 20 min and then they markedly decreased. AUCIs of the 1st part were higher in the ascending order of APP+LJH+GJE<control<APP+LJH=positive-control<APP (Fig. 3d). The AUCGIs of the 2nd part were highest in the control and APP groups, lower in the APP+LJH group and even lower in the APP+LJH+GJE and positive-control groups (Fig. 3d). Thus, the serum glucose levels in the 1st part were mainly associated with serum insulin levels.

Insulin resistance can be determined by ITT. In ITT, the decrease of serum glucose levels during the 1st part (0–30 min) of AUC was related to insulin resistance. In the 2nd part (40–90 min) of ITT, serum glucose levels were maintained and slowly increased. The AUC of serum glucose levels in the 1st and 2nd parts of the ITT were higher in the control group than the positive-control group and APP+LJH+GJE decreased the 1st part the most (Fig. 4). The 2nd part of AUC was higher in the control group than the positive-control group. APP, APP+LJH and APP+LJH+GJE had similar AUCs in the 2nd part as the positive-control during ITT (Fig. 4). These results indicated that APP+LJH+GJE reduced insulin resistance the most.

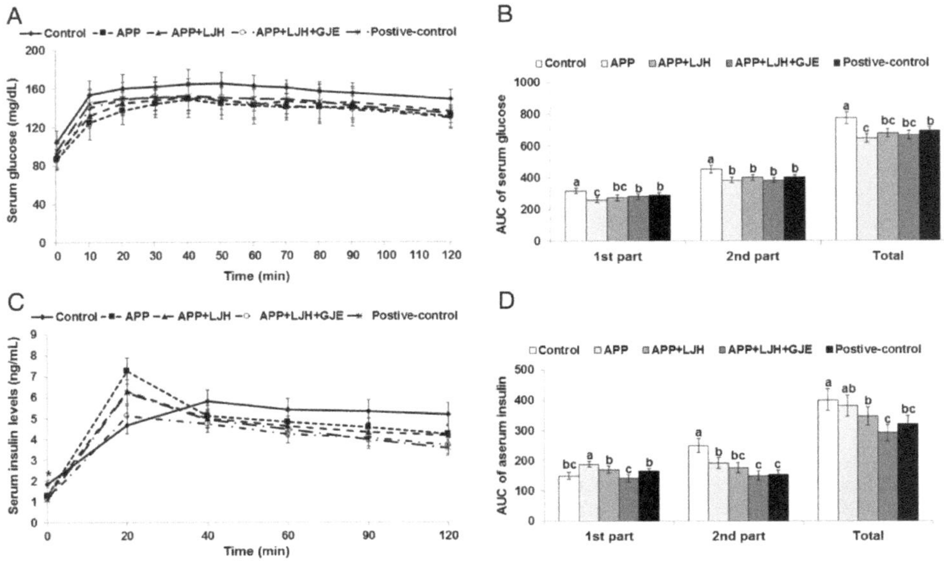

Figure 3: Serum glucose levels and area under the curve of glucose and insulin during oral glucose tolerance test (OGTT). Control, OVX rats fed a high-fat diet with 2 % dextrin; APP, OVX rats fed a high fat diet with 2 % Artemisia princeps Pamp water extract; LJH, OVX rats fed a high fat diet with 2 % Leonurus japonicas Houtt water extract; GJE, OVX rats fed a high fat diet with 2 % Gardenia jasminoides Ellis water extract; positive-control, OVX rats fed a high fat diet with 30 μg/kg body weight 17β-estradiol + 2 % dextrose. At 7th week, an OGTT was conducted in overnight-fasted animals by orally administering 2 g glucose/kg body weight. After glucose loading, blood samples were taken by tail bleeding at 0, 10, 20, 30, 40, 50, 60, 70, 80, 90, and 120 min to measure serum glucose levels (a) and serum insulin levels were measured at 0, 20, 40, 60, 90 and 120 min (c). Serum glucose levels had significant time and treatment effects but not interaction effects ($P < 0.05$) in two-way repeated measures ANOVA whereas serum insulin levels exhibited significant effect of time and treatment effects and their interaction effect ($P < 0.05$). During OGTT area under the curve (AUC) of serum glucose (b) and insulin (d) in the first part (0–40 min), second part (40–120 min) and total parts were given. Bars or dots and error bars represent means ± SD (n = 12). a,b,c Significantly different among all groups by Tukey's test at $p < 0.05$.

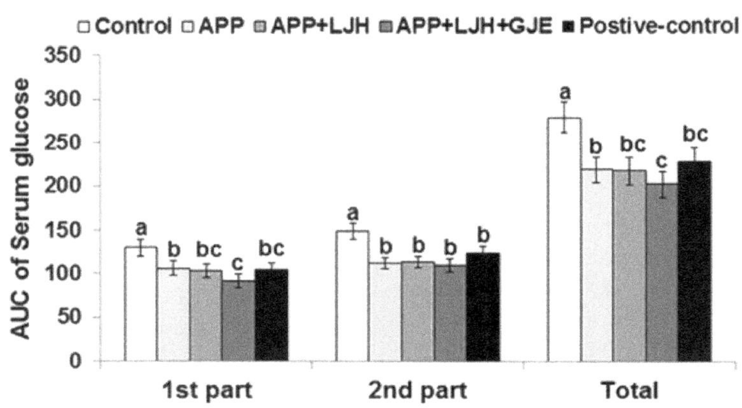

Figure 4: Serum glucose levels and area under the curve of glucose and insulin during insulin tolerance test (ITT). Control, OVX rats fed a high-fat diet with 2 % dextrin; APP, OVX rats fed a high fat diet with 2 % Artemisia princeps Pamp water extract; LJH, OVX rats fed a high fat diet with 2 % Leonurus japonicas Houtt water extract; GJE, OVX rats fed a high fat diet with 2 % Gardenia jasminoides Ellis water extract; positive-control, OVX rats fed a high fat diet with 30 μg/kg body weight 17β-estradiol + 2 % dextrose. At three days after OGTT, after the withdrawal of food for 6 h serum glucose levels were measured every 15 min for 90 min after intraperitoneal insulin injection (0.75 U/kg body weight). During ITT area under the curve (AUC) of serum glucose in the first part (0–30 min) and second part (30–90 min). Bars and error bars represent means ± SD (n = 12). * Significantly different among groups in one-way ANOVA at $p < 0.05$. [a,b,c] Significantly different among all groups by Tukey's test at $p < 0.05$

Lipid in the liver and skeletal muscles

Since a high fat diet induces insulin resistance by increasing intracellular triglyceride storage, intracellular triglyceride contents were measured in the liver and skeletal muscle. The contents of hepatic triglyceride storage were higher in the control group than in the positive-control group (Table 5). The storage was lower in the descending order of the control<APP=APP+LJH<APP+LJH+GJE=positive-control. The gastrocnemius and quadriceps muscles, the major skeletal muscles in the leg, stored more triglyceride in the control rats than in the positive-control rats. The lower triglyceride storage in skeletal muscles of rats given APP+LJH+GJE were similar to that of the positive-control group (Table 5).

Table 5: Triglyceride storage in the liver, gastrocnemius muscle and quadriceps muscles

	Control (n = 12)	APP (n = 12)	APP + LJH (n = 12)	APP + LJH + GJE (n = 12)	Positive-control (n = 12)
Liver (mg/g tissues)	0.86 ± 0.10^{a}	0.75 ± 0.09^{b}	0.74 ± 0.07^{b}	0.61 ± 0.09^{c}	0.57 ± 0.08^{c}
Gastrocnemius muscles (mg/g tissues)	1.25 ± 0.21^{a}	1.18 ± 0.18^{ab}	1.10 ± 0.22^{ab}	1.02 ± 0.17^{b}	0.93 ± 0.17^{b}
Quadriceps muscles (mg/g tissues)	2.62 ± 0.39^{a}	2.24 ± 0.36^{b}	2.18 ± 0.32^{bc}	2.01 ± 0.28^{c}	1.93 ± 0.29^{c}

Control, OVX rats fed a high-fat diet (OVX-CON) with 2 % dextrin; APP, OVX rats fed a high fat diet with 2 % *Artemisia princeps Pamp* water extract; LJH, OVX rats fed a high fat diet with 2 % *Leonurus japonicas Houtt* water extract; GJE, OVX rats fed a high fat diet with 2 % *Gardenia jasminoides* Ellis water extract; positive-control, OVX rats fed a high fat diet with 30 µg/kg body weight 17β-estradiol + 2 % dextrose. Values are means ± SD

a,b,c Significantly different among all groups by Tukey test at $p < 0.05$

Expression of Genes Related Fatty Acid Oxidation and Synthesis and Hepatic Insulin Signaling

Since triglyceride storage is the net of fatty acid oxidation and synthesis, the expression of genes involved in oxidation and synthesis were measured. The metabolism of fatty acids was associated with PGC-1α in the liver. The expression of PGC-1α was lower in the control rats than other treatment groups and APP+LJH+GJE increased PGC-1α expression in comparison to the control group (Fig. 5a). In parallel with the modulation of PGC-1α expression, hepatic expression of CPT-1, the mitochondrial transporter of fatty acids and a major regulator of fatty acid oxidation, was lower in control rats than in positive-control rats (Fig. 5a). This result indicated that fatty acid oxidation increased in APP+LJH+GJE compared to the control group. The expressions of hepatic SREBP-1c, FAS and ACC, which are regulatory enzymes of fatty acid synthesis, were higher in control rats than the positive-control rats (Fig. 5b). In OVX rats given APP+LJH+GJE CPT-1 expression was higher and FAS expression was lower; SREBP-1c and ACC to levels similar to those of the positive-control rats (Fig. 5b).

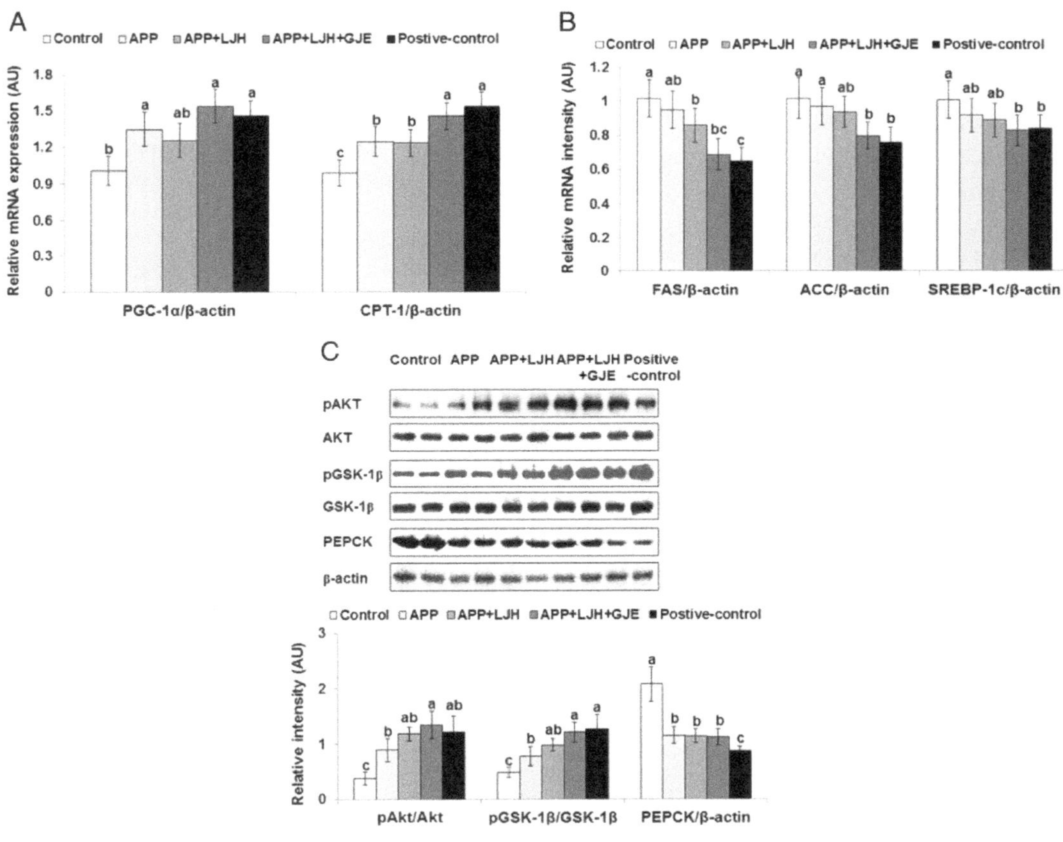

Figure 5: The mRNA expression of genes related to fatty acid metabolism and insulin signaling in the liver. Control, OVX rats fed a high-fat diet with 2 % dextrin; APP, OVX rats fed a high fat diet with 2 % Artemisia princeps Pamp water extract; LJH, OVX rats fed a high fat diet with 2 % Leonurus japonicas Houtt water extract; GJE, OVX rats fed a high fat diet with 2 % Gardenia jasminoides Ellis water extract; positive-control, OVX rats fed a high fat diet with 30 µg/kg body weight 17β-estradiol + 2 % dextrose. At end of the experimental period, mRNA levels of hepatic genes involved in fatty acid oxidation (a) and synthesis (b) were measured by real-time PCR. Insulin signaling (c) also measured by immunoblotting assay. Bars and error bars represent means ± SD (n = 6). a,b,c Significantly different among all groups by Tukey's test at $p < 0.05$

In hepatic insulin signaling, the phosphorylation of Akt was lower in the control group than in the positive-control and APP groups. APP+LJH and APP+LJH+GJE treatments increased phosphorylated Akt in OVX rats (Fig. 5c). In parallel with Akt phosphorylation, the phosphorylation of GSK-1β was lower in control rats than positive-control rats whereas APP+LJH+GJE elevated the phosphorylated GSK-1β (Fig. 5c).

DISCUSSION

Estrogen deficiency deteriorates energy, lipid, glucose and bone metabolism. Hormonal therapy is known to be lacking in terms of preventing the deterioration and has been reported to cause adverse effects [1, 4, 24]. Alternative treatments have been researched and some herbal extracts have been successfully employed [25]. In Korea, herbs such as APP and LJH were traditionally used for enhancing women's health [8, 10] and they can be beneficial for menopausal symptoms. APP, LJH, and GJE are traditionally used to improve women's health. The present study found that the herbal extracts attenuated various menopausal symptoms and their possible action mechanisms were examined. APP+LJH+GJE appeared to improved energy balance by increasing energy expenditure as much as estrogen treatment. Furthermore, APP+LJH+GJE treated OVX rats had less visceral and intramuscular fat and greater lean body mass and were similar to the positive-control. APP itself was sufficient to improve lipid profiles in the circulatory system and to elevate glucose-stimulated insulin secretion. APP+LJH+GJE also attenuated dyslipidemia while improving glucose intolerance and insulin resistance in OVX rats. The expressions of hepatic genes involved in fatty acid synthesis were lower in the APP+LJH+GJE group in comparison to the control group, which may improve lipid and glucose metabolism. These results suggested that the combination of APP+LJH+GJE may have potential as a therapeutic agent for the treatment of postmenopausal symptoms.

Estrogen plays an important role in maintaining energy, glucose, lipid and bone metabolism [2, 3]. As expected, estrogen deficiency offsets the estrogen effects in these processes, which leads to metabolic disorders such as insulin resistance, obesity, dyslipidemia, high blood pressure and hyperglycemia. Epidemiologic studies have demonstrated that overweight and obesity rates are increased in menopausal women and that obesity is closely related to the risk of metabolic syndromes [26, 27]. However, the cause of obesity in menopausal women remains unclear. Estrogen deficiency plays an important role in developing menopausal symptoms including gaining weight but it cannot explain them completely. Furthermore, estrogen treatment only partly reduces metabolic disturbances in menopausal women but exacerbates some cardiovascular diseases such as stroke [28]. Therefore, the discrepancy in the metabolic effects of hormone therapy needs to be explained. The present study suggested that estrogen deficient rats experienced the reduction of energy expenditure without changing food intake and the increase of visceral fats by decreasing fatty acid oxidation. In parallel with increasing visceral fat mass, dyslipidemia and hyperglycemia were exhibited in OVX rats in comparison to the positive-control. The previous studies have exhibited that sham-operated rats have a similar metabolism of energy, glucose, lipid and bone as the positive-control rats [7, 29]. Thus, the results in the present study indicated that OVX rats had impaired energy, glucose, lipid and bone metabolism and estrogen treatment attenuated the deterioration of the metabolism in OVX rats. As life expectancy increases, women are substantially influenced by menopausal symptoms for a significant period. As a result, estrogen deficiency may increase the chance to develop metabolic diseases in post-menopausal women. Estrogen increases PGC-1α expression via nitric oxide/cGMP signaling pathways through estrogen receptors which enhances lipid, energy and glucose metabolism [28, 30]. Overnutrition suppresses the expression of PGC-1α [31]. Thus, both overnutrition and estrogen deficiency suppress the transcription of genes related to metabolic and mitochondrial function such as nuclear respiratory factors, peroxisomal proliferator-activated receptors (PPARs) and cAMP responding element binding protein [28, 30, 31]. The inhibition of these signals decreases oxidative phosphorylation, lipid oxidation and insulin sensitivity [31]. Thus, the activation of PGC-1α signaling attenuates menopausal symptoms without direct activation of the estrogen receptor by estrogen. The present study showed similar results in OVX rats, as the expression of PGC-1α was lower, whereas it was higher with APP and APP+LJH+GJE treatments. The lower PGC-1α expression in OVX rats increased the expression of genes involved in fatty acid synthesis. Thus, APP and APP+LJH+GJE potentiated PGC-1α expression and improved dyslipidemia and hyperglycemia. These results indicated that the improvement of PGC-1α may attenuate menopausal symptoms.

APP, LJH and GJE have not been evaluated for their effects on anti-menopausal symptoms in experimental animal and human studies. APP and its essential oils have been reported to have anti-oxidant, anti-microbial (eucalyptol and α-terpineol) and anti-thrombotic (sulfated polysaccharides) activities [8, 9]. Some of these activities are associated with the inhibition of the expression of proinflammatory cytokines through the activation of NF-κB pathways [9]. In the present study, APP ameliorated the suppressed glucose metabolism due to estrogen deficiency in rats. APP also potentiated glucose-stimulated insulin secretion and attenuated insulin resistance measured by HOMA-IR. However, it did not inhibit the deterioration of energy and lipid metabolism in OVX rats as much as the positive-control. Thus, the combined mixture is beneficial for reducing various menopausal symptoms without increasing uterine proliferation.

In the estrogen deficient state, glucose metabolism is deteriorated by increased insulin resistance and menopausal women with low insulin secretion capacity are susceptible to type 2 diabetes [2, 3]. Insulin resistance develops mainly in the liver, skeletal muscle, and adipose tissue. In the insulin resistant state, myocytes in skeletal muscle exhibit reduced glycogen synthesis [32] and glycogenolysis and gluconeogenesis in the liver are increased even in hyperinsulinemic states [33]. Previous studies have demonstrated that OVX rats impaired glucose metabolism by increasing insulin resistance and deteriorating the regulation of insulin secretion in comparison to the sham-operated rats [14, 29]. Thus, estrogen deficiency develops insulin resistance even though overnight fasted serum glucose levels are normal. The present study indicated that OVX rats had insulin resistance. APP+LJH+GJE markedly lowered serum glucose levels more than the other treatments in the early part of ITT although rats with all treatments had lower HOMA-IR, an insulin resistance index. The results also indicated that APP+LJH+GJE treated and positive control rats had the best insulin sensitivity in the hyperinsulinemic state. This was consistent with the potentiation of Akt and GSK-1β phosphorylation and with the lower PEPCK expression. Therefore, APP+LJH+GJE can be a good candidate for improving insulin sensitivity in post-menopausal women.

LJH and GJE extracts were selected as additional herbs with APP for the purpose of attenuating menopausal symptoms. LJH has not been studied previously, although it is a folk medicine for reducing primary dysmenorrhea by enhancing blood circulation in Korea. In Chinese medicine, APP is often used with LJH to synergistically improve the function of both herbs [10]. Since APP itself did not reduce energy and lipid metabolism in OVX rats, GJE was added to the mixture in the present study. GJE and its major components such as geniposides, genipin and crocetin have been studied previously with respect to various functions [12,34–36]. They exhibited anti-inflammatory activities to ameliorate atopic dermatitis and arthritis and have anti-depressant, anti-oxidant, anti-hypertensive and hypoglycemic activities [35, 36]. Chen et al. [12] demonstrated that GJE activates PPAR-γ to induce adipocyte differentiation, decreases glucose release from the liver and enhances glycogen storage in skeletal muscle cells. This hypoglycemic effect is reported to be similar to thiazolidinediones, known as PPAR-γ activators, by stimulating of PPAR-γ activity and/or its expression to increase glucose uptake in adipose tissues [12]. In the present study, APP improved glucose tolerance by increasing insulin secretion in the first part of OGTT, but also lowered insulin sensitivity more than the control. However, APP+LJH+GJE increased insulin sensitivity more than APP alone during ITT. This was associated with the decrease of hepatic PEPCK expression, a regulatory enzyme for gluconeogenesis, potentiating the phosphorylation of GSK-1β, which is involved in insulin signaling. In addition, treating with both APP and APP+LJH+GJE resulted greater liver expression of PGC-1α, the upstream modulator of PPAR-γ. Not only GJE but also APP may improve glucose metabolism by potentiating hepatic insulin signaling. Therefore, APP+LJH+GJE may alleviate the impairment of hepatic glucose metabolism due to estrogen deficiency in menopausal women.

Potential toxicity of dietary and medicinal herbs need to be evaluated prior to their use. Although some plants of the genus *Artemesia* are known to be toxic, *Artemisia princeps* is a commonly used herb in traditional medicines, teas, and as a cooked green vegetable. It is rich in a variety of nutrients and is generally considered to a safe food [37]. *Leonurus japonicas* is a very commonly used herb in Oriental Medicine with a well-established record for safe use; one of its valuable properties is as a hepatoprotective intervention against liver toxins [38]. The ripe fruits of *Gardenia jasminoides* contain geniposide, an iridoid glycoside, which was determined in this study. A toxicology study established the LD50 of geniposide to be 1431.1 mg/kg body weight and liver toxicity was seen at 574 mg/kg [39]. When geniposide was administered at 24.3 and 72.9 mg/kg body weight for 90 days there was no indication of toxicity. The doses used in this study would only result in doses of about 1 mg/kg body weight. Therefore, it is apparent that the doses used in this study were within a very safe range. Nevertheless, it is important to be aware of its potential toxicity at higher doses, and possibly in individuals with greater sensitivity. In the present study no rats exhibited any side-effect of APP, LJH and GJE at the doses they were given. No rats died during the experimental period and no organs had apparent damage when they were examined. Serum ALT and AST levels in the treatment groups were lower than the control group, indicating no hepatotoxicity. Furthermore, these herbs have not been reported to have side effects and toxicity and they are registered as foods by the Korean Food and Drug Administration. However, since they have been demonstrated to cause apoptosis of carcinoma cells [37, 38], they might have some toxicity.

CONCLUSIONS

APP alone improved dyslipidemia and glucose intolerance by potentiating glucose-stimulated insulin secretion as much as APP+LJH+GJE in OVX rats. However, energy expenditure, fatty acid oxidation and insulin resistance was inhibited by APP+LJH+GJE as much as the positive-control. APP and APP+LJH+GJE potentiated PGC-1α expression in the liver and improved glucose tolerance in estrogen deficient rats. APP+LJH+GJE inhibited gluconeogenesis in the fed state by suppressing PEPCK expression. These results suggested that the combination of APP+LJH+GJE attenuated various menopausal symptoms in OVX rats. Thus, it may have potential as a therapeutic agent for the treatment of postmenopausal symptoms.

Funding

This work was supported by "Food Functionality Evaluation program" under the Ministry of Food, Agriculture, Forestry and Fisheries in Korea in 2014 and by Korea Food Research Institute (E0150302-02).

Availability of data and materials

All data and materials are contained and described within the manuscript.

Author's contributions

SP, HJY and DYK participated in designing the study and writing the manuscript. BRM and SK conducted biochemical experiments. MJK quantified individual components of herbs. SK and ARK participated by conducting the animal study. All authors read and approved the final manuscript.

Competing interests

The authors declare that there are no conflicts of interest.

Ethics approval

All surgical and experimental procedures were approved by Hoseo University Animal Care and Use Review Committee (2013–04), which reviewed the procedures based on NIH Guidelines.

ABBREVIATION

APP	*Artemisia princeps Pamp*
LJH	*Leonurus japonicas Houtt*
GJE	*Gardenia jasminoides Ellis*
OVX	ovariectomy
OGTT	oral glucose tolerance
VO_2	average oxygen consumption
VCO_2	average carbon dioxide production

PGC-1α	peroxisome proliferator-activated receptor-γ coactivator-1α
CPT-1	carnitine palmitoyltransferase-1
ACC	acetyl CoA carboxylase
SREBP-1c	sterol regulatory element-binding protein-1c
FAS	fatty acid synthase
En%	energy percent
ITT	insulin tolerance test
DEXA	dual-energy X-ray absorptiometry
BMD	bone mineral density
AST	aspartate aminotransferase
ALT	alanine aminotransferase
HOMA-IR	homeostasis model assessment estimate of insulin resistance
AUCG	area under the curves of glucose
AUCI	area under the curves of insulin
ANOVA	analysis of variance
PPAR	peroxisomal proliferator-activated receptor

REFERENCES

1. Nelson HD. Commonly used types of postmenopausal estrogen for treatment of hot flashes: scientific review. JAMA. 2004;291:1610–20. doi: 10.1001/jama.291.13.1610.
2. Jia G, Aroor AR, Sowers JR. Estrogen and mitochondria function in cardiorenal metabolic syndrome.Prog Mol Biol Transl Sci. 2014;127:229–49. doi: 10.1016/B978-0-12-394625-6.00009-X.
3. Mauvais-Jarvis F, Clegg DJ, Hevener AL. The role of estrogens in control of energy balance and glucose homeostasis. Endocr Rev. 2013;34:309–38. doi: 10.1210/er.2012-1055.
4. Rossouw JE, Anderson GL, Prentice RL, LaCroix AZ, Kooperberg C, Stefanick ML, et al. Risks and benefits of estrogen plus progestin in healthy postmenopausal women: principal results From the Women's Health Initiative randomized controlled trial. JAMA. 2002;288:321–33. doi: 10.1001/jama.288.3.321.
5. Roberts H, Lethaby A. Phytoestrogens for menopausal vasomotor symptoms: a Cochrane review summary. Maturitas. 2014;78:79–81. doi: 10.1016/j.maturitas.2014.04.004.
6. Ko BS, Lee HW, da Kim S, Kang S, Ryuk JA, Park S. Supplementing with Opuntia ficus-indica Mill and Dioscorea nipponica Makino extracts synergistically attenuates menopausal symptoms in estrogen-deficient rats. J Ethnopharmacol. 2014;155:267–76. doi: 10.1016/j.jep.2014.05.020.
7. Ko BS, Kim da S, Kang S, Ryuk JA, Park S. Prunus mume and Lithospermum erythrorhizon Extracts Synergistically Prevent Visceral Adiposity by Improving Energy Metabolism through Potentiating Hypothalamic Leptin and Insulin Signalling in Ovariectomized Rats. Evid Based Complement Altern Med.2013;2013:750986.
8. Ryu KR, Choi JY, Chung S, Kim DH. Anti-scratching behavioral effect of the essential oil and phytol isolated from Artemisia princeps Pamp. in mice. Planta Med. 2011;77:22–6. doi: 10.1055/s-0030-1250119.
9. Trinh HT, Lee IA, Hyun YJ, Kim DH. Artemisia princeps Pamp. Essential oil and its constituents eucalyptol and alpha-terpineol ameliorate bacterial vaginosis and vulvovaginal candidiasis in mice by inhibiting bacterial growth and NF-kappaB activation. Planta Med. 2011;77:1996–2002. doi: 10.1055/s-0031-1280094.
10. Shang X, Pan H, Wang X, He H, Li M. Leonurus japonicus Houtt.: ethnopharmacology, phytochemistry and pharmacology of an important traditional Chinese medicine. J Ethnopharmacol. 2014;152:14–32. doi: 10.1016/j.jep.2013.12.052.
11. Liang H, Liu P, Wang Y, Song S, Ji A. Protective effects of alkaloid extract from Leonurus heterophyllus on cerebral ischemia reperfusion injury by middle cerebral ischemic injury (MCAO) in rats. Phytomedicine.2011;18:811–8. doi: 10.1016/j.phymed.2011.01.020.
12. Chen YI, Cheng YW, Tzeng CY, Lee YC, Chang YN, Lee SC, et al. Peroxisome proliferator-activated receptor activating hypoglycemic effect of Gardenia jasminoides Ellis aqueous extract and improvement of insulin sensitivity in steroid induced insulin resistant rats. BMC Complement Altern Med. 2014;14:30. doi: 10.1186/1472-6882-14-30.
13. Higashino S, Sasaki Y, Giddings JC, Hyodo K, Sakata SF, Matsuda K, et al. Crocetin, a carotenoid from Gardenia jasminoides Ellis, protects against hypertension and cerebral thrombogenesis in stroke-prone spontaneously hypertensive rats. Phytother Res. 2014;28:1315–9. doi: 10.1002/ptr.5130.
14. Kojima K, Shimada T, Nagareda Y, Watanabe M, Ishizaki J, Sai Y, et al. Preventive effect of geniposide on metabolic disease status in spontaneously obese type 2 diabetic mice and free fatty acid-treated HepG2 cells. Biol Pharm Bull. 2011;34:1613–8. doi: 10.1248/bpb.34.1613.
15. Song X, Zhang W, Wang T, Jiang H, Zhang Z, Fu Y, et al. Geniposide plays an anti-inflammatory role via regulating TLR4 and downstream signaling pathways in lipopolysaccharide-induced mastitis in mice. Inflammation. 2014;37:1588–98. doi: 10.1007/s10753-014-9885-2.
16. Tenore GC, Ritieni A, Campiglia P, Novellino E. Nutraceutical potential of monofloral honeys produced by the Sicilian black honeybees (Apis mellifera ssp. sicula) Food Chem Toxicol. 2012;50:1955–61. doi: 10.1016/j.fct.2012.03.067.
17. Zhishen J, Mengcheng T, Jianming W. The determination of flavonoid contents in mulberry and their scavenging effects on superoxide radicals. Food Chem. 1999;64:555–9. doi: 10.1016/S0308-8146(98)00102-2.
18. Reeves PG, Nielsen FH, Fahey GC., Jr AIN-93 purified diets for laboratory rodents: final report of the American Institute of Nutrition ad hoc writing committee on the reformulation of the AIN-76A rodent diet. J

Nutr. 1993;123:1939–51.

19. Niwa H, Ogawa Y, Kido Y, Abe Y, Kobayashi M, Mori T, et al. The rate of lipid oxidation in septic rat models. Jpn J Surg. 1989;19:439–45. doi: 10.1007/BF02471625.

20. Lusk G. Animal calorimetry: Twenty-fourth paper. Analysis of the oxidation of mistures of carbohydrate and fat. J Biol Chem. 1924;59:41–2.

21. Tuomilehto J. Point: a glucose tolerance test is important for clinical practice. Diabetes Care.2002;25:1880–2. doi: 10.2337/diacare.25.10.1880.

22. Stumvoll M, Mitrakou A, Pimenta W, Jenssen T, Yki-Jarvinen H, Van Haeften T, et al. Use of the oral glucose tolerance test to assess insulin release and insulin sensitivity. Diabetes Care. 2000;23:295–301. doi: 10.2337/diacare.23.3.295.

23. Sebokova E, Klimes I, Moss R, Stolba P, Wiersma MM, Mitkova A. Muscle GLUT 4 protein levels and impaired triglyceride metabolism in streptozotocin diabetic rats. Effect of a high sucrose diet and fish oil supplementation. Ann N Y Acad Sci. 1993;683:218–27. doi: 10.1111/j.1749-6632.1993.tb35710.x.

24. Lambrinoudaki I. Progestogens in postmenopausal hormone therapy and the risk of breast cancer. Maturitas. 2014;77:311–7. doi: 10.1016/j.maturitas.2014.01.001.

25. Aidelsburger P, Schauer S, Grabein K, Wasem J. Alternative methods for the treatment of post-menopausal troubles. GMS Health Technol Assess. 2012;8:Doc03.

26. Davis SR, Castelo-Branco C, Chedraui P, Lumsden MA, Nappi RE, Shah D, et al. Understanding weight gain at menopause. Climacteric. 2012;15:419–29. doi: 10.3109/13697137.2012.707385.

27. Polotsky HN, Polotsky AJ. Metabolic implications of menopause. Semin Reprod Med. 2010;28:426–34. doi: 10.1055/s-0030-1262902.

28. Mirebeau-Prunier D, Le Pennec S, Jacques C, Gueguen N, Poirier J, Malthiery Y, et al. Estrogen-related receptor alpha and PGC-1-related coactivator constitute a novel complex mediating the biogenesis of functional mitochondria. FEBS J. 2010;277:713–25. doi: 10.1111/j.1742-4658.2009.07516.x.

29. Kim MJ, Park JH, Kwon DY, Yang HJ, da Kim S, Kang S, et al. The supplementation of Korean mistletoe water extracts reduces hot flushes, dyslipidemia, hepatic steatosis, and muscle loss in ovariectomized rats. Exp Biol Med. 2015;240:477–87. doi: 10.1177/1535370214551693.

30. Bourdoncle A, Labesse G, Margueron R, Castet A, Cavailles V, Royer CA. The nuclear receptor coactivator PGC-1alpha exhibits modes of interaction with the estrogen receptor distinct from those of SRC-1. J Mol Biol. 2005;347:921–34. doi: 10.1016/j.jmb.2005.01.048.

31. Handschin C, Spiegelman BM. Peroxisome proliferator-activated receptor gamma coactivator 1 coactivators, energy homeostasis, and metabolism. Endocr Rev. 2006;27:728–35. doi: 10.1210/er.2006-0037.

32. Kruszynska YT, Mukherjee R, Jow L, Dana S, Paterniti JR, Olefsky JM. Skeletal muscle peroxisome proliferator- activated receptor-gamma expression in obesity and non- insulin-dependent diabetes mellitus. J Clin Invest. 1998;101:543–8. doi: 10.1172/JCI1076.

33. Kovacs P, Stumvoll M. Fatty acids and insulin resistance in muscle and liver. Best Pract Res Clin Endocrinol Metab. 2005;19:625–35. doi: 10.1016/j.beem.2005.07.003.

34. Liu H, Chen YF, Li F, Zhang HY. Fructus Gardenia (Gardenia jasminoides J. Ellis) phytochemistry, pharmacology of cardiovascular, and safety with the perspective of new drugs development. J Asian Nat Prod Res. 2013;15:94–110. doi: 10.1080/10286020.2012.723203.

35. Zhang H, Xue W, Wu R, Gong T, Tao W, Zhou X, et al. Rapid Antidepressant Activity of Ethanol Extract of Gardenia jasminoides Ellis Is Associated with Upregulation of BDNF Expression in the Hippocampus. Evid Based Complement Altern Med. 2015;2015:761238.

36. Sung YY, Lee AY, Kim HK. The Gardenia jasminoides extract and its constituent, geniposide, elicit anti-allergic effects on atopic dermatitis by inhibiting histamine in vitro and in vivo. J Ethnopharmacol.2014;156:33–40. doi: 10.1016/j.jep.2014.07.060.

37. Mao F, Zhang L, Cai MH, Guo H, Yuan HH. Leonurine hydrochloride induces apoptosis of H292 lung cancer cell by a mitochondria-dependent pathway. Pharm Biol. 2015;53:1684–90. doi: 10.3109/13880209.2014.1001406.

38. Bang MH, Han MW, Song MC, Cho JG, Chung HG, Jeong TS, et al. A cytotoxic and apoptosis-inducing sesquiterpenoid isolated from the aerial parts of Artemisia princeps PAMPANINI (Sajabalssuk)Chem Pharm Bull. 2008;56:1168–72. doi: 10.1248/cpb.56.1168.

CHAPTER 8

REVEALING THE DYNAMICS OF THYLAKOID MEMBRANES IN LIVING CYANOBACTERIAL CELLS

Laura-Roxana Stingaciu, HughO'Neill, Michelle Liberton, Volker S. Urban, Himadri B. Pakrasi & MichaelOhl

Cyanobacteria dominated Earth's biosphere for billions of years and are the reason for the oxygen atmosphere on our planet[1]. Thylakoid membranes in cyanobacteria are not found in chloroplasts as in plants and eukaryotic algae[2], but typically occupy the peripheral region of the cell interior[3]. In Synechocystis sp. PCC 6803, the thylakoid membranes form concentric layers that tend to follow the shape of the cell envelope. Electron tomography[4] and other high-resolution structural studies like super-resolution microscopy with fluorescent tagging[5], fluorescence recovery after photobleaching[6] and NMR spectroscopy[7] have given reliable information about thylakoid membrane structure and architecture within cells.

Among non-invasive techniques, Small Angle Neutron Scattering (SANS) in combination with other methods[8,9,10,11,12] is able to provide spatially averaged structural information about membranes systems under different experimental conditions, in algal cells[8,9,10,11] and in isolated thylakoid membrane sistems[11,12]. The SANS data in Liberton et al. (2013) showed that within wild type cyanobacterial cells, the thylakoid membranes are arranged in various architectures, with center-to-center distances between 630 Å and 110 Å. This complex spacing corresponds to membrane pairs of different organizations: from pairs with sufficient space to accommodate light-harvesting complexes (phycobilisomes) within the interthylakoidal space, up to very closely appressed thylakoid membrane pairs that cannot accommodate light-harvesting antennas. These findings were acknowledged by[11] in their review article, and Nagy et al. (2011) observed similar architecture in eukaryotic algae and plant thylakoids[9]. Because SANS is a sensitive technique capable of detecting repeat distances in large populations of cells, even membrane architectures that are not readily seen in small samples of cells examined by TEM are revealed by SANS.

However, to date little information exists about the internal dynamics of thylakoids in terms of flexibility and mobility of the membranes in their native environment, since biophysical studies are typically carried out under defined conditions in dilute aqueous environments that do not reflect the complex crowded environment of a living cell. In this work, we observe the mobility of thylakoid membranes in live cyanobacterial cells by neutron spectroscopy. Neutron Spin-Echo[13](NSE) is a high-resolution spectroscopic technique that has already proved to be a successful method of studying the structure and dynamics of model bilayer lipid membranes[14,15,16,17] and proteins in solution[18] but has thus far not been used to study the internal dynamics of intact living cell components. This technique allows simultaneous observation of correlation times and length scales and thereby can directly deduce the geometry and modes of motion observed, such as membrane undulation or diffusion, which are not possible using purely energy-resolved spectroscopic techniques.

Citation: Stingaciu L-R, O'Neill H, Liberton M, Urban VS, Pakrasi HB, Ohl M. Revealing the Dynamics of Thylakoid Membranes in Living Cyanobacterial Cells. Scientific Reports. 2016;6:19627. doi:10.1038/srep19627.

RESULTS AND DISCUSSION

Assessment of the Internal Dynamics of Thylakoids

We used the NSE spectrometer at the Spallation Neutron Source[19] to study wild type cyanobacterial cells in a heavy water (deuterium oxide) solution. The high content of deuterium (D) in the heavy-water-filled spaces of the cyanobacterium provides excellent neutron contrast for the hydrogen-rich (H) lipid membranes and, therefore, the correlations between neighboring membranes dominate the coherent scattering signal in our probed window of observation, as shown previously[8,9,10,11,12]. Neutron data were collected at 20 °C during 12-hour light and dark alternating cycles to mimic the circadian clock. NSE measures the so-called intermediate scattering function which represents pair-correlation distances observed in reciprocal space (as in diffraction) and the time-dependent loss or relaxation of such correlations due to the mobility of the correlated spatial features. The reciprocal space coordinate or scattering vector magnitude, q, is inversely proportional to distances, d, in real space as $d=2\pi/q$, and the corresponding time information is expressed as either relaxation times or inversely as relaxation rates Γ.

The normalized intermediate scattering functions of cyanobacteria (Fig. 1) show identical relaxation modes for both light and dark states at q values higher than 0.07 $Å^{-1}$ ($d \leq 90$ Å). Conversely, at q values lower than 0.07 $Å^{-1}$ ($d \geq 90$ Å) considerable difference exists between the two states, with slower decay in measurements collected in light compared to measurements collected in dark. NSE data were fitted by a stretched exponential function with a stretching exponent of 2/3, as predicted by Zilman & Granek[20] for bilayer membranes (Table 1 and Methods). The spatial or q dependence of the relaxation rate Γ displayed in Fig. 2 is the major result of this study since its variations describe the dynamics of our system. Starting from the right side of Fig. 2 (high q values), a q^3 dependence of the relaxation rate Γ is observed, which is a typical signature of the undulation motions in lipid bilayer membranes[21] at length scales much smaller than the inter-membrane distance (the interthylakoidal space). In general, the tendency of decreasing magnitude of Γ with decreasing q is a common observation, as relaxations related to larger distances tend to take longer times. The absolute value of the relaxation rate at these high q values (short correlation distances) is similar for both light and dark measurements. More interestingly, we also observe a strong deviation from the $\Gamma(q^3)$ dependence when increasing the observed correlation distances to q values below 0.07 $Å^{-1}$ ($d \geq 90$ Å), peaks region in Fig. 2. This strong deviation from q^3 dependence represents significant excess mobility that exists in addition to the underlying q^3 dependent undulation dynamics. Between 0.07 $Å^{-1}$ and 0.03 $Å^{-1}$, Γ increases up to a maximum value at around 0.046 $Å^{-1}$ and then decreases again slowly. This corresponds to a speed-up of the internal motion up to a very fast relaxation (peak in Γ relaxation rate represents a minimum of the relaxation time) followed by a slow-down process. This effect is more pronounced during the dark cycle compared to the light cycle.

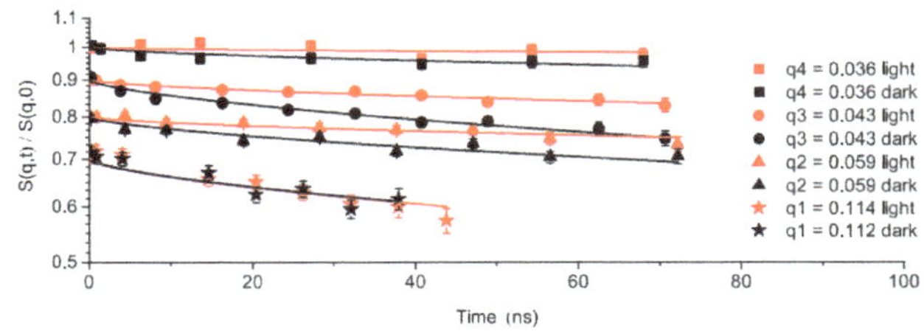

Figure 1. S(q, t)/S(q, 0) of cyanobacteria at 20 °C for light and dark states. All scattering functions start at unity and have been shifted for better visualization. Solid lines represent the stretched exponential function fitting with stretching exponent of 2/320; 0.035$Å^{-1}$, 0.05$Å^{-1}$ and 0.095$Å^{-1}$ minimum value of momentum transfer were measured according to the correlation peaks observed in the SANS experiment8 . Five different groupings of the time channels were applied, leading to 15 q values. Only four q values for each state are shown here in relation to Figs 2 and 3.

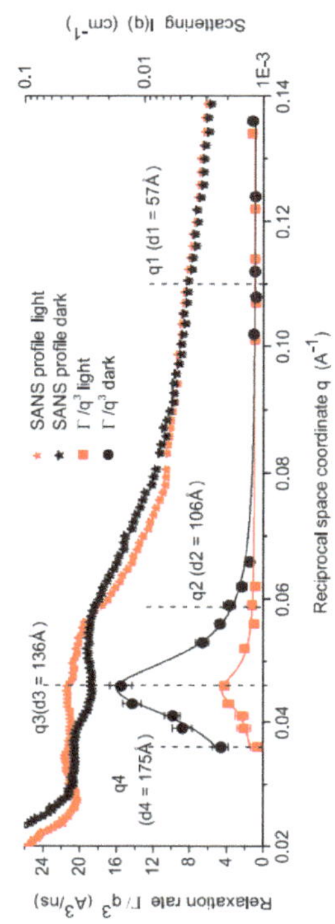

Figure 2: q^3 **dependence of the decay rate Γ for cyanobacteria during light and dark states.**

Table 1: Characteristics of membrane relaxation and bending fluctuations.

Sample	q ($Å^{-1}$)	Γ/q^3 ($Å^3$/ns)	Γ (1/ns)	τ (ns)	$\tilde{\kappa}$ (k_BT) **
PCC6803 light	0.114	0.92	1.37E-3	729.9	7655.45
PCC6803 dark	0.112	0.94	1.33E-3	751.9	7304.45
PCC6803 light	0.059	1.19	2.44E-4	4098.3	-
PCC6803 dark	0.059	3.69	7.59E-4	1317.5	-
PCC6803 light	0.043	3.75	2.98E-4	3355.7	-
PCC6803 dark	0.043	14.34	1.14E-3	877.2	-
PCC6803 light	0.036	0.61	3.35E-5	29850.7	-
PCC6803 dark	0.036	4.61	2.29E-4	4366.8	-

The q^3 dependent bending undulations describe correlations within a single membrane and are controlled by the elastic modulus with a higher bending coefficient or membrane stiffness corresponding to faster relaxation rates. From the q^3 dependence of the relaxation rate Γ at high q values effective bending coefficients of $7655k_BT$ for membrane structure in light and $7304k_BT$ for membrane structure in dark were calculated[20] (Table 1). The calculated bending coefficients show that during light measurement the membrane is stiffer compared to the dark state in which the bending coefficient shows a more flexible system. This is in agreement with a previous study on chloroplast remodeling within living cells of Chlamydomonas reinhardtii, which reports more stacked thylakoids with well-defined periodicity during the light harvesting state in comparison with a dark state in which the thylakoids are partially unstacked and undulated due to physiological responses of thylakoids to anaerobiosis[22].

Correlation of Dynamics with Membrane Flexibility

For model lipid bilayers, dynamic modes in addition to undulation have been interpreted as thickness fluctuations[14,15,16,17]. We propose a different explanation for our observations in cyanobacteria. The additional dynamics appear in the range of average length between 100 Å–180 Å that is rich with scattering intensity arising from correlations between very closely appressed thylakoid membrane pairs that cannot accommodate light harvesting antennas, as shown by[8,11] and observed also by[10] for other algal cells.

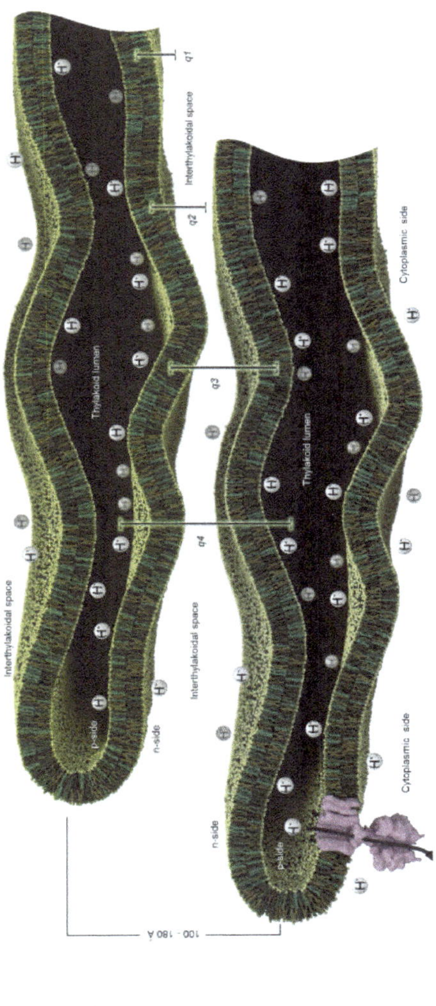

Figure 3: Thylakoid membrane sheet-like structure dynamics.

The dynamics at this length scale therefore should contain strong contributions from the relative motion of neighboring membranes across the interthylakoid space (Fig. 3). In contrast to the q^3 dependent bending undulations of individual membranes, motions of membrane pairs relative to each other in the direction across the interthylakoid space (large correlation distances) may be thought of as the local observation of a diffusive motion of large membrane areas. This diffusive component of the membranes motion is only observable through the time-dependent decay of correlations between membranes, and would not appear in the coherent signal of isolated single membranes. Importantly, although a stiffer membrane, submitted to a random undulation, relaxes faster to its flat state, this membrane is less efficient in exploring the available volume, in our case the interthylakoid space. The same space is sampled faster by a flexible membrane structure that is very efficient in exploring the available volume[20] (Table 1). The membrane exhibits a speed-up/slow-down of undulation motion with different amplitude when two adjacent membranes feel each other's influence in light or dark. Softer thylakoid membranes in the dark display three-to four-fold excess dynamics compared to membranes under light conditions. The more rigid membrane in light takes a longer time to relax which is the time necessary for the membrane structure to move through the available space. This represents the relaxation times we observe by NSE of about 2000 ns–3000 ns in light and only 600 ns–800 ns in the dark, for the peak position in Fig. 2.

The Relation between Excess Dynamics and Physiological Processes within Thylakoids

There could be several possible explanations for the difference in membrane mobility observed between light and dark. One reason could lie in the function of the photosynthetic apparatus in the thylakoid membrane. Absorption of light by photosystem II initiates water oxidation that results in transfer of photo-generated reducing equivalents to photosystem I. This results in an excess of protons in the lumenal space of the thylakoid membrane (Fig. 3) that is alleviated by moving an electrochemical gradient (H^+) through the membrane by chemiosmosis[23]. In light, photosynthesis maintains the presence of a proton concentration gradient. The increased density of protons within the lumenal space could result in increased pressure upon membranes. Such pressurized membranes will appear stiffer when observed during light and display restricted motility. The protons move across the thylakoid membrane by the function of ATP-Synthase. They are the motive force to generate ATP. In the dark cycle when light harvesting and photosynthesis stop, the difference in the proton density between the lumenal and the interthylakoidal space is finally, equilibrated over time. Pressure is released, and the membrane becomes more relaxed and flexible and can move easily within the available space.

Furthermore, refinements to the chemiosmotic hypothesis have proposed that localized domains of protons are confined to the region near the thylakoid membrane, rather than being delocalized within the lumenal space[24]. Such organization suggests that in addition to the increased pressure by the presence of protons, the inner leaflet of the thylakoid membrane bilayer (the p-side, Fig. 3) has a higher positive charge than the outer leaflet (the n-side). The negative outer leaflets would therefore repel each other when the undulating motion brought two neighboring membranes in close proximity. This repelling will constrict the excess undulation motion during light. The electrochemical gradient is dissipated by chemiosmosis. The repelling will stop and the membranes recover their higher flexibility during dark when the proton density equilibrates.

Another explanation of the observed differences in membrane excess dynamics between light and dark could be the dynamic control of protein diffusion within the thylakoid lumen[25]. In higher plant thylakoids, the lumenal space undergoes a significant expansion in light. In a study involving electron microscopy, light scattering and theoretical modeling of the lumenal space in Arabidopsis, it was shown that reversible changes in the thylakoid lumen induced by alteration of the light environment would modulate the diffusion of proteins within the restricted space of the lumenal compartment. This is done in order to facilitate PC-mediated electron transport and to repair photo-damaged photosystem II complexes[25]. If similar effects are present in cyanobacterial thylakoids, then the expansion of the lumen restricts the undulation motions of the membranes, i.e. membranes observed during light will appear again stiffer.

Is not certain whether the observed changes may constitute a mechanism by which cyanobacteria regulate photosynthetic activity in response to light intensity, or may be a byproduct of photosynthetic activity, e.g., through charge gradients across lipid membranes, or through an unknown, more indirect chain of events. However, this study is the first direct measurement of the undulation dynamics of photosynthetic membranes in living prokaryotic cells and adds to the open question of whether other components of the photosynthetic machinery exhibit internal dynamics and how these dynamics might affect the yield of photosynthesis. The question of how the plasticity and dynamics of thylakoid membranes can be regulated, and related to the functions of sub-cellular architectures, is now an open field in biological research. This holds promise for important discoveries and we are giving a first demonstration of the possibility to observe directly such complex dynamic relations.

METHODS

Sample Preparation

Synechocystis sp. PCC 6803 was grown in BG11 medium at 20 °C and 30 µmole photons·m^{-2}·s^{-1} light intensity. The cells were exchanged into a D_2O based BG11 medium by three sequential centrifugation and resuspension steps[8]. For the final resuspension step the cells were suspended in 4 ml of D_2O-based BG11 media (OD_{750nm}=41.6, 1 cm pathlength). A control sample from the same culture was maintained under conditions identical to the NSE sample (OD_{750nm}=36.5, 1 cm pathlength). There was no significant change in the OD_{750nm} of the sample over the time course of the experiment.

Neutron Spin Echo experiment

NSE experiments were performed using the NSE spectrometer at the Spallation Neutron Source[19], Oak Ridge National Laboratory. Measurements were carried out in 4 mm-path quartz cells. A closed sample environment was used to allow dark adaptation of the samples. The sample holder was equipped with a lighting apparatus that illuminated the samples with cool white LED lights set at 100 µmole photons·m^{-2}·s^{-1}. The NSE experiment was performed at incident wavelengths of 9 Å and 11 Å accessing a dynamical range of $0.1 \leq \tau_{max} \leq 130$ ns at different momentum transfers for each sample according to correlation peaks observed in the SANS experiment[8]. Solid angles of 0.035 Å^{-1}, 0.05 Å^{-1} and 0.095 Å^{-1} were measured (minimum value of momentum transfer). NSE spectra were taken at 20 °C under 12 h light and dark alternating cycles to mimic the circadian clock. One hour of equilibration was allowed between each cycle so that cells may adapt to the light or dark condition before data collection. Graphite foil was used as a standard elastic reference and solvent BG11 measurements were also necessary for proper data reduction. The entire experiment lasted 14 days: 7 days for cyanobacterial sample and 7 days for reference and background sample.

Data Analysis

The data analysis was performed with the standard reduction software package of the NSE-SNS instrument called ECHODET. Data were acquired with sufficient statistics so the entire wavelength spectra could be divided in 5 different solid angles for a better discretization of the q dependence. The intermediate scattering functions were fitted by a stretched exponential function with a stretching exponent of 2/3 predicted by Zilman & Granek[20] for single membrane (Fig. 1):

$$\frac{S(q,t)}{S(q,0)} = \exp[-(\Gamma t)^{2/3}] \tag{1}$$

were Γ is the relaxation rate. Using this approach for fitting data one might argue that the translational diffusion of the entire cyanobacteria cell has also a contribution to the overall dynamics of the system. However, the cyanobacteria cell has a diameter of about 3.8 – 5.2 µm and the diffusion coefficient of such a large cell is outside the NSE window we can access for measurements. Therefore, translational diffusion was neglected from further analysis. To account for the bending motion of the membranes we applied the single membrane fluctuation model of Zilman & Granek[20]. This model, based on mode coupling theory, explains the dependence with q^3 of the relaxation rate for systems where the hydrodynamic interactions dominate and the wavelengths are shorter than the characteristic correlation length, in our case 1/q « interthylakoidal distance[21]:

$$\Gamma = 0.025\alpha\sqrt{\frac{k_B T}{\tilde{\kappa}}} \cdot \frac{k_B T}{\eta_{D2O}} \cdot q^3 \tag{2}$$

with $\tilde{\kappa}$ as the effective bending modulus; α is a scaling factor equal unity; k_BT is the thermal energy; η_{D2O} is the viscosity of the solvent at T=20 °C. Γ/q^3 represents in fact the linear dependence observed at q values higher than 0.07 Å^{-1} (d ≤ 90 Å) in Fig. 2. The calculated values of bending coefficients are displayed in Table 1.

ACKNOWLEDGMENTS

Neutron beam time for this research has been allocated

by the Jülich Center for Neutron Science, JCNS1, Jülich Forschungszentrum GmbH, Germany, outstation to the Spallation Neutron Source (SNS), Oak Ridge, TN. The authors acknowledge Malcolm Cochran for the construction of the illumination setup and technical support during measurements, Dr. Piotr Zolnierczuk for providing automatic control of the illumination setup and software support during measurements, Rhonda Moody and Dr. Kevin Weiss from SNS biochemistry lab support, William B. O'Dell for assistance with Small-Angle Neutron Scattering (SANS) data, Dr. Jonathan Nickels, Prof. Dr. Dieter Richter and Dr. Michael Monkenbusch for discussion on data interpretation. The generation of neutrons for this research at ORNL's Spallation Neutron Source and High Flux Isotope Reactor was sponsored by the Scientific User Facilities Division, Office of Basic Energy Sciences, Office of Biological and Environmental Research, U.S. Department of Energy. M.L., H.B.P., H.O.N. and V.S.U. were supported by the Photosynthetic Antenna Research Center (PARC), an Energy Frontier Research Center funded by the U.S. Department of Energy, Office of Science and Office of Basic Energy Sciences, under Award Number DE-SC 0001035.

REFERENCES

1. Herrero A. & Flores E. (eds.) The Cyanobacteria: Molecular Biology, Genomics and Evolution.ISBN: 978-1-904455-15-8 (Caister Academic Press, 2008).
2. Liberton M. & Pakrasi H. B. in The Cyanobacteria: Molecular Biology, Genomics, and Evolution(Herrero A. & Flores E. eds) 217–287 (Horizon Scientific Press, Norwich, United Kingdom, 2008).
3. Mullineaux C. W. The thylakoid membranes of cyanobacteria: structure, dynamics and function. Aust. J. Plant Physiol. 26, 671–677 (1999).
4. Liberton M., Austin J. R. 2nd, Berg R. H. & Pakrasi H. B. Unique thylakoid membrane architecture of a unicellular N_2-fixing cyanobacterium revealed by electron tomography. Plant Physiol. 155, 1656–1666 (2011).
5. Manley S., Gunzenhauser J. & Olivier N. A starter kit for point-localization super-resolution imaging.Curr. Opin. Chemical Biol. 15, 813–821 (2011).
6. Mullineaux C. W., Tobin M. J. & Jones G. R. Mobility of photosynthetic complexes in thylakoid membranes. Nature 390, 421–424 (1997).
7. Tochio H. Watching protein structure at work in living cells using NMR spectroscopy. Curr. Opin. Chemical Biol. 16, 609–613 (2012).
8. Liberton M. et al. . Organization and flexibility of cyanobacterial thylakoid membranes examined by neutron scattering. J. Biol. Chem. 288(5), 3632–3640 (2013).
9. Nagy G. et al. . Reversible membrane reorganizations during photosynthesis in vivo: revealed by small-angle neutron scattering. Biochem. J. 436, 225–230 (2011).
10. Nagy G. et al. . Modulation of the multilamellar membrane organization and of the chiral macrodomains in the diatom Phaeodactylum tricornutum revealed by small-angle neutron scattering and circular dichroism spectroscopy. Photosyn. Res. 111, 71–79 (2012).
11. Ünnep R., Nagy G., Markó M. & Garab G. Monitoring thylakoid ultrastructural changes in vivo using small-angle neutron scattering. Plant Physiology and Biochemistry 81, 197–207 (2014).
12. Nagy G. et al. . Kinetics of structural reorganizations in multilamellar photosynthetic membranes monitored by small-angle neutron scattering. Eur. Phys. J. E Soft Matter 36(7), 69–80 (2013).
13. Mezei F. ed. Neutron Spin Echo. Lecture Notes in Physics. Berlin, Heidelberg, New York: Springer128, 1980).
14. Yi Z., Nagao M. & Bossev D. Bending elasticity of saturated and monosaturated phospholipid membranes studies by neutron spin echo technique. J. Phys. Condens. Matter 21, 155104 (2009).
15. Nagao M. Observation of local thickness fluctuations in surfactant membranes using neutron spin echo. Phys. Rev. E. 80, 031606 (2009).
16. Nagao M. Temperature and scattering contrast dependencies of thickness fluctuations in surfactant membranes. J. Chem. Phys. 135, 074704 (2011).
17. Woodka A. C. et al. . Lipid Bilayers and Membrane Dynamics: Insight into Thickness Fluctuations.Phys. Rev. Lett. 109, 058102 (2012).
18. Biehl R. & Richter D. Slow internal protein dynamics in solution. J. Phys. Condens. Matter 26, 503103 (2014).
19. Ohl M. et al. . The spin-echo spectrometer at the Spallation Neutron Source (SNS). Nuclear Instruments and Methods A 696, 85–99 (2012).
20. Zilman A. G. & Granek R. Undulations and dynamic structure factor of membranes. Phys. Rev. Lett.77, 4788–4791 (1996).
21. Ohta T. & Kawasaki K. Mode coupling theory of dynamic critical phenomena for classical liquids I.Progress of Theoretical Physics 55(5), 1384–1395 (1976).
22. Nagy G. et al. . Chloroplast remodeling during state transitions in Chlamydomonas reinhardtii as revealed by noninvasive techniques in vivo. PNAS 13(111), 5042–5047 (2014).
23. Mitchell P. Coupling of phosphorylation to electron and hydrogen transfer by a chemi-osmotic type of mechanism. Nature 191(4784), 144–148 (1961).
24. Dilley A. & Ewy R. G. Distinguishing between Luminal and Localized Proton Buffering Pools in Thylakoid Membranes. Plant Physiol. 122, 583–595 (2000).
25. Kirchhoff H. et al.. Dynamic control of protein diffusion within the granal thylakoid lumen. PNAS108(50), 20248–20253 (2011).

CHAPTER 9

A SINGLE-LEVEL TUNNEL MODEL TO ACCOUNT FOR ELECTRICAL TRANSPORT THROUGH SINGLE MOLECULE- AND SELF-ASSEMBLED MONOLAYER-BASED JUNCTIONS

Alvar R.Garrigues, LiYuan, LejiaWang, Eduardo R. Mucciolo, Damien Thompon, Enrique del Barco and ChristianA. Nijhuis

INTRODUCTION

The last decade has led to substantial advances in the understanding of electrical conduction through molecular systems, mostly due to an improved control of the different measuring techniques employed to study molecular junctions, including electromigrated single-electron transistors (SETs)[1,2,3,4,5,6,7,8], scanning tunneling microscopy (STM) break-junctions[9,10], and techniques enabling measuring self-assembled monolayers (SAMs) of molecules sandwiched between two macroscopic electrodes[11,12,13,14,15,16,17,18,19,20]. By far most fabrication techniques produce two-terminal junctions in the form of electrode-molecule-electrode, with either a single molecule or a self-assembled monolayer (SAM) as the active component. However, the lack of an electrical gating capability in two-terminal junctions complicates determination of the energy level alignment. STM break-junctions can be used to obtain statistics on electron transport at an individual molecular level but the supramolecular and the electronic structure of the junctions cannot be investigated independently. By contrast, it is possible to measure both the supramolecular and electronic structure of the SAMs immobilized on one of the electrodes (usually referred to as the bottom-electrode) using standard surface characterization techniques before fabrication of the top-contact. For most systems, however, it is not clear how much the energy level alignment changes once the molecules are in contact with the top-electrodes[21] and therefore models are often used to extract the relevant transport parameters. From a technological point of view, SAM-based junctions are promising candidates for molecular electronics technologies, with rectification ratios rapidly approaching values of commercial semiconductor devices[22,23]. On the other hand, three-terminal SETs electrically gate the molecular electronic states of an individual molecule, enabling the full spectroscopic resolution of the molecular energy landscape[1,2,3,4,5,6,7,8]. Comparative studies using these different techniques are required to establish the principles governing electrical conduction through molecular tunnel junctions under a wide range of experimental conditions.

Despite the increasing number of experimental reports, and to a lesser extent theoretical works, dedicated to the topic of electrical conduction through molecular junctions (see refs 24, 25, 26 and references therein), there remains the challenge of extracting fundamental characteristic junction parameters from the experiments in a consistent way. The complexity of the real devices, with large variations in the parameters that affect electrical conduction, and the different possible transport regimes, make the rigorous treatment of electron transport through junctions extremely difficult. In this

Citation: Garrigues, A. R. et al. A Single-Level Tunnel Model to Account for Electrical Transport through Single Molecule- and Self-Assembled Monolayer-based Junctions. Sci. Rep. 6, 26517; doi: 10.1038/srep26517 (2016).

context, first-principles calculations have the advantage that one can account for structural details of the junctions27·28. However, they are not generally applicable (i.e., they need to be tailored to the specific junction under study) and may require very long calculation times even on high-performance computing platforms. Alternatively, analytical models dealing mainly with the energetics of the junction, such as the Landauer formalism for coherent tunneling (or the rate equations for incoherent tunneling), are easy to employ but rely on rough approximations of some critical parameters and may miss some of the physics involved in the process, particularly when conduction is driven or assisted by interactions with the molecular surroundings (e.g., reorganization energies, many-body interactions, energy level renormalization, image charges, etc.).

There are ongoing theoretical efforts focused on solving the problem by integrating different effects to formulate a non-equilibrium theory in terms of molecular states that includes many-body correlations when computing electron transport through molecular junctions (for a detailed discussion on this see ref. 26). This is far from the goal of the present work, where we do not attempt to develop a new theory but instead aim to test the applicability of a well-known single-level tunneling model and variations thereof29·30·31·32, to capture the essential physics of molecular junctions. The key metric for success of our model is the ability to easily implement it using experimentallyobtained input parameters. In this article, we test this single-level tunneling model using properties of molecular diodes measured in two different test-beds: (i) Three-terminal SET experiments performed at different temperatures on individual $S(CH_2)_4Fc(CH_2)_4S$ molecules; and (ii) experiments in SAM-based EGaIn junctions of $S(CH_2)_{11}Fc_2$ molecules where two distinct molecular levels participate in charge transport22. In addition, we employ the model to interpret experiments performed on SAM-based EGaIn junctions of $S(CH_2)_nFcC_{13-n}$ molecules where the ferrocene (Fc) unit is placed at different positions (n) within the alkyl chain33. Fitting our model to the experimental data enables the determination of the electrostatic potential profile in the junction, which we find to be strongly non-linear, indicating the presence of significant electrical screening in the junction. Comparison between the different junctions shows that the single-level tunneling model satisfactorily describes both single molecule and SAM based junctions. The model captures the essential physics, including the asymmetry of the J(V) curves, the shape of the electrostatic potential profile and the temperature dependence of the transport excitations. Finally, we use this model to discuss in detail the range of validity of coherent versus incoherent theoretical formulations that have been used to rationalise some recent experimental results.

RESULTS

The single-level Transport Model: Coherent Versus Incoherent Tunneling

In the simplest approximation, a molecular junction can be represented by the schematic diagrams shown in Fig. 1, where the molecule sandwiched between two electrodes (three in the case of a SET) is represented by a discrete set of levels (with N representing the number of electrons in each level) separated from the electrostatic potential in the electrodes by tunnel barriers. The molecular levels closest to the electrostatic potentials of the leads at zero bias are known as the HOMO (highest occupied molecular orbital, N) and LUMO (lowest unoccupied molecular orbital, N+1) levels and are well separated in energy by a ~2–5 eV conductance gap in solid state junctions. The level closest to the Fermi energy of the electrodes governs the electrical conduction through the junctions for moderate bias potentials.

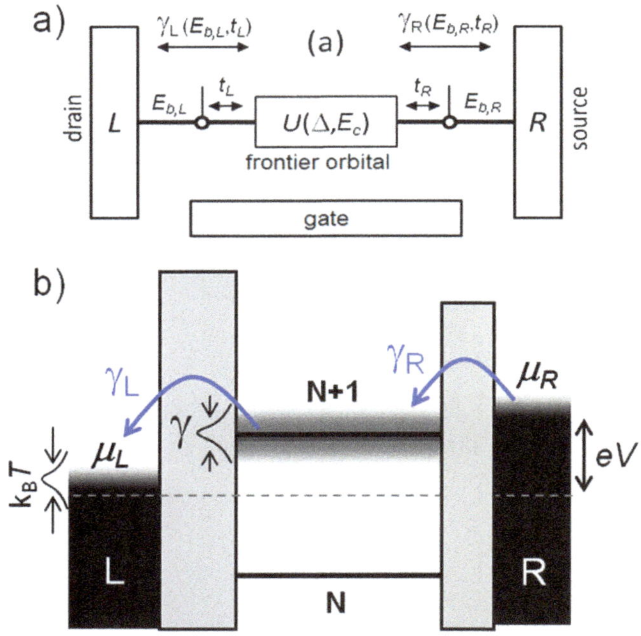

Figure 1: A molecular tunnel junction. Schematics of the electrical coupling (**a**) and corresponding energy level alignment (**b**) in a molecular junction composed of a molecule with multiple electrostatic levels sandwiched in between two electrodes (L (left): drain and R (right): source), including all the functional parameters that govern electron tunneling through the junction. These are the tunneling rates, $\gamma_{L/R}$, the binding energies, $E_{b,L/R}$, and the intermolecular couplings, $t_{L,R}$, between the molecule and left and right electrodes, with $\gamma=\gamma_L+\gamma_R$ being the level width. In addition, Δ represents the intramolecular level spacing, E_c the charging energy, and $\mu_{L/R}$ the electrostatic potentials of the leads.

Conduction through a molecule can, in principle, be understood in terms of sequential tunneling (either coherent or incoherent) involving one molecular level. In such a process the electron tunnels from the source into the molecule at a rate γ_R and then out into the drain lead at a rate γ_L. The ratio between these rates and the characteristic intramolecular relaxation times determines whether the electron transport occurs coherently or incoherently. This means that if the electron spends sufficient time inside the molecule to significantly couple with internal/external degrees of freedom (i.e., short relaxation time and short decoherence time relative to the tunneling rates) then the process will become incoherent, with an associated loss of phase in the case of elastic tunneling, and loss of both phase and energy in the case of inelastic tunneling. In wet electrochemical molecular charge-transfer processes, energy relaxation is usually facilitated by intramolecular vibrational modes and polaron excitations in the surrounding solvent molecules. In terms of Marcus theory34, these are known as inner- and outer-sphere molecular reorganization processes, respectively, and are characterized by an activation energy arising from the classical energy barrier governing the process. At sufficiently high temperatures, these processes lead to an exponential increase of the charge transfer (accompanied by an Arrhenius law and incoherent tunneling). However, in the junctions discussed in this work, where the molecules are present in solid-state devices and charge transport is studied in the absence of solvent, outer-sphere reorganization processes do not likely play an important role in the electrical conduction through the junction. This is not necessarily the case in all SAM-based junctions, since even in the absence of solvent molecules, neighboring molecules can play this role, particularly when considering complex molecules with high degrees of polarizability.

We expect that inner-sphere reorganization will always be important, and can be related to intra-molecular vibrational modes with energies usually below a few tens of millielectronvolts (<30 meV), which is smaller than or comparable to other energy scales relevant to the conduction process, such as thermal broadening at high temperatures ($k_BT = 13–86$ meV in the range 100–1000 K). Thus, when only one level participates in the conduction through the junction (which may be the case in many molecular junctions), the vibrational modes can be formally incorporated into an effective temperature-dependent level broadening (the higher the temperature the more vibrational modes will be sampled), which means that it is reasonable to still treat the transport process with simple single-level transport models, as we do in the present work. Note the different causes of level broadening, which we can separate into two main contributions: (i) broadening due to the coupling of the molecule to the electrodes, with tunneling rate $\gamma = \gamma_L + \gamma_R$, and, (ii) broadening arising from the coupling of the electron to other degrees of freedom of the molecule, γ_0. Of course, this is just a simplification (virtual transitions to high-energy molecular states take place during the conduction process, leading to renormalized parameters) and should be taken as such. However, this picture can be justified from a more fundamental calculation involving a non-equilibrium quantum mechanical treatment of the response of the molecule to its coupling to leads. When a single energy level is active in the molecule, and the coherent interactions between electrons in the molecule and those in the leads are negligible, the non-equilibrium expression for the current reduces to an expression similar to the Landauer-Büttiker formula30, namely,

$$I = \frac{2q}{h}\int dE\left[f\left(\frac{E-\mu_L}{k_BT}\right) - f\left(\frac{E-\mu_R}{k_BT}\right)\right] Tr\left\{\frac{\gamma_L(E)\gamma_R(E)}{\gamma_L(E)+\gamma_R(E)} Im[G^r(E)]\right\}, \quad (1)$$

where $f(x) = 1/(e^x+1)$, μ_L and μ_R are the chemical potential at the leads, $\gamma_L(E)$ and $\gamma_R(E)$ are the partial level widths due to the coupling to right and left leads, respectively, and G^r is the molecule's single-particle retarded Green's function, which contains a self-energy that takes into account the coupling to leads and to other degrees of freedom as an imaginary part, namely, a total level width. In fact, this has been the starting point of several successful theoretical descriptions of molecular electronic transport experiments (see, e.g., refs 35,36).

We acknowledge that there may be other processes playing important roles in the transport through solid-state junctions, such as intermolecular interactions or electrostatic potential changes due to electron charging or image-charge effects27, among others. Some of these could be incorporated into simple single-level transport models. As we show below, for a single-level junction system the coherent and incoherent tunneling treatments provide identical predictions of the current at all temperatures, provided that one incorporates into the level broadening the two effects discussed above concerning outer-sphere reorganization and vibrational modes. Therefore, even a single-level coherent transport model can explain the exponential thermal enhancement of the conductance through a molecular junction without the need to invoke Marcus relaxation processes. This is because the temperature dependence of the conductance naturally arises from the thermal broadening of the energy of the electrons in the leads (see, e.g., the work by van der Zant et al.37). In other words, a transition between a plateau-like to temperature-dependent conductance in a molecular junction should not be taken as conclusive proof of inelastic incoherent tunneling. Sequential tunneling may well be coherent and one simply cannot distinguish them when transport involves a single energy level without considering the ratio of the relaxation rates involved.

As mentioned before, the transition between coherent and incoherent tunneling is determined by the tunneling rates in and out of the molecule (when compared with the intrinsic molecular relaxation rates). As explained by Moth-Poulsen and Bjørnholm in their review article38, the tunneling rates for a complex molecule usually depend on the strength of the bond of the molecule to the electrodes, $E_{b,i}$, and on the intramolecular coupling, t_i, between the electrode and the molecular frontier orbital (see Fig. 1a), and are, in general, difficult to engineer. As mentioned above, the molecule-electrode couplings result in broadening of the molecular levels ($\gamma = \gamma_L + \gamma_R$). The capacitive couplings of the frontier orbital with the electrodes determine the electric potential drops at both sides of the active conduction unit (V_L and V_R) and provide asymmetry to the junctions when $L_L \neq L_R$. This asymmetry is parameterized by the dimensionless division parameter $\eta = V_R/(V_L + V_R)$, which gives the ratio of the voltage drop between the molecule and the right electrode with respect to the total voltage drop in the junction.

The molecular frontier orbital follows the average effective potential within the molecule and its energy can be expressed as $E_m = \mu_R(V) + \varepsilon + \eta qV$, where ε is the zero-bias energy offset between the molecular orbital and the electrodes, q is the electron charge, and μ_R is the electrochemical potential of the right electrode. The behaviour of E_m, μ_L and μ_R with respect to the bias potential V is shown in Fig. 2 for different values of η. For $\eta \sim 0.5$ the energy of the level remains constant as the bias potential is increased, while for $\eta \sim 0$ or ~ 1, the molecular level closely follows the right or left potentials, respectively.

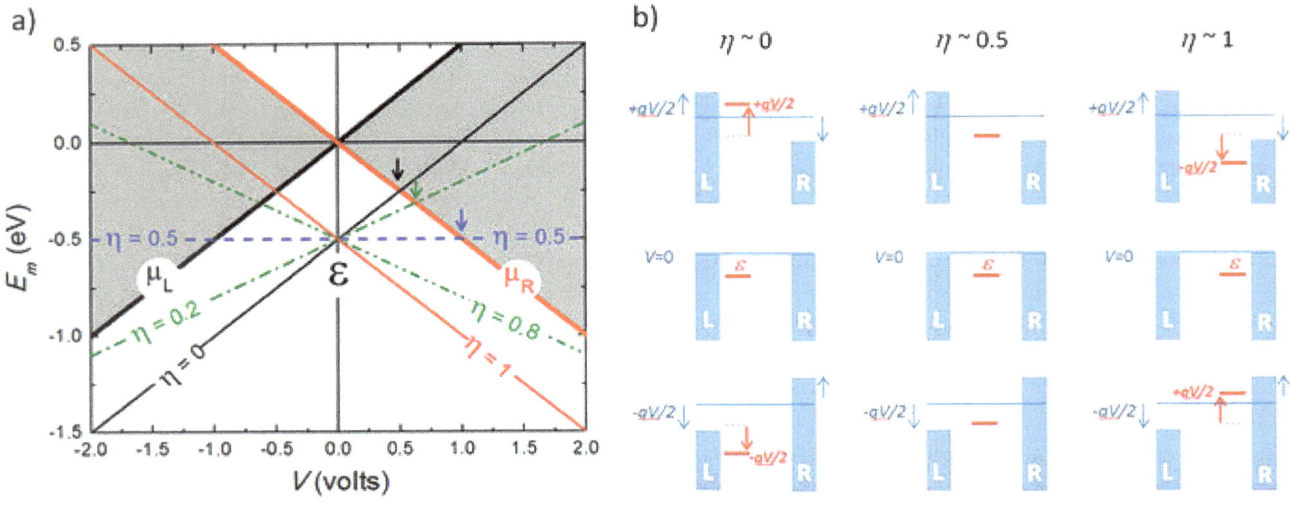

Figure 2: Electrostatic coupling in a tunnel junction. (**a**) Voltage dependence of the energy of the molecular frontier orbital E_m for different values of the voltage division parameter η, with respect to the electrochemical potentials μ_L and $\mu_R$of the electrodes in the junction. ε is the zero-bias energy offset between the molecular orbital and the electrodes. The arrows show when the molecular level enters the conduction window (grey areas). (**b**) Represent the energetic configurations of the molecular level and the electrochemical potential of the electrodes for three representative values of the voltage division parameter, i.e., η=0, 0.5 and 1.

Finally, thermal broadening of the energy of the electrons in the leads also greatly affects conduction through the junction and needs to be taken into account. When a bias potential is applied, two distinct Fermi distributions need to be used, one for each lead:

$$f_{L,R}(E) = \frac{1}{1 + \exp[(E - \mu_{L,R})/k_B T]} \tag{2}$$

In the case of a single conduction level, the electrical current can be calculated by solving the rate equations for each side of the molecule (see Supplemental Information for a generalized treatment of the rate equations for a molecule with multiple levels). In the limit of very small broadening ($\gamma \ll k_B T, |\varepsilon|, |\mu_L - \mu_R|$ one obtains the following electrical current expression (see detailed discussion in Supplemental Information and in refs 24,39):

$$I_{L,R}(E) = (-q)\frac{\gamma_{L,R}}{\hbar}(f_{L,R}(E) - N), \tag{3}$$

where N is the occupation number or average number of electrons in the steady state. By definition, in this state there is no net transfer of electrons in or out of the junction (i.e., $I_L + I_R = 0$), from which one can extract an expression for N:

$$N = \frac{\gamma_L f_L(E) + \gamma_R f_R(E)}{\gamma_L + \gamma_R} \tag{4}$$

Substituting this functional for N in Eq. (3) one obtains the steady-state current through the junction (without taking into account the spin) as

$$I = I_L = - I_R = \frac{2q}{\hbar}\frac{\gamma_L \gamma_R}{\gamma_L + \gamma_R}[f_L(E) - f_R(E)]. \tag{5}$$

Note that this treatment accounts for sequential tunneling of the electron from one of the leads into the molecule and further into the opposite lead and is incoherent in nature. In other words, the rate equations formulation assumes that the electron's phase memory is completely lost once it tunnels into the molecule. That can only happen through relaxation processes —purely elastic in the case of a single level (no other levels to transit into), but possibly even inelastic. Apart from this, the first conclusion that one can extract from this simple expression is that current will only flow through the junction if a molecular level lies between the electrochemical potentials of the left and right electrodes (i.e., within the conductance window). Of course, this approximation fails when the level width surpasses such energy separation, since it will lie partially out of the conductance window. In other words, the current will not increase indefinitely by simply increasing the level width (i.e., increasing the coupling to the electrodes). On the other hand, the resistance has a quantum mechanical limit, which for a single transport channel translates into a universal quantum of conductance, G_0, as originally predicted by Landauer40.

Interestingly, it is possible to relax the condition of very small broadening $\gamma \ll k_B T, |\varepsilon|, |\mu_L - \mu_R|$ and incorporate the effect of a finite level broadening into the rate equation calculation by introducing a broadened density of states (DOS) in the shape of a Lorentzian centered at the energy level ε,

$$D_\varepsilon(E) = \frac{\gamma/2\pi}{(E - \varepsilon)^2 + (\gamma/2)^2}. \tag{6}$$

The result is

$$I = \frac{q}{h}\int_{-\infty}^{\infty} dE\, D_\varepsilon(E)\frac{\gamma_L \gamma_R}{\gamma_L + \gamma_R}[f_L(E) - f_R(E)], \tag{7}$$

This expression matches exactly the one derived by Jauho, Wingreen, and Meir in 199430 using a fully coherent formulation based on the Keldysh Green's function formalism (see Supplemental Information for a detailed discussion of this point). This may seem remarkable, but we note that for a single channel and a single level in the molecule, interference plays no role in the electron conduction through the molecule. As a result, coherent and incoherent sequential tunneling cannot be distinguished at a formal level. Both models provide the same level of accuracy in describing transport at low bias through a single level, as was verified in a pioneering experiment by the Saclay group41. In the standard derivation of Eq. (7) in the context of coherent tunneling, the level broadening is entirely given by the sum of the partial broadenings due to leakage through the leads ($\gamma=\gamma_R+\gamma_L$). In the derivation based on rate equations (incoherent sequential tunneling) one can introduce an additional source of broadening unrelated to leakage, namely, $\gamma=\gamma_R+\gamma_L+\gamma_0$. Other than that, the two results are formally identical at any temperature or bias voltage42.

Landauer derived an equation similar to Eq. (7) under the assumption that the entire system (leads and "molecule") were one dimensional. But now we know that the validity of Eq. (7) does not depend on the spatial dimensions, but rather on the number of channels involved. Namely, it is a correct description only in the case of a single channel in and out of the molecule on each lead. The derivation presented here is also only valid in the case of spinless charge carriers. It is straightforward to extend it to include spin ½ carriers; provided that coherence does not extend beyond the molecule, both rate equation and fully-coherent formulations yield again identical results. When the charging energy is strong and forbids double occupancy of the molecular level (that is, $E_c \gg k_B T, |\mu_L - \mu_R|$, which is typically the case for solid-state molecular junctions), the expression for the current becomes

$$I = \frac{2q}{h}\int_{-\infty}^{\infty} dE\, D_\varepsilon(E)\frac{\gamma_L\gamma_R}{\gamma_L+\gamma_R+\Gamma_{0\to1}}[f_L(E) - f_R(E)], \tag{8}$$

where the level width associated to the level lifetime is given by

$$\Gamma_{0\to1} = \int_{-\infty}^{\infty} dE\, D_\varepsilon(E)\,[\gamma_L f_L(E) + \gamma_R f_R(E)]. \tag{9}$$

In the linear regime and when the molecular level falls within the bias voltage window, $\Gamma_{0\to1} \approx \gamma_R + \gamma_L$ and one finds that the current matches closely that of Eq. (7), with the prefactor of 2 cancelling out. Notice that, in the opposite limit, when charging energy is weak and double occupancy is permitted, one can simply adopt Eq. (7) and account for spin degeneracy in the conduction process by multiplying the right-hand side by a factor of 2 (see Supplemental Information for details).

Thus, it is our claim that Eq. (7) can be formally taken as a good approximation to describe conduction through a molecular junction at all temperatures provided that: (i) Electrical conduction through the junction is mainly governed by a single level (no transitions to high-energy states); and, (ii) that intra- and inter-molecular interactions can be incorporated into the level width (elastic processes). These conditions are likely to be applicable to solid state molecular junctions, where interaction with solvent molecules is not possible and Marcus-like energy relaxation processes are unlikely.

It is easy to see from Eq. (7) that at low temperature ($f_L(E) - f_R(E) = 1$ if $\mu_L<E<\mu_R$, $=0$ otherwise) and low bias ($\mu_L \sim \mu_R$), the maximum conductance (occurring when the energy level coincides with the average electrochemical potential, $\varepsilon=\mu$) will be

$$G_{\max} = \frac{I}{V} = \frac{q^2}{h}\frac{4\gamma_L\gamma_R}{(\gamma_L+\gamma_R)^2} \tag{10}$$

this expression coincides with the universal conductance $G_0 = \frac{q^2}{h}$, if the two rates are equal ($\gamma_L=\gamma_R$) [for the spinless case].

As a side note, one can break up the Landauer conductance into two contributions, since

$$G^{-1} = \frac{h}{q^2 M}\frac{I}{T} = \frac{h}{q^2 M} + \frac{h}{q^2 M}\frac{I-T}{T} = G_C^{-1} + G_T^{-1}, \tag{11}$$

with M representing the number of channels and the characteristics of the junction (see Eq. (8) when $\gamma_L \neq \gamma_R$). This is convenient in order to separate the contact resistance (G_C^{-1}) from the tunneling resistance (G_T^{-1}) in a molecular junction, and particularly relevant for AC impedance transport measurements, where one can distinguish experimentally between capacitive and resistive contributions in the transport through the junction43.

An interesting exercise within the context of the discussion in this article is to solve the asymptotic limit of Eq. (7) at high temperatures. To be more explicit, this pertains to the case when $\gamma \ll |E_m| \ll k_B T \ll qV$ (seeSupplemental Information for a detailed derivation). In this case, Eq. (7) reduces to:

$$I = \frac{2q}{\hbar}\frac{\gamma_L\gamma_R}{\gamma_L+\gamma_R}\left(\frac{qV}{k_B T}\right)e^{-E_m/k_B T}. \tag{12}$$

The current in this case shows an activation behavior, with the "activation energy" ε being the offset between the energy level in the molecule and the Fermi energy in the leads. Note that a similar behaviour is expected from classical Marcus charge transfer processes, for which the activation energy has a completely different origin, as described above. We note that the activation energies arising from these two very different mechanisms may display different dependencies on the electric potentials applied to the junction, which therefore become powerful diagnostic tools to discern charge transport mechanisms. From Eq. (12) it follows that the activation energy ε would decrease with increasing bias voltage, as the electrostatic potential of the lead approaches the molecular level. This does not need to be the case in, e.g., polaron-assisted conduction, where polarization of neighbouring molecules may be affected by the applied bias in intricate ways, resulting from its direct effect on the polaron dynamics.

Figure 3 shows tunneling calculations using Eqs (5) and (7), to solve for the electrical current through a single molecular level described with the following characteristic parameters: $\varepsilon=0.4$ eV, $\eta=0.5$, $V_G=0$, and $\gamma_L=\gamma_R=1$ meV (in Fig. 3a–c) and 10 meV (in Fig. 3d) and with different bias voltages: $V \in 0.4-0.7$ V. As can be clearly seen, there are two distinct regimes: decaying and saturated currents. The temperature of the transition between the two regimes (namely, the inflexion point where the plateau starts to develop) depends on the bias voltage: the lower the bias the higher the transition temperature (compare Fig. 3a–c). The same correspondence is observable for the slope in the thermally activated regime. As determined from Eq. (12), the "activation energy" decreases with increasing bias (since the separation ε between the molecular level and the electrochemical potential of the leads decreases with increasing bias). In addition, one can observe that increasing the level width smears out the difference between the two regimes (compare Fig. 3b,d). Important for our discussion is the fact that one can explain temperature-dependent electrical conductance in a molecular junction with the same functionality (exponential) and within the common temperature range of existing experiments without the need for a thermal activation derived from Marcus-like energy relaxation processes in the molecules (also possible but less likely in many solid state junctions). Indeed, this approach was successfully employed to explain the temperature dependence of the conductance through an individual sulfur end-functionalized tercyclohexylidene molecule in a SET device by van der Zant and collaborators in 200637. In that case, the width of the molecular level had to be adjusted for different bias voltages, from a few meV at low bias up to 30 meV as the system was brought into resonance at large bias. As discussed above, this effect can be understood in terms of a broadening of the molecular level width as a result of internal molecular vibrations excited by the increasing current when the system is brought closer to resonance.

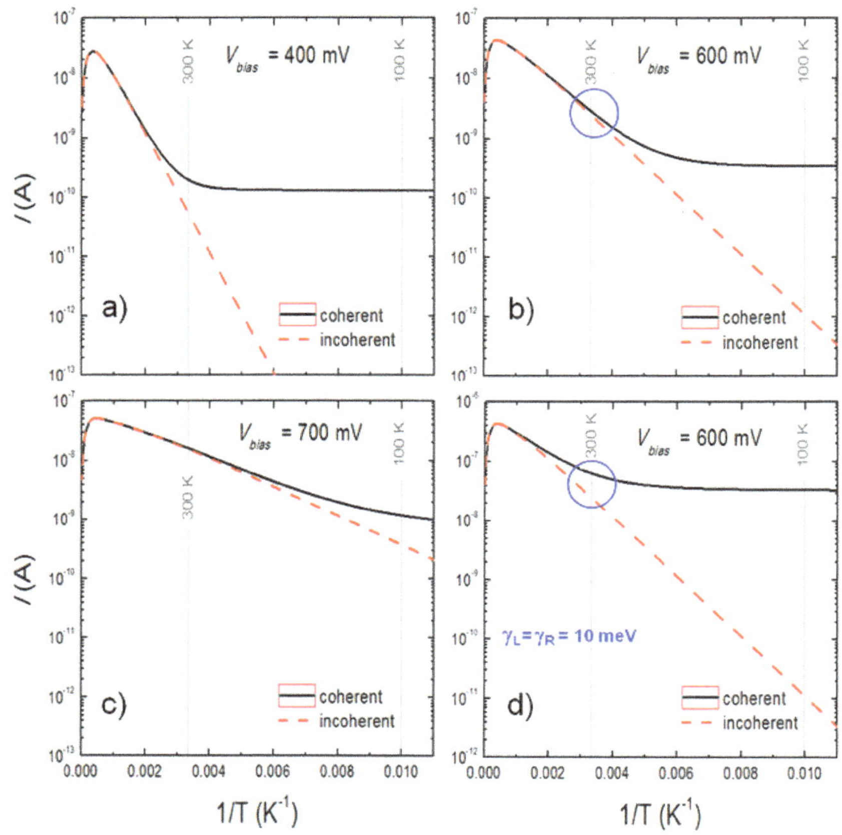

Figure 3: Coherent vs. Incoherent tunnelling. The electrical current through a single-level molecular junction as expressed by Eqs (7) (broadened molecular level) and (5) (zero-width molecular level). The parameters used in the calculations are $\varepsilon=0.4$ eV, $\eta=0.5$, $V_G=0$, and $\gamma_L=\gamma_R=1$ meV for (**a–c**) and $\gamma_L=\gamma_R=10$ meV for (**d**), which are typical values in the molecular junctions studied in this work. The plateau at low temperatures is a direct consequence of the broadening of the molecular level as given by Eq. (6), which sets a maximum for a conductance at low temperatures.

Three-Terminal single-Electron Transistor Measurements of a Ferrocene-Based Junction

Figure 4a shows the electrical current at T = 80 K as a function of gate voltage for a bias of V = 10 mV through an individual S-$(CH_2)_4$-Fc-$(CH_2)_4$-S molecule placed in between the leads of an electromigrated three-terminal Al_2O_3/Au SET (see sketch in Fig. 4) fabricated according to the procedure given in ref. 44. Two peaks in the current are clearly visible at V = −0.3 V and +1.7 V. These peaks separate three consecutive charge states (N − 1, N and N + 1) of the molecule, when no current is allowed to flow at low bias (i.e., Coulomb blockade). Details of the transport results and the synthesis of the molecule are given in ref. 45. From I vs. $V-V_g$ measurements, the actual separation between these two charge points is 160 meV, from which we extract a gate capacitance factor (i.e., $V_g=cV$) of c = 0.08. The data in Fig. 4a (solid black circles) can be well resolved (continuous red line in Fig. 4a) using the single-level model in Eq. (7) including two conduction levels (1 and 2) with the following characteristic parameters: $\varepsilon_1 = -25$ meV, $\varepsilon_2 = 135$ meV ($\varepsilon_1 - \varepsilon_2 = 0.160$ eV), $\eta = 0.5$, and $\gamma_{L1} = \gamma_{R1} = 0.21$ meV and $\gamma_{L2} = \gamma_{R2} = 0.06$ meV.

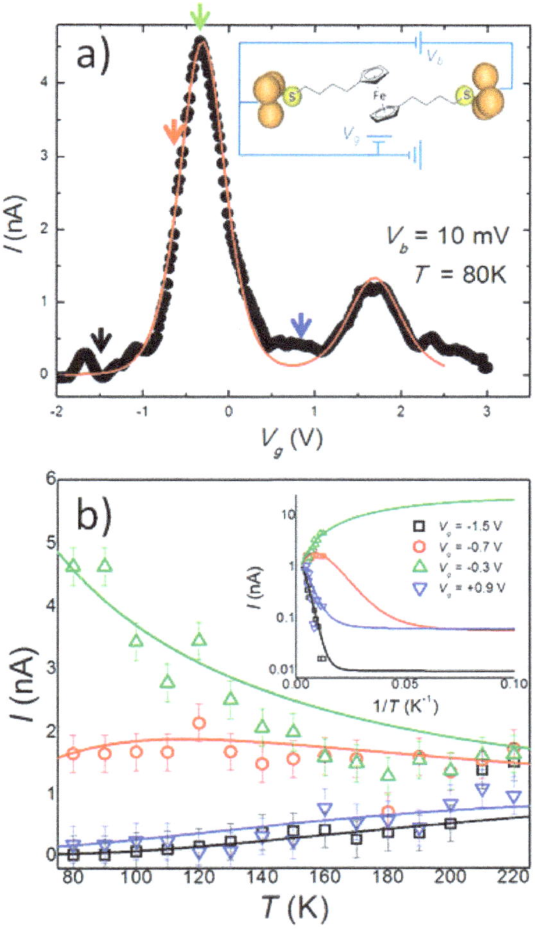

Figure 4: Temperature dependent tunneling in a single molecule junction. **(a)** Electrical current through a single $S(CH_2)_4Fc(CH_2)_4S$ molecule in a three-terminal SET as a function of gate voltage obtained at bias voltage V = 10 mV and temperature T = 80 K. Two charge points at $V_g = -0.3$ and +1.7 V separate three different charge states in the molecule,i.e., N − 1, N and N + 1 electrons (Coulomb blockade). Inset: Schematic illustration of $S(CH_2)_4Fc(CH_2)_4S$ in a three-terminal Al_2O_3/Au single-electron transistor. **(b)** Experimental (data points) and theoretical (lines) temperature behavior of the tunnel current for four different gate voltages in the temperature range 80–220 K (marked with coloured arrows in panel a). The calculated behaviour of the junction was obtained using Eq. (7) with the following parameters: $\varepsilon_1 = -25$ meV, $\varepsilon_2 = 135$ meV ($\varepsilon_1 - \varepsilon_2 = 0.160$ eV), $\eta = 0.5$, and $\gamma_{L1} = \gamma_{R1} = 0.21$ meV and $\gamma_{L2} = \gamma_{R2} = 0.06$ meV.

The measurement in Fig. 4a was repeated at different temperatures up to T = 220 K. The behavior of the tunnel current with temperature is shown in Fig. 4b for four different gate voltages ($V_g = -1.5$, −0.7, −0.3, and 0.9 V) characteristic of the different conduction regimes in this junction, as indicated by the arrows in Fig. 4a. The agreement with the theoretical expectation is significant for all voltages, whether the conduction is in the Coulomb blockade regime ($V_g = -1.5$ and +0.9 V) or in the transport regime ($V_g = -0.7$ and −0.3 V).

The inset to Fig. 4b shows fits of current against inverse temperature, from which one can clearly discern the overall behavior for the different conduction regimes, including the transition into a temperature independent regime (quantum coherent plateau) at sufficiently low temperatures. Both the transition temperature and the "activation energy" depend on the gate voltage, as expected from Eq. (12) and discussed above.

We note that the same temperature behaviour has been observed in several other studies of this family of ferrocene-based molecules in EGaIn junctions15, illustrating the robustness of the theoretical analysis used in this work to determine the characteristic parameters of solid-state molecular junctions, and serving as a contextualization for the following theoretical analysis of experiments performed in SAM-based molecular junctions.

A two-level SAM-Based Molecular Junction

Let us focus now on some very recent experiments in SAM-based molecular junctions where a record-high rectification ratio of three orders of magnitude has been achieved22. Figure 5 shows a sketch of the molecular junction formed by a SAM of $S(CH_2)_{11}Fc_2$ molecules (with Fc_2representing a biferrocenylene (Fc=Fc) head group) sandwiched in between GaO_x^{cond}/ EGaIn (top) and Ag (bottom) electrodes (see refs 22,46for synthesis and experimental details). In these junctions, the Fc_2 lies at the end of a long insulating alkyl chain, providing a non-covalent contact (van der Waals coupling) to the top electrode. This results in a highly asymmetric drop of electric potential at both sides of the Fc_2(where the frontier orbitals are located) enabling large rectification ratios. In particular, the Fc_2 presents two distinct conduction levels (HOMO and HOMO −1, see Fig. 5) separated by an energy of 0.6 eV, as measured by cyclovoltammetry (CV) in solution and ultra-violet photoemission spectroscopy (UPS) in vacuum, which is much smaller than the HOMO-LUMO gap (which is approx. 2.0–2.5 eV). Figure 5shows the room-temperature rectification ratio (defined as R = |I(−V)|/|I(V)|) as a function of the bias voltage applied to the junction. Apart from the high rectification ratio of 1.1×10^3 reached at V = 0.875 V, the R(V) plot shows two clear kinks at around 0.35 and 0.75 volts which were ascribed to the HOMO and HOMO-1 levels subsequently entering the conduction window defined by the difference in electrochemical potential of the top and bottom electrodes22.

These results can be easily explained by incorporating a second level into the single-level transport model to account for the two Fc_2 levels,i.e., the HOMO and the HOMO-1. Specifically, the addition of two different Lorentzian functions, as given in Eq. (6), weighted by the corresponding tunneling rates is used in Eq. (7) to represent the two contributing levels. The results of the calculation using Eq. (7) are shown in Fig. 5 (continuous red line) for the following characteristic parameters: $\varepsilon_1 = 0.35$ eV, $\varepsilon_2 = 0.75$ eV, $\eta = 0.95$, and $\gamma_{1,L} = 3$ meV, $\gamma_{1,R} = 1$ meV,$\gamma_{2,L} = 5$ meV, $\gamma_{2,R} = 2$ meV. Note: the same results are obtained by Eq. (5) at room-temperature. The agreement is significant given the approximations in the model and the inherent degree of dispersion in the SAM-based molecular junctions. The theory recovers the voltage position of the two kinks, supporting their association with the HOMO and HOMO-1 levels of the Fc_2 group, as well as the overall value of the rectification for most of the bias voltage range, including the maximum rectification of three orders of magnitude. As discussed in ref. 22, the separation between the levels in the R-V plot (0.4 eV, Fig. 5) is smaller than that measured by CV and UPS (0.6 eV), which can be ascribed to an energy-renormalization of the molecular levels as a result of charge-image effects by the electrodes. This effect should also decrease the HOMO-LUMO gap (estimated to be >2 eV), allowing it to play some role in the conduction at large bias voltages (reverse polarity), which is likely responsible for the decrease of the rectification ratio above 1 V (see Fig. 5).

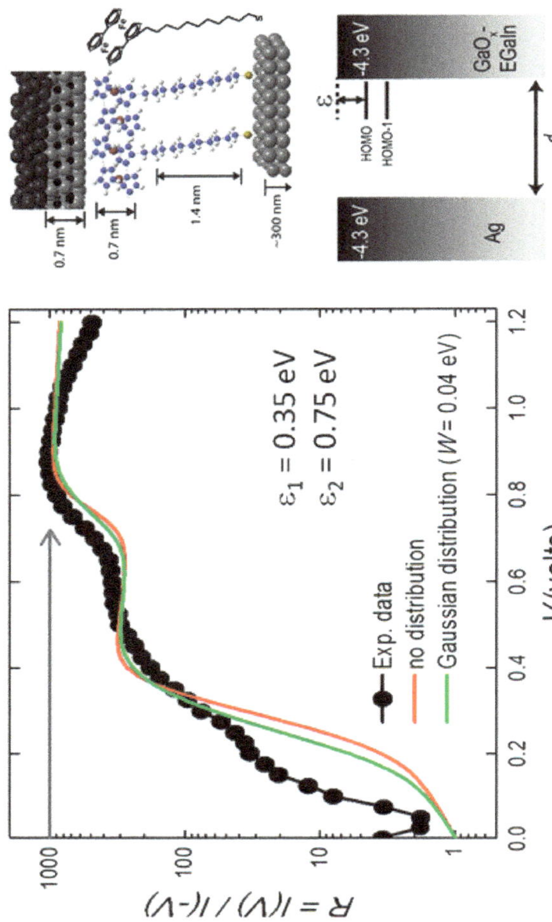

Figure 5: A two-level molecular junction. Room-temperature electrical current rectification as a function of bias voltage in a SAM-based junction of $S(CH_2)_{11}Fc_2$ molecules. Two distinct jumps in the rectification at 0.35 and 0.75 volts can be understood in terms of the HOMO and HOMO-1 levels entering the conduction bias window (as depicted in the sketch on the right), after which a record-high rectification ratio of 10^3 is achieved22. The solid lines are fits to a double-level transport model given by Eqs (5) and (7), with (green) and without (red) the energy distribution of ε_1 and ε_2 given by Eq. (13).

Given the fact that molecules are not all identically placed with respect to the electrodes and that their disposition may actually change during the course of a complete measurement, we have used Gaussian distributions for the energies of the HOMO and HOMO-1 levels to account for the degree of dispersion in the system. To do so, we modify Eq. (7) as follows (note once again that Eq. (5) can be treated in the same way to give the same results at high temperature):

$$I = \frac{q}{h}\int\int_{-\infty}^{\infty} dE\,dE'\,D_\varepsilon(E)\frac{\gamma_L\gamma_R}{\gamma_L+\gamma_R}[f_L(E)-f_R(E)]g_\varepsilon(E'), \qquad (13)$$

where,

$$g_\varepsilon(E) = \exp\left[\frac{(E-\varepsilon)^2}{2W^2}\right], \qquad (14)$$

The best results are obtained for a width of the Gaussian distribution W=0.04 eV for each level (green line in Fig. 5). This small dispersion in the position of the energy of the molecular levels (<10%) is indicative of the high degree of electrical stability in these SAM-based molecular junctions.

Maximum Rectification Ratio for a Single-Level Junction

We note that the rectification ratio of three orders of magnitude observed for this molecular junction (with typical level widths $\gamma\sim$1–10 meV) is at the theoretical maximum limit expected from a single-level transport mechanism for characteristic positions of the frontier orbital with respect to the Fermi energy ($\varepsilon\sim$0.5–1 eV) and operating voltages (V ~ 1 volt). This is easy to deduce by looking in Fig. 6 at the current and rectification ratios calculated from Eq. (7) for a molecular junction, where the parameters have been chosen to be representative of these kinds of junctions (i.e., derived from experimental data). In particular, the calculations have been obtained for a single-level positioned at ε=0.5 eV, and with a maximally asymmetric voltage divider parameter η=0, to maximize the rectification ratio. Obviously, the rectification will be maximum for a bias voltage larger than −0.5 V, which is the forward voltage needed to bring the molecular level into resonance when η=0 (as obtained from the definition of E_m and illustrated in Fig. 2).

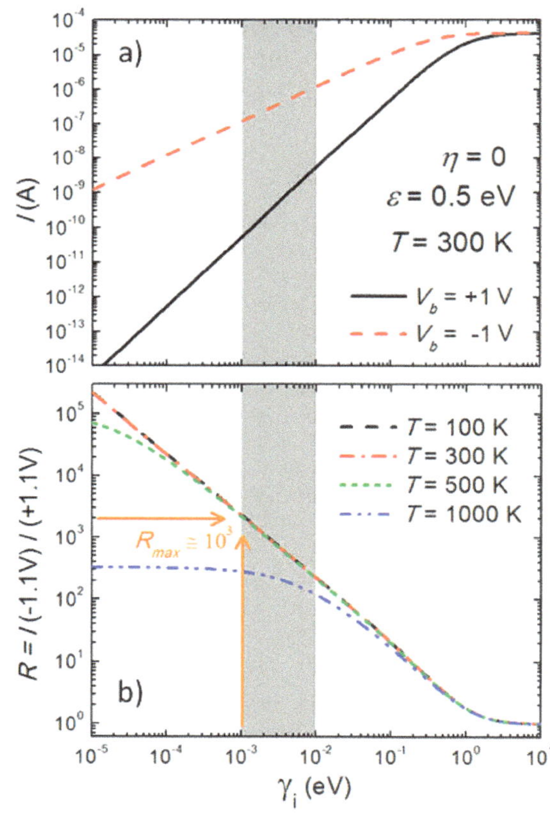

Figure 6: Limits on Molecular Tunneling Rectification. (**a**) Electrical current and (**b**) rectification ratio as a function of γ_i (with $\gamma_L=\gamma_R$) calculated solving Eq. (7) for a single-level

molecular junction with the following parameters: $\varepsilon=0.5\,eV$ and $\eta=0$. For the typical smallest values of $\gamma_I=1\,meV$, the rectification originated from a single level (e.g., the HOMO orbital) in a molecular junction is theoretically limited to about three orders of magnitude.

Figure 6a shows the behaviour of the room-temperature electrical current with γ_i (with $\gamma_L=\gamma_R$) for both polarities of the bias voltage ($V=\pm1$ volt). As expected, the difference between the forward and reverse currents increases as the tunneling rates (and consequently the level widths) decrease. Both currents reach the same saturation value at sufficiently high tunneling rates (i.e., $\gamma_i \gtrsim 1\,eV$), when the level width is larger than the separation between electrochemical potentials in the leads and the conductance becomes constant, as discussed above. The corresponding R for these voltages is shown in Fig. 6b for different temperatures, to emphasise that it does not vary below room-temperature. In the range $10^{-5}<\gamma_i \lesssim 1\,eV$ the electrical rectification decreases exponentially at room-temperature and below, saturating at$R=1$ (no rectification) for $\gamma_I \gtrsim 1\,eV$, corresponding to the behaviour of the current in Fig. 6a. From these results one can conclude that in molecular junctions of this kind, where the γ_i takes values typically above 1 meV, the rectification ratio will be limited to three orders of magnitude when a single electrostatic molecular level contributes to the conductance through the junction, which is the value found in the experiments reported in ref. 22 and marked by the arrows in Fig. 6.

We want to stress again that this result is found for voltage conditions that maximize the rectification efficiency (i.e., $\eta=0$) but with an experimentally relevant energy location of the frontier orbital ($\varepsilon=0.5\,eV$) and typical values of γ_i. Larger ε values and, consequently, larger voltages will increase this limit, but not substantially. Evidently, if more than one molecular level is involved in charge transport the rectification can be improved by a factor proportional to the total number of levels involved. However, since it is unlikely to find molecular junctions with more than a few levels in the vicinity of the electrochemical potential of the leads, it is unwise to expect substantially higher electrical rectification ratios originating from the molecular level structure. Imaginative ways should be explored in order to overcome this limit in practice, perhaps by engineering junctions that mix high single-level rectification properties with conformational changes in the molecule that can build up to increase the overall rectification of the junction.

Determination of the Electrostatic Potential Profile in a Molecular Junction

Some of the authors in this article have recently reported a study of electrical rectification in EGaIn SAM-based junctions where the Fc unit was placed at 14 different positions within the SAM. The molecules have the form of S-C_n-Fc-C_{13-n}, where C_n represents the number of aliphatic carbons (CH_2 or a terminal CH_3), with n=0 to 1333. Controlling the position of the Fc unit (where the molecular HOMO level is localized) along the alkyl chain for different values of n (see sketches in Fig. 7) allows us to quantify the rectification response for different energetic symmetries in the junction (values of η), since the electric potential at both sides of the conduction unit (i.e., Fc) will strongly depend on its position along the chain. This enables sampling of the electrostatic potential profile along the junction gap, which was found to be highly non-linear in this particular molecular junction33.

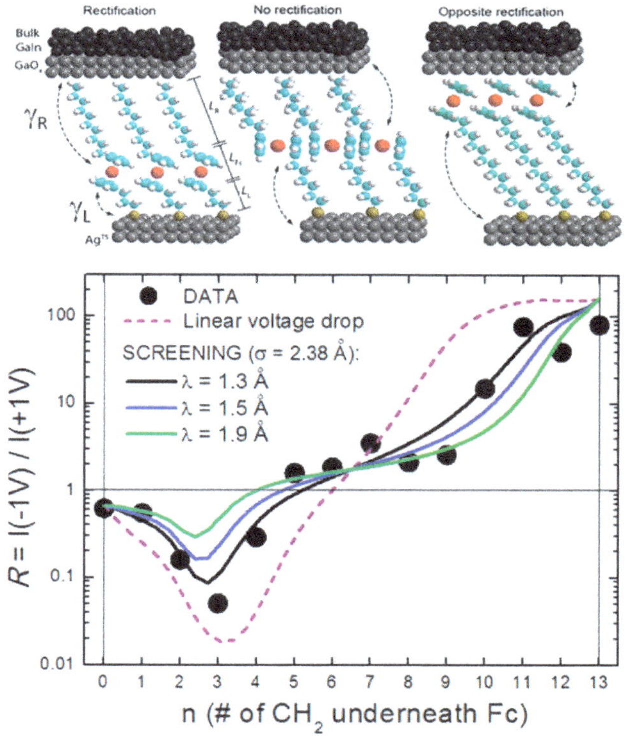

Figure 7: Electrostatic Potential Profile: Experiment. Sketch) Representation of the EGaIn SAM-based junctions formed by S-C_n-Fc-C_{13-n}molecules for three different positions of the Fc unit within the alkyl chain. **Main panel)**Measured rectification ratio of the junctions in ref. 33 as a function of the position n of the Fc unit within the chain (solid circles). The lines represent fittings to the data using the single-level transport model in Eq. (7) using different shapes for the electrostatic potential profile in the junction, including linear (dashed purple line) and non-linear as derived from a model that assumes the molecule a cylinder of radius $\sigma=2.38$ Å and takes into account screening of the electric field by the molecule (with screening length λ).

Figure 7 shows R as a function of n measured at room-temperature and calculated by comparing the electrical currents at $V=\pm1\,V$. As can be clearly observed, the rectification behavior is highly non-linear, with several distinct areas, and with the maximum rectification values achieved when the Fc unit is placed close to one of the electrodes (i.e., n=3 and n>10), as expected from the asymmetry in the potential drops at both sides of the Fc. For n<3 the strong hybridization between the Fc and the bottom electrode (chemisorbed contact) leads to large level broadenings and leakage currents, which in turn results in lower rectification ratios (see ref. 33 for more details). A similar effect (i.e., Rsaturates), although weaker, arises for n>10 due to proximity to the top electrode, where the coupling between the molecule and the lead is non-covalent (van der Waals interactions). The intermediate range of Fc positions (4<n<11) is the most interesting in the context of the analysis of the electrostatic potential profile in this article, since coupling between the Fc unit and the electrodes is non-covalent, and the molecular level responsible for conduction is well defined, with a width that can be considered constant, as we discuss below. The results in Fig. 7 have been fitted to the single-level model given in Eq. (7) with characteristic energy parameters defined in the following. We reiterate once again that Eq. (5) will give the same exact results for this temperature and electric potential conditions. Before getting into the details of the calculations, we emphasise that the fitting results show that the distinct behaviour of R for intermediate n values, and particularly its small slope for n=5–9 (Fig. 7), can only be explained by taking into account screening effects, i.e., a non-linear electrostatic potential profile. As we show below, single-level transport models together with a functional treatment to account for electric field screening proposed by Nitzan and collaborators47 allows determination of the electrostatic potential profile and extraction of a quantitative estimate of the screening length in a molecular junction, which has not been possible to date.

In the following we describe the analytical models that have been employed to fit the rectification data in Fig. 7 by means of Eq. (7). First, for the distance dependence of the energy of the HOMO level, ε, the following "synthetic" function has been employed: $\varepsilon(n)=\varepsilon(n\sim13)-Ae^{-Bn}$, with $\varepsilon(n\sim13)=0.75\,eV$ (see solid black line in Fig. 8b). The rationale behind this selection is based on the n-dependence of ε^{exp} obtained from UPS data (see solid data in Fig. 8a), which departs from its linear behavior for distances below n~5. The solid black squares in Fig. 8b, closely following the function used in the calculations (solid black line), represent the result of subtracting a linear function (dashed line in Fig. 8a) from the experimentally obtained value of ε^{exp}. The linear behavior is associated with screening by the metallic electrodes and depends on the Fc-distance (thus the linearity). Therefore, for n>5 the energy offset ε(n) is expected to be constant (i.e., independent of n). Departures from linearity (as observed for n<5) are ascribed to a real change in the distance between the molecular level and the Fermi energy of the electrode, which in this case decreases. Figure 8b shows the experimental data of Fig. 8a after subtraction of the linear slope (black squares), together with the function used to fit the data (black line).

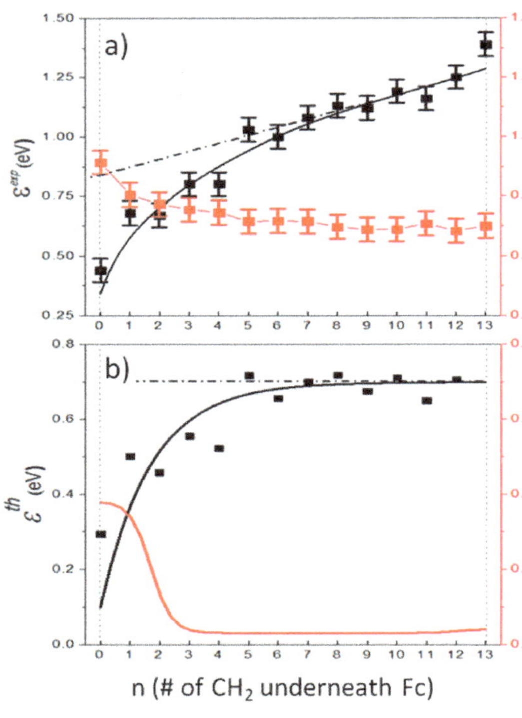

Figure 8: Molecule-Electrodes coupling. (**a**) Experimental value of ε^{exp} (left axis) and γ of the S-C_n-Fc-C_{13-n} SAMs on Ag^{TS} as a function of n obtained from UPS experiments reported in ref. 39. The lines are guides to the eye, with the dashed line representing the linear background expected from screening by the metallic electrodes, which is proportional to the Fc-electrode distance. (**b**) Functions employed to represent ε and γ for different positions of the Fc within the alkyl chain (black and red lines, respectively). The solid squares are the data in panel ***a*** after subtraction of the observed linear slope in the UPS data caused by electron-hole screening effects.

Distance-dependence functions have been chosen to represent the coupling energies $\gamma_L(n)$ and γ_R (n), and consequently the HOMO level width $\gamma(n)=\gamma_L(n)+\gamma_R(n)$, and ε(n) in order to account for the effect of the coupling to the respective electrodes. Figure 8b shows the values of γ(n) and ε(n) used in the calculations (solid curves). The broadening of the HOMO level in proximity with the electrode is expected to increase exponentially with decreasing Fc-electrode distance. Fermi's golden rule gives $\gamma_{L,R}=t^2_{L,R}/D_{L,R}$, where $t_{L,R}$ is the effective coupling matrix and $D_{L,R}$ the electron bandwidth in the electrodes48. In general, $\gamma_{L,R}\propto e^{-ekL_{L,R}}$, where k is the tunneling attenuation coefficient, proportional to $\sqrt{E_{barrier}-E_F}$, which for alkyl chains on Au or Ag is on the order of a few eV. As can be observed in Fig. 8b, the functions chosen to mimic this effect for γ_L and γ_R are not purely exponential with n close to the electrodes but saturate for distances below two CH_2 units (n<2 and n>12) from the respective electrodes. This fitting is based on the spectroscopic data33 that shows little differences observed in the spectroscopic results between SAMs with n=0 and 1 (and 2 to some extent). As shown in Fig. 8b (red curve), the used dependence for the level width reaches maximum values of $\gamma_L(n<2)=360\,meV$ (due to strong hybridization with the Ag electrode) and $\gamma_R(n>11)=40\,meV$, and a base value of $\gamma=30\,meV$ (with $\gamma_L=\gamma_R=15\,meV$) for intermediate distances, as obtained from the fitting of the rectification curves (see below). Note that taking the full width at half maximum from UPS data to estimate the molecular level broadening results in an overestimation of γ due to limitations of the technique (see ref. 33 for details) and therefore we only can derive the relative values of γ.

According to Eqs (5) and (7), the only other ingredient for calculating the electrical current through the molecular junctions is the functional describing the electrostatic potential profile in the junction, which is represented by η(n). We have used two different models to obtain this parameter: (i) A simple linear electrostatic potential profile; and, (ii) a correction to account for electrical screening in the alkyl chain.

Linear Electrostatic Potential Profile

Assuming a linear dependence of the voltage drop as a function of the distance between the Fc and the respective electrodes (i.e., $V_L \alpha L_R$), one can rewrite $\eta=V_R/(V_L+V_R)$ as $\eta\approx L_R/(L_L+L_R)$. To account for the voltage drop at the Fc, the expression above can be rewritten assuming a correction L_{Fc} for the respective lengths49, and the expression for the dimensionless division parameter is:

$$\eta(L_R) = \frac{L_R + L_{Fc}}{L_L + L_R + 2L_{Fc}} + \eta_{shift}, \tag{14}$$

where η_{shift} is an effective shift introduced to account for the asymmetry generated by different couplings of the molecule with the two different electrodes and to explain the small rectification ratios (R<2) observed in equivalent alkyl chains without the Fc unit, or when the Fc is placed in the middle of the chain. The corresponding η in the case of our molecule is given in Fig. 9 (dashed purple line). Note that 8.7% [$=L_{Fc}/(L_A+2L_{Fc})$] of the voltage drops in the Fc and remains unavailable for use in the rectification process. We note that the strength of this correction is proportional to the size of the active unit employed to sample the electrostatic potential profile, which in our case is substantially larger than the adjacent CH_2 units. An ideal unit to sample the electrostatic profile would be one of negligible size (which is not possible, of course). That would also guarantee that the sampled molecule would remain undistorted when moving the active unit within the chain.

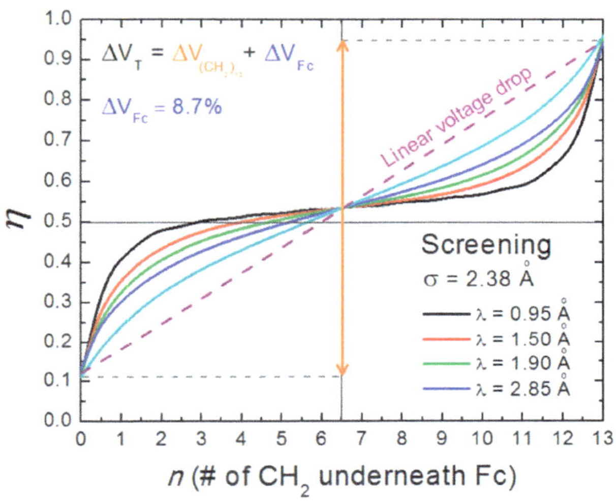

Figure 9: Electrostatic Potential Profile: Theory. Electrostatic potential profiles in the molecular junctions as represented by the dimensionless voltage division parameter η. The curves represent the voltage drop within the alkyl chain on both sides of the Fc unit (totalling 91.3%), which is not the total voltage drop in the junction, since a substantial fraction decays in the Fc (8.7%). Also, the curves have been shifted from η=0.5 in order to account for the different binding energies of the molecule with the left and right electrodes, which leads to a small rectification even in the absence of the Fc unit (or when it is placed at the center of the junction). The linear profile that would be expected in vacuum (red curve) is modified when accounting for electrical screening by the alkyl chain, idealized as a cylinder of diameter σ and electrical screening length λ.

The resulting rectification curve obtained using the linear model defined by Eq. (14) (dashed purple line in Fig. 7) does not give an overall good fit to the data but can explain some features. It quantitatively explains the high rectification ratios for 2<n<5 and n>9 and accounts for the abrupt decrease of rectification for n<3 (resulting from a substantial level broadening due to hybridisation with the left Ag electrode). However, it fails to explain the non-monotonic behaviour of the rectification for intermediate n values, and, particularly, the small slope for 4<n<10 (almost constant). This flat region is one of the main results arising from the sampling procedure to extract the exact shape of the electrostatic potential profile in these junctions, enabled by the capacitive nature of the Fc-electrodes coupling in this range of n values. This discrepancy presents a clear indication that the voltage does not drop linearly within the molecule, as it would be the case in vacuum.

Correction for the Screening of the Electric Field by the Molecule

The electrochemical potential in electrode-conductor-electrode junctions drops mainly at the contacts, but this is not applicable to the electrostatic potential, which may show profiles extending well into the low-dimensional conductor (e.g., molecule) for sufficiently long electric screening lengths, λ, which is the case in semiconductors and molecular insulators (see Liang et al.48 and references therein for a more detailed discussion). It is therefore necessary to account for electrical screening effects when trying to explain our observations. For this, we have followed the theoretical formalism developed by Nitzan et al.47, in which the molecule is approximated by a cylindrical conductor of diameter σ, and electric field screening length λ. As discussed by the authors, the screening length in molecular conductors is an open question, and systems with small HOMO-LUMO gaps are expected to screen well. Assuming that screening over the characteristic length of our system can be described by a Poisson formalism, the voltage division parameter describing the electrostatic potential profile can be written as:

$$\eta(L_R) = \eta_0(L_R) + \frac{A}{\pi}\sum_{m=1}^{\infty}\frac{F_m}{m(1+F_m)}\sin\left(\frac{2\pi m}{L_L+L_R}L_R\right) + \eta_{shift}, \tag{15}$$

where $\eta_0(z)$ is the linear profile given in Eq. (14). A is a factor to correct for the voltage drop in the Fc. The coefficients F_m can be expressed analyticaly as:

$$F_m = \frac{1}{2}\left(\frac{\sigma}{\lambda}\right)^2 e^{\chi}\int_{\chi}^{\infty} du\frac{e^{-u}}{u}, \tag{16}$$

with,

$$\chi = \frac{1}{2}\left(2\pi m\frac{\sigma}{\lambda}\right)^2 \tag{17}$$

Using σ=2.38 Å ($=0.125(L_L+L_R)$), which is taken as the effective diameter of the alkyl chain, the electrostatic potential profiles for different screening lengths (λ=0.95, 1.50, 1.90 and 2.85 Å) are shown in Fig. 9. It can be clearly observed that the departure from linear behavior is more pronounced as the screening length decreases. In the limit λ→0, the electrostatic potential will follow the electrochemical potential and drop entirely within the vicinity of the contacts.

The results of the corresponding fittings of the rectification data are shown in Fig. 7 for three screening lengths: λ=1.30 Å (black line), λ=1.50 Å (blue line) and λ=1.90 Å (green line). From the good quantitative agreement with the experimental results, where the small slope in the rectification ratio can be well accounted for at intermediated values of n, one can estimate the characteristic screening length to be approximately λ=1.50±0.2 Å. These values of the screening length are comparable to the molecular diameter σ, indicating that the electric field lines from vacuum penetrate well into the molecule.

DISCUSSION

We have shown the validity of a simple but formal analytical single-level tunneling model that uses input parameters to calculate I(V) curves that can be experimentally obtained (tunneling rate γ, the dimensionless division parameter $\eta = V_R/(V_L + V_R)$, and the zero-bias energy offset between the molecular orbital and the electrodes ε). This model was tested against a well-characterized molecular diode and faithfully recovered the rectification ratio and the temperature dependent behavior of the junctions, from which we conclude that it is applicable to describe transport data from both single molecule and SAM-based junctions where intra-molecular collective behavior does not significantly affect the electronic structure of the junction.

One of the main conclusions from our theoretical treatment of the rectification ratio is that the highly non-linear electrostatic potential profile in typical molecular tunneling junctions is the result of a sizeable screening length in the molecule ($\lambda = 1.50 \pm 0.2$ Å, for ferrocene-alkanethiol). Fitting our equations to the experimental results shows that the electrostatics of the molecular junction may be sampled simply by placing the redox Fc unit at different positions along the junction. These results confirm the expectation that potential profiles in insulating molecules are neither linear (as in vacuum) nor do they follow the electrochemical potential profile, which would lead to flat profiles within the entire length of the molecule and would prevent the attainment of high rectification ratios by capacitive coupling with the active transport unit. It is important thus to identify molecular ligands that provide long screening lengths in order to achieve linear-like potential profiles and extend the range of lengths for which the molecular diode would be an efficient rectifier of electrical current.

In addition, we have shown that simple single-level transport models (to describe incoherent and coherent tunneling) can account for the typical temperature dependencies of the conductance in solid state junctions and provide a powerful analysis tool to extract the characteristic parameters governing tunneling in these junctions in a consistent way.

ACKNOWLEDGEMENTS

A.R.G., E.D.B. and E.R.M. acknowledge support from the National Science Foundation (grants NSF-ECCS #1402990 and #1518863). D.T. acknowledges the Science Foundation Ireland (Grant No. 11/SIRG.B2111), and computing time at the Irish Centre for High-End Computing. C.A.N acknowledges support from the National Research Foundation, Prime Minister's Office, of Singapore under its Medium sized centre programme, and the NRF fellowship award No. NRF-RF 2010-03.

REFERENCES

1. Park, H. et al. Nanomechanical oscillations in a single-C60 transistor. Nature 407, 57–60 (2000).
2. Park, J. et al. Coulomb blockade and the kondo effect in single atom transistors. Nature 417, 722–725 (2002).
3. Liang, W. J., Shores, M. P., Bockrath, M., Long, J. R. & Park, H. Kondo resonance in a single-molecule transistor. Nature 417, 725–729 (2002).
4. Kubatkin, S. et al. Single-electron transistor of a single organic molecule with access to several redox states. Nature 425, 698–701 (2003).
5. Zhetenev, N. B., Meng, H. & Bao, Z. Conductance of small molecular junctions. Phys. Rev. Lett. 88, 226801 (2002).
6. Yu, L. H. et al. Inelastic electron tunneling via molecular vibrations in single-molecule transistors. Phys. Rev. Lett. 93, 266802 (2004).
7. Haque, F., Langhirt, M., del Barco, E., Taguchi, T. & Christou, G. Magnetic field dependent transport through a Mn4 single-molecule magnet. J. Appl. Phys. 109, 07B112 (2011).
8. O'Neill, K., Osorio, E. A. & van der Zant, H. S. J. Self-breaking in planar few-atom Au constrictions for nanometer-spaced electrodes. Appl. Phys. Lett. 90, 133109 (2007).
9. Díez-Pérez, I. et al. Rectification and stability of a single molecular diode with controlled orientation. Nat Chem. 1, 635–641 (2009).
10. Venkataraman, L., Klare, J. E., Nuckolls, C., Hybertsen, M. S. & Steigerwald, M. L. Dependence of single-molecule junction conductance on molecular conformation. Nature 442, 904–907 (2006).
11. Chiechi, R. C., Weiss, E. A., Dickey, M. D. & Whitesides, G. M. Eutectic gallium–indium (EGaln): a moldable liquid metal for electrical characterization of self-assembled monolayers. Angew. Chem. Int. Ed. 47, 142–146 (2008).
12. Reus, W. F., Thuo, M. N., Shapiro, N. D., Nijhuis, C. A. & Whitesides, G. M. The SAM, not the electrodes, dominates charge transport in metal-monolayer//Ga2O3/gallium-indium eutectic junctions. ACS Nano 6, 4806–4822 (2012).
13. Nijhuis, C. A., Reus, W. F., Barber, J. & Whitesides, G. M. Comparison of SAM-based tunneling junctions with Ga2O3/EGaIn topelectrodes to other large-area junctions. J. Phys. Chem. 116, 14139–14150 (2012).
14. Fracasso, D., Valkenier, H., Hummelen, J. C., Solomon, G. & Chiechi, R. C. Evidence for quantum interference in SAMs of arylethynylene thiolates in tunneling junctions with eutectic Ga-In (EGaIn) top-contacts. J. Am. Chem. Soc. 133, 9556–9563 (2011).
15. Nijhuis, C. A., Reus, W. F., Barber, J., Dickey, M. D. & Whitesides, G. M. Charge transport and rectification in arrays of SAM-based tunneling junctions. Nano Lett. 10, 3611- 3619 (2010).
16. Fracasso, D., Muglali, M. I., Rohwerder, M., Terfort, A. & Chiechi, R. C. Influence of an atom in EGaIn/Ga2O3 tunneling junctions comprising self-assembled monolayers. J. Phys. Chem. 117, 1136–11376 (2013).
17. McCreery, R. L. & Bergren, A. Progress with molecular electronic junctions: meeting experimental challenges in design and fabrication. J. Adv. Mater. 21, 4303–4322

(2009).

18. Haj-Yahia, A. E. et al. Substituent variation drives metal/monolayer/semiconductor junctions from strongly rectifying to ohmic behavior. Adv. Mater. 25, 702–706 (2013).
19. Bowers, C. M. et al. Characterizing the metal–SAM interface in tunneling junctions. Nano Lett. 9, 1471–1477 (2015).
20. Zhang, Y., Zhao, Z., Fracasso, D. & Chiechi, R. C. Bottom-Up molecular tunneling junctions formed by self-assembly. Isr. J. Chem. 54, 513–533 (2014).
21. Perrin, M. L. et al. Large tunable image-charge cffects in single-molecule junctions. Nat. Nanotechnol. 8, 282–287 (2013).
22. Yuan, L., Breuer, R., Jiang, L., Schmittel, M. & Nijhuis, C. A. Molecular diode with a statistically robust rectification ratio of three orders of magnitude. Nano Lett. 15, 5506–5512 (2015).
23. Capozzi, B. et al. Single-molecule diodes with high rectification ratios through environmental control. Nat. Nanotechnol. 10, 522–527 (2015).
24. Datta, S. Electrical resistance: an atomistic view. Nanotechnology 15, S433–S451 (2004).
25. Galperin, M., Ratner, M. A. & Nitzan, A. Molecular transport junctions: vibrational effects. J. Phys. Condens. Matter 19, 103201 (2007).
26. Migliore, A., Schiff, P. & Nitzan, A. On the relationship between molecular state and single electron pictures in simple electrochemical junctions. Phys. Chem. Chem. Phys. 14, 13746–13753 (2012).
27. Kornilovitch, P. E., Bratkovsky, A. M. & Williams, R. S. Current rectification by molecules with asymmetric tunneling barriers. Phys. Rev. B 66, 165436 (2002).
28. Liu, R., Ke, S., Yang, W. & Baranger H. U. Organometallic molecular rectification. J. Chem. Phys. 124, 024718 (2006).
29. Davis, J. H., Hershfield, S., Hyldgaard, P. & Wilkins, J. W. Current and rate equations for resonant tunneling. Phys. Rev. B 47, 4603 (1992).
30. Jauho, A. P., Wingreen, N. S. & Meir, Y. Time-dependent transport in interacting and noninter-acting resonant-tunneling systems. Phys. Rev. B 50, 5528 (1994).
31. Baldea, I. & Koppel, H. Sources of negative differential resistance in electrical nanotransport. Phys. Rev. B 81, 193401 (2010).
32. Baldea, I. Ambipolar transition voltage spectroscopy: Analytical results and experimental agreement. Phys. Rev. B 85, 035442 (2012).
33. Yuan, L. et al. Controlling the direction of rectification in a molecular diode. Nat. Commun. 6, 6324 (2015).
34. Marcus, R. A. On the theory of oxidation-reduction reactions involving electron transfer I. J. Chem. Phys. 24, 966 (1956).
35. Datta, S. et al. Current-voltage characteristics of self-assembled monolayers by scanning tunneling microscopy. Phys. Rev. Lett. 79, 2530 (1997).
36. Heurich, J., Cuevas, J. C., Wenzel, W. & Schön, G. Electrical transport through single-molecule junctions: from molecular orbitals to conduction channels Phys. Rev. Lett. 88, 256803 (2002).
37. Poot, M. et al. Temperature dependence of three-terminal molecular junctions with sulfur end-functionalized tercyclohexylidenes. Nano Lett. 6, 1031–1035 (2006).
38. Moth-Poulson, K. & Bjornholm, T. Molecular electronics with single molecules in solid-state devices. Nat. Nanotechnol. 4, 551–556 (2009).
39. Datta, S. Lessons from Nanoelectronics: A New Perspective on Transport Vol. 1 (World Scientific, Singapore, 2012).
40. Landauer, R. Can a length of perfect conductor have a resistance. Phys. Rev. A 85, 91–93 (1981).
41. Kergueris, C. et al. Electron transport through a metal-molecule-metal junction. Phys. Rev. B 59, 12505 (1999).
42. Due to the absence of electron-electron interactions, the single-level model is exactly solvable. Thus, formally one derives Eq. (7) from scattering theory and from that Eq. (5) in the limit of weakly coupled leads; see Appendix B in Koch, J., von Oppen, F. & Andreev, A. V. Theory of the Franck-Condon blockade regime. Phys. Rev. B 74, 205438 (2006).
43. Suchand Sangeeth, C. S., Wan, A. & Nijhuis, C. A. Equivalent circuits of a self-assembled monolayer-based tunnel junction determined by impedance spectroscopy. J. Am. Chem. Soc. 136, 11134–11144 (2014).
44. Henderson, J. J., Ramsey, C. M., del Barco, E., Mishra, A. & Christou, G. Fabrication of nanogapped single-electron transistors for transport studies of individual single-molecule magnets. J. Appl. Phys. 101, 09E102 (2007).
45. Garrigues, A. R., Lejia, W., del Barco, E. & Nijhuis, C. A. Electrostatic control over temperature-dependent tunneling across a single molecule junction. Nat. Commun. 7:11595 doi: 10.1038/ncomms11595 (2016).
46. Breuer, R. & Schmittel, M. Redox-stable SAMs in water (pH 0–12) from 1,1′-biferrocenylene-terminated thiols on gold. Organometallics 31, 6642–6651 (2012).
47. Nitzan, A., Galperin, M., Ingold, G.-L. & Grabert, H. On the electrostatic potential profile in biased molecular wires. J. Chem. Phys. 117, 10837–10841 (2002).
48. Liang, G. C., Ghosh, A. W., Paulsson, M. & Datta, S. Electrostatic potential profiles of molecular conductors. Phys. Rev. B 69, 115302 (2004).
49. Kornilovitch, P. E., Bratkovsky, A. M. & Williams, R. S. Current rectification by molecules with asymmetric tunneling barriers. Phys. Rev. B 66, 165436 (2002).

CHAPTER 10 MANGANESE DIOXIDE-MODIFIED CARBON PASTE ELECTRODE FOR VOLTAMMETRIC DETERMINATION OF RIBOFLAVIN

Eda Mehmeti, Dalibor M. Stanković, Sudkate Chaiyo, Ľubomir Švorc, Kurt Kalcher

Keywords: Differential pulse voltammetry, Cyclic voltammetry, Vitamin B_2, Electrooxidation

INTRODUCTION

Vitamins are the important group of organic compounds and they are necessary for human health. They are required in the diet and to human body for normal growth and nutrition where their absence can lead to many diseases [1]. Riboflavin or Vitamin B_2 is a water soluble vitamin and an essential component of flavoenzymes which plays a significant role in biochemical reactions of the human body [2]. It cannot be synthesized in human body therefore has to be obtained dietary from the sources such as liver, cheese, milk, meat, eggs, wines, and tea [3] therefore its insufficiency is associated with eye lesions and skin disorders [4].

Up to the date several analytical methods are used for the determination of Vitamin B_2 such as HPLC [5] chemiluminescence [6], fluorescence [7] and capillary electrophoresis [8]. These methods provide highly sophisticated instrumentation setup and they are more time consuming [9]. Therefore, electrochemical methods have received great attention due to their simple, sensitive, low cost and rapid response time [10–12]. The use of chemically modified electrodes offers a tool for improving the performance of electrodes and several functional materials are used as a modifier for vitamin sensors [13, 14].

Carbon paste electrodes (CPE) are widely used as an electrode material for the development of various electrochemical sensors and biosensors and they can be simply modified [15]. The main advantages of carbon paste electrode are due to easy way of preparation, the new reproducible surface and low residual current in wide potential windows [16]. The modification of carbon paste electrodes with catalytic metals, has received also considerable attention [11]. There are already reported several carbon electrodes with numerous types of modifiers such as MnO_2, Fe_3O_4, FeO, SnO_2, CuO, Fe_2O_3 which are found to be very sensitive to acids and basis media, oxides of platinum group metals, complexes of copper, nickel, iron and chromium or/and also nanocomposites of these modifiers. The purpose of the use of these modifiers is due to the lowering overpotential for the oxidation or reduction of analytes, in comparison with unmodified electrodes. The obtained analytical responses are significantly higher and with high reproducibility of the electrode performance. MnO_2 based electrodes are very popular for all mentioned characteristics with high catalytic effect at an appropriate potential for sensing the target analyte [17]. The aim of this work was to find a simple and cheap procedure for the determination of Vitamin B_2 at low concentrations based on a manganese dioxide bulk-modified carbon paste electrode (MnO_2/CPE).

Citation: Mehmeti E, Stanković DM, Chaiyo S, Švorc Ľ, Kalcher K. Manganese dioxide-modified carbon paste electrode for voltammetric determination of riboflavin. Mikrochimica Acta. 2016;183:1619-1624. doi:10.1007/s00604-016-1789-4.

EXPERIMENTAL

Chemicals

Boric acid, sodium hydroxide, acetic acid, phosphoric acid and manganese(IV) oxide (99.99 %, diameter approximately 5 μm), were purchased from Sigma–Aldrich (https://www.sigmaaldrich.com/) and used as received without any further purification. Calibration solutions were prepared from the stock solution (1 mM) by appropriate dilution with supporting electrolyte. Britton–Robinson buffer was prepared in usual way by mixing of 40 mM of all necessary components (phosphoric acid, acetic acid and boric acid). The pH of different Britton–Robinson buffer was adjusted with sodium hydroxide (0.2 M).

Working solutions of vitamin B_2 (VB_2) were freshly prepared on the day of the experiment by appropriate dilution with the supporting electrolyte. All other chemicals were of analytical reagent grade. Deionized water with a resistivity of 18 MΩ cm (Millipore Milli-Q system) was used for the preparation of all the solutions.

Apparatus

Cyclic voltammetric (CV) measurements and differential pulse voltammetric (DPV) measurements were performed using an Autolab PGSTAT 302 N (http://www.metrohm.com/de-at) potentiostat/galvanostat controlled by corresponding software (Nova 1.10). The electrochemical cell (total volume of 10 mL) consisted of a glass vessel equipped with the Ag/AgCl (3 M KCl, Metrohm 6.0733.100) as a reference electrode, platinum wire as a auxiliary electrode and carbon paste as a working electrode. All of the pH values were measured using a pH meter (Orion, model 1230) with a combined electrode (glass-reference electrodes), which was calibrated weekly with standard buffer. All potentials given in the text are versus the Ag/AgCl reference electrode at room temperature.

Preparation of a Carbon Electrode Modified with Manganese Dioxide (MnO_2/CPE)

Plain carbon paste was prepared by carefully hand mixing 380 μL of paraffin oil with 1 g of graphite powder in a mortar with a pestle. After standing overnight a portion of the resulting paste was packed into the end of a Teflon tube (an inner diameter 5 mm, outer diameter 10.15 mm) and the surface was polished using a PTFE plate or wet filter paper. The carbon paste was modified by adding 5 % (m/m) of MnO_2 as received. The amount of modifier was selected according to our experience and previously described articles, where it is found that modification with 5 % of MnO_2 gives best analytical response [11].

Whenever regeneration was required, a layer of the surface was removed and replaced by fresh paste. Electrical contact was made with a copper wire through the center of the tube.

Procedures

Cyclic voltammetry with a scan rate of 0.1 V/s (if not stated otherwise) was used for characterizing the electrochemical behavior of the analyte at the unmodified and modified electrode surface. The investigated solutions were transferred into the voltammetric cell and the voltammograms (usually 5 cycles) were recorded in a potential range between −0.5 V and +0.5 V.

Differential pulse voltammetry with optimized parameters in the potential range from −0.5 V to +0.3 V (pulse amplitude of 0.12 V and pulse time of 0.04 s) was used for the quantification of VB_2.

Interference Studies

Oxidation behavior of some possible interferences such as vitamins B_1, B_6 and B_{12}, ascorbic acid and glucose, were tested in concentrations of 1 μM under optimized experimental conditions. The changes of the peak current of 1 μM VB_2 were compared in the absence and in the presence of selected interferences It was considered that tested compounds strongly interfere with the determination of riboflavin if gives signal changes more than ±10 %.

Sample Analysis

Vitamin B_2 tablets (4.55 g) were dissolved in 10 mL of water and an aliquot (10 μL) was added to 10 mL of buffer at pH 2.0 and recorded by DPV under optimized experimental conditions. The concentration of $VB_2$was evaluated from calibration curve. All experiments were performed in triplicate.

RESULTS AND DISCUSSION

Electrochemical Behavior of Vitamin B_2 on MnO_2/CPE

Cyclic voltammetry was applied to study the electrochemical behavior of VB_2 on a MnO_2/CPE. All necessary factors influencing the current response of VB_2 were carefully studied to explore the best conditions at which the best analytical performance was achieved. The electrochemical behavior of the MnO_2/CPE towards VB_2 was compared to the unmodified CPE electrode (Fig. (Fig.1).

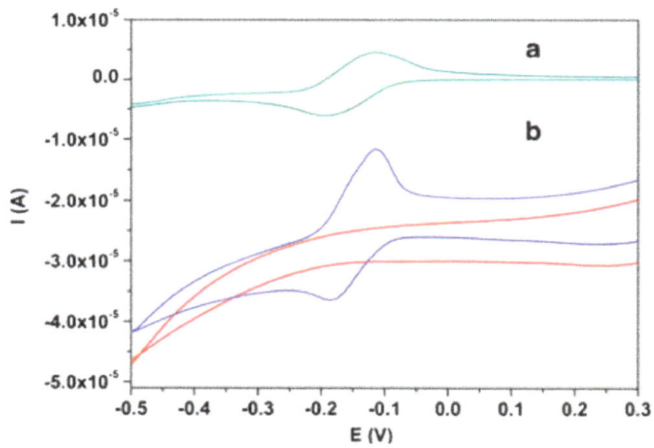

Figure 1: Cyclic voltammograms of 0.1 mM Vitamin B_2 on A) CPE unmodified and B) MnO_2/CPE in buffer at pH 2.0, scan rate of 0.1 $V \cdot s^{-1}$

The unmodified CPE (A) gave a small current response with well-defined oxidation peak at around −0.15 V and reduction peak at around −0.2 V. The modified electrode CPE /MnO_2 (B) gives a well manifested oval-shape signal response at −0.15 V in the anodic direction. In the reverse scan also the reduction is observable at around −0.2 V indicating that the oxidation and reduction of the analyte during the reaction is electrochemically reversible. It is noted that the current corresponding to MnO_2/CPE electrode is at about two time higher value when compared to the unmodified electrode. This is mainly attributed to the higher active surface area of MnO_2 particles present on the surface of modified CPE electrodes. The current response obtained for the modified electrode approves the effect of MnO_2 in the electrode structure.

Effect of pH Value of Supporting Electrolyte

Effect of pH on peak current and peak potential was investigated in the range from 2 to 6 using buffer. The peak current decreased considerably beyond pH 2.0. The peak potential of VB_2 was shifted to more negative potentials linearly with increasing of the pH of supporting electrolyte. Based on these facts pH 2.0 was chosen for further experiments. When considering pH from 2 to 6 the peak potential shift to the more negative values occurs with the corresponding equation $Ep\ (V) = -0.052 \times pH - 0.01272$.

The slope of 52 mV per pH unit is close to the ideal value of 59 mV which might indicate that the number of protons and electrons involved in the electrochemical reaction is in the ratio 1:1. In Fig.2 the dependence of the peak current (Ip) and of the peak potential (Ep) on the pH of buffer is represented. Obtained proton/electron ratio is same as those previously described in the literature and in accordance with the oxidation reaction of riboflavin where two electrons and two protons are involved.

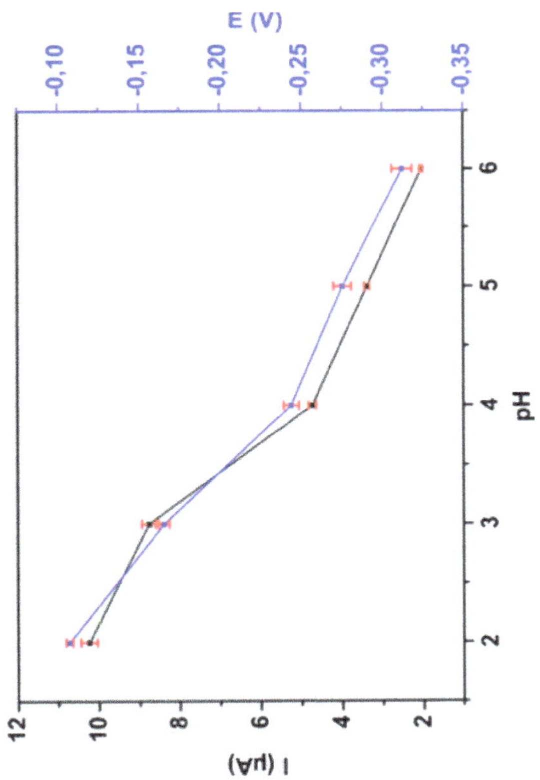

Figure 2: Effect of pH on the peak potential (■) and peak current (■) of 0.1 mM Vitamin B_2 in buffer at pH 2.0 on MnO_2/CPE using CV at scan rate of 0.1 $V \cdot s^{-1}$

Effect of Scan Rate

In order to study the nature of the electrochemical reaction of VB_2 on MnO_2/CPE the effect of different scan rate in the range from 0.01 $V \cdot s^{-1}$ to 0.5 $V \cdot s^{-1}$ on the peak current and peak potential was investigated by CV (Fig. 3) in buffer at pH 2.0. The peak current of VB_2 increased practically linearly with square root of the scan rate indicating that the oxidation and reduction process on the electrode surface is controlled by diffusion rather than by adsorption. Inset of Fig. 3 the linear dependence can be expressed by the equation: $Ip(\mu A) = 1.8760 \ x\ v^{1/2}\ (mV \cdot s^{-1}) - 4.662$ ($R^2 = 0.9927$). Increase of the scan rate does not cause significant changes in redox peak potentials (ΔEp ~70 mV). These results indicate reversible process for the nature of electrochemical reaction.

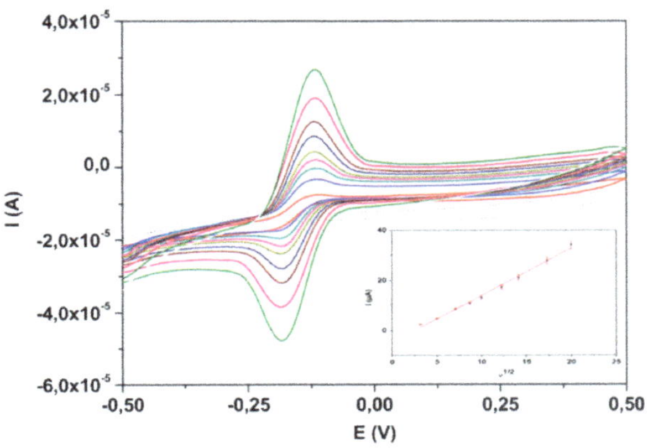

Figure 3: Cyclic voltammograms of 0.1 mM Vitamin B_2 in buffer at pH 2.0 on MnO_2/CPE at various scan rates from 0.01 $V \cdot s^{-1}$ to 0.5 $V \cdot s^{-1}$. The peak current (Ip) as a function of $v^{1/2}$ for the oxidation peak of Vitamin B_2 is shown in the inset

Optimization of DPV Parameters

For the quantitative determination of VB_2, DPV was used as a suitable electroanalytical technique due to the low background currents and low detection limits. The parameters for DPV such as pulse amplitude and pulse time were optimized to find the best experimental setup for the quantification of VB_2. The optimization was carried out in previously selected buffer at pH 2.0 with the concentration of 0.1 mM VB_2. During this optimization procedure one investigated parameter was varied while the others were kept fixed. When the pulse time was varied from 0.01 to 0.1 s, the peak current increased up to value of 0.04 s and with further increase of the pulse time obtained current was decreasing. The most appropriate peak currents was observed at 0.04 s. Varying the pulse amplitude in the range of 0.01–0.35 V the peak currents increased with concomitant broadening of the peaks; finally a value of 0.12 V of pulse amplitude was chosen which was found to be most appropriate with respect to the current response and peak shape of VB_2. All other experiments such as interference studies, calibration curve and sample analysis were carried out under these optimized parameters.

Analytical Performance

Calibration curve for determination of VB_2 on MnO_2/CPE was obtained using DPV under the optimized experimental conditions and was constructed by plotting estimated oxidation peak current versus known VB_2concentrations. FigureFigure4 shows a typical DP voltammograms obtained for different concentrations of VB_2. The obtained currents were linear with logarithm of concentration in the range from 0.02 to 9 μM.

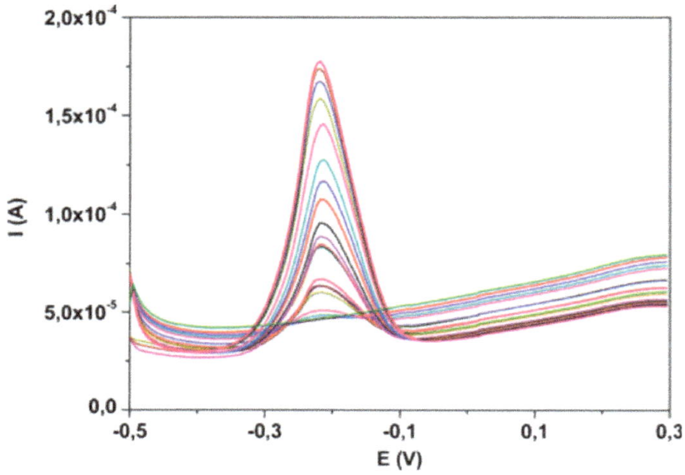

Figure 4: DP voltammograms for different concentrations of Vitamin B_2 from 0.02 to 9 μM in buffer at pH 2.0 on MnO_2/CPE at optimized DPV parameters

The graph shows a dynamic range for concentrations from 0.02 to 9 μM, with a corresponding linear equation: I (μA) = *55.51723 x* log_C μM + *92.31389*. Correlation coefficient of R^2 is equal to 0.9976. The detection limit ($3\sigma_{c=0.06\mu M}$/slope) was estimated as 15 nM. The repeatability (*n* = 4 measurements, c = 0.06 μM VB_2) was calculated as ±2 %. RSD, and the reproducibility of the electrode preparation, based on measurements of the 0.06 μM of VB_2 with four separately prepared electrodes, was estimated to be ±3 %, which improves our statement that this electrode can be satisfactory replacement for commercial electrodes.

This sensor offers low detection limit, wide linear range with a good sensitivity and reproducibility and in comparison with previously reported data this sensor possess comparable or better characteristics for the quantification of VB_2.

Table Table1 shows a comparison of the MnO_2/CPE with other electrodes described recently in the literature. The method presented in this work has a comparable or better performance with wide linear range and low detection limit. The advantages of our method are based on simplicity of the electrode preparation.

Table 1: Recently reported electrodes for the determination of riboflavin

Electrode	Modifier	Method	Limit of Detection	Linear Range	Ref.
Gold Electrode	homoadenine single-stranded DNA/ molybdenum disulfide–graphene nanocomposite	DPV	20 nM	0.025–2.25 μM	[18]
Carbon-paste electrode	zeolite	CV	0.71 μM	1.7–34 μM	[19]
Copper	Bi-film	SWAdSV	100 nM	0.3–0.8 μM and 1.0–9.0 μM	[2]
Pencil graphite electrode	DNA	DPV	0.9 μM	1–186 μM	[20]
Carbon paste	MnO_2	DPV	15 nM	0.02 to 9 μM	This work

Interference Studies

In order to evaluate the selectivity of the method toward VB_2, the effect of possible interfering agents was investigated under optimized conditions. Some possible interfering compounds were tested, such as vitamin B_1, B_6, B_{12}, ascorbic acid and glucose. These compounds in concentration of 1 μM, in absence of VB_2, practically do not provide electrochemical activity in the tested potential range (Fig. 5a). The presence of these interferences in same concentration level as VB_2 (1 μM) do not causes changes in the peak current obtained for VB_2 (Fig. 5 b). Hydrogen peroxide, dopamine and uric acid can be expected as possible interferences in the human body fluid samples (urine and blood serum). According previously published data [11, 21, 22] in strongly acidic media these compounds provides oxidation peaks at higher potentials compared to VB_2 (~ −0.2 V). Based on these results this method has a good selectivity for the electrochemical determination of VB_2.

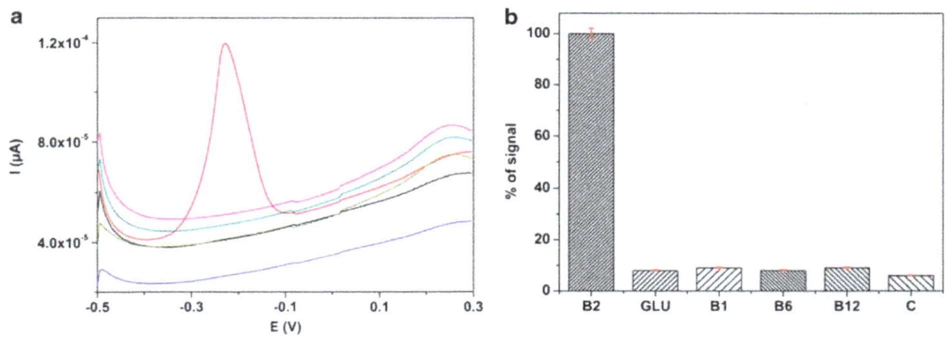

Figure 5: **a** DPVs of all tested compounds in concentration of 1 µM in absence of VB_2 and **b** Signals of tested compounds in the presence of 1 µM VB_2, expressed as relative signals of VB_2 on MnO_2/CPE in buffer at pH 2.0 at scan rate of 0.1 $V{\cdot}s^{-1}$

Analytical Application

To test the practical applicability of the method it was applied for the analysis of VB_2 content in the pharmaceutical formulations. The content of VB_2 was determined from the calibration curve by optimized DPV method. The samples were prepared as it is previously described. The mean value of the concentration obtained by the calibration curve as 0.24 µM corresponds quite well to the labeled value of the commercial pharmaceutical formulation 0.26 µM. Standard addition of different amount of VB_2 caused current increments at the sample potential (Fig. 6)) which allows the evaluation of the recovery values.

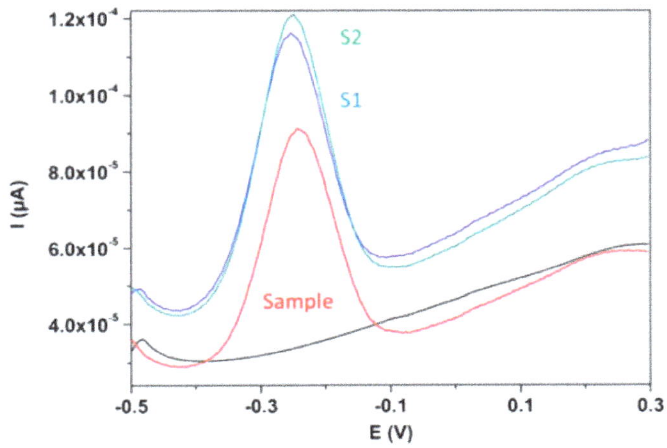

Figure 6: DP voltammograms obtained for determination of Vitamin B_2 (S1-S2 standard addition 1 and 2) at MnO_2/CPE electrode, in buffer at pH 2.0, under optimized experimental parameters

The found values are in good agreement with added amount of VB_2 and the accuracy was evaluated with recovery experiments (Table .2). The results are confirming that the sensor was applied for the determination of the concentration of VB_2 in pharmaceutical formulations.

Table 2: Results for determination of VB_2 in pharmaceutical formulations

Sample/Tablet	Labeled/µM	Found/µM	Added(S1)/µM	Found(S1)/µM	Recovery/%	Added(S2)/µM	Found(S2)/µM	Recovery/%
	0.24	0.26	0.10	0.35	97	0.2	0.53	95

CONCLUSIONS

Carbon paste electrode modified with manganese dioxide was described for the determination of the vitamin B_2 by differential pulse voltammetry. The observed results showed that incorporating of MnO_2 in the structure of carbon paste electrode increases its affinity towards determination of riboflavin with a good reproducibility and very low detection limit. Based on the simplicity of the electrode preparation, its sensitivity and selectivity we propose simple, inexpensive electrochemical sensor which can be used for application on the field of riboflavin analysis.

ACKNOWLEDGMENTS

E.M wishes to acknowledgement Higher KOS Stipendien, financed by ADA and MEST and Austrian Agency for International Cooperation in Education and Research (OeAD-GmbH), Centre for International Cooperation & Mobility (ICM). D.M.S. wishes to acknowledgement the Ministry of Education and Science of the Republic of Serbia (project No. OI 172030).

REFERENCES

1. Revin SB, John SA. Simultaneous determination of vitamins B_2, B_9 and C using a heterocyclic conducting polymer modified electrode. Electrochim Acta. 2012;75:35–41. doi: 10.1016/j.electacta.2012.04.056.
2. Sá SÉ, Silva SP, Jost LC, Spinelli A. Electrochemical sensor based on bismuth-film electrode for voltammetric studies on vitamin B_2 (riboflavin) Sensors Actuators B. 2015;209:423–430. doi: 10.1016/j.snb.2014.11.136.
3. Hua L, Yang X, Wang C, Yuan H, Xiao D. Determination of riboflavin in urine and beverages by capillary electrophoresis with in-column optical fiber laser-induced fluorescence detection. J Chromatogr B.2007;856:245–251. doi: 10.1016/j.jchromb.2007.06.011.
4. Qi H, Cao Z, Hou L. Electrogenerated chemiluminescence method for the determination of riboflavin at an ionic liquid modified gold electrode. Spectrochim Acta A. 2011;78:211–215. doi: 10.1016/j.saa.2010.09.023.
5. Vidović S, Stojanović B, Veljković J, Prazi AL, Roglić G, Manojlović D. Simultaneous determination of some water-soluble vitamins and preservatives in multivitamin syrup by validated stability-indicating high-performance liquid chromatography method. J Chromatogr A. 2008;1202:155–162. doi: 10.1016/j.chroma.2008.06.039.
6. Wang M, Zhao L, Liu M, Lin JM. Determination of riboflavin by enhancing the chemiluminescence intensity of peroxomonosulfate–cobalt(II) system. spectrochim. Acta A. 2007;66:1222–1227. doi: 10.1016/j.saa.2006.06.010.
7. Shumyantseva VV, Bulko VT, Petushkova AN, Samenkova FN, Kuznetsova PG, Archakov IA. Fluorescent assay for riboflavin binding to cytochrome P450 2B4. J Inorg Biochem. 2004;98:365–370. doi: 10.1016/j.jinorgbio.2003.10.024.
8. Safavi A, Maleki N, Ershadifar H, Tajabadi F. Development of a sensitive and selective riboflavin sensor based on carbon ionic liquid electrode. Anal Chim Acta. 2010;674:176–181. doi: 10.1016/j.aca.2010.06.012.
9. Stankovic DM, Kalcher K. The immunosuppressive drug rapamycin electroanalytical sensing using boron- doped diamond electrode. Electrochim Acta. 2015;168:76–81. doi: 10.1016/j.electacta.2015.03.200.
10. Svorc L, Stankovic DM, Mehmeti E, Kalcher K. Sensitive electrochemical determination of yohimbine in primary bark of natural aphrodisiacs using boron-doped diamond electrode. Anal Methods. 2014;6:4853–4859. doi: 10.1039/C4AY00704B.
11. Schachl K, Alemu H, Kalcher K, Jezkova J, Svancara I, Vytras K. Amperometric determination of hydrogen peroxide with a manganese dioxide-modified carbon paste electrode using flow injection analysis. Analyst. 1997;122:985–989. doi: 10.1039/a701723e.
12. Taba Z, Wang J. Electrocatalysis and flow detection at a glassy carbon electrode modified with a thin film of oxymanganese species. Electroanalysis. 1991;3:215–219. doi: 10.1002/elan.1140030313.
13. Nie T, Xu JK, Lu LM, Zhang KX, Bai L, Wen YP. Electroactive species-doped poly(3,4-ethylenedioxythiophene) films: enhanced sensitivity for electrochemical simultaneous determination of vitaminsB_2, B_6 and C. Biosens Bioelectron. 2013;50:244–250. doi: 10.1016/j.bios.2013.06.057.
14. Qijin W, Nianjun Y, Haili Z, Xinpin Z, Bin X. Voltammetric behavior of vitamin B2 on the gold electrode modified with a self-assembled monolayer of L-cysteine and its application for the determination of vitamin B2 using linear sweep stripping voltammetry. Talanta. 2001;55:459–467. doi: 10.1016/S0039-9140(01)00437-4.
15. Kalcher K. Chemically modified carbon paste electrodes in voltammetric analysis. Electroanalysis.1990;2:419–433. doi: 10.1002/elan.1140020603.
16. Tashkhourian J, Nami-Ana SF. A sensitive electrochemical sensor for determination of gallic acid based on SiO_2 nanoparticle modified carbon paste electrode. Mater Sci Eng C. 2015;52:103–110. doi: 10.1016/j.msec.2015.03.017.
17. Zbiljić J, Guzsvány V, Vajdle O, Prlina B, Agbaba J, Dalmacija B, Kónya Z, Kalcher K. Determination of H_2O_2 by MnO_2 modified screen printed carbon electrode during Fenton and visible light-assisted photo-Fenton based removal of acetamiprid from water. J Electroanal Chem. 2015;755:77–86. doi: 10.1016/j.jelechem.2015.07.027.
18. Wang Y, Zhuang Q, Ni Y. Fabrication of riboflavin electrochemical sensor based on homoadenine single-stranded DNA/molybdenum disulfide–graphene nanocomposite modified gold electrode. J Electroanal Chem. 2015;736:47–54. doi: 10.1016/j.jelechem.2014.10.028.
19. Nezamzadeh Ejhieh A, Pouladsaz P. Voltammetric determination of riboflavin based on electrocatalytic oxidation at zeolite-modified carbon paste electrodes. J Ind Eng Chem. 2014;20:2146–2152. doi: 10.1016/j.jiec.2013.09.044.
20. Ensafi AA, Heydari-Bafrooei E, Amini M. DNA-functionalized biosensor for riboflavin based electrochemical interaction on pretreated pencil graphite electrode. Biosens Bioelectron. 2012;31:376–381. doi: 10.1016/j.bios.2011.10.050.
21. Sljukic B, Compton RG. Manganese dioxide graphite composite electrodes formed via a low temperature method: detection of hydrogen peroxide, ascorbic acid and nitrite. Electroanalysis. 2007;19:1275–1280. doi: 10.1002/elan.200703878.
22. Zhang L, Yuan SM, Yang LM, Fang Z, Zhao GC. An enzymatic glucose biosensor based on a glassy carbonvelectrode modified with manganese dioxide nanowires. Microchim Acta. 2013;180:627–633. doi: 10.1007/s00604-013-0968-9.

CHAPTER 11

CADMIUM SULPHIDE-REDUCED GRAPHENE OXIDE-MODIFIED PHOTOELECTRODE-BASED PHOTOELECTROCHEMICAL SENSING PLATFORM FOR COPPER(II) IONS

I Ibrahim, H. N Lim, N. M Huang, A Pandikumar

INTRODUCTION

Copper (II) ions (Cu^{2+}) are the third most abundant transition metal ion and an essential trace element in the human body. Although Cu^{2+} ions are physiologically essential for various reasons, they are also an important environmental pollutant. The recommended allowance of copper is 0.8–0.9 mg/day for normal adults [1] The drinking water quality guidelines by the World Health Organization (WHO) suggests that the amount of copper in drinking water should be limited to 1.3 mg/L (~20 μM) [2, 3] Aberrant levels of copper exhibit significant deleterious effects on living organisms, such as Wilson's disease, pre-menstrual syndrome, adult polycystic kidney disease, fibroids, and infertility [3, 4] As a result, there has been ongoing interest in developing better sensors for Cu^{2+} ion detection, because of their significance in the environment and biological systems. Although conventional methods such as chromatography-mass spectrometry, inductively coupled plasma mass spectrometry, and atomic absorption spectrometry are common detection tools, expensive instrumentation, extremely time-consuming processes, and the necessity of sophisticated operation have retarded their progress. Hence, the discoveries of sustainable techniques are needed to overcome the limitations. Recently, photoelectrochemical (PEC) has attract much attention among researcher as a sensing platforms for Cu^{2+} ion detection because its offered sensitive, fast, reproducible, simple, low-cost, and accurate potential alternatives to the above conventional methods.

PEC sensing is an approach that relies on photo-illumination to transfer electrons between an analyte, a semiconductor, and an electrode [5, 6] A PEC sensor harnesses light to stimulate the photoactive species on the electrode, and utilizes a photocurrent as a means to quantify the changes. Furthermore, it has numerous inherent benefits such as a simple electronic circuit and easy fabrication, and it can use a low-power light source, which results in a very low instrumentation cost [7] In order to further enhance the performance of PEC sensors, triethanolamine, ascorbic acid, and sodium sulphide are often added to the PEC detection cell as electron donors [8] Semiconductors such as TiO_2 [9] ZnO [10] and CdS [11] can be used as the photoactive materials in the PEC sensor.

The photoactive material plays a major role in the PEC sensing performances for heavy metal ions. A PEC sensing system with high sensitivity can be achieved by amplifying the photocurrent intensity and minimizing the electron–hole recombination. Cadmium sulfide (CdS) semiconductor nanoparticles are the best candidate to achieve this target because of their application potential, based on size-tunable optical properties and high mobility [12] CdS possesses a

Citation: Ibrahim I, Lim HN, Huang NM, Pandikumar A (2016) Cadmium Sulphide-Reduced Graphene Oxide-Modified Photoelectrode-Based Photoelectrochemical Sensing Platform for Copper(II) Ions. PLoS ONE 11(5): e0154557. doi:10.1371/journal.pone.0154557.

band gap of ~2.4 eV, and has been considerably investigated for PEC sensing, as well as for the other visible-light-driven PEC applications [13] In order to heighten the photocurrent generated by the CdS nanoparticles, it is essential to hold back the rapid recombination of the photogenerated electron–holes using CdS on support materials with large surface areas or by hybridizing CdS with conductive matrices such as carbon materials [14,15]

As a flexible atom-thin 2D carbon material, graphene has emerged as a rising star because of its outstanding properties and excellent performance, including its very large surface area (theoretically, 2630 $m^2 g^{-1}$ for a single-layer of graphene) [16], extraordinary electrical and thermal conductivities [17, 18], excellent mechanical strength [19] and good biocompatibility [20]. Considering its remarkable physical, chemical, and structural properties, materials based on graphene and its derivatives have sprung up quickly and been given tremendous attention across many disciplines, including novel nanoelectronics, high-frequency electronics, energy conversion and storage, biomedical devices, and sensors. Graphene-based electrochemical studies have received much interest because of its important role in accelerating electron transfer, with the merits of a high carrier mobility and exceptional electrical conductivity [21], which subsequently provide new opportunities to develop high-performance electrochemical sensors. To date, most graphene-based electrochemical sensors have used reduced graphene oxide (rGO) instead of graphene oxide (GO) because (i) it is rich in structural defects and its chemical groups facilitate the charge transfer process and thus enhance its electrochemical activity; (ii) the abundant chemical moieties on the rGO surface create availability and flexibility for various surface functionalizations, which boost the sensing performance; (iii) excellent tunable chemical and electrical properties; and (iv) efficient charge transport compared to non-conductive GO. The aforementioned benefits of rGO in electrochemical sensing can be enhanced by assembling CdS nanoparticles on its surface to intensify the photocurrent performance through the synergic effects between CdS and rGO [21] in addition to addressing the photocorrosion problems of CdS nanoparticles, making this nanocomposite suitable for photoelectrochemical sensing. The present research focused on the development of an easy-to-use, economical, mobile, selective, and sensitive analytical approach for the detection of Cu^{2+} ions based on a photoelectrochemical technique. CdS semiconductor nanoparticles were synthesized on an indium tin oxide (ITO) electrode with an aerosol-assisted chemical vapor deposition (AACVD) procedure using $Cd(CH_3COO)_2{\cdot}2H_2O$ and CH_4N_2S as the precursors to Cd^{2+} and S^{2-}, respectively. The rGO was then assembled on the CdS nanoparticles using a dip-coating method. The as-prepared ITO/CdS-rGO electrode was used for the detection of Cu^{2+} ions in an aqueous solution, in the presence of an electron donor, triethanolamine (TEA). It was found that the nanocomposites enhanced the photocurrent response up to 63 μA as a result of the rapid interfacial electron-transfer process. The results of this study demonstrated that the ITO/CdS-rGO photoelectrode has the potential to exhibit an excellent photocurrent response in 0.1 M of KCl under light illumination, which could provide a remarkable response upon the addition of Cu^{2+} ions. Based on the significant quenching in the photocurrent signal, an unconventional strategy was developed for the PEC detection of Cu^{2+}ions, with a linear concentration range of 0.5–120 μM, which gives a LoD of 16 nM.

EXPERIMENTAL METHODS

Materials

Graphite flakes were purchased from Asbury Graphite Mills Inc., USA. Cadmium acetate dihydrate ($Cd(CH_3COO)_2{\cdot}2H_2O$) and copper(II) sulfate pentahydrate ($CuSO_4.5H_2O$) were purchased from Hamburg Chemical. Thiourea (CH_4N_2S) was purchased from Fisher scientific. Methanol (CH_3OH, 99.9%) and acetone ($(CH_3)_2CO$, 99.5%) were purchased from Friendemann Schmidt. Ethanol ($CH_3CH_2OH_2$, 95%) was purchased from Systerm. Potassium chloride (KCl) was purchased from Merck. Triethanolamine (TEA, 99%) was received from R & M Chemicals. Indium tin oxide (ITO) conducting glass slides (7 Ωsq^{-1}) were commercially supplied by Xin Yan Technology Limited, China. Stock solutions of Cu^{2+}, Ni^{2+}, Co^{2+}, Fe^{3+}, Mn^{2+}, Na^+, K^+, Ba^{2+}, Zn^{2+}, Mg^{2+}, and Al^{3+} were prepared by dissolving suitable amounts of the $CuSO_4$, $NiSO_4$, $CoSO_4$, $Fe(NO_3)_3$, $MnCl_2$, NaCl, KCl, $BaCl_2$, $ZnSO_4$, $MgCl_2$, and $AlCl_3$ compounds, respectively, in Milli-Q water with a resistivity larger than 18 MΩ cm. Each sample solution was 10 mL. The entire experiment employed deionized water. The reagents and materials were used as received without further purification.

Fabrication of CdS thin film by AACVD method

A CdS thin film was prepared on an ITO glass substrate (1 cm × 1.5 cm) using an in-house AACVD assembly process. The synthesis strategy was reported earlier by our group [22, 23] The ITO glass was ultrasonically cleaned in acetone, ethanol, and deionized water before being used. Then, the CdS films were prepared from 0.05 M $Cd(CH_3COO)_2{\cdot}2H_2O$ and 0.1 M CH_4N_2S in 35 mL of a methanol solution that was generated into an aerosol to be deposited as CdS on the ITO in the reaction chamber.

Fabrication of Photoelectrochemical Sensor Electrode

The deposition of 0.1 M GO, which was produced using a simplified Hummer's method [24] on the surface of the AACVD-prepared CdS thin film was carried out using a dip-coating method, as reported by our group [22]. The CdS-GO thin film was placed in a furnace at 150°C, and argon gas was flowed for 15 min at a flow rate of 50 mL/min for the reduction process. The electrode was labeled as CdS-rGO.

Characterization Techniques

The surface morphologies of the nanocomposites were analyzed using a field emission scanning electron microscope (FESEM, FEI Quanta SEM Model 400 F) equipped with an energy dispersive X-ray (EDX) accessory. The crystalline phase of the CdS-rGO nanocomposite was studied using a Philips X'pert system X-ray powder diffractometer with Cu Kα radiation (λ = 1.5418 Å), and a Raman spectral analysis was carried out using a Renishaw inVia Raman microscope with laser excitation at λ = 514 nm.

Photoelectrochemical Studies

Photoelectrochemical measurements were performed using a homemade photoelectrochemical system containing a 150-W halogen lamp (Halloid) as the illumination source. The photocurrent signal was measured on a computer-controlled Versa-STAT-3 electrochemical analyzer from Princeton Applied Research. A CdS-rGO-modified photoelectrode with an active area of 1 $cm^2$was used as the working electrode, and a Pt wire and saturated Ag/AgCl were used as the counter and reference electrodes, respectively. The working electrode potentials stated in this paper are with reference to the Ag/AgCl, unless otherwise stated. All the photocurrent measurements were conducted by dipping the CdS-rGO-modified photoelectrode into a mixture of 0.1 M KCl and 0.5 M TEA at a constant potential of 0 V vs. Ag/AgCl.

RESULTS AND DISCUSSION

Morphological Studies

Fig 1(A) shows a low-magnification FESEM image of the CdS obtained using the AACVD method at 400°C for 180 min. This image indicates that the sphere-like morphology of pure CdS was composed of homogenous nanospheres with a narrow diameter distribution of about 350–400 nm. These nanospheres were dispersed with good monodispersity (**Fig 1(A), inset**), although a few seemed to agglomerate. Upon the addition of 0.1 mg/ml GO, a transparent ultrathin GO layer was clearly observed on the surface of the CdS nanoparticles (**Fig 1(B)**). The GO nanosheets were curled and corrugated, and CdS nanospheres were distributed over the GO sheets. The GO was then reduced by sintering at a temperature and pressure of 150°C and 50 mL/min, respectively. **Fig 1(C)** shows that the rGO formed a continuous network with CdS nanoparticles, resulting in excellent hybridization of the rGO and CdS. The rGO sheets blanketed the CdS nanoparticles, which enhanced the photo-generated carriers separation, and consequently elevated the PEC performance [25]. The corresponding histograms of the particle size distribution (**Fig 1(D)**) are also shown, along with the FESEM images. The size distribution was determined based on the measurement of 180 CdS nanospheres. It is worth noting that aggregates consisting of small particles were not taken into account in the determination of the particle crystal size. It was found that most of the particles had sizes of <400 nm, and the average particle size was 350–400 nm, as indicated in the histogram.

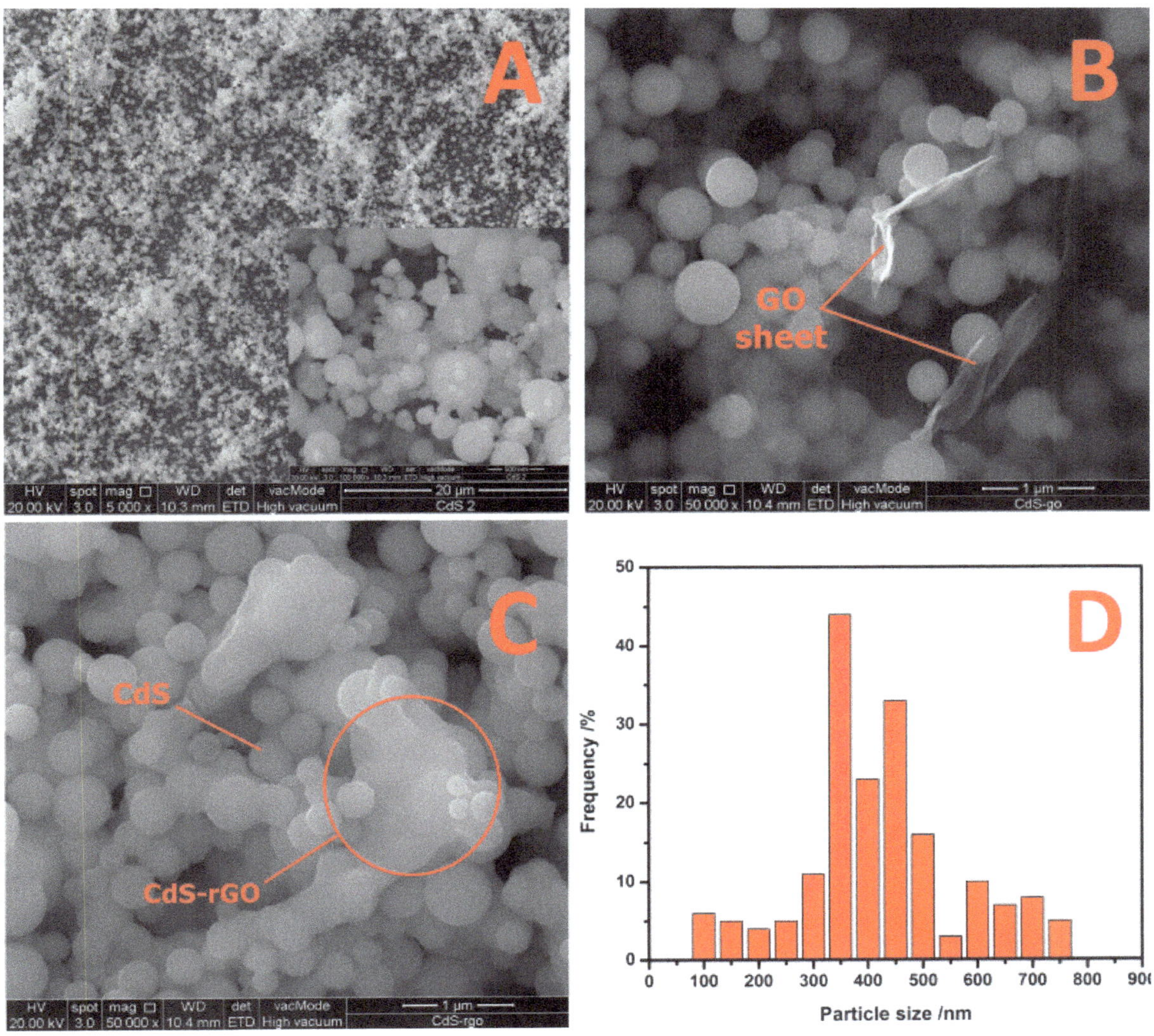

Figure 1. FESEM images of (A) CdS nanospheres, (B) CdS-GO, and (C) CdS-rGO, as well as (D) histogram showing size distribution of CdS nanoparticles in CdS-rGO nanocomposite.

X-ray Diffraction Analysis

XRD patterns revealed the effect of the graphene on the crystallinity of the CdS nanoparticles for the GO, CdS, and CdS-rGO films. Their diffraction patterns are given in **Fig 2.** The XRD pattern of GO portrays a distinguishable peak at 2θ = 10.6° at the index of (002). The characteristic peaks at 245°, 26.7°, 26.8°, 30.3°, 35.2°, 43.5°, 50.6°, 51.9°, and 60.2° respectively correspond to the (100), (002), (111), (101) (102), (110), (201), (112), and (202) planes of the hexagonal crystal structure of CdS (JCPDS card no.: 02–0549). The XRD pattern of the CdS-rGO nanocomposite only shows the peaks for the hexagonal CdS phase. However, the nanocomposite shows a peak with lower intensities than those for pure CdS. This would be caused by the graphene wrapped around the surface of the CdS particles, which could induce crystal diffraction. Moreover, no obvious peak is observed at the (002) index for the CdS-rGO composite, which corresponds to the oxygen-containing functional groups located between the layers of GO, suggesting that the GO was reduced to rGO during the heating treatment, and the interlayer spacing was decreased [26].

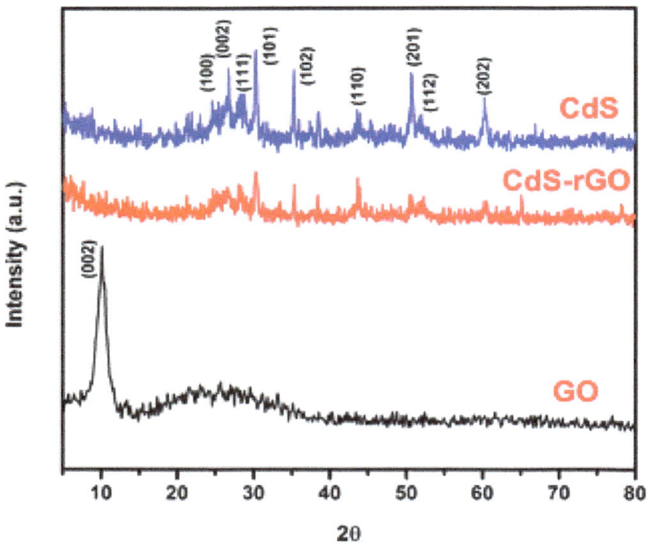

Figure 2. XRD patterns of as-prepared GO, CdS, and CdS-rGO samples.

Raman Spectral Analysis

Raman spectroscopy was performed on the GO, rGO, CdS, CdS-GO, and CdS-rGO nanocomposites, and the results are shown in **Fig 3.** The Raman profile of GO the typical D and G bands assigned to the κ-point phonons of the A_{1g} symmetry and E_{2g} phonon of the sp^2carbon at 1353 and 1593 cm^{-1}, respectively. These two bands shifted to lower wavelengths of 1340 and 1577 cm^{-1} after the reduction. The chemical reduction of GO enhanced the intensity ratio of the D to G bands (I_D/I_G) from 0.974 to 0.986. The increase in the I_D/I_G ratio for rGO was due to the restoration of the sp^2 network and the formation of unrepaired defects after the removal of numerous oxygen functionalities [27, 28]. Hence, it gives a clear indication of the effective reduction of GO. Furthermore, the appearance of the 2D band at 2662 cm^{-1} in the rGO spectrum indicates the reduction of GO to rGO [27]. The characteristic peaks of CdS presented at 301, 599, and 905 cm^{-1} correspond to the first-, second-, and third-order longitudinal optical phonon modes (1LO, 2LO, and 3LO), respectively. For the CdS-rGO nanocomposite, all the Raman bands for CdS and rGO can be found. When CdS nanoparticles are deposited on graphene, the intensity of the D band increases relative to the G band, and both bands shift to higher wavenumbers [29]. In the present study, the D and G bands for the CdS/rGO composite appeared at 1355 and 1593 cm^{-1}, respectively. The intensity ratios of the D and G bands for CdS–GO and CdS-rGO were 0.989 and 0.991, respectively. The I_D/I_G ratios for these two nanocomposites were higher than those for GO, indicating that a significant number of structural defects were introduced to the graphene lattice in the reaction. Conclusively, the larger I_D/I_G ratios in the composite materials compared to GO indicate an increase in the amount of smaller sp^2 domains and the revival of graphene network conjugation (re-aromatization) [30]. Moreover, the revived graphene network size was smaller than that of the GO starting material. This effect gave rise to an increased I_D/I_G ratio in the CdS/rGO composite materials [31].

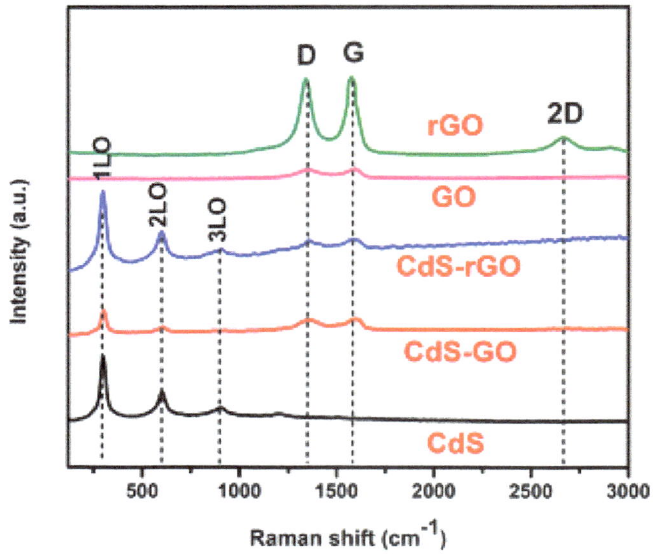

Figure 3. Raman spectra of GO, rGO, CdS, CdS-GO, and CdS-rGO.

Electrochemical Impedance Spectroscopy Analysis

The electron transfer characteristics and recombination processes at the electrolyte and electrode interfacial surface were investigated using the electrochemical impedance spectra (EIS) in 0.1 M KCl and 0.5 M TEA, as shown by Nyquist plots (**Fig 4**). The EIS were found in a frequency range of 10 kHz to 0.01 Hz to evaluate the frequency response of the nanocomposites. The equivalent series resistance (ESR) was extracted from the x-intercept of the Nyquist plot to evaluate the resistance of the KCl and TEA mixture solution, intrinsic resistance of the nanocomposite, and contact resistance at the interface between the electrode and current collector [32, 33]. The ESR values for the CdS, CdS-GO, CdS-rGO, and CdS-rGO-Cu were 89.14, 88.24, 72.37, and 351.22 Ω, respectively. The CdS-rGO electrode (**Fig 4,** curve c) shows a lower resistance value, indicating that the presence of rGO improved the transfer of the charge performance of the CdS electrode. Meanwhile, the larger resistance value of the CdS-rGO-Cu electrode (curve d) was due to the presence of Cu^{2+} ions on the CdS-rGO electrode, which hindered the electron transfer to the as-prepared electrode, increasing the intrinsic resistance of the nanocomposite. In the high-frequency range, the sizable semi-circle indicates the high charge transfer resistance (R_{ct}) caused by the weak transfer of the generated electrons and the recombination properties [10]. In the Nyquist plots, CdS-rGO exhibits a smaller semi-circle of 10.85 Ω compared to CdS-GO (25.33 Ω), suggesting that the electrochemically unstable oxygenous groups were removed during the reduction, hence overcoming the pseudocapacitance [34]. The CdS-GO (curve b) appears to have a large R_{ct} because of excessive oxygenated species, which produced an insulating behavior and obstructed its electrochemical characteristics [35]. Obviously, bare CdS electrodes (curve c) appear to have the largest semi-circle of 8841.2 Ω, particularly because of the absence of rGO, which behaved as an electron transfer medium for CdS excitation [36]. The CdS-rGO-Cu shows a slightly high Rct value of 7700.4 Ω, demonstrating that the Cu^{2+} ions were successfully incorporated on the surface of the sensor electrode [36], which provided a sensitive sensing platform for Cu^{2+} ion detection. Additionally, the low-frequency region portrays a straight line caused by a diffusion-limiting process that is more vertical for CdS-rGO, implying even more diffusion routes for electrons within the CdS-rGO electrode [34], and hence rapid electrolyte diffusion on the electrode–electrolyte interfaces [37].

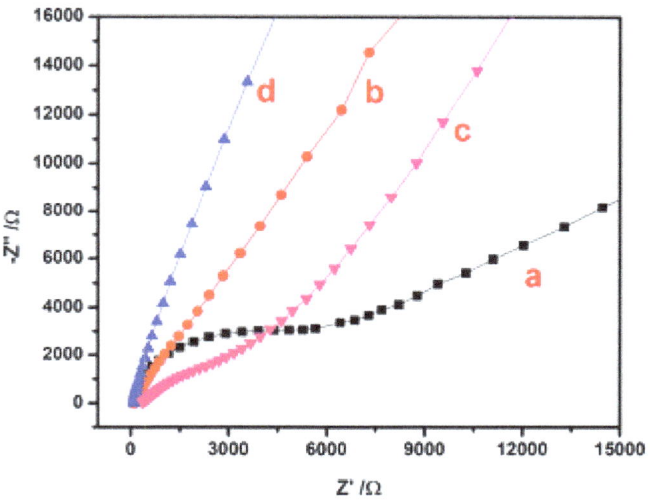

Figure 4. Nyquist plots obtained for (a) CdS, (b) CdS-GO, (c) CdS-rGO, and (d) CdS-rGO-Cu after addition of 4 μM Cu^{2+} ions into 0.1 M KCl and 0.5 M TEA solution.

Photoelectrochemical Performance

S1(A) Fig shows the linear sweeps of the CdS-rGO electrode in a potential range of -0.6 to 1.0 V versus Ag/AgCl in 0.1 M KCl electrolyte. Under irradiation, the CdS-rGO electrode achieved a maximal current of 100.07 μA at an applied bias of 0.96 V versus Ag/AgCl. Meanwhile, the current was obviously low in the absence of light irradiation, and it had a maximum current of 89.87 μA. The light alternated in 30-s intervals (light on and off) showed the comparable and instantaneous photocurrent response under light and dark conditions due to the PEC effect [38]. Likewise, **S1(B) Fig** appears to have a similar LSV to that in **S1(A) Fig** but the photocurrent response for **S1(B) Fig** is much more pronounced. The current response upon light illumination was 163.03 μA at 0.70 V versus Ag/AgCl, while in the dark condition, it had a current response of 140.47 μA with the same applied bias. The enhanced photocurrent was due to the introduction of 0.5 M TEA in the electrolyte. This is because TEA is a suitable sacrificial electron donor, leading to rapid electron–hole separation, and thus magnifying the photoelectrochemical responses and minimizing the over potential [39]. Conclusively, the CdS-rGO electrode worked extremely well in the presence of TEA, thus giving a stronger current response because more electrons were brought together from the PEC reaction, which suggests an improved separation efficiency [40]. Upon the addition of 4 μM of Cu^{2+} ions, the photocurrent dropped even more dramatically (**S1(C) Fig**). The current was reduced to 67.41 μM under light irradiation and 44.96 μM in the dark condition, both at an applied bias of 0.60 V versus Ag/AgCl. The photocurrent decrease corresponded to the extensive carrier recombination rate [41]. The obvious increase in cathodic current at potentials of -0.20 to -0.10 V was a result of the reduction in Cu^{2+} ions, corresponding to the following reaction: $Cu^{2+} + e^- \rightarrow Cu^+$. Further, the influence of the photocurrent performance was investigated in the absence and presence of 4 μM Cu^{2+} ions with 0.1 M KCl and 0.5 M TEA under the light "on-off" condition (**S1(D) Fig**). It was clear for both the absence and presence of Cu^{2+} ions that the current was higher under light illumination and declined when the light was cut off. Upon the addition of Cu^{2+} ions into the solution, there was a reduction in the photocurrent compared to the former value. This suggests that the photocurrent quenching was due to the photocatalytic reduction of Cu^{2+} to Cu^+ ions. From the results obtained, the photocurrent increased as the electrode potential was made more positive, suggesting that an anodic photocurrent appeared, with electrons as its majority carriers [42]. Thus, the ITO/CdS-rGO electrode behaved as an n-type semiconductor electrode. The n-type material induced oxidation on the electrolyte species at the semiconductor-electrolyte during the PEC performance [42].

Subsequently, the PEC response was investigated by employing different materials to decorate the ITO sheet used as a photoanode. The photocurrent of each prepared sample was calculated in a 0.1 M KCl solution containing 0.1 M TEA under "on-off" light illumination cycles at a bias of 0 V vs. Ag/AgCl, as shown in **Fig 5(A)**. The photoresponse for rGO was non-existent because rGO could not be excited, as shown in curve a [43]. Similarly, for GO (curve b), no obvious photocurrent was observed. This was typically due to the nature of graphene oxide, which is electrically insulating and requires further reduction steps to generate a photocurrent signal [44]. After modification by the AACVD process, the ITO/CdS electrode increased the photocurrent to 9.96 μA (curve c). For the ITO/CdS-rGO electrode, this could boost the photocurrent by at least 6-fold, reaching a value of 64.35 μA (curve e). CdS nanoparticles were bound to the surface of the rGO nanosheets through the van der Waals force [45]. The binding force enabled the CdS nanoparticles to integrate intimately with the ITO electrode. The encapsulation of CdS by the rGO improved the interfacial contact, which facilitated the electron transfer between the CdS and ITO, thus preventing possible current leakage in the ITO/CdS-rGO electrode. Besides, rGO is also capable of dividing and accumulating the photogenerated electrons, which minimizes the recombination rate of the photogenerated electrons and holes, and boosts the photocurrent production [46, 47]. Meanwhile, for the ITO/CdS-GO electrode (curve d), a slightly lower signal (16.27%

or 53.88 µA) was revealed compared to CdS-rGO. This was because the electrical conductivity of rGO is higher than that of GO [48].

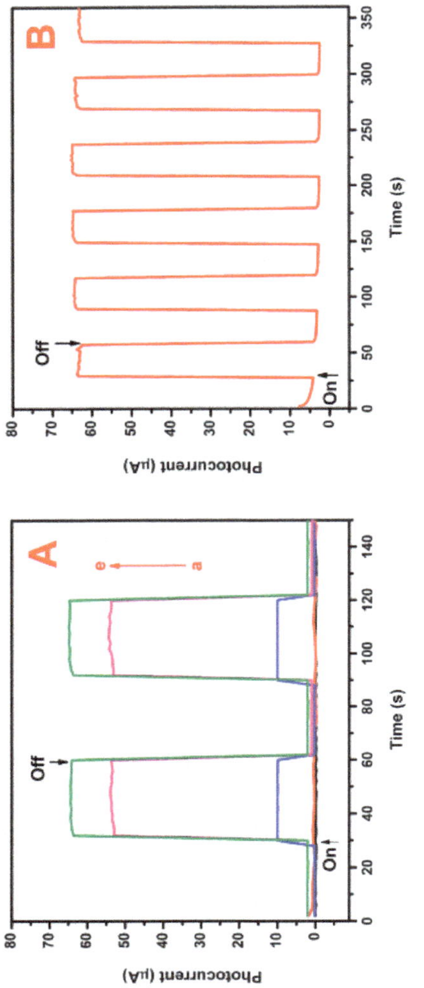

Figure 5. (A) Photocurrent responses of (a) rGO-, (b) GO-, (c) CdS-, (d) CdS-GO-, and (e) CdS-rGO-modified electrodes and (B) time-based photocurrent response of ITO/CdS-rGO in 0.1 M KCl and 0.5 M TEA under chopped irradiation.

The stability of the amperometric response of the CdS-rGO-modified photoelectrode in the presence of 0.5 M TEA under the light "on-off" condition is given in **Fig 5(B)**. This graph displays the photocurrent response of the ITO/CdS-rGO based on time, which was repeatedly measured six times at 30-s intervals under visible irradiation. No current was observed in the dark, which clearly suggests that no photoinduced charge separation took place. Interestingly, when the light was turned on, the photocurrent intensity was significantly increased to 64.35 µA. This may have been due to the photoinduced electron–hole separation at the CdS, in which the holes were scavenged by the TEA, and rGO acted as an electron transfer medium. Thus, the electrons were transported to the ITO electrode, resulting in photocurrent generation. The electrode showed a pronounced and stable photocurrent response during the light "on-off" condition, which it maintained. The relative standard deviation was only 1.02%, which implied that the constructed PEC sensor had a stable photocurrent activity.

Photoelectrochemical Detection of Copper(II) Ions

As shown in **Fig 6(A)**, the PEC performance of the CdS-rGO was evaluated by the sensing of Cu^{2+} ions. The photocurrent density of the ITO/CdS-rGO decreased with increasing concentration of Cu^{2+} ions. **Fig 6(B)** shows the increase in the photocurrent ΔI ($\Delta I = I - I_0$) and the Cu^{2+} ions concentration, where I and I_0 are the photocurrent intensities of CdS-rGO in the presence and absence of Cu^{2+} ions, respectively. The inset of **Fig 6(B)** shows an excellent linear relationship of R^2 = 0.9915. The regression equation was ΔI (µA) = 13.89 + 6.97 log [Cu^{2+}] (µM) between the value of ΔI and the logarithm of the Cu^{2+} ions concentration of 0.5–120.0 µM. The limit of detection (LoD) was evaluated using $3\sigma/S$ and was found to be 0.016 µM, where σ is the standard deviation of a blank signal, and S is the slope of the linear calibration plot. A possible schematic mechanism for the PEC discernment of Cu^{2+} ions is shown in **Fig 7**. Under illumination, the photogenerated electrons are transferred from the valence band (VB) to the conduction band (CB) of a CdS nanoparticles present in the CdS-rGO nanocomposite, and then holes are generated in the VB. The photogenerated holes are scavenged by TEA, and the electrons are promoted to the CB of CdS. Hence, when CdS was in contact with graphene, the photogenerated electrons in the CB of the CdS were transferred to the two-dimensional carbon sheets, which hindered the charge recombination and prolonged the lifetime of the photogenerated carriers. The photocurrent intensity was greatly decreased upon the addition of Cu^{2+} ions. The addition of Cu^{2+} ions to CdS resulted in the binding of $Cu^{2+}$with S^{2-}, and the reduction of Cu^{2+} to Cu^{+} ions under irradiation. Consequently, Cu_xS (x = 1, 2) was formed on the surface of CdS because Cu^{2+} chemically displaced Cd^{2+} because Cu_xS has a lower solubility than CdS. The development of recombination centers (Cu_xS) on the CdS surface generated a low band gap energy, which effectively recombined excited electrons in the CB and holes in the VB [49–52]. In addition, to prove the quenching of the photocurrent upon the addition of Cu^{2+} ions, FESEM images were recorded for the CdS-rGO nanocomposite after the addition of Cu^{2+} ions to the photoelectrochemical cell. In **Fig 6(C)**, a small hole is observed in some of the spherical CdS structures due to the presence of 1.15% Cu content in the CdS-rGO nanocomposite. The transparent ultrathin rGO sheets were seen to retain their role of linking the spherical CdS, even after the addition of 4 µM Cu^{2+} ions under an illuminated condition. A comparison of this FESEM image with **Fig 6(B)** clearly indicates a photocurrent quenching due to the formation of Cu_xS, *i.e.*, analyte-induced morphological changes caused by photocurrent quenching. The presence of CdS in the composite was demonstrated by the Cd and S peaks in the EDX profile (**Fig 6(D)**). The spectrum reveals an atomic Cd to S ratio of about 1:1, which agrees with the stoichiometric CdS. The C came from graphene, and the O was contributed by the residual oxygenous groups on the rGO. Thus, the result indicates that the CdS was successfully deposited on the rGO sheets. In contrast, the small peak observed in the EDX spectrum implies the existence of 0.55 atomic % of Cu^{2+} (**Fig 6(D), inset**). Therefore, the photocurrent intensity of the CdS-rGO decreased in the Cu^{2+} ion solution, making the nanocomposite suitable for the selective determination of Cu^{2+} ions.

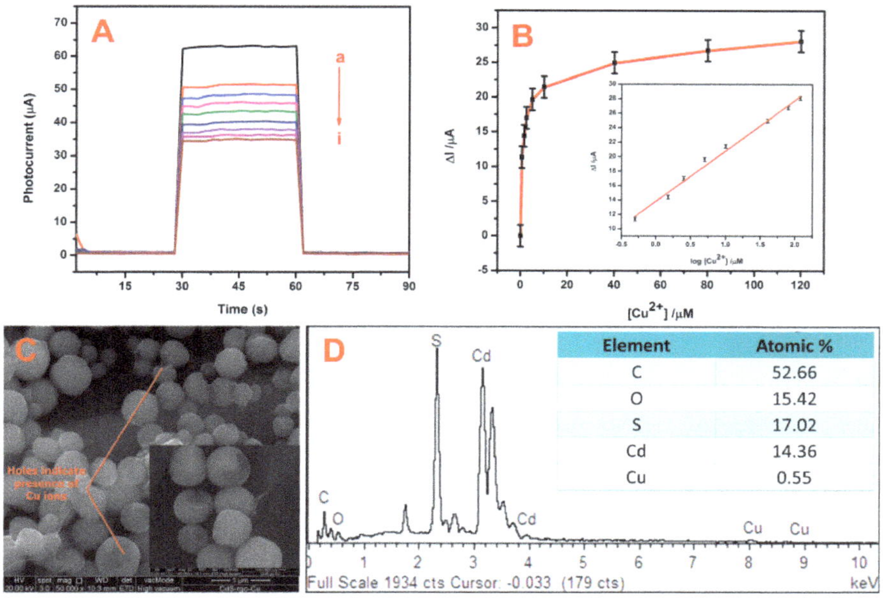

Figure 6. (A) Photocurrent responses of ITO/CdS-rGO electrode at different concentrations of Cu^{2+} ions: (a) 0, (b) 0.5, (c) 1.5, (d) 2.5, (e) 5.0, (f) 10.0 (g) 40.0, (h) 80.0, and (i) 120 μM, and (B) photocurrent change against Cu^{2+} ions concentration. Inset: corresponding calibration curve. (C) Low-magnification image of CdS-rGO-Cu, inset: high-magnification image of CdS-rGO-Cu, and (D) EDX spectrum of CdS-rGO-Cu, inset: weight percentage of elements found in spectrum.

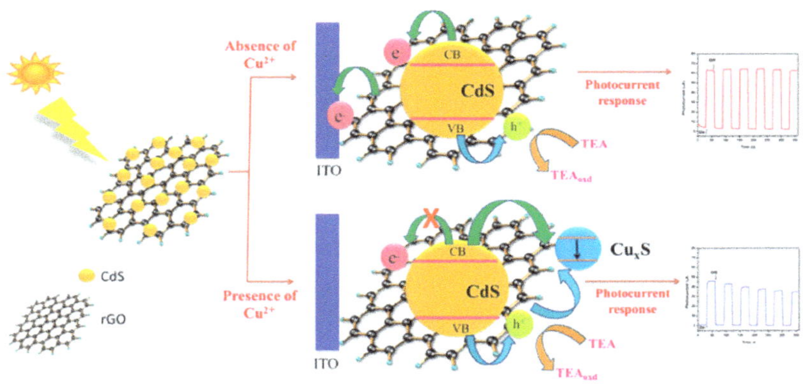

Figure 7. Mechanism for sensing Cu^{2+} **ions based on ITO/**CdS-rGO nanocomposites.

Table 1. Comparison of various Cu^{2+} **ions sensor based on various detection methods.**

Detection method	Electrode/substrate	Electrode preparation	Linear range (μM)	LoD (μM)	Ref.
Photoelectrochemical sensor	$Cd_xZn_{1-x}S$–rGO on GCE	facile one-pot reaction	0.02–20	0.067	[46]
Photoelectrochemical sensor	SnO_2/CdS heterostructural films on FTO	SILAR	1.0–38.0	0.55	[11]
Electrochemiluminescence	MPA-capped CdTe/CdS	In-situ	0.1–10	0.02	[53]
Luminescent chemosensor	CdS cluster, $[Cd_{10}S_4(SR)_{12}]$	Ligand-exchange reaction	10	0.07	[54]
Luminescent response	L-cysteine and thioglycerol-capped CdS QD	Reflux	1600	1	[50]
Fluorescence resonance energy transfer (FRET)	Phenol formaldehyde resin PFR-CdTe QDs	Hydrothermal	0.80–5.36	0.16	[55]
Adsorption stripping voltametry (AdSV)	CPE modified with Ac-Phos SAMMS	Adsorption of monolayers	0.160–3.147	0.0786	[56]
Potentiometric method using ion-selective electrodes	CPE modified with MWCNT	In-situ	0.01–10000	1.07	[57]
Differential pulse voltammetry measurement	Carbon tip electrode modified with mercury film	Electrochemical deposition	9.44–157.37	1.57	[58]
Photoelectrochemical sensor	CdS-rGO on ITO	AACVD	0.5–120	0.016	Present work

doi:10.1371/journal.pone.0154557.t001

Even though much research has been reported on the PEC sensing of Cu^{2+} ions, our current work demonstrated a facile, environmentally friendly, and low-cost method, especially in relation to the synthesis of a modified electrode. There has been no work reported based on the AACVD synthesis of CdS using cadmium acetate and thiourea as the precursors. In addition, an

ITO/CdS-rGO electrode holds the promise of high sensitivity toward Cu^{2+} ions, without requiring a ternary hybrid such as $Cd_xZn_{1-x}S$–rGO [46] and MPA-capped CdTe/CdS [53]. Some of the Cu^{2+} ions sensors with various detection methods are presented in **Table 1**. As shown in **Table 1**, the LoD obtained from the ITO/CdS-rGO-modified electrode was the lowest compared to the other reported works. Our LoD of 0.016 μM is far below the WHO's guideline value of ~20 μM that is allowed for Cu^{2+} ions in drinking water. This indicates that the proposed PEC sensor is ultra-sensitive and holds promise for the detection of a trace amount of Cu^{2+} ions in drinking water.

Interference studies

The selectivity of a Cu^{2+} ions assay was determined by investigating the influences of various kinds of metal ions at the same concentration of 120 μM, including Ni^{2+}, Co^{2+}, Fe^{3+}, Mn^{2+}, Na^{+}, K^{+}, Ba^{2+}, Zn^{2+}, Mg^{2+}, and Al^{3+}, on the photocurrent response of the prepared ITO/CdS-rGO electrode when the concentration of Cu^{2+} ions was at 0.5 μM. As shown in **Fig 8**, even with a 240-fold excess of another metal ion (120 μM vs. 0.5 μM Cu^{2+} ions), no significant photocurrent change can be seen in comparison with Cu^{2+} ions. In addition, it was observed that the photocurrent intensity of the ITO/CdS-rGO electrode was also decreased by Ni^{2+}, Co^{2+}, Fe^{3+}, Na^{+}, Zn^{2+}, Mg^{2+}, and Al^{3+}. This was due to the displacement of Cd^{2+}, leading to the formation of MS (for M = Ni, Co, Fe, Na, Zn, Mg, Al) on the CdS surface [31]. The formed MS behaved as sites for recombining photogenerated electron–hole pairs due to its defects on the surface and lower band energy. However, the ITO/CdS-rGO electrode was significantly sensitive to Cu^{2+} ions compared to the other metal ions because CuS has a notably lower K_{sp} value than NiS, CoS, FeS, and ZnS [31]. Hence, the prepared ITO/CdS-rGO electrode was sensitive and selective toward Cu^{2+} ions.

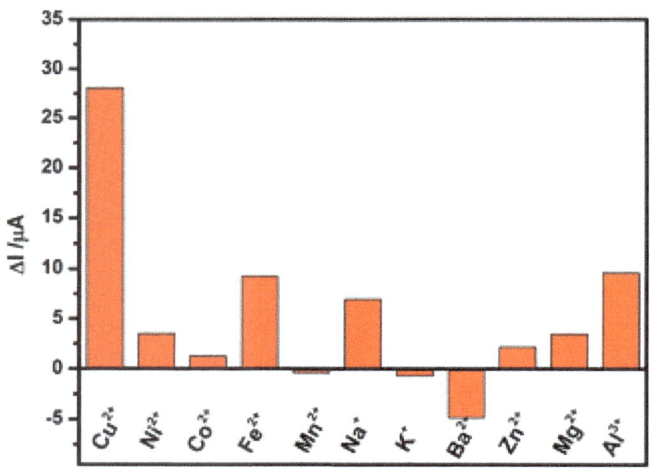

Figure 8. Photocurrent changes in prepared ITO/CdS-rGO electrode toward 0.5 **μM** Cu^{2+} **ions against 120 μM** of other metal ions.

CONCLUSION

A PEC sensor based on the CdS-rGO nanocomposite was successfully fabricated using AACVD and dip-coating strategies. The CdS-rGO nanocomposite showed the highest PEC activity compared to that of bare GO, rGO, CdS, and CdS-GO electrodes under visible-light illumination. This was because the presence of rGO facilitated the transfer of electrons from the conduction band of the CdS, thus manifesting the photocurrent performance of the nanocomposite. The photocurrent intensity of the CdS-rGO nanocomposite was effectively suppressed by the displacement of Cd^{2+} by Cu^{2+}. This phenomenon resulted in a PEC sensor with heightened sensitivity and selectivity toward Cu^{2+} ions, manifesting its potential in real applications.

Supporting Information

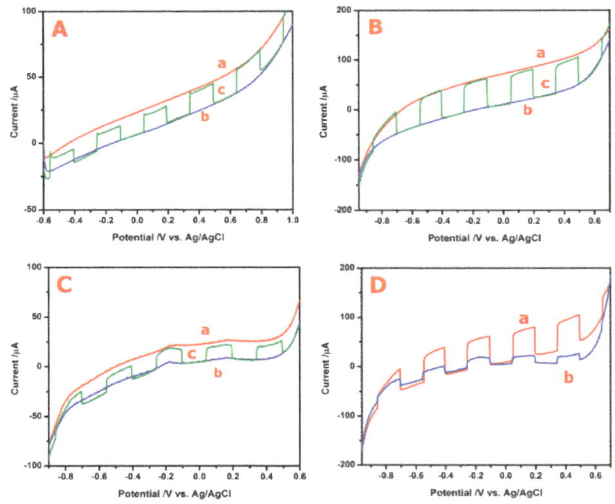

S1 Fig. LSV obtained for ITO/CdS-rGO photoelectrode dipped into (A) 0.1 M KCl, (B) mixture of 0.1 M KCl and 0.5 M TEA, and (C) 0.1 M KCl, 0.5 M TEA, and 4 μM Cu (II) under (a) light irradiation, (b) dark condition, and (c) light "on-off" condition. (D) LSV responses in (a) absence and (b) presence of 4 μM Cu^{2+} ions with 0.1 M KCl and 0.5 M TEA under light "on-off" condition at scan rate of 0.1 Vs^{-1}.

doi:10.1371/journal.pone.0154557.s001

Acknowledgments

This research work was supported by a Wighton Titular Fellowship in Engineering (6385600–10501) and a High Impact Research Grant (UM.C/625/1/HIR/MOHE/SC/21) from the Ministry of Higher Education.

Author Contributions

Conceived and designed the experiments: II HNL NMH AP. Performed the experiments: II. Analyzed the data: II HNL. Contributed reagents/materials/analysis tools: HNL NMH. Wrote the paper: II HNL. Designed the homemade PEC system and experimental method: II HNL NMH AP. Gave an idea and advice on every problem that occurred while doing this research: II HNL NMH AP. Gave ideas on Cu2+ metal ion sensing: II HNL NMH AP. Contribution on research fund: HNL NMH.

REFERENCES

1. Trumbo P, Yates AA, Schlicker S, Poos M. Dietary reference intakes: vitamin A, vitamin K, arsenic, boron, chromium, copper, iodine, iron, manganese, molybdenum, nickel, silicon, vanadium, and zinc. Journal of the American Dietetic Association. 2001;101(3):294–301. pmid:11269606 doi: 10.17226/10026

2. Guidelines for drinking-water quality: recommendations: World Health Organization; 1993.
3. Lan GY, Huang CC, Chang HT. Silver nanoclusters as fluorescent probes for selective and sensitive detection of copper ions. Chemical Communications. 2010;46(8):1257–9. doi: 10.1039/b920783j. pmid:20449269
4. Lee SH, Sumranjit J, Tongkate P, Chung BH, Lee HJ. Voltammetric Studies of Cu (II) Ion Transfer Reaction with Picolinamide-phenylenevinylene across Liquid/liquid Interfaces and Their Sensing Applications. Electrochimica Acta. 2014;123:198–204. doi: 10.1016/j.electacta.2014.01.003
5. Li H, Hao W, Hu J, Wu H. A photoelectrochemical sensor based on nickel hydroxyl-oxide modified n-silicon electrode for hydrogen peroxide detection in an alkaline solution. Biosensors and Bioelectronics. 2013;47:225–30. doi: 10.1016/j.bios.2013.03.028. pmid:23584227
6. Wang P, Dai W, Ge L, Yan M, Ge S, Yu J. Visible light photoelectrochemical sensor based on Au nanoparticles and molecularly imprinted poly (o-phenylenediamine)-modified TiO2 nanotubes for specific and sensitive detection chlorpyrifos. Analyst. 2013;138(3):939–45. doi: 10.1039/c2an36266j. pmid:23232561
7. Lu W, Jin Y, Wang G, Chen D, Li J. Enhanced photoelectrochemical method for linear DNA hybridization detection using Au-nanopaticle labeled DNA as probe onto titanium dioxide electrode. Biosensors and Bioelectronics. 2008;23(10):1534–9. doi: 10.1016/j.bios.2008.01.011. pmid:18294836
8. Li H, Li J, Wang W, Yang Z, Xu Q, Hu X. A subnanomole level photoelectrochemical sensing platform for hexavalent chromium based on its selective inhibition of quercetin oxidation. Analyst. 2013;138(4):1167–73. doi: 10.1039/c2an36605c. pmid:23282732
9. Gao P, Ma H, Yang J, Wu D, Zhang Y, Du B, et al. Anatase TiO_2 based photoelectrochemical sensor for the sensitive determination of dopamine under visible light irradiation. New Journal of Chemistry. 2015. doi: 10.1039/c4nj01909a
10. Shen Q, Zhao X, Zhou S, Hou W, Zhu JJ. ZnO/CdS hierarchical nanospheres for photoelectrochemical sensing of Cu^{2+}. The Journal of Physical Chemistry C. 2011;115(36):17958–64. doi: 10.1021/jp203868t
11. Huang F, Pu F, Lu X, Zhang H, Xia Y, Huang W, et al. Photoelectrochemical sensing of Cu^{2+} ions with SnO_2/CdS heterostructural films. Sensors and Actuators B: Chemical. 2013;183:601–7. doi: 10.1016/j.snb.2013.04.047
12. Liu S, Chen Z, Zhang N, Tang ZR, Xu YJ. An Efficient Self-Assembly of CdS Nanowires–Reduced Graphene Oxide Nanocomposites for Selective Reduction of Nitro Organics under Visible Light Irradiation. The Journal of Physical Chemistry C. 2013;117(16):8251–61. doi: 10.1021/jp400550t
13. Pawar RC, Lee CS. Single-step sensitization of reduced graphene oxide sheets and CdS nanoparticles on ZnO nanorods as visible-light photocatalysts. Applied Catalysis B: Environmental. 2014;144:57–65. doi: 10.1016/j.apcatb.2013.06.022
14. Wang P, Jiang T, Zhu C, Zhai Y, Wang D, Dong S. One-step, solvothermal synthesis of graphene-CdS and graphene-ZnS quantum dot nanocomposites and their interesting photovoltaic properties. Nano Research. 2010;3(11):794–9. doi: 10.1007/s12274-010-0046-0
15. Sheeney Haj Ichia L, Basnar B, Willner I. Efficient generation of photocurrents by using CdS/carbon nanotube assemblies on electrodes. Angewandte Chemie International Edition. 2005;44(1):78–83. doi: 10.1002/anie.200461666
16. Chen D, Feng H, Li J. Graphene oxide: preparation, functionalization, and electrochemical applications. Chemical reviews. 2012;112(11):6027–53. doi: 10.1021/cr300115g. pmid:22889102
17. Novoselov K, Geim AK, Morozov S, Jiang D, Katsnelson M, Grigorieva I, et al. Two-dimensional gas of massless Dirac fermions in graphene. nature. 2005;438(7065):197–200. pmid:16281030 doi: 10.1038/nature04233
18. Balandin AA, Ghosh S, Bao W, Calizo I, Teweldebrhan D, Miao F, et al. Superior thermal conductivity of single-layer graphene. Nano letters. 2008;8(3):902–7. doi: 10.1021/nl0731872. pmid:18284217
19. Bunch JS, Verbridge SS, Alden JS, Van Der Zande AM, Parpia JM, Craighead HG, et al. Impermeable atomic membranes from graphene sheets. Nano letters. 2008;8(8):2458–62. doi: 10.1021/nl801457b. pmid:18630972
20. Allen MJ, Tung VC, Kaner RB. Honeycomb carbon: a review of graphene. Chemical reviews. 2009;110(1):132–45. doi: 10.1021/cr900070d
21. Zhao WW, Ma ZY, Yu PP, Dong XY, Xu JJ, Chen HY. Highly sensitive photoelectrochemical immunoassay with enhanced amplification using horseradish peroxidase induced biocatalytic precipitation on a CdS quantum dots multilayer electrode. Analytical chemistry. 2011;84(2):917–23. doi: 10.1021/ac203184g. pmid:22148581
22. Jumeri F, Lim H, Zainal Z, Huang N, Pandikumar A. Titanium dioxide-reduced graphene oxide thin film for photoelectrochemical water splitting. Ceramics International. 2014;40(9):15159–65. doi: 10.1016/j.ceramint.2014.06.130
23. Lim SP, Pandikumar A, Huang NM, Lim HN. Silver/titania nanocomposite-modified photoelectrodes for photoelectrocatalytic methanol oxidation. International Journal of Hydrogen Energy. 2014;39(27):14720–9. doi: 10.1016/j.ijhydene.2014.07.066
24. Huang N, Lim H, Chia C, Yarmo M, Muhamad M. Simple room-temperature preparation of high-yield large-area graphene oxide. International journal of nanomedicine. 2011;6:3443. doi: 10.2147/IJN.S26812. pmid:22267928
25. Liu X, Pan L, Lv T, Zhu G, Sun Z, Sun C. Microwave-assisted synthesis of CdS–reduced graphene oxide composites for photocatalytic reduction of Cr (vi). Chemical Communications. 2011;47(43):11984–6. doi: 10.1039/c1cc14875c. pmid:21960206

26. Wang DW, Li F, Wu ZS, Ren W, Cheng HM. Electrochemical interfacial capacitance in multilayer graphene sheets: Dependence on number of stacking layers. Electrochemistry Communications. 2009;11(9):1729–32. doi: 10.1016/j.elecom.2009.06.034
27. Vusa CSR, Berchmans S, Alwarappan S. Facile and green synthesis of graphene. RSC Advances. 2014;4(43):22470–5. doi: 10.1039/c4ra01718h
28. Rao LG, Guns E, Rao AV. Lycopene: its role in human health and disease. AgroFood Industry hi-tech. 2003;15:25–30.
29. Subrahmanyam K, Manna AK, Pati SK, Rao C. A study of graphene decorated with metal nanoparticles. Chemical Physics Letters. 2010;497(1):70–5. doi: 10.1016/j.cplett.2010.07.091
30. Song S, Gao W, Wang X, Li X, Liu D, Xing Y, et al. Microwave-assisted synthesis of BiOBr/graphene nanocomposites and their enhanced photocatalytic activity. Dalton Transactions. 2012;41(34):10472–6. doi: 10.1039/c2dt31088k. pmid:22825685
31. Pan S, Liu X. ZnS–graphene nanocomposite: synthesis, characterization and optical properties. Journal of Solid State Chemistry. 2012;191:51–6. doi: 10.1016/j.jssc.2012.02.048
32. Aravinda L, Nagaraja K, Nagaraja H, Bhat KU, Bhat BR. ZnO/carbon nanotube nanocomposite for high energy density supercapacitors. Electrochimica Acta. 2013;95:119–24. doi: 10.1016/j.electacta.2013.02.027
33. Haldorai Y, Voit W, Shim JJ. Nano ZnO@ reduced graphene oxide composite for high performance supercapacitor: Green synthesis in supercritical fluid. Electrochimica Acta. 2014;120:65–72. doi: 10.1016/j.electacta.2013.12.063
34. Yang J, Gunasekaran S. Electrochemically reduced graphene oxide sheets for use in high performance supercapacitors. Carbon. 2013;51:36–44. doi: 10.1016/j.carbon.2012.08.003
35. Chen Y, Zhang X, Zhang D, Yu P, Ma Y. High performance supercapacitors based on reduced graphene oxide in aqueous and ionic liquid electrolytes. Carbon. 2011;49(2):573–80. doi: 10.1016/j.carbon.2010.09.060
36. Sun B, Zhang K, Chen L, Guo L, Ai S. A novel photoelectrochemical sensor based on PPIX-functionalized WO_3–rGO nanohybrid-decorated ITO electrode for detecting cysteine. Biosensors and Bioelectronics. 2013;44:48–51. doi: 10.1016/j.bios.2013.01.014. pmid:23391706
37. Rajagopalan B, Chung JS. Reduced chemically modified graphene oxide for supercapacitor electrode. Nanoscale research letters. 2014;9(1):1–10. doi: 10.1186/1556-276x-9-535
38. Ye H, Park HS, Bard AJ. Screening of electrocatalysts for photoelectrochemical water oxidation on W-doped $BiVO_4$ photocatalysts by scanning electrochemical microscopy. The Journal of Physical Chemistry C. 2011;115(25):12464–70. doi: 10.1021/jp200852c
39. Han L, Guo S, Xu M, Dong S. Photoelectrochemical batteries for efficient energy recovery. Chem Commun. 2014;50(87):13331–3. doi: 10.1039/c4cc06708h
40. Yu H, Wang J, Zhang S, Li X, Zhao H. Layered Fe (III) doped TiO_2 thin-film electrodes for the photoelectrocatalytic oxidation of glucose and potassium hydrogen phthalate. Chinese Science Bulletin. 2011;56(23):2475–80. doi: 10.1007/s11434-010-4502-8
41. Xie R, Su J, Li M, Guo L. Structural and photoelectrochemical properties of Cu-doped CdS thin films prepared by ultrasonic spray pyrolysis. International Journal of Photoenergy. 2013;2013. doi: 10.1155/2013/620134
42. Ikeda S, Nakamura T, Lee SM, Yagi T, Harada T, Minegishi T, et al. Photoreduction of water by using modified CuInS2 electrodes. ChemSusChem. 2011;4(2):262–8. doi: 10.1002/cssc.201000169. pmid:21328555
43. Zang Y, Lei J, Hao Q, Ju H. "Signal-On" Photoelectrochemical Sensing Strategy Based on Target-Dependent Aptamer Conformational Conversion for Selective Detection of Lead (II) Ion. ACS applied materials & interfaces. 2014;6(18):15991–7. doi: 10.1021/am503804g
44. Bonaccorso F, Sun Z, Hasan T, Ferrari A. Graphene photonics and optoelectronics. Nature photonics. 2010;4(9):611–22. doi: 10.1038/nphoton.2010.186
45. Cao A, Liu Z, Chu S, Wu M, Ye Z, Cai Z, et al. A Facile One-step Method to Produce Graphene–CdS Quantum Dot Nanocomposites as Promising Optoelectronic Materials. Advanced materials. 2010;22(1):103–6. doi: 10.1002/adma.200901920. pmid:20217706
46. Yan J, Wang K, Liu Q, Qian J, Dong X, Liu W, et al. One-pot synthesis of $Cd_xZn_{1-x}S$–reduced graphene oxide nanocomposites with improved photoelectrochemical performance for selective determination of Cu^{2+}. RSC Advances. 2013;3(34):14451–7. doi: 10.1039/c3ra41118d
47. Pawar RC, Lee CS. Sensitization of CdS nanoparticles onto reduced graphene oxide (RGO) fabricated by chemical bath deposition method for effective removal of Cr (VI). Materials Chemistry and Physics. 2013;141(2):686–93. doi: 10.1016/j.matchemphys.2013.05.062
48. Neogi A, Karna S, Shah R, Phillipose U, Perez J, Shimada R, et al. Surface plasmon enhancement of broadband photoluminescence emission from graphene oxide. Nanoscale. 2014;6(19):11310–5. doi: 10.1039/c4nr03055a. pmid:25139261
49. Wang GL, Xu JJ, Chen HY. Selective detection of trace amount of Cu^{2+} using semiconductor nanoparticles in photoelectrochemical analysis. Nanoscale. 2010;2(7):1112–4. doi: 10.1039/c0nr00084a. pmid:20648335
50. Chen Y, Rosenzweig Z. Luminescent CdS quantum dots as selective ion probes. Analytical Chemistry. 2002;74(19):5132–8. pmid:12380840 doi: 10.1021/ac0258251
51. Wang P, Ma X, Su M, Hao Q, Lei J, Ju H. Cathode photoelectrochemical sensing of copper (II) based on analyte-induced formation of exciton trapping. Chem Commun. 2012;48(82):10216–8. doi: 10.1039/c2cc35643k

52. Isarov AV, Chrysochoos J. Optical and photochemical properties of nonstoichiometric cadmium sulfide nanoparticles: surface modification with copper (II) ions. Langmuir. 1997;13(12):3142–9. doi: 10.1021/la960985r

53. Wang J, Jiang X. Anodic near-infrared electrochemiluminescence from CdTe/CdS core small/shell thick quantum dots and their sensing ability of Cu^{2+}. Sensors and Actuators B: Chemical. 2015;207:552–5. doi: 10.1016/j.snb.2014.10.112

54. Konishi K, Hiratani T. Turn-On and Selective Luminescence Sensing of Copper Ions by a Water-Soluble Cd10S16 Molecular Cluster. Angewandte Chemie International Edition. 2006;45(31):5191–4. doi: 10.1002/anie.200601491

55. Yang P, Zhao Y, Lu Y, Xu QZ, Xu XW, Dong L, et al. Phenol formaldehyde resin nanoparticles loaded with CdTe quantum dots: a fluorescence resonance energy transfer probe for optical visual detection of copper (II) ions. ACS nano. 2011;5(3):2147–54. doi: 10.1021/nn103352b. pmid:21344860

56. Yantasee W, Lin Y, Fryxell GE, Busche BJ. Simultaneous detection of cadmium, copper, and lead using a carbon paste electrode modified with carbamoylphosphonic acid self-assembled monolayer on mesoporous silica (SAMMS). Analytica Chimica Acta. 2004;502(2):207–12. doi: 10.1016/j.aca.2003.10.001

57. Soleimani M, Afshar MG. Potentiometric Sensor for Trace Level Analysis of Copper Based on Carbon Paste Electrode Modified With Multi-walled Carbon Nanotubes. Int J Electrochem Sci. 2013;8:8719–29.

58. Kudr J, Nguyen HV, Gumulec J, Nejdl L, Blazkova I, Ruttkay-Nedecky B, et al. Simultaneous Automatic Electrochemical Detection of Zinc, Cadmium, Copper and Lead Ions in Environmental Samples Using a Thin-Film Mercury Electrode and an Artificial Neural Network. Sensors. 2014;15(1):592–610. doi: 10.3390/s150100592. pmid:25558996

CHAPTER 12

INVESTIGATION OF HEMOGLOBIN/GOLD NANOPARTICLE HETEROLAYER ON MICRO-GAP FOR ELECTROCHEMICAL BIOSENSOR APPLICATION

Taek Lee, Tae-Hyung Kim, Jinho Yoon, Yong-Ho Chung, Ji Young Lee and Jeong-Woo Choi

Keywords: Electrochemical biosensor, hemoglobin, gold nanoparticle, cyclic voltammetry, Au micro-gap

INTRODUCTION

In the field of nanobiotechnology, the electrochemical study of biomolecules plays an important role in understanding the electron transfer mechanism of living organisms [1,2]. Recently, the electrochemical study of biomolecules has made advances in the form of nano-biotechnologies, such as the fabrication of biochips, biosensors, and bioelectronic applications [3,4]. The electrochemical analysis of biomolecules not only provides an understanding of their electrochemical behavior, but also furthers applications to biomedical and environmental biosensors. In particular, an electrochemical-based biosensor composed of protein can be a very effective tool to characterize a target analyte [5]. Various types of electrochemical-based biosensor composed of protein such as antibody and enzyme have been reported for environmental, biomedical, and food industry applications [6,7,8,9].

In particular, the detection of hydrogen peroxide (H_2O_2) is a quite significant chemical compound in the clinical, environmental, pharmaceutical, and cosmetic industries [10,11]. Within a living organism H_2O_2 is a reactive oxygen by-product which related with key elements as a regulating diverse biological stress. The oxidative stress related with H_2O_2 has been linked to cytotoxic effects and immune cell activation, as well as intracellular thermogenesis [12]. For these reasons the detection of H_2O_2 at low concentration is considered to be an important analyte. Usually, various enzymes such as peroxidases, lipoxygenase, and catalases are used to detect H_2O_2 for biosensor applications [13,14,15]. However, these enzymes are too sensitive, unstable, and expensive. Recently, metalloproteins have been introduced to detect H_2O_2 because they can provide a direct transfer of electrons and improved sensitivity [16].

Hemoglobin (Hb) is a well-investigated metalloprotein with a molecular weight of ~64.5 kDa, and has four electroactive ferrous ions (Fe^{2+}) in a heme group that are oxidized to ferric ions (Fe^{3+}). It has been studied in the context of the life sciences, pharmaceutical sciences, and medicine [17,18]. In living organisms, hemoglobin transports oxygen in red blood cells [12]. Recently, hemoglobin has been used to develop an enzymeless H_2O_2 biosensor for peroxidase activity. Because the hemoglobin provides the direct electron transfer (DET) to achieve the fast electron transfer with high sensitivity [19,20].

Several groups have studied the hemoglobin-based biosensor on a conventional electrode surface [12,21,22,23,24,25], while others introduced nanoparticles [21], graphene [22] and other heavy metal materials [23] on an

Citation: Lee T., Kim, T.-H., Yoon J., Chung, Y.-H., Lee, J.Y., Choi, J.-W., Investigation of Hemoglobin/Gold Nanoparticle Heterolayer on Micro-Gap for Electrochemical Biosensor Application. Sensors 2016, 16, 660.

electrode to facilitate the electron transfer between H_2O_2 and hemoglobin. However, such heavy metal-based nanomaterials are toxic and possess low biocompatibility. A gold nanoparticle was introduced to facilitate the electron transfer between hemoglobin and Au electrode. The gold nanoparticle is toxic at low levels and widely used to fabricate the electrochemical biosensor [26,27]. Furthermore, the electrochemical-based biosensor has been studied on a bulk scale electrode. Some groups proposed miniaturized biosensor to detect the ErbB2 protein [28], PSA [29] and DNA hybridization [30]. Those miniaturized biosensors provide (1) defined distance between working and counter electrodes; (2) detect the low concentration of sample. To develop the advanced electrochemical biosensor based on direct electron transfer, the miniaturized biosensor with high sensitivity will be needed.

In the present study, the hemoglobin/gold nanoparticle heterolayer was fabricated on the micro-gap for signal-enhanced electrochemical H_2O_2 biosensor applications. To determine the H_2O_2 detection property, the hemoglobin/gold nanoparticle (Hb/GNP) was immobilized onto the Au electrode and Au micro-gap electrode through self-assembly method. The immobilization of the hemoglobin/gold nanoparticle heterolayer was confirmed by atomic force microscopy (AFM) and surface-enhanced Raman spectroscopy (SERS). The performance of the hemoglobin/gold nanoparticle heterolayer biosensor was investigated by cyclic voltammetry (CV). Figure 1 shows the schematic diagram of fabricated hemoglobin/gold nanoparticle immobilized on the Au micro-gap.

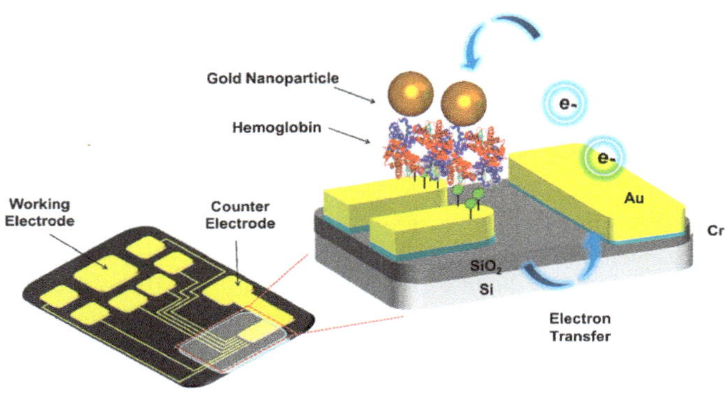

Figure 1. Schematic diagram of fabricated hemoglobin/gold nanoparticle heterolayer immobilized on the micro-gap for H_2O_2 detection.

MATERIALS AND METHODS

Materials

The Au substrates (Au (43 nm)/Cr (2 nm)/SiO_2 (200 nm) Si (p-type) wafers) were prepared to carry out the AFM and CV experiments to manufacture the working electrode (G-MEK, Seoul, Korea). Negative photoresist (Su-82), developer, and stripper for the photolithographic process were purchased from Microchem (Westborough, MA, USA). Pt wire for the counter electrode and the Ag/AgCl reference electrode were purchased from BAS (West Lafayette, IN, USA) to conduct the electrochemical experiments. The 6-Mercaptohexanoic acid (6-MHA) and hemoglobin extracted from *horse heart* were purchased from Sigma-Aldrich (Saint Louis, MO, USA). A 0.2 mg/mL hemoglobin solution was prepared and dilluted in 10 mM HEPES buffer at pH 7.2. For the electrochemical experiment, hydrogen peroxide (H_2O_2) was purchased from Daejung Chemical and Metals Co. Ltd. (Siheung-si, Korea). The gold nanoparticles (20 nm and 60 nm) were purchased form BBI for electrochemical experiment and SERS experiment, respectively (Cardiff, UK).

Fabrication of Micro-Gap

The Si substrate (100) with a thermally oxidized SiO_2 was initially cleaned with piranha solution composed of H_2SO_4(Daejung Chemical and Metals Co. Ltd., Siheung-si, Korea) and H_2O_2 (Duksan Pure Chemical and Metals Co. Ltd., Seoul, Korea) with a volume ratio of 5:1 at 70 °C for 10 min, and baked on the hot plate at 200 °C for 20 min for adhesion promotion with photoresist. The pre-pattern for metal deposition was fabricated by the standard photolithographic process using negative photoresist. Initially, Su-82 photoresist was spin-coated in cleaned substrate with a speed of 3000 rpm, and then baked at 65 °C for 1 min and 95 °C for 1 min. After exposure of 110 mJ/cm^2 (i-line) by MA6 mask aligner (SUSS MicroTec., Garching, Germany), substrate was post-baked by the same steps with pre-baking and developed for 70 s. Metal deposition of Cr (5 nm) and Au (50 nm) was operated with an electron beam evaporator. Finally, a lift-off process using remover PG was performed at 70 °C for 2 h. The fabricated substrates were cleaned by piranha solution with the same conditions. The working chamber, which has the volume of 400 μL for electrochemical measurements, was attached on the prepared substrate by using PDMS linkage. After attachment of the chamber on the multi-electrode, substrates were immersed in acetone and isopropyl alcohol (IPA) for 20 min sequentially to remove organic contaminant, which could be induced in the curing process of PDMS at high-temperature chamber.

Fabrication of Hemoglobin/Gold Nanoparticle Heterolayer

To fabricate a micro-gap-based electrochemical sensor, the prepared Au micro-gap was treated with piranha solution at 70 °C for 3 min, washed with deionized water, and dried under N_2 gas to remove dust and residue before the self-assembly process, [31,32]. Then, 3 μL of 10 mM 6-MUA was added onto the Au surface for 3 h. The thiol group of the 6-MHA molecule was bridged as a terminal group to the Au surface by covalent bonding during the immobilization period. The remaining 6-MHA on the surface was rinsed thoroughly with DI water and ethanol to remove excess residue. Then, 5 μL of the 0.2 mg/mL hemoglobin solution was added onto the self-assembled 6-MHA layer over 6 h. During this step, the free amine group of hemoglobin was bound to the carboxyl groups of 6-MHA by EDC/NHS reaction. Additionally, the modified substrates were cleaned with deionized water and dried under an Ar gas stream. Then, 50 mM solution of 1-Octadecanethiol was added for 6 h, and the prepared 0.2 mg/mL of gold nanoparticle solution (3 μL) was added onto the hemoglobin self-assembled substrate for 6 h. Finally, the modified substrates were cleaned with deionized water and dried under an Ar gas stream [33]. Figure 2a–c show the schematic diagram, the optical image of fabricated micro-gap electrode with working chamber for H_2O_2 and optical image of zoomed Au micro-gap electrode, respectively.

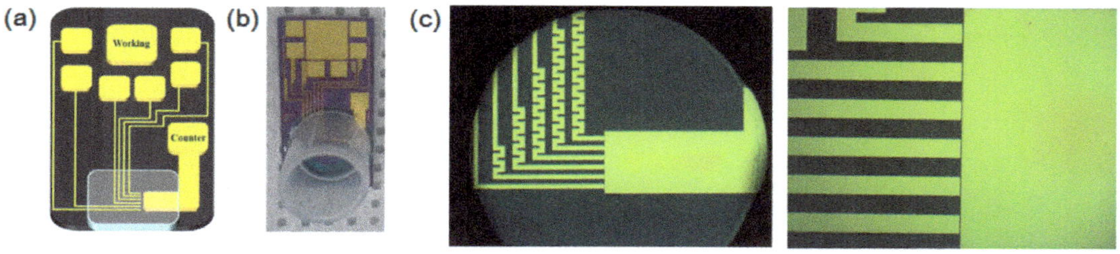

Figure 2. (**a**) Schematic diagram of micro-gap electrode; (**b**) Optical image of fabricated micro-gap electrode with working chamber for H_2O_2 biosensor application; (**c**) Optical image of zoomed micro-gap.

Investigation of Hemoglobin/Gold Nanoparticle Heterolayer Using AFM and SERS

To confirm the immobilization of the Hb/GNP hybrid layer, the surface topography of the Hb/GNP heterolayer was investigated through AFM. The tapping-mode AFM was operated with a Nanoscope IV/Multimode from Digital Instruments. The tips used for the AFM measurement were phosphorous (n-type doped Si), and the resonance peaks in the frequency response of the cantilever were selected at around of 230–305 kHz. Spring constants of 20–80 N/m were used. Before scanning the sample, the set point was adjusted to optimize the force between the tip and the biomolecule.

For SERS measurement, indium tin oxide (ITO)-coated glass (G-mek, Seoul, Korea) was prepared by sonication for 30 min in 1% Triton X-100, DIW, and ethanol, sequentially. The preparation of SERS substrate was followed by our previous study. Then, the hemoglobin was immobilized via 6-MHA linker, and the gold nanoparticle was dropped onto the hemoglobin-modified substrate for SERS measurement [34].

Electrochemical Analysis of Hemoglobin-Gold Nanoparticle Heterolayer

An electrochemical H_2O_2 detection test of the Hb/GNP heterolayer was performed using a CHI660A electrochemical workstation (CH Instruments, Austin, TX, USA) and a three-electrode system. The Hb/GNP immobilized on the working electrode in the micro-gap. And the counter electrode part in micro-gap was used as a counter electrode and an Ag/AgCl (saturated KCl) electrode was used for a reference electrode. The electrochemical analysis for the H_2O_2 sensitivity was performed in 10 mM PBS buffer solution (pH = 7.4) at room temperature [16]. The electrochemical experiment was repeated with five samples.

RESULTS AND DISCUSSION

The Surface Investigation of Hemoglobin/Gold Nanoparticle Hybrid by AFM and SERS

AFM measurement was conducted to investigate the surface morphology of fabricated Hb/GNP heterolayer. The Hb self-assembled on 6-MHA layer and Hb/GNP hybrid layer were investigated. In the case of the hemoglobin surface, small lumps of protein clusters were oriented onto the Au substrate via 6-MHA linker. The average lump sizes were around 20–30 nm and the vertical distance of each lump was 4–7 nm (Figure 3a). However, the surface of the Hb/GNP heterolayer, the lump size, and vertical distance showed a different size and shape compared to hemoglobin's lump. The size of the hemoglobin lumps were around 40–55 nm and vertical size around 15–20 nm (Figure 3b). The hemoglobin/gold nanoparticle clusters were larger than the lumps of Hb. Presumably, the result showed the gold nanoparticle was adsorbed on the hemolgobin molecule because the hemoglobin molecule was self-assembled onto the 6-MHA-modified Au substrate. As a result, the hemoglobin/gold nanoparticle was well oriented onto the Au substrate.

Moreover, the surface roughness of the Hb/GNP and Hb were analyzed using AFM Nanoscope software. The surface roughness values of the the Hb self-assembled on 6-MHA layer and the Hb/GNP self-assembled on 6-MHA layer are shown in Figure 3c, respectively. In the case of the Hb, the R_a value is 0.298 ± 0.174 nm, RMS roughness (R_q) shows 1.094 ± 0.275 nm, and R_{max} is 1.315 ± 0.333 nm. Also, the average vertical distance of hemoglobin is 4.685 ± 1.221 nm. In the case of the hemoglobin/gold nanoparticle, the average R_a, R_q, R_{max} and vertical distance values are 1.253 ± 0.418 nm, 5.747 ± 1.199 nm, 4.865 ± 0.795 nm, and 17.206 ± 3.337 nm, respectively. Based on these results, the Hb/GNP heterolayer was well immobilized to the Au surface for further experiment.

Figure 3d displays the SERS spectra of the hemoglobin layer (Brown line) and the Hb/GNP heterolayer (Blue line), respectively. The Raman peaks of the Hb were monitored near the region of 1126 cm^{-1}, 1214 cm^{-1}; the 1126 cm^{-1} peak is assigned to the pyrrole half-ring stretching mode [35]. The intense peak at 1216 cm^{-1} was attributed to the C_mH out-of-plane deformation stretching mode. And the Raman spectra of Hb was recorded near 651 cm^{-1}, which had peaks assigned to the C-S (cys) stretching. The intense peak monitored near 1347 and 1356 cm^{-1} correspond to the anti-symmetric υ(pyr half-ring)$_{sym}$. And the intense band near the region of 924 cm^{-1} was previously assigned to modes involving C-COO^- stretching [36]. Those peaks are monitored to both of SERS spectra. Compared to Hb, the Hb/GNP heterolayer possesses a greater SERS intensity. Based on this result, the Hb/GNP heterolayer was well-oriented onto the gold nanoparticle-coated ITO surface.

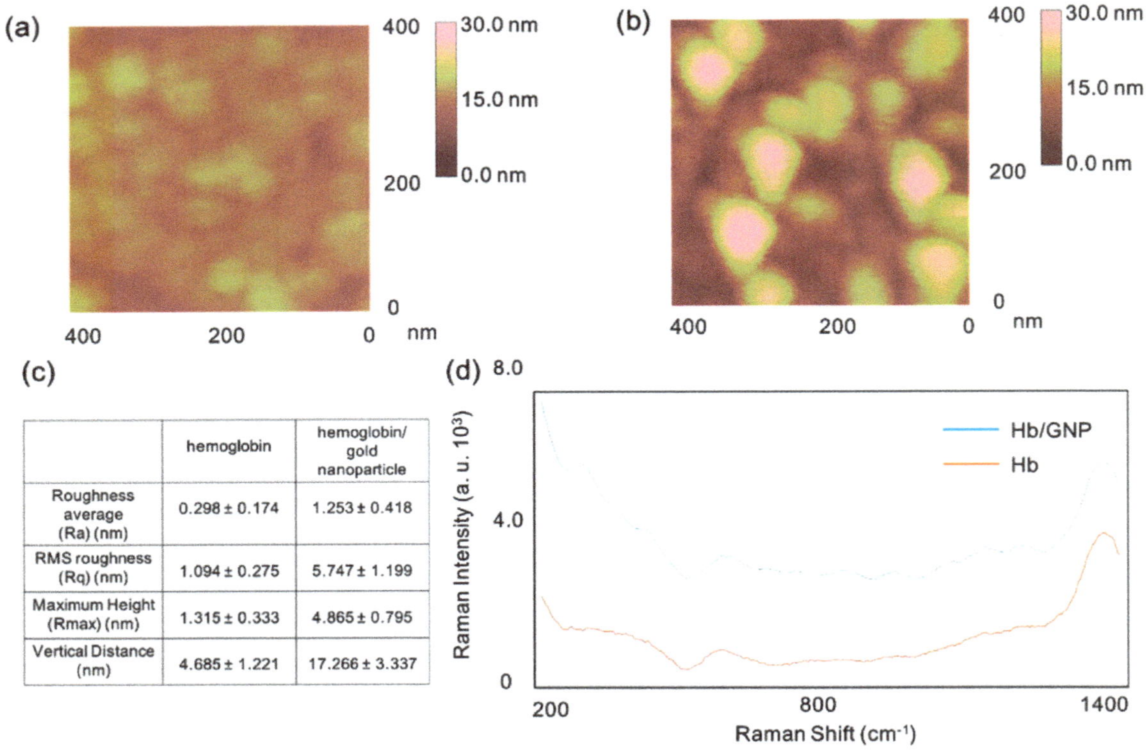

	hemoglobin	hemoglobin/ gold nanoparticle
Roughness average (Ra) (nm)	0.298 ± 0.174	1.253 ± 0.418
RMS roughness (Rq) (nm)	1.094 ± 0.275	5.747 ± 1.199
Maximum Height (Rmax) (nm)	1.315 ± 0.333	4.865 ± 0.795
Vertical Distance (nm)	4.685 ± 1.221	17.266 ± 3.337

Figure 3. Surface morphology investigation of (**a**) hemoglobin; (**b**) hemoglobin/gold nanoparticle on 6-mercaptohexanoic acid (6-MHA) layer; (**c**) Surface roughness analysis of the hemoglobin, hemoglobin/gold nanoparticle; (**d**) Raman spectra of hemoglobin (Brown line); (**b**) hemoglobin/gold nanoparticle on 6-MHA layer (Blue line).

Cyclic Voltammetric Behavior of Hemoglobin/Gold Nanoparticle

A cyclic voltammetry was conducted to compare the redox properties of the Hb and Hb/GNP heterolayer-immobilized on a conventional Au electrode (1 × 2 cm) and Au micro-gap electrode. As shown in Figure 4a, the redox property of the Hb showed a quasi-reversible redox peaks (Brown line) because of the $Fe^{3+/2+}$ redox center that provides the electron exchange. The anodic peak potential (E_{pa}) and cathodic peak potential (E_{pc}) were showed at 103 mV and 67 mV (*vs.* Ag/AgCl), respectively, with a peak-to-peak separation (ΔE_p) of 36 mV. The redox property of the Hb/GNP showed a signal-enhanced redox peak (Purple line) because the GNP facilitates electron transfer between Hb and H_2O_2. The anodic peak potential (E_{pa}) and cathodic peak potential (E_{pc}) were 233 and 181 mV (*vs.* Ag/AgCl), respectively, with a peak-to-peak separation (ΔE_p) of 52 mV. This result shows that the gold nanoparticles provide the higher surface roughness and active area which yields to a higher coverage [37].

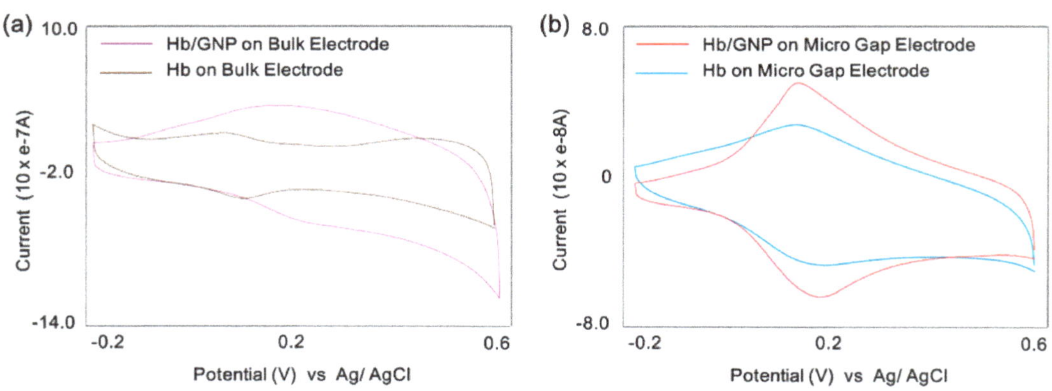

Figure 4. Cyclic voltammogram of (**a**) hemoglobin (Brown line) and hemoglobin/gold nanoparticle (Purple line) immobilized on bulk Au electrode, respectively; (**b**) hemoglobin (Red line) and hemoglobin/gold nanoparticle (Blue line) immobilized on micro-gap Au electrode, respectively.

Moreover, a CV test of Hb and Hb/GNP was carried out on the micro-gap electrode. The Figure 4b shows the cyclic voltamogram of Hb and Hb/GNP. In the case of Hb on the micro-gap electrode displayed a couple of well-defined quasi-reversible redox peaks (Blue line) because of the $Fe^{3+/2+}$ redox center. The redox current signal was shown to be less than 10 times the bulk Au electrode, the anodic peak potential (E_{pa}) and cathodic peak potential of Hb were determined to be 0.213 and 0.196 V (*vs.* Ag/AgCl) (Blue line) with a peak-to-peak separation (ΔE_p) of 17 mV, which was narrow compared to the bulk Au electrode.

The redox peaks of Hb/GNP showed a signal-enhanced redox peak (Red line) because the GNP facilitate electron transfer between Hb and H_2O_2 similar to the bulk Au electrode. The anodic peak potential (E_{pa}) and cathodic peak potential (E_{pc}) were 224 mV and 187 mV (*vs.* Ag/AgCl), respectively, with a peak-to-peak separation (ΔE_p) of 37 mV, which was narrow compared to the bulk Au electrode. This result shows that the Hb/GNP heterolayer maintained the facilitation of the electron transfer on the micro-gap electrode.

Electrocatalytic Properties of Hemoglobin/Gold Nanoparticle on the Micro-Gap Electrode

Cyclic voltammograms of the Hb/GNP-modified micro-gap electrode were obtained with different scan rates from 0.01 to 0.05 V/s (Figure 5a). The result displays well-defined and slightly shifted cathodic and anodic peaks at each scan rate. Also, the CV result shows a linear plot for cathodic and anodic peak currents against the scan rate (Figure 5b), respectively. The linear regression equations were $i_p/10 \times e^{-8}$ A = 0.0067 v(mV/s) − 0.0229 for cathodic currents and $i_p/10 \times e^{-8}$ A = −0.0728 v(mV/s) − 0.0266 for anodic currents. The linearity of the CV results corresponded to the scan rate (v) show an increment of scan rate yields a slight shift of the cathodic peak to a more negative potential, while the anodic peak was moved to a more positive potential. Plotting E_{pa} and E_{pc} *vs.* log(scan rate) yielded two straight lines with a slope equal to −2.3 RT/αnF for the cathodic peak, and a slope of 2.3 RT/(1−α)nF for the anodic peak. The charge transfer coefficient was calculated (0.27) from the slopes of the straight lines based on the following equation [38]:

$$\log\frac{k_a}{k_c} = \log\left[\frac{\alpha}{(1-\alpha)}\right] \; or \; \frac{k_a}{k_c} = \frac{\alpha}{1-\alpha} \quad (1)$$

where k_a is the slope of the line derived from E_{pa} = f(log scan rate); k_c is the slope of the line derived from; E_{pc} = f(log α).

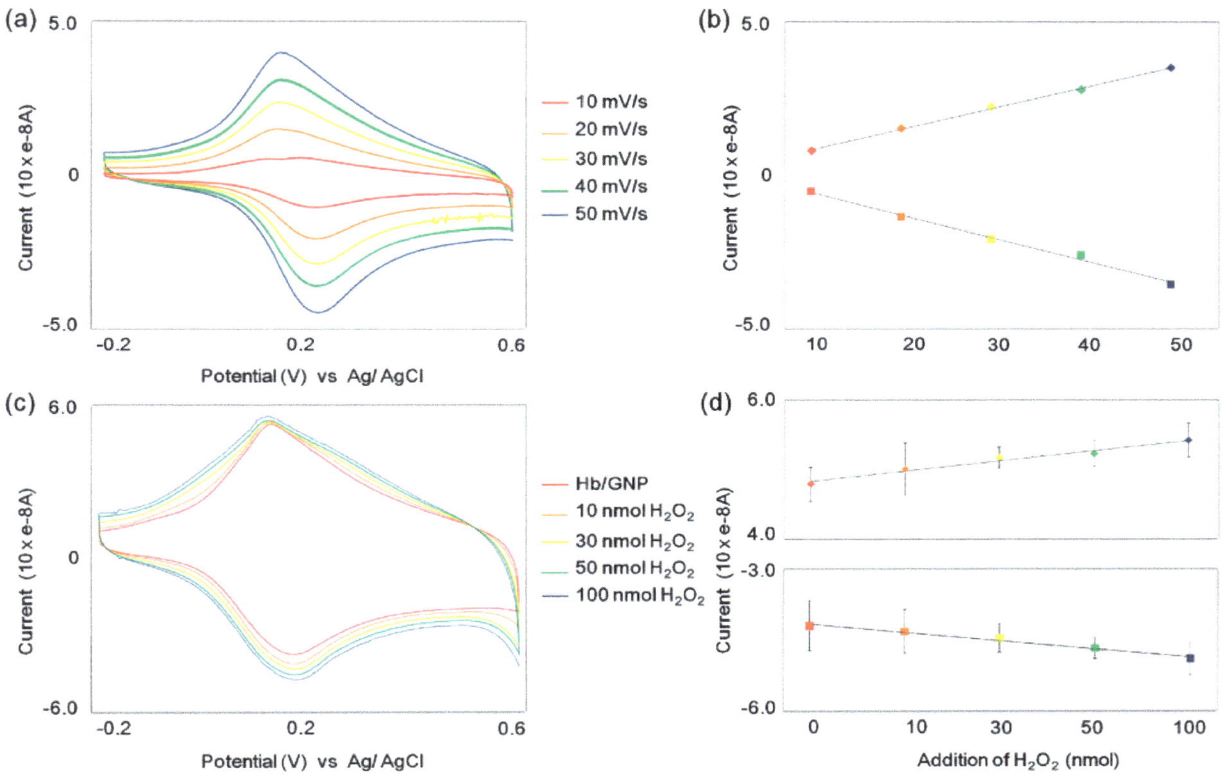

(e)

	Hb/GNP	10 nmol	30 nmol	50 nmol	100 nmol
Cathodic Current (10 x e-8 A)	4.465 ±0.308	4.732 ±0.477	4.936 ±0.192	5.010 ±0.233	5.252 ±0.311
Anodic Current (10 x e-8A)	-4.114 ±0.651	-4.258 ±0.555	-4.426 ±0.382	4.698 ±0.275	-4.970 ±0.423

Figure 5. (**a**) Cyclic voltammogram of hemoglobin/gold nanoparticle (Blue line) immobilized on Au micro-gap electrode in 10 mM PBS (pH = 7.4) at different scan rate (mV/s) (Red line: 10 mV/s, Orange line: 20 mV/s, Yellow line: 30 mV/s, Green line: 40 mV/s, Blue line: 50 mV/s); (**b**) Plots of anodic and cathodic peaks currents *vs.* scan rates; (**c**) Cyclic voltammogram of hemoglobin/gold nanoparticle (Blue line) immobilized on Au micro-gap electrode containing (Red line: 0, Orange line: 10 nmol, Yellow line: 30 nmol, Green line: 50 nmol, Blue line: 100 nmol H_2O_2 at 50 mV/s); (**d**) Plots of anodic and cathodic peaks currents *vs.* addition of H_2O_2; (**e**) Table of current values corresponding to H_2O_2 concetrations.

The electron-transfer rate constant (k_s) for electron transfer between the micro-gap electrode and the surface Hb layers was also 0.35 s^{-1} based on the following equation [16,38];

tAlso, the *Hb* shows electrocatalytic activity because of its similarity with peroxidase for catalysis of H_2O_2 [9,20]. The electrocatalytic activity of the Hb/GNP to H_2O_2 on the micro-gap electrode was investigated by CV. The principle of catalyzing H_2O_2 of Hb was explained using the following equations:

$$Hb[Fe(III)] + e^{-} \rightarrow Hb[Fe(II)] \quad (3)$$

$$2\,[Hb - Fe(II)] + 2H^{+} + H_2O_2 \rightarrow 2\,[Hb - Fe(III)] + 2H_2O \quad (4)$$

The overall reaction would be:

$$H_2O_2 + 2H^{+} + 2e^{-} \rightarrow 2\,H_2O \quad (5)$$

Figure 5c shows the electrochemical signal-enhancement effect of anodic and cathodic currents corresponding to the successive addition of H_2O_2. The redox currents of Hb/GNP were gradually increased according to the addition of 10 nmol H_2O_2, 30 nmol H_2O_2, 50 nmol H_2O_2, 100 nmol H_2O_2. This result shows the electrocatalytic reduction process of the Hb/GNP heterolayer on the micro-gap. The anodic peak potential (E_{pa}) and cathodic peak potential (E_{pc}) were 0.207 and 0.205 V (*vs.*Ag/AgCl), respectively. As shown in Figure 5d,e, cathodic and anodic current peaks were obtained for the Hb/GNP heteroalyer on micro-gap electrode after the addition of 20 μL aliquots of 10 mM H_2O_2 in 1 mL of 10 mM HEPES at pH 7.0. The result shows a linear plot for anodic and cathodic peak currents against the addition of H_2O_2, and the sensor calibration plot was linear with $R^2 = 0.97$ (anodic peak) and 0.98 (cathodic peak), respectively. The slope values of anodic and cathodic peak currents against the addition of H_2O_2, $I_{pa} = 0.1862X + 4.3234(10 \times e^{-8}$ A) and $I_{pc} = -0.2152X - 3.8476(10 \times e^{-8}$ A). And the 17.6% of cathodic peak current and 20.8% of anodic peak current were increased corresponding to the addition of H_2O_2. Based on this analysis, an increase of H_2O_2 concentration shifts the oxidation peaks to a more positive potential, while the reduction peaks shifted to a more negative potential on the micro-gap.

CONCLUSIONS

In summary, the Hb/GNP heterolayer on the micro-gap was fabricated to detect the H_2O_2 with electrochemical-signal enhanced effect. The fabricated Hb/GNP hybrid layer was investigated by AFM and SERS. Additionally, the redox properties of Hb/GNP hybrid layer on bulk electrode and micro-gap were confirmed using CV. Also, the H_2O_2 detection test of Hb/GNP heterolayer on the micro-gap was conducted. As a result, the fabricated Hb/GNP heterolayer on the Au micro-gap was detected the low concentration (10 nmol) of H_2O_2 with high fidelity. These results show the electrochemical signal-enhancement effect of Hb/GNP heterolayer can be directly applied to a protein-based biosensor system at the micro-scale. This metalloprotein-nanoparticle hybrid layer can be used for miniaturized H_2O_2 biosensor applications.

Acknowledgments

This research was supported by Nanobioelectronic Laboratory in Sogang University.

Author Contributions

T.L. and T.-H.K. conceived and designed the experiments; T.L. and J.Y. performed the experiments; Y.-H.C. and J.Y.L. analyzed the data; T.L., T.-H.K. and J.-W.C. wrote the paper.

Conflicts of Interest

The authors declare no conflict of interest.

Abbreviations

Hb	Hemoglobin
GNP	Gold Nanoparticle
AFM	Atomic Force Microscopy
SERS	Surface-Enhanced Raman Spectroscopy
6-MHA	6-Mercaptohexanoic Acid
LbL	layer-by-layer
CV	Cyclic Voltammetry

REFERENCES

1. Bartlett, P.N. *Bioelectrochemistry: Fundamentals, Experimental Techniques and Applications*, 1st ed.; John Wiley & Sons Inc.: West Sussex, UK, 2008; pp. 3–38.
2. Kulys, J.; Gorton, L.; Domingues, E.; Emnéus, J.; Jarskog, H. Electrochemical characterization of carbon pastes modified with proteins and polycations. *J. Electroanal. Chem.* **1994**, *372*, 49–55.
3. Bănică, F. *Chemical Sensors and Biosensors: Fundamentals and Applications*; John Wiley & Sons Inc.: Chichester, UK, 2012; pp. 133–139.
4. Bartlett, P.N.; Birkin, P.R.; Wang, J.H.; Palmisano, F.; Benedetto, G.D. An Enzyme Switch Employing Direct Electrochemical Communication between Horseradish Peroxidase and a Poly(aniline) Film. *Anal. Chem.* **1998**, *70*, 3685–3694.
5. Wang, J.; Lin, Y.; Chen, L. Organic-phase biosensors for monitoring phenol and hydrogen peroxide in pharmaceutical antibacterial products. *Analyst* **1993**, *118*, 277–280.
6. Gu, T.; Hasebe, Y. DNA–Cu(II) poly(amine) complex membrane as novel catalytic layer for highly sensitive amperometric determination of hydrogen peroxide. *Biosens. Bioelectron.* **2006**, *21*, 2121–2128.
7. Chen, S.; Yuan, R.; Chai, Y.; Zhang, L.; Wang, N.; Li,

X. Amperometric third-generation hydrogen peroxide biosensor based on the immobilization of hemoglobin on multiwall carbon nanotubes and gold colloidal nanoparticles. *Biosens. Bioelectron.* **2007**, *22*, 1268–1274.

8. Du, D.; Ding, J.; Cai, J.; Zhang, A. Determination of carbaryl pesticide using amperometric acetylcholinesterase sensor formed by electrochemically deposited chitosan. *Colloid Surface B* **2007**, *58*, 145–150.

9. Yagati, A.K.; Choi, J.-W. Protein Based Electrochemical Biosensors for H2O2 Detection towards Clinical Diagnostics.*Electroanalysis* **2014**, *26*, 1259–1276.

10. Rojking, R.; Dominguez-Rosales, J.-A.; Nieto, N.; Greenwel, P. Role of hydrogen peroxide and oxidative stress in healing responses. *CMLS Cell. Mol. Life. Sci.* **2002**, *59*, 1872–1891.

11. Lean, J.M.; Jagger, C.J.; Kirstein, B.; Fuller, K.; Chambers, T.J. Hydrogen Peroxide is Essential for Estrogen-Deficiency Bone Loss and Osteoclast Formation. *Endocrinology* **2005**, *146*, 728–735.

12. Tan, X.-C.; Zhang, J.-L.; Tan, S.-W.; Zhao, D.-D.; Huang, Z.-W.; Mi, Y.; Huang, Z.-Y. Amperometric Hydrogen Peroxide Biosensor Based on Immobilization of Hemoglobin on a Glassy Carbon Electrode Modified with Fe_3O_4/Chitosan Core-Shell Microspheres. *Sensors* **2009**, *9*, 6185–6199.

13. Lei, C.-X.; Hu, S.-Q.; Gao, N.; Shen, G.-L.; Yu, R.-Q. An amperometric hydrogen peroxide biosensor based on immobilizing horseradish peroxidase to a nano-Au monolayer supported by sol-gel derived carbon ceramic electrode.*Bioelectrochemistry* **2004**, *65*, 33–39.

14. Sassolas, A.; Blum, L.J.; Leca-Bouvier, B.D. Immobilization strategies to develop enzymatic biosensors. *Biotechnol. Adv.***2012**, *30*, 489–511.

15. Aghebati-maleki, L.; Salehi, B.; Behfar, R.; Saeidmanesh, H.; Ahmadian, F.; Sarebanhassanabadi, M.; Negahdary, M. Designing a Hydrogen Peroxide Biosensor Using Catalase and Modified Electrode with Magnesium Oxide Nanoparticles. *Int. J. Electrochem. Sci.* **2014**, *9*, 257–271.

16. Yagati, A.K.; Lee, T.; Min, J.; Choi, J.-W. Electrochemical performance of gold nanoparticle-cytochrome c hybrid interface for H_2O_2 detection. *Colloid. Surf. B* **2012**, *92*, 161–167.

17. Yip, R. Significance of an abnormally low or high hemoglobin concentration during pregnancy: Special consideration of iron nutrition. *Am. J. Clin. Nutr.* **2000**, *72*, 272–279.

18. Schechter, A.N. Hemoglobin research and the origins of molecular medicine. *Blood* **2008**, *112*, 3927–3938.

19. Han, X.; Cheng, W.; Zhang, Z.; Dong, S.; Wang, E. Direct Electron Transfer between Hemoglobin and a Glassy Carbon Electrode Facilitated by Lipid-Protected Gold Nanoparticles. *Biochim. Biophys. Acta Bioenerg.* **2002**, *1556*, 273–277.

20. Ramanavicius, A.; Ramanaviciene, A. Hemoproteins in design of biofuel cells. *Fuel Cells* **2009**, *9*, 25–36.

21. Xuan, J.; Jia, X.-D.; Jiang, L.-P.; Abdel-Halim, E.S.; Zhu, J.-J. Gold nanoparticle-assembled capsules and their application as hydrogen peroxide biosensor based on hemoglobin. *Bioelectrochemistry* **2012**, *84*, 32–37.

22. Xie, L.; Xu, Y.; Cao, X. Hydrogen peroxide biosensor based on hemoglobin immobilized at graphene, flower-like zinc oxide, and gold nanoparticle nanocomposite modified glassy carbon electrode. *Colloids Surf. B* **2013**, *107*, 245–250.

23. Akhtar, N.; El-Safty, S.A.; Khairy, M. Simple and Sensitive Electrochemical Sensor-Based Three-Dimensional Porous Ni-Hemoglobin Composite Electrode. *Chemosensors* **2014**, *2*, 235–250.

24. Tian, H.; Liang, F.; Jiao, J.; Hu, J. Direct Electrochemistry of Hemoglobin on Nickel Ion Implanted-Modified ITO Electrode and Biosensing for H_2O_2. *J. Electrochem. Soc.* **2013**, *160*, B125–B131.

25. Xu, M.-Q.; Wu, J.-F.; Zhao, G.-C. Direct Electrochemistry of Hemoglobin at a Graphene Gold Nanoparticle Composite Film for Nitric Oxide Biosensing. *Sensors* **2013**, *13*, 7492–7504.

26. German, N.; Ramanavicius, A.; Ramanaviciene, A. Electrochemical Deposition of Gold Nanoparticles on Graphite Rod for Glucose Biosensing. *Sens. Actuators B Chem.* **2014**, *203*, 25–34.

27. German, N.; Kausaite-Minkstimiene, A.; Ramanavicius, A.; Semashko, T.; Mikhailova, R.; Ramanaviciene, A. The Use of Different Glucose Oxidases for the Development of an Amperometric Reagentless Glucose Biosensor based on Gold Nanoparticles Covered by Polypyrrole. *Electrochim. Acta* **2015**, *169*, 326–333.

28. Seo, S.M.; Kang, T.J.; Kim, Y.; Kim, N.; Ahn, J.; Kim, T.W.; Kim, Y.H.; Ryu, S.H.; Park, Y.J. Electrode Asymmetry Driven Self-gating Effect on the Electrical Detection of Protein. *Sens. Actuators B Chem.* **2014**, *191*, 800–805.

29. Chornokur, G.; Arya, S.K.; Phelan, C.; Tanner, R.; Bhansali, S. Impedance-BasedMiniaturized Biosensor for Ultrasensitive and Fast Prostate-Specific Antigen Detection. *J. Sens.* **2011**, *983751*, 1–7.

30. Ben-Yoav, H.; Dykstra, P.H.; Bentley, W.E.; Ghodssi, R. A Microfluidic-Based Electrochemical Biochip for Label-Free Diffusion-Restricted DNA Hybridization Analysis. *Biosens. Bioelectron.* **2012**, *38*, 114–120.

31. Lee, T.; Kim, S.-U.; Min, J.; Choi, J.-W. Multi-Level Biomemory Device Consisting of Recombinant Azurin/Cytochrome c. *Adv. Mater.* **2010**, *22*, 510–514.

32. Chung, Y.-H.; Lee, T.; Park, H.J.; Yun, W.S.; Min, J.; Choi, J.-W. Nanoscale biomemory composed of recombinant azurin on a nanogap electrode. *Nanotechnology* **2013**, *24*, 365301.

33. Lee, T.; Yoo, S.-Y.; Chung, Y.-H.; Min, J.; Choi, J.-W. Signal enhancement of Electrochemical Biomemory Device Composed of Recombinant Azruin/Gold Nanoparticale. *Electroanalysis* **2011**, *23*, 2023–2029.

34. Lee, T.; Chung, Y.-H.; Yoon, J.; Min, J.; Choi, J.-W. Fusion Protein-based Dual-level Biomemory Device Consisted of Recombinant Azurin-Myoglobin. *Appl.*

Surf. Sci. **2014**, *329*, 448–454.

35. Drescher, D.; Büchner, T.; McNaughton, D.; Kneipp, J. SERS reveals the specific interaction of silver and gold nanoparticles with hemoglobin and red blood cell components. *Phys. Chem. Chem. Phys.* **2013**, *15*, 5364–5373.

36. Kang, Y.; Si, M.; Zhu, Y.; Miao, L.; Xu, G. Surface-enhanced Raman scattering (SERS) spectra of hemoglobin of mouse and rabbit with self-assembled nano-silver film. *Spectrochim. Acta Mol. Biomol. Spectrosc.* **2013**, *108*, 177–180.

37. Pingarron, J.M.; Yanez-Sedeno, P.; Gonzalez-Cortes, A. Gold Nanoparticle-Based Electrochemical Biosensors. *Electrochimica Acta* **2008**, *53*, 5848–5866.

38. Laviron, E. General Expression of the Linear Potential Sweep Voltammogram in the Case of Diffusionless Electrochemical Systems. *J. Electroanal. Chem.* **1979**, *101*, 19–28.

CHAPTER 13

A HIGHLY CONTROLLABLE ELECTROCHEMICAL ANODIZATION PROCESS TO FABRICATE POROUS ANODIC ALUMINUM OXIDE MEMBRANES

Yuanjing Lin, Qingfeng Lin, Xue Liu, Yuan Gao, Jin He, Wenli Wang, and Zhiyong Fan

Keywords: Anodic aluminum oxide, Nanoporous structure, Integrated charge density, Controllable electrochemical anodization

BACKGROUND

Metal anodization has been broadly used in industry as a surface treatment technique to render materials with resistance against uncontrolled oxidation, abrasion, and corrosion. Although this technique has been developed for a long time, it was until 1990s that researchers discovered that highly ordered nanoporous structures can be achieved by properly tuning anodization conditions including electrolyte composition and concentration, temperature, as well as anodization voltage [1]. Among all valve metals that can be anodized, aluminum (Al) and titanium (Ti), particularly Al, can be anodized into nanoporous structures with well-controlled diameter, pitch, and depth. Membranes consist of these nanostructures, i.e., anodic titanium oxide (ATO) and anodic aluminum oxide (AAO), have wide nanoengineering applications that have attracted enormous attention. For example, AAO membranes have been used as templates to directly assemble semiconductor nanowires and nanorods for photodetection [2] and solar energy conversion [3–5]. A large internal surface area of these oxide nanostructures can also be harnessed to build high-performance energy storage devices such as Li-ion batteries [6] and supercapacitors [7, 8]. Meanwhile, it is worth pointing out that AAO structures can be engineered into a number of variants via proper combination of wet chemical etching and anodization processes. These variants include nanowells [9], inverted nanocones [10, 11], nanobowls [12], nanospikes [13–15], and the integrated nanopillar-nanowell structures [16]. These structures have been used as scaffold of energy harvesting and storage devices in our past works.

There are certainly a number of advantages of using porous anodic nanostructures, particularly AAO membranes for nanoengineering applications, such as large surface area, high regularity and scalable low-cost production [17]. In many applications, precise control of nanostructure shape and geometry is critical. It is known that the geometric features of porous AAO membranes are widely tunable via electrochemical anodization process [10]. Factors such as applied voltage, acid type, and concentration contribute to the formation of the porous nanostructures with various barrier layer thicknesses (D_b), interpore distances (D_{int}), and periodicity [18–20]. Meanwhile, it is noteworthy that the growth rate of porous AAO nanostructures is not constant during the entire growth process, even when the growth starts with a fixed voltage and given electrolyte. This can be attributed to two competing factors. On one hand, AAO growth rate is sensitive to electrolyte temperature. In general, higher temperature expedites growth and lower temperature slows down growth. Therefore, environmental temperature fluctuation leads to the

Citation: Lin Y, Lin Q, Liu X, et al. A Highly Controllable Electrochemical Anodization Process to Fabricate Porous Anodic Aluminum Oxide Membranes. Nanoscale Research Letters. 2015;10:495. doi:10.1186/s11671-015-1202-y.

variation of growth rate [19]. Besides, anodization current flow through electrolyte causes temperature increase. This should be also incorporated into consideration as well, particularly for high voltage/high-current anodization. On the other hand, during the anodization process, electrolyte composition is being gradually changed. Specifically, Al cations will be injected into electrolyte and hydrogen evolution at the cathode that reduces proton concentration in electrolyte [21]. The composition and concentration change of electrolyte inevitably affects anodization rate. All these complicated factors pose a challenge to precisely control anodization process, especially when there is a stringent requirement on final membrane thickness. In this work, we revisited the electrochemical reactions during Al anodization. Then a generic electric charge integral approach was developed to monitor the growth pore depth (D_p) of AAO in real time. This approach is based on the discovered linear correlation between measured total electric charge flow through the circuit during stable anodization process and the molar amount of anodic alumina grown in the process. It was found that this linear correlation is rather insensitive to anodization voltage and composition of the electrolyte. This suggests that for a given anodization process, AAO pore depth can be predicted in real time regardless of the temperature variation and electrolyte concentration change. Therefore, a program was coded to visualize the AAO growth process. In this case, not only the AAO growth pore depth can be monitored, but also a target pore depth can be set and the anodization process can be terminated when the projected pore depth reaches the set point. Overall, the revealed correlation between the quantities of electric charge and anodic material is of scientific and practical importance, and it facilitates precise control of AAO anodization in a wide voltage range for a broad applications.

METHODS

The fabrication of porous AAO membranes with nanostructures mainly follows a two-step electrochemical anodization process [22] after proper pretreatments [23, 24]. The aluminum foils are initially covered by natural oxide layers and surface roughness caused by thermal and mechanical processes [25]. To minimize the impact of these defects on the fabrication of highly ordered porous AAO membranes, pretreatments such as pressing and electropolishing play an essential role in removing particles and smoothing the surface of aluminums. Normally, the two-step anodization of aluminum leads to formation of AAO nanopores with short-range hexagonal ordering [26]. Typically, D_{int} is simply proportional to the applied voltage (U) with a linear constant of 2.5 nm/V, namely Dint(nm)=2.5(nmV)×U(V) [18]. Meanwhile the depth of the nanopores and their diameter can be controlled by anodization time and the subsequent wet chemical etching. In some applications, highly ordered structures are required. For example in nanophotonic applications, both ordering and periodicity may affect light-structure interaction [9]. Moreover, when utilizing nanopores to directly integrate semiconducting nanowires for nanoelectronic applications, each nanowires needs to be individually addressable which also requires perfect ordering. In the past, we have intensively explored fabrication of AAO membranes with perfect hexagonal and square orderings with an area up to tens of centimeter square, and the periodicity ranging from 500 nm to 3 μm [9, 10, 13, 14, 27]. In order to achieve long-range regular structures, hexagonally or squarely ordered nanohole arrays are produced on electrochemically polished aluminum surface with nanoimprint, using a silicon stamp mold with ordered short pillars on the surface. Then the aluminum sample undergoes anodization process yielding a perfectly ordered AAO nanopore array on the surface. Note that it is important to satisfy a matching condition between the periodicity of the nanoimprinted nanoholes and the anodization voltage, governed by the linear constant of 2.5 nm/V. Figure 1a,bb shows the scanning electron microscopic (SEM) images of an as-fabricated AAO membrane demonstrating ideally regular configuration of the straight and parallel holes. In this case, the Al foil was anodized with an applied voltage of 400 V in 230 ml 1:1(v/v) mixture of 4 wt.% citric acid ($C_6H_8O_7$) and ethylene glycol (EG) with an extra 15 ml 0.1 wt.% phosphoric acid (H_3PO_4) at a temperature of 10 °C for 30 min.

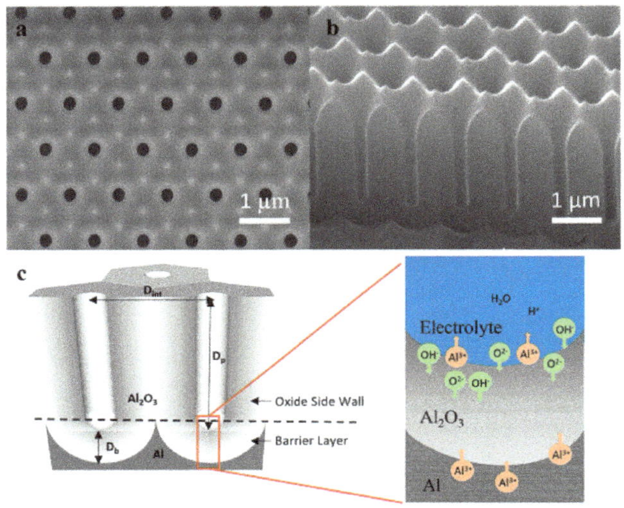

Figure 1: SEM images of imprinted AAO and schematic of AAO structure and formation mechanism. a Top-view SEM image of imprint AAO with Dint of 1 μm. b Cross-sectional SEM image of imprint AAO with Dint of 1 μm. c Schematic of AAO nanoporous structure and major features of AAO formation

In order to have a more precise control of the anodization process in this work, the voltage and current of the anodization process are monitored and recorded by Keithley 2410 SourceMeter, controlled by a home-built computer program based on LabView. The voltage was increased in a linear manner from zero to the target value with a ramping rate of 10 V/min, so that the current will not rise too fast to avoid overheating of the electrolyte and burning aluminum samples. Meanwhile, we have also discovered that a ramping process which lasted too long will result in forming a fairly smooth anodic oxide film without nanoporous structure. The unnecessarily slow voltage ramping cannot provide high enough electric field for expelling Al^{3+} into electrolyte, and the Al^{3+} is retained in oxide bulk. Thus, the oxidization is fairly stable without variation in electric field distribution on the oxide layer and therefore hinders the formation of hemispherical oxide layer and nanoporous structure [25]. Once the ramping process is done, the program maintains target voltage while it is observed that the anodization current falls naturally down to a stable level, which is an indication of inception of stable anodization [28]. In our experiment, a two-step anodization approach is adopted. In the first step, an initial AAO layer was formed after anodization in electrolyte for over 12 h. The first layer of AAO was then removed in a mixture of 6 wt.% phosphoric acid (H_3PO_4) and 1.8 wt.% chromic acid (H_2CrO_4) at 98 °C for 30 min, leaving a highly periodic nanoconcave structure on the surface of the aluminum substrate, which forms the initiation sites for the formation of pores in the second anodization step. Interestingly, this nanoconcave structure can be used to enhance light absorption and power conversion efficiency of thin-film solar cells which was reported by us previously [29]. Furthermore, AAO membrane with highly ordered porous nanostructure is produced in the same electrolyte during second anodization step. It is well known that an AAO membrane contains two parts of oxide layers, namely barrier layer and the oxide side wall, as shown in Fig. 1c. AAO is mainly formed at the AAO/Al interface when OH^- and O^{2-}diffuse through the barrier layer driven by the electric field [25]. The thickness of the barrier layer remains constant for a particular anodization voltage during the stable anodization process, while the length/height of the oxide side wall continues to grow. In the first step anodization, the nanoimprinted nanoholes on the aluminum surface are the centers for concentrated electric field. During the initial voltage ramping process, an oxide barrier layer with hemispherical shape is formed due to radial alignment of electric field line vectors. When the oxide barrier layer is thin, electric field is strong; thus, the field driven diffusion of OH^- and O^{2-} to AAO/Al interface is fast leading to fast oxidation of Al and increase of barrier layer thickness. At the AAO/electrolyte interface, field-assisted etching of AAO and thinning of the barrier layer occur since dissolution of Al_2O_3 in pH <5 electrolyte [28]. It is obvious that the above barrier layer growth and etching processes are the two competing processes. The etching rate is not only a function of local electric field intensity, local pH value, but also highly depends on temperature. In practice, the temperature is kept at 5~10 °C to slow down the barrier layer dissolution. Generally, the barrier layer growth rate is largely determined by electric field intensity. In the beginning of the barrier layer growth, small oxide thickness leads to high electric field thus, the overall barrier layer thickness increases over the time. However, after a certain period of time, the electric field intensity drops to a level so that a balanced barrier oxide layer dissolution and growth is achieved. Thereafter, the barrier layer thickness remains constant. As a natural outcome of this dynamic process, the AAO/Al interface moves deeper and deeper into Al bulk, leading to a sustainable growth of pore depth. This formation mechanism explains tubular structure of each AAO nanochannel, and the side-wall thickness is roughly twice of that of the barrier layer, since half of the side wall comes from the barrier layer. It is interesting to notice that the hexagonal nanopore arrangement is typically observed even without assistance of nanoimprint, though in a short range. This can be explained by the volumetric expansion of nanopores during formation of oxide which induces stress between neighboring pores [30]. In order to achieve minimal system energy, a close-pack hexagonal ordering is naturally formed.

RESULTS AND DISCUSSION

Figure 2a shows measured current density-time (J-t) and voltage-time (V-t) curves recorded during a 400 V anodization process. According to the AAO growth mechanism discussed above, ionic current is the dominant contributor to the observed current flow while electronic current can be ignored since AAO itself is an insulator [31]. The transportation of anions (O^{2-}, OH^-) and cations (Al^{3+}) contributes to most of the ionic current. It has been reported that the ionic current density is related to electric field strength (E) with an exponential function [21]:

$$J_{\text{ion}} = Ane^{BE} = Ane^{\frac{BU}{D_b}} \quad (1)$$

where both A and B are constants related to electrolyte temperature, n is the surface density of mobile ions, U is the potential drop across the barrier layer, and D_b is the thickness of the barrier layer. At the moment (t_0) when an equilibrium state of dissolution and oxidization is established, the typical barrier thickness (D_{b0}) is proportional to applied voltage with a linear constant (λ_b) approximately ranges from 1.0 to 1.4 nm/V depends on electrolyte components and temperature [21], namely, Db0(nm)=λb(nmV)×V(V). The growth rate of the barrier layer can be defined as [32]:

$$\frac{dD_b}{dt} = \begin{cases} G_0 - \frac{D_b}{\tau} & (D_b < D_{b0}) \\ 0 & (D_b = D_{b0}) \end{cases} \quad (2)$$

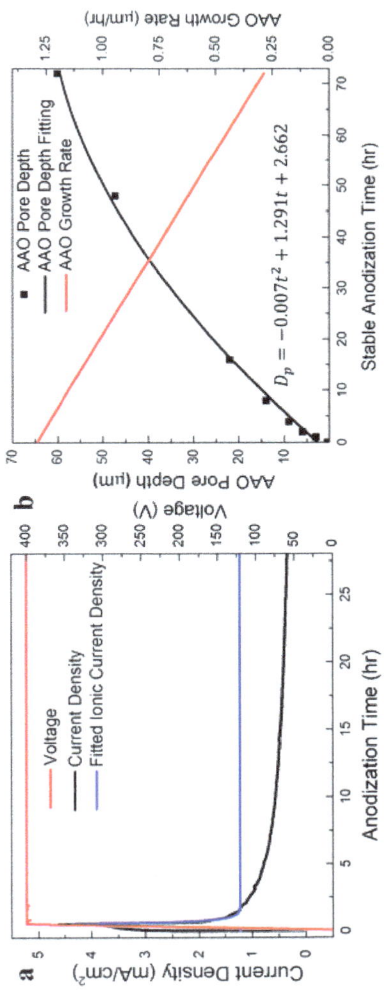

Figure 2: Temporal response of anodization current and voltage and AAO growth rate. **a** Current density and voltage curve versus time. Ionic current is the dominated contributor to the current flow. **b** AAO pore depth and growth rate over stable anodization time

Where G_0 is the initial growth rate at the start of a stable applied anodization voltage, τ is a time constant. After integral, formula (2) can be rewritten as:

$$D_b = \begin{cases} G_0\tau\left(1-e^{-\frac{t}{\tau}}\right) & (t < t_0) \\ D_{b0} & (t \geq t_0) \end{cases} \qquad 3$$

Therefore, after combining Eqs. 1 and 3, the ionic current density can be expressed as:

$$J_{\text{ion}} = \begin{cases} \dfrac{BU}{G_0\tau\left(1-e^{-\frac{t}{\tau}}\right)} \; Ane & (t < t_0) \\ \dfrac{BU}{D_{b0}} \; Ane & (t \geq t_0) \end{cases} \qquad 4$$

The current density curve can be mathematically fitted with the Eq. 4 as shown in Fig. 2a, and the fitted expression is:

$$J_{\text{ion}} = \begin{cases} 1.47 \times 10^{-4} \dfrac{9.03}{(1-e^{-4.18t})} & (t < t_0) \\ 1.22 & (t \geq t_0) \end{cases} \qquad 5$$

The fitting proved that the ionic current is the main contributor to the growth of AAO, and the ionic current should remain unchanged during stable anodization process. However, we notice that the measured current density has a decaying nature in stable anodization process. This can be ascribed to the gradually reduced concentration of H^+ in electrolyte and as a result, the dissolution rate together with oxidation rate is continuously slowed down.

To investigate the growth rate of AAO pore depth, seven AAO membranes were grown with different anodization time under an applied voltage of 400 V and a temperature of 10 °C. Figure 2b shows AAO pore depth, which is roughly AAO thickness, versus anodization time with polynomial fitting. It is obvious that the growth rate of AAO is not constant even without fluctuation in environmental conditions. As a result, a declining growth rate can be derived and plotted, as shown in Fig. 2b. One of the effective ways to recover the growth rate after a certain thickness of AAO has grown is to replenish acid during the reaction to compensate proton loss. As shown in Fig. 3a, immediate current rise can be recorded upon admittance of 2 and 4 ml 0.1 wt.% H_3PO_4 after 7 and 12.5 h, respectively, which indicates an accelerated oxidation rate of AAO growth. With an electric field established in the electrolyte, additional protons from the replenished acid quickly migrate to cathode leading to moving extra amount of electrons from Al anode through external circuit, which is indicated by the fast current rise. Once the new charge distribution is established, the current declines. Overall, the amount of protons at the proximity of cathode and oxygen containing anions (PO_4^{3-}) at the AAO/electrolyte interface increased and accelerated the anodization reaction. This explains the observation of a current spike followed by a step increase of anodization current. It is obvious that fluctuation of anodization conditions brings in difficulties in predicting the real-time AAO pore depth during the anodization process. Considering the fact that the measured electric current in external circuit is equal to the ionic current in magnitude, the electric current can be used to indicate the growth rate of AAO pore depth. Since the integral of grow rate over time leads to the pore depth, the integral of current density in principle can also be used to indicate AAO pore depth. As a matter of fact, the relationship between AAO pore depth and the integrated charge density is shown in Fig. 3b, which clearly suggests a linear relationship between the pore depth and charge density. Therefore, this linear relationship can be utilized to monitor AAO pore depth in real time.

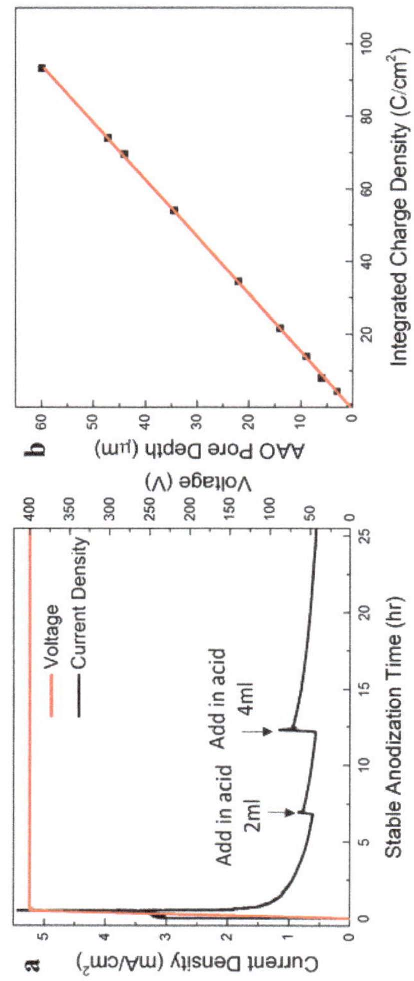

Figure 3: Temporal response of anodization current and voltage response with perturbation on anodization conditions. a Current density and voltage temporal response with adding acid twice during the process. b Linear fitting of AAO pore depth dependence on integral of stable anodization charge density

The validity of this linear correlation was further verified with anodization under applied voltage ranging from 20 to 600 V in different electrolytes (see Additional file 1: Figure S1). The slope of the linear fitting was defined as growth constant (G), which has a physical meaning of increment of AAO membrane volume per unit charge and can be theoretically estimated. In experiments, we have observed small intercepts on the AAO pore depth (Y) axis with value ranging from 30~500 nm. This is caused by initial unstable anodization of Al before the constant thickness of the barrier layer is established. Specifically, the volume of AAO (V_A) can be expressed as:

$$V_A = \frac{n \times M}{\rho} \quad 6$$

where ρ is the mass density of Al_2O_3, n is the molar amount of Al_2O_3 produced in oxidation, and M is the molar mass. In an ideal case, all the charges come from the oxidation of Al and are irrelevant to the dissolution of Al_2O_3 and thus, the charge quantity (q) has a linear relationship with n as:

$$q = 6eN_A \times n \quad 7$$

where e is the elementary charge and N_A is the Avogadro constant. Therefore, combining Eqs. 6 and 7, the theoretical growth constant can be calculated by using Eq. 8:

$$G = \frac{dV_A}{dq} = \frac{M}{6eN_A \times \rho} \quad 8$$

It can be seen from Eq. 8 that in theory, G is a constant independent of anodization condition. Specifically, the growth constant G (mm^3/C) is theoretically independent of variations on factors such as acid concentration, temperature, and AAO pore diameter. While the growth rate (μm/h) of AAO membrane can be accelerated by higher acid concentration and electrolyte temperature, the integral of charge quantity from Al oxidation is independent of oxidation rate. Therefore, the linear correlation will not be affected even when there is a fluctuation in acid concentration and temperature. However, in experiments, we have discovered that excessive temperature leads to high current and burning of AAO due to uncontrollable fast anodization. The integral of charge density is not affected by variation of pore diameter and the porosity of AAO membrane, which indicates the linear correlation could be applied in fabrication of both non-imprint and imprint AAO membranes. However, the measured G obtained from Additional file 1: Figure S1a and b shows variation for different anodization voltage, as it can be clearly seen from Fig. 4a. We have attributed the variation of growth constants under different anodization voltages to mass density (ρ) deviation of AAO membrane obtained with different voltages. It is known that incorporation of anions and water molecules from the electrolyte into the oxide layer occurs during anodization process which changes mass density of AAO [33, 34]. It is reported that minor constituents such as anions from the electrolyte can be as much as 20 % of the AAO wall structure in volume depending upon the electrolytes and the growth conditions [35] and the amount of incorporation depends on electric field and acid concentration, etc.

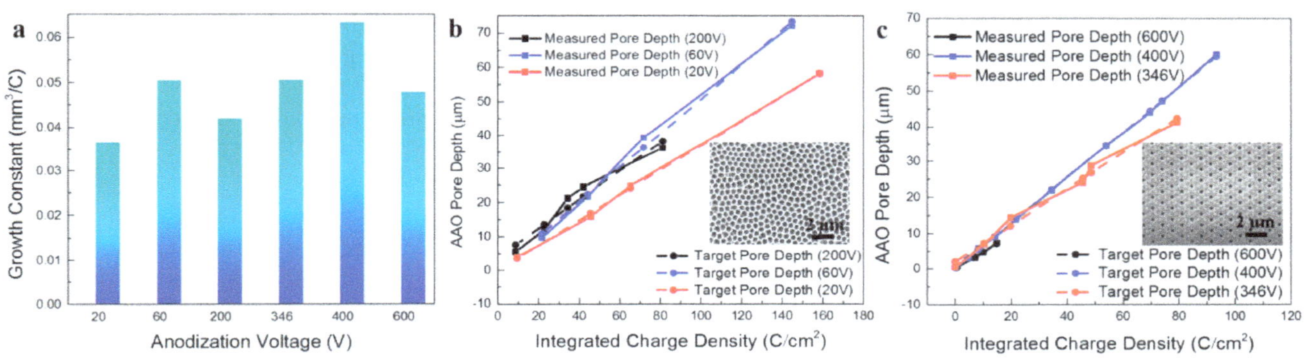

Figure 4: Growth constant and controlled AAO growth. a The measured growth constant of AAO membrane under different applied voltages. b Measured pore depth of non-imprinted AAO is consistent with target values. c Measured pore depth of imprinted AAO is consistent with target values

With the achieved understanding on growth constant, the control of AAO membrane thickness can be realized with a programmable source-meter unit. Here, we have performed a set of AAO anodization with various target pore depths. Utilizing the measured growth constant G, pore depth is monitored by a computer connected with Keithley source-meter unit in real time and anodization processes were terminated automatically when reaching target AAO thickness. Figure 4b, c indicates that in a large voltage range and regardless of pore diameters, with this charge integration approach, the AAO pore depths can be precisely controlled with high consistency with the actual thickness measured with SEM. The marginal deviations may possibly come from: (1) inaccuracy in measuring the AAO pore depth with SEM, mainly from viewing angle deviation for cross-section measurement; and (2) deviation in fitting method. All the curves were fitted in concise models such as linear and exponential fitting so as to indicate the primary trend and thus, the measured growth rates derived from the fitting result could have a reasonable deviation.

CONCLUSIONS

In this work, a linear correlation between AAO pore depth and integrated charge density during electrochemical anodization process was discovered and utilized to implement precise control of AAO growth process. This method proved to be repeatable and robust regardless of anodization voltage, temperature, and electrolyte composition and concentration variation. Thus, a home-built automatic AAO fabrication setup combining a source-meter unit and software was developed to realize real-time monitoring and automatic anodization termination when reaching a target pore depth. The good reliability in control of the as-fabricate AAO membrane thickness has been verified in a voltage range from 20 to 600 V, indicating its practicality in a broad nanoengineering applications.

ACKNOWLEDGEMENTS

This work was partially supported by the General Research Fund (612113) from Hong Kong Research Grant Council, ITS/362/14FP from Hong Kong Innovation Technology Commission, Fundamental Research Project of Shenzhen Science & Technology Foundation JCYJ20130402164725025, National Natural Science Foundation of China under Grants (61574005), and the Priority Academic Program Development of Jiangsu Higher Education Institutions [PAPD].

Additional file

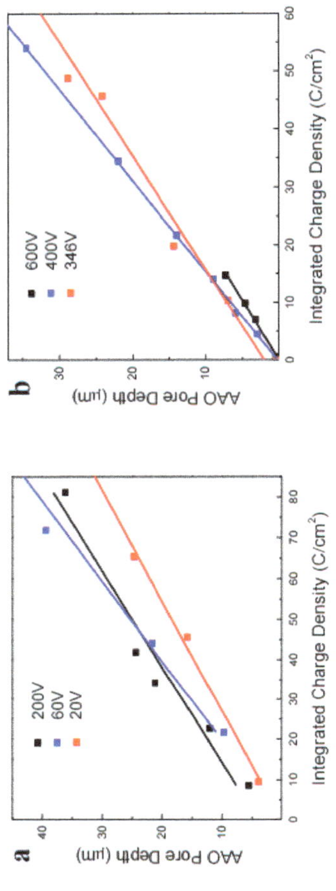

Figure S1: Linear correlation between AAO pore depth and integrated charge density in a large voltage range (a) 20–200 V; (b) 346–600 V.

Authors' Contributions

YL prepared the AAO membranes, carried out the SEM characterization, data analysis, and drafted the manuscript. QL fabricated the silicon mold for imprint AAO and revised the manuscript. XL helped on software design. YG and JH provided suggestions on data analysis. ZF and WW provided the idea and completed the manuscript. All authors read and approved the final manuscript.

REFERENCES

1. Masuda H, Yamada H, Satoh M, Asoh H, Nakao M, Tamamura T. Highly ordered nanochannel-array architecture in anodic alumina. Appl Phys Lett. 1997;71:2770–2772. doi: 10.1063/1.120128.
2. Fan ZY, Dutta D, Chien CJ, Chen HY, Brown EC, Chang PC, Lu JG. Electrical and photoconductive properties of vertical ZnO nanowires in high density arrays. Appl Phys Lett. 2006;89:213110–213112. doi: 10.1063/1.2387868.
3. Fan ZY, Kapadia R, Leu PW, Zhang XB, Chueh YL, Takei K, Yu K, Jamshidi A, Rathore AA, Ruebusch DJ, Wu M, Javey A. Ordered arrays of dual-diameter nanopillars for maximized optical absorption. Nano Lett. 2010;10:3823–3827. doi: 10.1021/nl1010788.
4. Gu L, Liu X, Kwon K, La C, Lee MH, Yu K, Chueh Y, Fan Z. In situ doping control and electrical transport investigation of single and arrayed CdS nanopillars. Nanoscale. 2013;5:7213–7218. doi: 10.1039/c3nr01975f.
5. Ergen O, Ruebusch DJ, Fang H, Rathore AA, Kapadia R, Fan Z, Takei K, Jamshidi A, Wu M, Javey A. Shape-controlled synthesis of single-crystalline nanopillar arrays by template-assisted vapor–liquid–solid process. J Am Chem Soc. 2010;132:13972–13974. doi: 10.1021/ja1052413.
6. Liu C, Gillette EI, Chen X, Pearse AJ, Kozen AC, Schroeder MA, Gregorczyk KE, Lee SB, Rubloff GW. An all-in-one nanopore battery array. Nat Nanotechnol. 2014;9:1031–1039. doi: 10.1038/nnano.2014.247.
7. Xu J, Wu H, Lu L, Leung S, Chen D, Chen X, Fan Z, Shen G, Li D. Integrated photo-supercapacitor based on Bi-polar TiO2 nanotube arrays with selective one-side plasma-assisted hydrogenation. Adv Funct Mater. 2014;24:1840–1846. doi: 10.1002/adfm.201303042.
8. Wu H, Xu C, Xu J, Lu L, Fan Z, Chen X, Song Y, Li D. Enhanced supercapacitance in anodic TiO2 nanotube films by hydrogen plasma treatment. Nanotechnology. 2013;24:455401. doi: 10.1088/0957-4484/24/45/455401.
9. Leung SF, Yu M, Lin Q, Kwon K, Ching KL, Gu L, Yu K, Fan Z. Efficient photon capturing with ordered three-dimensional nanowell arrays. Nano Lett. 2012;12:3682–3689. doi: 10.1021/nl3014567.
10. Lin Q, Leung S, Tsui K, Hua B, Fan Z. Programmable nanoengineering templates for fabrication of three-dimensional nanophotonic structures. Nanoscale Res Lett. 2013;8:268. doi: 10.1186/1556-276X-8-268.
11. Lin Q, Leung S, Lu L, Chen X, Chen Z, Tang H, Su W, Li D, Fan Z. Inverted nanocone-based thin film photovoltaics with omnidirectionally enhanced performance. ACS Nano. 2014;8:6484–6490. doi: 10.1021/nn5023878.
12. Qiu Y, Leung S, Zhang Q, Mu C, Hua B, Yan H, Yang S, Fan Z. Nanobowl optical concentrator for efficient light trapping and high-performance organic photovoltaics. Sci Bull. 2015;60:109–115. doi: 10.1007/s11434-014-0693-8.
13. Gao Y, Jin H, Lin Q, Li X, Tavakoli MM, Leung S, Tang WM, Zhou L, Chan HLW, Fan Z. Highly flexible and transferable supercapacitors with ordered three-dimensional MnO 2/Au/MnO 2 nanospike arrays. J Mater Chem A. 2015;3:10199–10204. doi: 10.1039/

C5TA01960E.

14. Leung S, Gu L, Zhang Q, Tsui K, Shieh J, Shen C, Hsiao T, Hsu C, Lu L, Li D, Lin Q, Fan Z. Roll-to-roll fabrication of large scale and regular arrays of three-dimensional nanospikes for high efficiency and flexible photovoltaics. Sci Rep. 2014;4:4243. doi: 10.1038/srep04243.
15. Leung S, Tsui K, Lin Q, Huang H, Lu L, Shieh J, Shen C, Hsu C, Zhang Q, Li D. Large scale, flexible and three-dimensional quasi-ordered aluminum nanospikes for thin film photovoltaics with omnidirectional light trapping and optimized electrical design. Energy Environ Sci. 2014;7:3611–3616. doi: 10.1039/C4EE01850H.
16. Lin Q, Hua B, Leung S, Duan X, Fan Z. Efficient light absorption with integrated nanopillar/nanowell arrays for three-dimensional thin-film photovoltaic applications. ACS Nano. 2013;7:2725–2732. doi: 10.1021/nn400160n.
17. Lee MH, Lim N, Ruebusch DJ, Jamshidi A, Kapadia R, Lee R, Seok TJ, Takei K, Cho KY, Fan ZY, Jang H, Wu M, Cho GJ, Javey A. Roll-to-roll anodization and etching of aluminum foils for high-throughput surface nano-texturing. Nano Lett. 2011;11:3425. doi: 10.1021/nl201862d.
18. Masuda H, Yada K, Osaka A. Self-ordering of cell configuration of anodic porous alumina with large-size pores in phosphoric acid solution. Jpn J Appl Phys. 1998;37:L1340–L1342. doi: 10.1143/JJAP.37.L1340.
19. Stępniowski WJ, Bojar Z. Synthesis of anodic aluminum oxide (AAO) at relatively high temperatures study of the influence of anodization conditions on the alumina structural features. Surf Coat Tech.2011;206:265–272. doi: 10.1016/j.surfcoat.2011.07.020.
20. Belwalkar A, Grasing E, Van Geertruyden W, Huang Z, Misiolek W. Effect of processing parameters on pore structure and thickness of anodic aluminum oxide (AAO) tubular membranes. J Membr Sci.2008;319:192–198. doi: 10.1016/j.memsci.2008.03.044.
21. Diggle JW, Downie TC, Goulding C. Anodic oxide films on aluminum. Chem Rev. 1969;69:365–405. doi: 10.1021/cr60259a005.
22. Masuda H, Fukuda K. Ordered metal nanohole arrays made by a two-step replication of honeycomb structures of anodic alumina. Science. 1995;268:1466–1468. doi: 10.1126/science.268.5216.1466.
23. Wu M, Leu I, Hon M. Effect of polishing pretreatment on the fabrication of ordered nanopore arrays on aluminum foils by anodization. J Vac Sci Technol B. 2002;20:776–782. doi: 10.1116/1.1467657.
24. Chu S, Wada K, Inoue S, Isogai M, Yasumori A. Fabrication of ideally ordered nanoporous alumina films and integrated alumina nanotubule arrays by high-field anodization. Adv Mater. 2005;17:2115–2119. doi: 10.1002/adma.200500401.
25. Poinern GEJ, Ali N, Fawcett D. Progress in nano-engineered anodic aluminum oxide membrane development. Materials. 2011;4:487–526. doi: 10.3390/ma4030487.
26. Li Y, Ling Z, Chen S, Wang J. Fabrication of novel porous anodic alumina membranes by two-step hard anodization. Nanotechnology. 2008;19:225604. doi: 10.1088/0957-4484/19/22/225604.
27. Qiu Y, Leung S, Zhang Q, Hua B, Lin Q, Wei Z, Tsui K, Zhang Y, Yang S, Fan Z. Efficient photoelectrochemical water splitting with ultra-thin film of hematite on three-dimensional nanophotonic structures. Nano Lett. 2014;14:2123–2139. doi: 10.1021/nl500359e.
28. Patermarakis G, Moussoutzanis K. Electrochemical kinetic study on the growth of porous anodic oxide films on aluminium. Electrochim Acta. 1995;40:699–708. doi: 10.1016/0013-4686(94)00347-4.
29. Huang H, Lu L, Wang J, Yang J, Leung S, Wang Y, Chen D, Chen X, Shen G, Li DD, Fan Z. Performance enhancement of thin-film amorphous silicon solar cells with low cost nanodent plasmonic substrates. Energy Environ Sci. 2013;6:2965–2971. doi: 10.1039/c3ee41139g.
30. Jessensky O, Muller F, Gosele U. Self-organized formation of hexagonal pore arrays in anodic alumina.Appl Phys Lett. 1998;72:1173–1175. doi: 10.1063/1.121004.
31. Lee I, Jo Y, Kim Y, Tak Y, Choi J. Electrochemical thinning for anodic aluminum oxide and anodic titanium oxide. Bull Korean Chem Soc. 2012;33:1465. doi: 10.5012/bkcs.2012.33.5.1465.
32. Chong B, Yu D, Jin R, Wang Y, Li D, Song Y, Gao M, Zhu X. Theoretical derivation of anodizing current and comparison between fitted curves and measured curves under different conditions. Nanotechnology. 2015;26:145603. doi: 10.1088/0957-4484/26/14/145603.
33. Sato T. Mechanism of electrophoretic deposition of organic coatings on anodized aluminium. Trans Inst Met Finish. 1982;60:25–30.
34. Le Coz F, Arurault L, Fontorbes S, Vilar V, Datas L, Winterton P. Chemical composition and structural changes of porous templates obtained by anodising aluminium in phosphoric acid electrolyte. Surf Interface Anal. 2010;42:227–233. doi: 10.1002/sia.3199.
35. Li AP, Muller F, Birner A, Nielsch K, Gosele U. Hexagonal pore arrays with a 50–420 nm interpore distance formed by self-organization in anodic alumina. J Appl Phys. 1998;84:6023–6026. doi: 10.1063/1.368911.

CHAPTER 14

REGULATION OF ELECTRON TRANSFER PROCESSES AFFECTS PHOTOTROPHIC MAT STRUCTURE AND ACTIVITY

Phuc T. Ha, Ryan S. Renslow, Erhan Atci, Patrick N. Reardon, Stephen R. Lindemann, James K. Fredrickson, Douglas R. Call and Haluk Beyenal

Keywords: Electron transfer, Hot Lake, microbial mats, current, microelectrodes, mat structure, metabolite analysis, microbial community

INTRODUCTION

Phototrophic microbial mats are remarkable self-sustaining natural ecosystems, being composed of highly interactive species that completely cycle energy and elements within them (Guerrero et al., 2002; Des Marais, 2003; Bender and Phillips, 2004). These systems undergo daily cycling of redox potential caused by variations in light energy input and metabolic interactions among the microbial species (Frund and Cohen, 1992; Burow et al., 2012). Light also drives mat structure and activity by providing a diverse energy spectrum to sustain photosynthesis in photosynthetic microorganisms, which fix carbon and produce organic materials needed by other microorganisms in the community. Numerous studies have characterized the variations in physicochemical parameters, the elemental cycles, and the diversity and interactions of the microbial community members within the mat (Des Marais, 1995; Decker et al., 2005; van der Meer et al., 2005; Burow et al., 2012; Lindemann et al., 2013; Pages et al., 2014). Mat systems provide an excellent ecological model that can be used to investigate how microbial populations associate and interact as well as how electrons are used for energy transfer (Guerrero et al., 2002; Babauta et al., 2014). In this ecological model, electrons have to be transferred to convert and carry captured light energy to the deeper layers of the mat. The mat structure, inclusive of its physical structure and pore distribution, may play a critical role in these mass and energy transport processes (Moran et al., 2014). Therefore, it is critical to understand the effects of electron transfer processes on mat structure and activity. Energy transfer in a microbial mat starts with light energy being absorbed in the photic zone of the mat, which allows fixation of carbon dioxide by photoautotrophs. This energy transfer is mediated by diverse processes, such as the synthesis of organic compounds (e.g., carbohydrates) and their diffusion to deeper, and frequently hypoxic, strata of the mat where they may be consumed by fermenters and/or sulfate reducers (Paerl et al., 2000). Many of these reactions involve extracellular electron transfer processes, which can occur either directly or through electron carriers (mediators). We hypothesized that regulating the electron transfer processes in a mat system would, in turn, change its structure and activity. "Structure" refers to the physical structure or architecture of the mat, and "activity" refers to the metabolites and energy exchanged by the mat. According to our previous work and other work in the literature, electron transfer processes can be controlled by controlling the potential of an inert solid electrode and using it as an electron donor or acceptor in a biological system (Renslow et al., 2013). Use of polarized electrodes to study electron transfer by microorganisms has become an important tool for filling the knowledge gap about

Citation: Ha PT, Renslow RS, Atci E, Reardon PN, Lindemann SR, Fredrickson JK, Call DR and Beyenal H (2015) Regulation of electron transfer processes affects phototrophic mat structure and activity. Front. Microbiol. 6:909. doi: 10.3389/fmicb.2015.00909.

microbial electrochemical activities (Rabaey et al., 2007). This technique offers a unique strategy for controlling the activity of a microbial population by accepting or donating electrons from/to the local environment. The measured electron transfer rate can be used to interpret the response of microbial activities to the perturbation of electron transfer processes. Some microbial species respond by altering their metabolism, activating a defense mechanism, or reducing their growth rates. Other microbial species benefit because of their ability to transfer or accept electrons from the electrodes (Babauta et al., 2014). Thus, by increasing the metabolic activities of certain species present near the electrode surface, the electrode potential will not only select the species enriched on it but also regulate the structure, activity, and composition of the local microbial community. To the best of our knowledge, the responses of mat structure, activity, and electron transfer rates when solid electrodes are employed to accept or donate electrons have not yet been quantified.

The goal of this work is to characterize the response of a microbial mat to a solid electron acceptor and a donor. The phototrophic microbial mats used in this study were derived from Hot Lake, a hypersaline, epsomitic lake located near Oroville (Washington, USA). The seasonal characteristics of mat and lake conditions were described detail by Lindemann et al. (2013). The electrodes were placed at the bottom of the mat to create the mat electrochemical systems. The electrodes that provided electrons to the mat were polarized at -700 $mV_{Ag/AgCl}$ (hereafter called the "CAT mat"). The ones that worked as an electron sink were polarized at +300 $mV_{Ag/AgCl}$ (hereafter called the "AN mat"). The electron transfer rates from mat-to-electrode and from electrode-to-mat were continuously monitored. We quantified the diel cycling of AN and CAT mats. To understand the factors affecting the diel cycle, (1) we examined the influences of temperature and light condition, separately, to electron transfer rate; (2) we used microelectrodes to measure the depth profiles of hydrogen sulfide and oxygen concentrations; and (3) we introduced acetate and sulfide into AN and CAT mats and quantified their influence on electron transfer rates, separately. To understand how AN or CAT processes regulate mat structure and activity, we used nuclear magnetic resonance (NMR) imaging for morphology and NMR metabolite analysis to quantify the activity of the mats. Finally, we employed amplicon sequencing across the V4 region of the 16S gene to determine the effect of electrochemical regime upon mat community structure.

MATERIALS AND METHODS

Field Site and Sample Collection

Hot Lake is located near Oroville, North Washington, CO, USA. Photosynthetic microbial mats and Hot Lake water were sampled on June 6th and September 26th, 2014. Each mat sample was between 1.5 and 2 cm in depth and about 200 cm^2 in size. The samples were kept in petri dishes and placed in a cold ice box when carried to our laboratory at Washington State University, Pullman, WA, USA. Upon arrival, the mat was paced in glass jars and covered with Hot Lake water from the field site. Mats were kept under natural solar conditions and at room temperature for 2 days before being transferred to the mat electrochemical systems.

Construction of Mat Electrochemical Systems and Current Monitoring

We used an acrylic aquarium with a working volume of 1.5 L to construct the mat electrochemical systems. The reactors mimicked the lake condition without flow (Figure 1). For each reactor, approximately 3 cm of sediment was added to the bottom of the reactor. Then the hollow polycarbonate plate on which the graphite cloth electrode (2.5c m × 2.5 cm) was embedded was placed on top of the sediment. The mat was carefully placed on top of the electrode. The electrode placed under the mat was used as the working electrode. The counter electrode was constructed by attaching another piece of graphite cloth (2.5 cm × 2.5 cm) to a hollow plastic dish and placing it in the water phase of the reactor. Ti wire was woven into each electrode and secured with nylon bolts. The part of the Ti wire that was exposed to Hot Lake water was covered by marine sealant and inserted into a sealed silicon tube. During operation of the reactor, deionized (DI) water was added periodically to maintain the water level and salinity in the reactor. All the reactors were operated under natural solar conditions: they were incubated near a laboratory window. For some experiments, the reactors were moved to an incubator or to a water bath with controlled temperature, as described in the relevant sections.

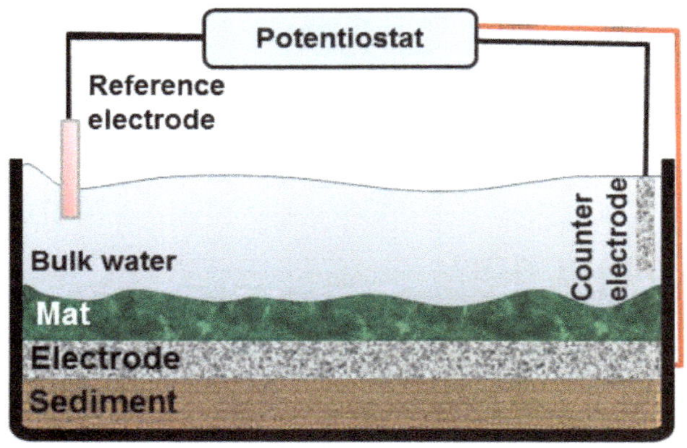

Figure 1. Schematic diagram of the mat electrochemical systems. The working electrode was placed right below the mat, and the counter electrode was located in the bulk solution. A potentiostat was used to control the working electrode potential vs. the Ag/AgCl reference electrode.

A 6-channel custom-built potentiostat was used to control the desired potential of the working electrodes under the mat (Renslow et al., 2011). The custom-made Ag/AgCl reference electrodes were used (~+210 mV vs. SHE). We operated reactors in which the graphite electrodes were polarized at potentials of -700 $mV_{Ag/AgCl}$ (CAT mat) and +300 $mV_{Ag/AgCl}$ (AN mat). These potentials are in the range that was previously used to enrich AN and CAT communities in bioelectrochemical systems (Wagner et al., 2010; Tremblay and Zhang, 2015). Moreover, due to the presence of sulfide in the mat we polarized cathode at lower potential than the redox potential of sulfide ($E^{o}_{H2S/So}$ = -490 $mV_{Ag/AgCl}$) to prevent the deposition of elemental sulfur on the electrode. We also operated a mat system under open circuit condition as the control mat [open circuit potential (OCP) mat]. Two technical replicates with two biological replicates were performed for all experiments.

Effects of Light Intensity, Temperature and Electron Donor/Acceptor Addition on the Mat

To examine the effects of light/dark conditions with AN and CAT currents, we operated our mat systems in an incubator. An incandescent light bulb was used as the light (light intensity of 6.4 ± 0.2 μmol/m²/s). The temperature was maintained by the incubator at the same value as under the light (22.5 ± 1°C). To examine the effect of temperature on current, the mat electrochemical systems were placed in a temperature-controlled water bath. The water bath was covered to prevent light penetration when needed. The effects of acetate and sulfide on the electron transfer process were examined by carefully injecting 2 mL of 1 M acetate solution or 0.3 mL of 0.1 M $Na_2S.9H_2O$ solution into the bottom of the mat. The mat systems were operated under constant light (6.4 ± 0.2 μmol/m²/s) and stable temperature (22.5 ± 1°C) to minimize the acetate and sulfide production by the mat itself.

Dissolved Oxygen and H_2S Microelectrodes and Measurements

The DO microelectrodes were constructed according to the procedure described by Lewandowski and Beyenal (2014). The final tip diameter of the DO microelectrodes was ~15 μm. A Keithley 6517 A electrometer/high-resistance meter was used as both the voltage source and an ammeter. The DO microelectrodes were polarized to -800 $mV_{Ag/AgCl}$ during the operation. Prior to use, the DO microelectrodes were calibrated using two-point calibration: in the air-saturated solution and in Na_2SO_3 solution (zero oxygen concentration). Microelectrode movements were administered using a Mercury Step motor controller PI M-230.10S Part No. M23010SX (Physik Instrumente, Auburn, MA, USA). Each microelectrode was positioned ~2000 μm above the mat surface and stepped down in 50-μm increments using a custom microprofiler. For the DO profiles, the measurements were taken in the air first (~1 mm) and in the bulk above the mat (~1 mm) until the microelectrode touched the mat surface; then ~3 mm of profile was measured in the mat. Before the DO profiles were measured, some of the bulk liquid above the mat was discharged.

The hydrogen sulfide microelectrodes were constructed and operated according to Jeroschewski et al. (1996). The hydrogen sulfide microelectrodes had tip diameters of ~20 μm. A Gamry Interface 1000TM potentiostat (Gamry® Instruments, Warminster, PA, USA) was used to polarize the hydrogen peroxide microelectrodes at +100 mV against a Pt counter electrode. The microelectrode was calibrated in standard solutions of various hydrogen sulfide concentrations prepared by dissolving Na_2S in anoxic phosphate buffer (16 mM, pH 7). The response of the microelectrodes was linear in the range of hydrogen sulfide concentrations from 0 to 570 μM. The hydrogen sulfide microelectrode was placed ~1 mm above the mat surface, and the microelectrode was stepped down in 1-mm increments through 13 mm of the mat thickness using the manual control. After the depth profiles were measured, the microelectrode was left at the 13-mm depth to monitor the hydrogen sulfide concentration over 24 h. Each microelectrode used in the measurements was also calibrated for verification after the measurements. A Zeiss Stemi 2000 stereomicroscope was used to determine the locations of the microelectrode tip and the mat surface during all microelectrode experiments.

Morphology Analysis using NMR Micro-Imaging

Pulsed-field gradient NMR (PFG-NMR) was used to observe mat morphology and determine intra-mat porosity and diffusion coefficients. The techniques used were similar to those of Renslow et al. (2013). Briefly, the NMR imaging experiments were conducted at 500.40 MHz for proton (1H) detection using a 89-mm-wide bore 11.7-T magnet with a Bruker Avance III digital NMR spectrometer and ParaVision 5.1 imaging software (Bruker Instruments, Billerica, MA, USA). Each mat sample was placed in a 15-mm NMR tube on a support bed of 2% agar gel. Experiments performed included 2D magnetic resonance imaging (mic_flash), diffusion tensor imaging for determining diffusion coefficients (DtiStandard; Renslow et al., 2013), and chemical shift selective imaging for generating porosity measurements.

Metabolite Analysis using NMR Spectroscopy

Sections of microbial mat were dissected out such that the full thickness of the mat was retained. Mat samples were frozen at -80°C immediately following dissection and lyophilized before their dry weight was measured. Dried samples were coarsely ground in glass vials using a small metal spatula. Metabolite extraction was performed using a 1:1:1 volume of water:methanol:chloroform. Ice cold solvents were sequentially added to the samples in the order listed. Samples were allowed to extract and phase separate at -20°C for ~24 h. Samples were spun at 4000 × g for 10 min, and the hydrophilic supernatant was removed. Samples were spun again at 20800 × g for 10 min, and 1 mL of supernatant was retained. The retained samples were dried under vacuum in a speedvac overnight. Dried samples were dissolved in 50 mM sodium phosphate at pH 7, with 10% D_2O, 448 uM DSS, and 10 mM imidazole.

Nuclear magnetic resonance spectroscopy was carried out using an Agilent VNMRS 600 MHz (proton frequency) spectrometer equipped with a cryogenically cooled HCN triple resonance probe. One-dimensional NOESY experiments were performed using the following parameters, 28896 complex points (acquisition time of 4 s), a recycle delay of 1 s, a sweep width of 12 ppm and 512 scans per experiment for a total acquisition time of ~48 min per sample. The temperature was held constant at 25°C. The NMR spectra were Fourier transformed, phased, baseline corrected and apodized (0.5-Hz line broadening) using nmrPipe. Spectra were referenced to DSS at 0 ppm. Data were imported into Chenomx NMR Suite (Chenomx, Inc., Edmonton, AB, Canada) for subsequent metabolic profiling. Profiling was carried out independently by two operators to reduce operator bias. The DSS internal standard was used to determine metabolite concentrations. Imidazole peaks were used to correct for pH differences.

Community Structure Analysis using 16S rRNA Sequencing

Mat community structure changes were examined using 16S rRNA amplicon sequencing. Two samples of mat (~4 cm²) were collected for each condition: (OCP, AN and CAT). These samples were further subdivided into nine subsamples. Genomic DNA (gDNA) was extracted as described inLindemann et al.

(2013) with the following modifications: prior to extraction, samples were washed three times with washing buffer containing 0.5 M EDTA and 0.55 M NaCl at pH 8.0 and suspended in lysis buffer (50 mM Tris/25 mM EDTA pH 8.0). Samples were then transferred to Lysing Matrix E tube (MP Biomedicals, Santa Ana, CA, USA) and beat for 2 min. After centrifugation (16,000 rpm for 90 s) the aqueous portions were incubated at 85° C for 5 min to inactivate native nucleases and slowly cooled to 37°C. Chemical lysis then was initiated by adding 70 μl of 10% SDS and proteinase K to the final concentration of 0.2 mg/ml prior to incubation at 56°C for 1 h. Subseqently, 100 μl of NaCl 5 M and 100 μl of CTAB/NaCl solution were added to samples and incubated at 65°C for 10 min. Post-lysis, DNA was extracted with phenol-chloroform-isoamyl alcohol (25:24:1) and then chloroform-isoamyl alcohol (24:1) before being treated with RNAase (10 μg) at 37°C for 30 min. DNA was precipitated with addition of 1:10 vol of sodium acetate 3 M (pH 5.5) and ice-cold 100% ethanol, washed in 70% ethanol, dried and resuspended in TE buffer (10 mM Tris-HCl at pH 8.0, 1 mM EDTA). Similarly, gDNA was extracted from fresh mats harvested at Hot Lake at the same time as those used in incubation studies but immediately frozen in 2.3 M sucrose and held at -80°C prior to extraction. gDNA was amplified and sequenced as described previously (Caporaso et al., 2012). Briefly, the V4 region of the 16S rRNA gene was amplified using primers 515F and 806R with 0–3 random bases and the Illumina sequencing primer binding site (Caporaso et al., 2011) and amplicons were sequenced on an Illumina MiSeq using MiSeq Reagent Kit MS-102-2001 {[v2 chemistry over 500 total cycles (2 × 250); Illumina, San Diego, CA, USA]}. Sequences were demultiplexed and processed using mothur v. 1.34.1 (Schloss et al., 2009) according to the MiSeq SOP (http://mothur.org/wiki/MiSeq_SOP, accessed 7/6/15) except for the following modifications: (1) sequences were aligned against a SILVA reference alignment to which full-length 16S clones from Hot Lake had been added (Lindemann et al., 2013), and (2) only sequences that could not be placed within any of the domains of life were removed using remove.lineage(taxon = unknown). Sequences were clustered using a 0.03 average neighbor cutoff using cluster.split at the Order level (taxlevel = 4) and subsampled to 6283 sequences per sample; all samples containing fewer sequences were discarded from the analysis. Bray–Curtis dissimilarity (Bray and Curtis, 1957) was calculated for all pairwise combinations of samples using dist.shared (calc = braycurtis) and a neighbor-joining dendrogram was constructed using the tree.shared() command.

RESULTS AND DISCUSSION

Electron Transfer Rates in Anodic and Cathodic Mat Systems

Current generated from AN and CAT mats over the first 20 days under ambient light conditions is shown in Figure 2. Increasing AN current over time (Figure 2A) indicated that electrons were transferred from the mat to the solid electrode during the first 15 days. The current generation started immediately after the electrode was polarized. It developed in the range from 10 μA to about 70 μA with cyclic fluctuation. We hypothesized that these electrons may be transferred to the anode either (1) through direct sulfide oxidation (Rabaey et al., 2006; Babauta et al., 2014) and/or (2) through microbial exogenous electron transfer (Lovley, 2008; Logan, 2009). In the first case, the abiotic oxidation of sulfide generates electrons and elemental sulfur is deposited on the electrode (Rabaey et al., 2006). Sulfide is commonly produced in phototrophic mats by sulfate-reducing bacteria. Because the sulfate concentration in the Hot Lake environment is extremely high, ranging from ~200 mM to 1.8 M (Lindemann et al., 2013), the rate of sulfide production depends on the availability of substrates for sulfate reducers such as H_2 and acetate. In the case of exogenous electron transfer, electrogenic bacteria act as a catalyst which oxidizes substrates and transfers electrons to the electrode. The rate of electron transfer is governed by the enrichment of the electrogens and the concentration of the available substrates in the mat. Thus, in both cases the current generated by an AN microbial mat is regulated by substrate availability at the bottom of the mat.

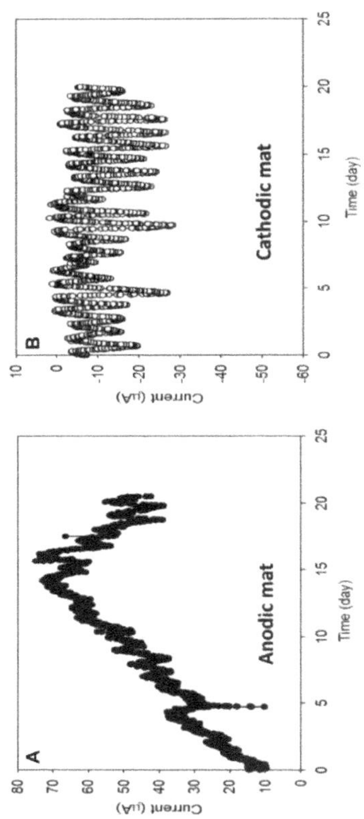

Figure 2. Electron transfer rates (A) from the microbial mat to the electrode in the anodic (AN) mat and (B) from the electrode to the microbial mat in the cathodic (CAT) mat. Both systems showed the cyclic fluctuation of current which is also referred to as the diel cycle.

A recent study reported that green sulfur bacteria were a dominant population on a polarized AN microelectrode (Babauta et al., 2013). Sulfur cycling in combination with the activities of green sulfur bacteria was hypothesized to be the cause for the cycling of current. In another study,Badalamenti et al. (2013) used a large-size graphite electrode to enrich AN biofilm from a microbial mat inoculum (Badalamenti et al., 2013). The authors found that the phototrophically enriched anode biofilms were dominated by green sulfur bacteria. These bacteria were reported to have the ability to respire on electrodes. Thus, it is possible that the phototrophic community is also involved in the current production from our AN mat.

We also observed that the negative current generation from the CAT mat started immediately after the electrode was polarized (Figure 2B). Like the AN condition, mat under CAT conditions showed cyclic current fluctuations. Among the cycles, the lowest current ranged from approximately -0 μA to approximately -5 μA while the highest ranged from about -10 μA to about -30 μA. The generation of negative current indicates the consumption of electrons from the solid electrode by the mat. These electrons could be consumed by (1) an abiotic oxidant such as oxygen or (2) biocathodic microbes (Lovley and Nevin, 2011; Rosenbaum et al., 2011). In the first case, the current would be controlled by the DO concentration near the electrode surface. Oxygen is generated in the mat during the day by oxygenic phototrophs in the illuminated zone near the surface. Thus, the abiotic CAT current will be ultimately governed by photosynthetic activity and the diffusion of the oxygen through the mat. In the latter case, the current would be governed by the enrichment of microbes that are capable of using the electrode as an electron source for their metabolism.

In the literature, CAT current has been obtained from mixed CAT cultures enriched from wastewater (Marshall et al., 2012), activated sludge (Su et al., 2013), freshwater (Babauta et al., 2013), and pure bacterial cultures (Gregory et al., 2004; Nevin et al., 2011; Ueki et al., 2014). Our study demonstrates that CAT current can also be generated by the phototrophic mat community. Microbes that can potentially consume electrons from solid electrodes include methanogens (Villano et al., 2010), acetogens (Nevin et al., 2011), phototrophs (Bose et al., 2014) and oxygen reducers (Clauwaert et al., 2007; Debuy et al., 2015).

Diel Cycle of Cathodic and Anodic Current

To understand the reasons for the cyclic fluctuation seen during the development of AN and CAT current, we superimposed individual cycles of each current as a function of time. Figure 3 shows that both AN and CAT current followed an oscillating pattern in tandem with the diel cycle. It is known that phototrophic mats generally undergo a diel cycling of redox potential and speciation of redox-active elements caused by diel variations in light intensity and temperature and hence metabolic interactions among the microbial species (Finke and Jorgensen, 2008; Burow et al., 2012). Thus, the diel cycling of our current is likely to have been caused by these diel variations.

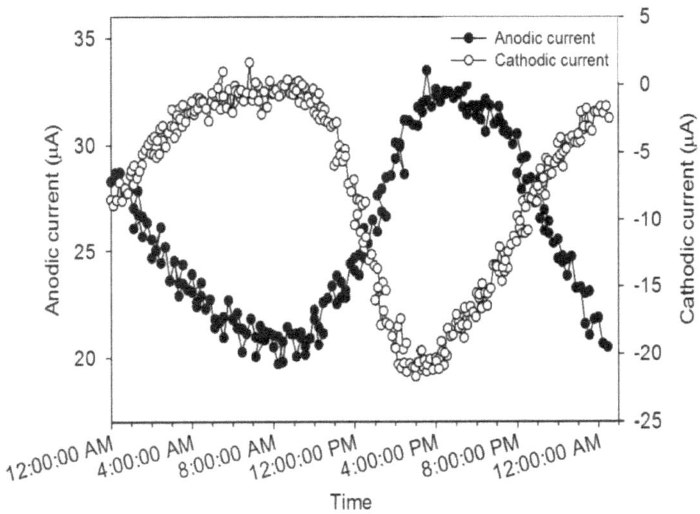

FIGURE 3. Diel cycle of AN and CAT current generated from microbial mat systems.

Interestingly, the lowest and highest points in the diel cycles of AN and CAT current occurred at the same time (Figure 3). The minimum values of AN and CAT current both occurred at about 8:00 a.m. (20 μA for AN current and -1 μA for CAT current). Similarly, the AN mat generated its highest current (33 μA) simultaneously with the CAT mat generating its highest current (-22 μA), at around 4:00 p.m. Thus, the CAT mat achieved its highest reducing potential at the same time that the AN mat achieved its highest oxidizing potential. The similarity in diel pattern suggested that the CAT and AN mat currents were governed by the same factors. However, the contrary redox characteristics of the two mat systems indicated that (1) there was an element or a microbial group in the mat that could function as both electron donor and electron acceptor to/from a solid electrode or (2) the characteristic of the mat was changed by the polarized potential of the solid electrode.

Effect of Temperature on Electron Transfer Process

Under natural conditions, temperature and light intensity co-vary. Therefore, to isolate the effect of temperature change on the current, we operated the system in the dark. The AN and CAT currents were compared with the maximum current obtained at 4:00 p.m. from the closest diel cycle. At that time, the temperature was about 21.3 ± 0.2°C.

Figure 4 shows that in the dark the AN current increased as temperature increased. This agrees with the observations made when the system was operated under diel temperature changes. The current was highest at 4:00 p.m., when the temperature reached a maximum value of about 23°C. The effect of temperature on AN current may be due to an increase of chemical and biological activities that contribute electrons to the electrode. Because of the temperature dependence, we believe that the decrease of AN current after 15 days of operation under ambient temperature (Figure 2A) was caused by temperature differences owing to seasonal change.

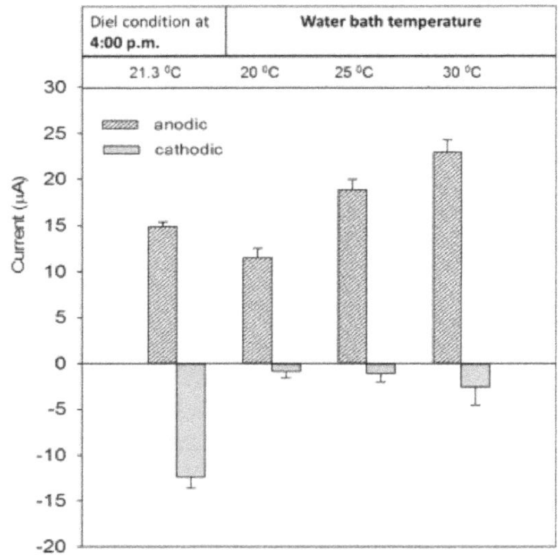

Figure 4. Performance of AN and CAT currents at various temperatures. The currents were compared with the maximum current obtained at 4:00 p.m. under the diel condition.

In the dark with a constant temperature, the CAT current did not oscillate as was observed under diel cycling. Despite an increasing temperature, the current was significantly lower than the maximum obtained during diel cycling (at 4:00 p.m.; -15 μA vs. > -3.5 μA). Although the CAT current also increased with increasing temperature the change was not significant compared with the difference under a diel temperature change. This suggests that the temperature change was not the main factor governing the diel cycling of CAT current.

Effect of Light Energy on Electron Transfer Processes

The effect of the light on the AN and CAT currents was further investigated under constant temperature. Figure 5 shows the currents obtained from the AN and CAT mats under light and dark conditions. Both the AN and the CAT current changed in response to light. The AN current increased when the system was operated in the dark (10–23 μA). In contrast, the CAT current was higher in the light than in the dark (-23 μA vs. -2.5 μA). Thus, the two exhibit an opposite responses to light.

Because of the association of light intensity and temperature, our recorded diel temperature showed that the highest temperature was reached when the light intensity was highest (at 4:00 p.m.). Thus, it is clear from the diel cycle that the highest AN current occurred when the highest light intensity was presented to the mat (Figure 3). This is the opposite of the observation in Figure5, in which the AN current was higher in the dark than in the light. The increase of AN current during the dark might be because: when illuminated, phototrophic microbes will convert light energy to chemical energy typically stored in organic polymers such as glycogen and in the dark depolymerized to form monomers that fermented to easily respired compounds such as H_2 and acetate. These compounds will diffuse through the mat (Des Marais, 2003; Burow et al., 2012). Fermentation products could increase sulfide production from sulfate reduction which, in consequence, increase the current caused by sulfide oxidation on electrode. Fermentation products could also be used as substrates for electrogenic bacteria which use the electrode as an electron acceptor. Thus, the inconsistency between the behavior of AN current in light/dark tests and that observed during the diel cycle suggests that a diffusion process may control the transport of fermentation products from the phototrophic layer to the bottom of the AN mat. Therefore, the highest AN current was delayed from the darkest time in the diel cycle.

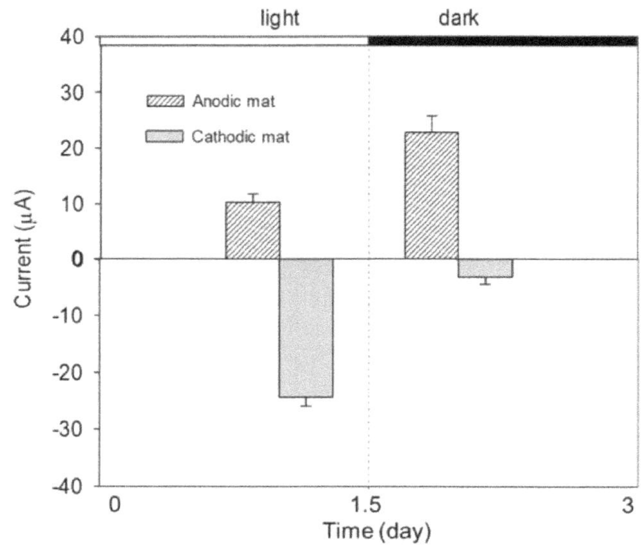

Figure 5. Performance of AN and CAT currents with a change of light condition. The AN and CAT mats showed opposite responses to the change of light condition.

The significant increase of CAT current in the light reveals that electron consumption in a CAT mat is light-dependent (Figure 5). This is consistent with the observation from diel cycle results in which the CAT current was highest when the highest light intensity was present. The light-dependent behavior of CAT current also explains why the CAT current in the dark was lower than that obtained at 4:00 p.m. in the diel condition in spite of increased temperature at this time (Figure 4). One hypothesis to explain this is that the CAT current was caused by the abiotic reduction of oxygen by electrode which diffused from the aerobic phototrophic layer. If this is the case, oxygen must have diffused through the mat fast enough that the CAT current cycle synchronized with the light intensity cycle.

Effect of Sulfide and Acetate Additions on the Electron Transfer Process

As previously mentioned, AN current is regulated by the availability of an electron source in the surrounding environment. In this experiment, we examined the response of current to two different electron sources: sulfide and acetate. Acetate is a common end product of fermentation while sulfide is a common product of anaerobic respiration in phototrophic mats (Grotzschel et al., 2002).

Figure 6A shows that the AN current increased immediately after an injection of sodium sulfide (5 mg). However, the current decreased quickly to the background value, indicating that sulfide was rapidly oxidized in this system. Since sulfide is normally produced from sulfate reduction in a microbial mat, these results demonstrate that sulfide could be abiotically oxidized by the AN electrode.

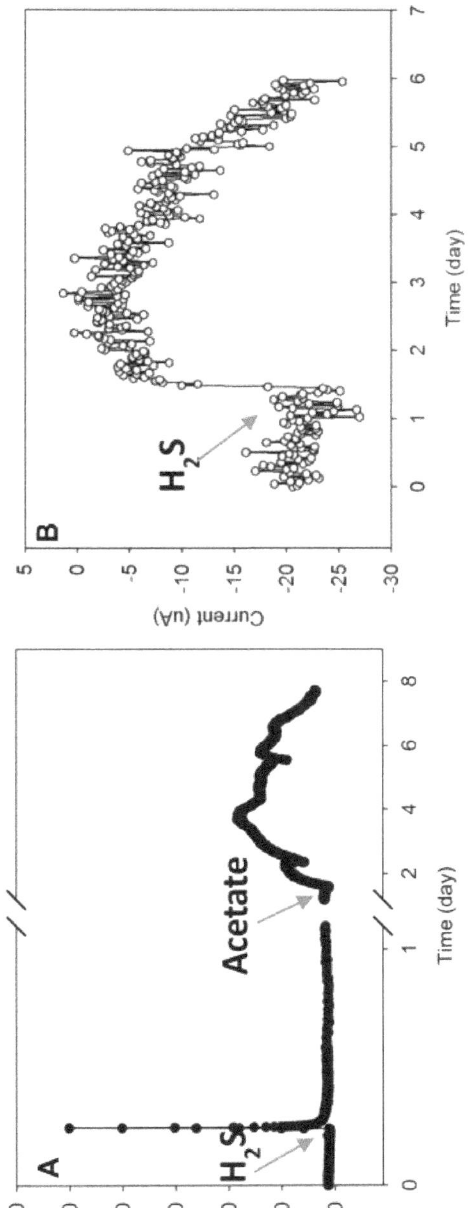

FIGURE 6. Effects of sulfide and acetate addition to the current from (A) the AN mat and (B) the CAT mat.

The AN current slowly increased after acetate was added (Figure 6A). After about 1.5 days, the current reached a maximum value of about 30 μA, then slowly decreased during the next 4 days. The gradual increasing and decreasing of AN current after acetate addition indicates that biological processes were likely involved in current generation from acetate. Acetate addition probably increased the rate of sulfate reduction at the bottom of the mat; the produced sulfide could then be oxidized on the electrode surface, driving increases in current. In addition, acetate could be used as a substrate for electrode-respiring microbes which can transfer electrons to an electrode. Thus the increase in AN current due to acetate addition may occur by both biotic and abiotic mechanisms.

The addition of the same amount of acetate (0.2 mmol) to the CAT mat did not alter current (data not shown). However, the CAT current decreased when sulfide was added (Figure 6B). If the CAT current was due to direct oxygen reduction on the electrode, it is possible that the introduction of sulfide reduced the oxygen available at the electrode, decreasing the current. Sulfide at high concentration is known to inhibitory or toxic to microbial metabolism including anaerobic and aerobic bacteria (Karhadkar et al., 1987; Chen et al., 2008). If these microbes are involved in the CAT current generation, it is possible that sulfide inhibited the consumption of electrons.

Oxygen and Sulfide Profiles in the Mats over the Diel Cycle

Figures 7A,B shows the DO and sulfide depth profiles measured from the tops of the mats at two time points (at 8:00 a.m. and 5:00 p.m.) during a diel cycle. We tried to measure the profiles at time points close to when the current peaks occurred, as shown in Figure 3. Similar DO profiles were obtained from AN mat (Figure 7A) and CAT mat (data not shown). The DO profile suggests that the oxygen concentration inside the mat decreased with the depth. At 3.5 mm bellow the mat surface, the DO concentration was close to zero, both for the 8:00 a.m. and 5:00 p.m. measurements. Since the thickness of the mat used in this study was 1.5–2 cm, this suggests that the environment surrounding the electrodes, which were placed at the bottom of the mat, was anoxic. To further confirm that the CAT electrode was in an anoxic environment, we switched the electrode potential of the cathode from -700 mV vs. Ag/AgCl to +300 mV vs. Ag/AgCl (i.e., to an AN condition). This switch occurred at about 4:00 p.m, when CAT current was highest. Positive current (AN current) was immediately generated; suggesting that an anoxic condition at the bottom of the mat, therefore allowed electrons transfer to the electrode.

Figure 7B shows the change in total sulfide concentration in the mat with depth for different measurement times. The sulfide concentration increased with the depth for both 8:00 a.m. and 5:00 p.m. measurements. However, at 8:00 a.m. the increase was only 3.6-fold, from 25 μM at the surface of the mat to about 90 μM at a depth of 1.3 cm. In contrast to the profile observed at 8:00 a.m., sulfide concentration increased significantly with the depth at 5:00 p.m. The total sulfide concentration at 1.3 cm beneath the mat surface was about 18 times higher than that observed at the surface, and about 5 times higher than that obtained at the same depth at 8:00 a.m. (460 μM vs 90 μM). This sulfide profile is similar to a previous observation in a different mat (van der Meer et al., 2005; Pages et al., 2014)

Since the sulfide concentration continued to increase with depth, we left the sulfide microelectrode near the colonized electrode (at 13 mm below the mat surface) to measure sulfide concentration over the diel cycle. This suggests that the electron transfer processes between the electrode and the mat have a strong relationship with the sulfur cycling in the mat.

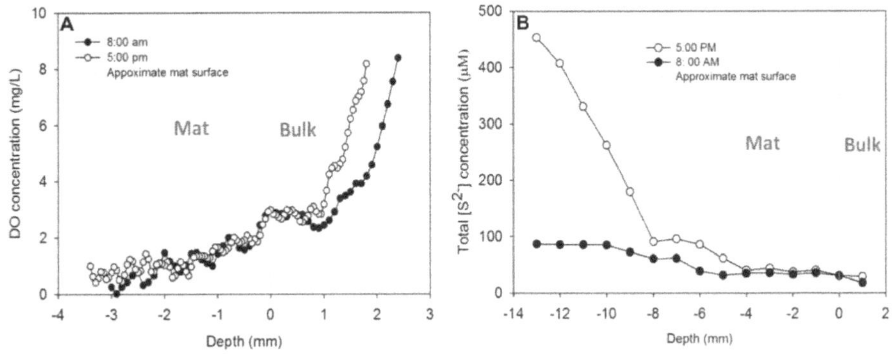

Figure 7. **(A)** Dissolved oxygen (DO) and **(B)** sulfide depth profiles measured from the tops of the mats at two time points (at 8:00 a.m. and 5:00 p.m.).

Microbial Mat Morphology and Structural Change under Different Polarized Conditions

Nuclear magnetic resonance imaging analysis showed that the internal architecture of the CAT mat system was very different from the AN mat (Figure 8). The darker regions in Figure 8 qualitatively represent denser biomass and regions with less porosity. The CAT mat had dispersed biomass, and therefore was thicker and more porous than the more compact AN mat. These results corroborate the visible and tactile form of the mat samples during the harvesting period.

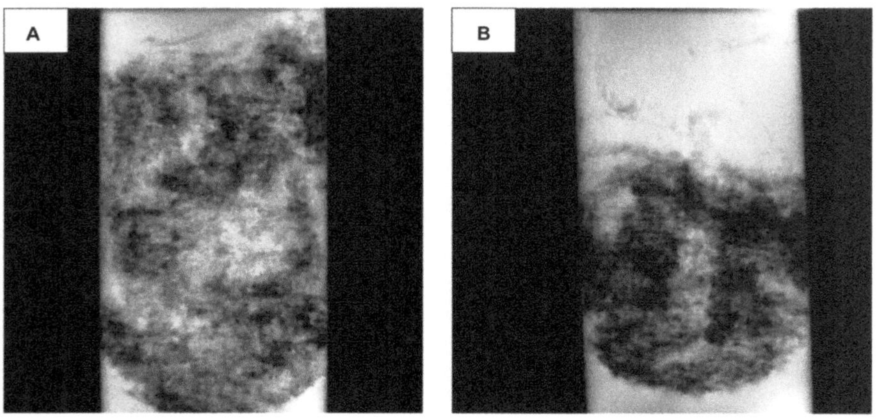

Figure 8. Representative nuclear magnetic resonance images (97.7 **μm** × 78.1 **μm** resolution, 5-mm slice) of (A) the CAT mat and (B) the AN mat.

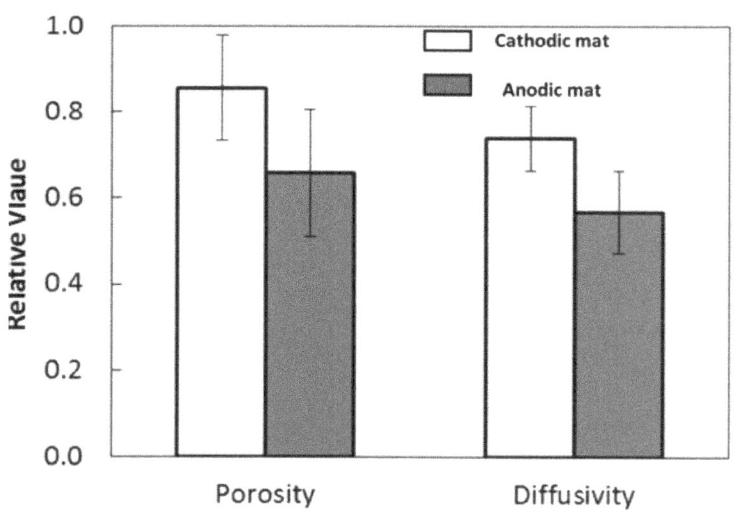

Figure 9. Porosity and diffusivity of the an and cat mats.

Nuclear magnetic resonance imaging showed that the quantitative porosity and diffusion coefficients were higher in the CAT mat than in the AN mat (*P*-value < 0.0001; Figure 9). These structural differences may have resulted from changes in microbial community composition selected under the different electrochemical conditions. Thus, this result indicates that the electrode potential regulated the development of the mat, leading to differences in morphology and structure.

The higher porosity and lower density of biomass of the CAT mat promote relatively rapid oxygen diffusion through the CAT mat, resulting in higher current. However, Figure 2B shows that the CAT mat was able to generate current from the initiation of the experiment, when the mat had a structure similar to that of the original sample as well as to that of the mat used in the AN system. This suggests that the electron-consuming processes in the CAT system were active in the native mat and not solely caused by the increasing porosity and diffusion coefficient of the mat.

We further used NMR spectroscopy for quantitative metabolite analysis of mat samples from our AN and CAT mats. The results were compared with data obtained from the control mat, which developed under OCP condition (OCP mat). The concentrations of major metabolites detected in the microbial mats under AN and CAT conditions. Generally, most of the significant metabolites (with concentration >10 ug/g sample) were found at higher concentrations in the CAT mat than in the AN mat (Table 1). The metabolite analysis results also showed that the AN mat and OCP mat were relatively similar while the CAT mat were clearly distinguished from the other mat samples.

Table 1. percent increase and average approximate concentration increase of primary metabolites with significant and consistently higher concentrations in the cathodic (cat) mat than in the anodic (an) mat.

	June 6th collected mat	September 26th collected mat	Ave. Approx. Increase in CAT mat
Trehalose	52.7%	162.3%	+439 µg/g
Betaine	58.8%	80.5%	+26 µg/g
Glutamate	35.1%	64.7%	+17 µg/g
Oxypurinol	143.4%	113.4%	+18 µg/g

Most of the significant metabolites—trehalose, betaine, and glutamate—are known osmolytes. Trehalose is known to be a stress molecule, protecting cells from injuries imposed by high osmolarity, heat, oxidation, and salinity (Purvis et al., 2005; Reina-Bueno et al., 2012). Similar functions have been found for betaine (McNeil et al., 1999; Kapfhammer et al., 2005) and glutamate (Feehily and Karatzas, 2013). Glutamate is also involved in the bacterial response to acid stress. Thus, the significantly higher concentrations of these compounds in the CAT mat may indicate that the CAT potential (-700 mV vs. Ag/AgCl) caused stress to the mat which led to a high response of the microbial mat community. It is possible that the addition of electrons from electrode to the mat alleviated the energy limitations allowing the mat to maintain large carbon pools of osmolytes. It is possible that some microorganisms in the mat such as acetogens and anaerobic phototrophs used electrons from the electrode to drive carbon fixations which also increased the total carbon available to the mat.

Community Structure Changes due to Electrochemical Potential

Analysis of the community structure of mats revealed that variation in the electrochemical conditions under which mats were generated significantly impacted the relative abundances of mat community members (Figure 10). Average within-treatment Bray–Curtis dissimilarity was 0.367 (with a pooled *SD* of 0.106) compared with 0.501 (*SD* = 0.130) for all pairwise combinations. All incubated samples were significantly different from the original mat, indicating a strong effect of electrode-based cultivation on community composition. This result was expected, considering the significant differences in environmental conditions between the field and laboratory cultivation conditions. Mat incubated under OCP formed a broad clade with AN mat; suggesting the additional effect of the anode exerted only modest impacts on microbial community structure beyond the general effects of cultivation.

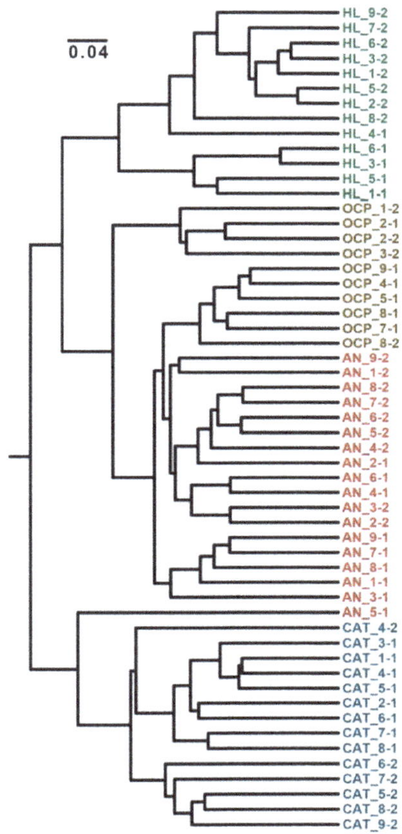

Figure 10. Dendrogram of Bray–Curtis dissimilarity between mats cultivated under divergent electrochemical conditions. Sample names denote the sample type (HL, initial mat from Hot Lake; OCP, control mat without polarization (called open circuit potential mat); AN, anodic mat; and CAT, cathodic mat), the subsample (1–9) and the sample (of two) from that condition. Mat frozen after initial collection from Hot Lake (green) forms a distinct subclade within a larger clade that encompasses open current potential (gold) and AN (red) mats. CAT mat (blue) forms a distinct clade, exhibiting a distinct community structure compared with the other conditions.

These data agree with the relatively small differences in mat metabolites observed between mat cultivated without polarization (OCP condition) and AN condition. Furthermore, this subclade was more similar to the original mat than were the CAT mats. In contrast, apart from one outlying sample of the AN mats, CAT mats formed a clade distinct from the initial, OCP, and AN mats. These community structure data are consistent with our metabolite analysis results that suggested greater changes in CAT mat metabolism compared with OCP and AN mat. Although it is not possible to determine whether changes in metabolism or community structure lead the CAT mat phenotype, our data strongly suggest that donation of electrons via a solid electrode is capable of strongly altering phototrophic mat morphology, community composition, and metabolic function.

CONCLUSION

We quantified the diel variations of electron transfer rates to and from the solid electrode in the mat electrochemical systems. We observed the diel cycling of current in which the lowest and highest magnitudes of AN current occurred simultaneously with those of CAT current. We also characterized the influence of physicochemical conditions (temperature, light, sulfide, and DO gradients) within the phototrophic mat upon electron transfer processes. AN current was higher in the dark and it increased with temperature. In contrast, CAT current increased with the illumination but did not change significantly with increased temperature. Comparing the behavior of AN current under constant light and dark conditions vs. diel-cycling mats revealed that AN current reached maximum under photic conditions under diel cycling conditions but was greater under constant dark conditions. This suggested diffusion control of the fermentable substrates from the photic layer to the bottom of the AN mat. We propose that the generated diel cycle data can be used to develop model of energy flow in the mat to predict diffusion controls. DO and sulfide gradients over diel cycle suggested that the current cycles of CAT and AN mats related to sulfur cycling in the mat. The congruence of community structure data with metabolite profiles suggested that variation in the electrochemical conditions under which mats were generated significantly impacted the relative abundances of mat members and mat metabolism. It remains unclear whether the electron transfer to and from the electrode to the mat was mediated by microbial metabolism of selectively enriched community on the electrode or was caused by abiotic reactions on electrode. These findings suggest that it is possible to electrochemically regulate the morphology, community composition, and activities of phototrophic mats and may represent a new paradigm for designing and manipulating microbial communities (Fredrickson, 2015).

Conflict of Interest Statement

The authors declare that the research was conducted in the absence of any commercial or financial relationships that could be construed as a potential conflict of interest.

ACKNOWLEDGMENTS

This research was supported by the Genomic Science Program (GSP) of the Office of Biological and Environmental Research (OBER), U.S. Department of Energy (DOE), and is a contribution of the Pacific Northwest National Laboratory (PNNL) Foundational Scientific Focus Area. A portion of the research was performed using EMSL, a DOE Office of Science User Facility sponsored by the Office of Biological and Environmental Research and located at Pacific Northwest National Laboratory. The authors are grateful for the assistance of Mr. Beau Morton in purifying extracted genomic DNA and Mrs. Sarah Fansler for 16S rRNA V4 amplification and MiSeq operation. The authors would further like to acknowledge the Wenatchee Field Office of the U.S. Bureau of Land Management for their assistance in authorizing this research and providing access to the Hot Lake Research Natural Area. R. Renslow was supported by a Linus Pauling Distinguished Postdoctoral Fellowship at PNNL.

REFERENCES

1. Babauta, J. T., Atci, E., Ha, P. T., Lindemann, S. R., Ewing, T., Call, D. R., et al. (2014). Localized electron transfer rates and microelectrode-based enrichment of microbial communities within a phototrophic microbial mat. *Front. Microbiol.* 5:11. doi: 10.3389/fmicb.2014.00011
2. Babauta, J. T., Nguyen, H. D., Istanbullu, O., and Beyenal, H. (2013). Microscale gradients of oxygen, hydrogen peroxide, and ph in freshwater cathodic biofilms. *Chemsuschem* 6, 1252–1261. doi: 10.1002/cssc.201300019
3. Badalamenti, J. P., Torres, C. I., and Krajmalnik-Brown, R. (2013). Light-responsive current generation by phototrophically enriched anode biofilms dominated by green sulfur bacteria. *Biotechnol. Bioeng.* 110, 1020–1027. doi: 10.1002/bit.24779
4. Bender, J., and Phillips, P. (2004). Microbial mats for multiple applications in aquaculture and bioremediation. *Bioresour. Technol.* 94, 229–238. doi: 10.1016/j.biortech.2003.12.016
5. Bose, A., Gardel, E. J., Vidoudez, C., Parra, E. A., and Girguis, P. R. (2014). Electron uptake by iron-oxidizing phototrophic bacteria. *Nat. Commun.* 5:3391. doi: 10.1038/ncomms4391
6. Bray, J. R., and Curtis, J. T. (1957). An ordination of the upland forest communities of southern wisconsin. *Ecol. Monogr.* 27, 326–349. doi: 10.2307/1942268
7. Burow, L. C., Woebken, D., Bebout, B. M., McMurdie, P. J., Singer, S. W., Pett-Ridge, J., et al. (2012). Hydrogen production in photosynthetic microbial mats in the Elkhorn Slough estuary. *Monterey Bay. ISME J.* 6, 863–874. doi: 10.1038/ismej.2011.142
8. Caporaso, J. G., Lauber, C. L., Walters, W. A., Berg-Lyons, D., Huntley, J., Fierer, N., et al. (2012). Ultra-high-throughput microbial community analysis on the Illumina HiSeq and MiSeq platforms. *ISME J.* 6, 1621–1624. doi: 10.1038/ismej.2012.8
9. Caporaso, J. G., Lauber, C. L., Walters, W. A., Berg-Lyons, D., Lozupone, C. A., Turnbaugh, P. J., et al. (2011). Global patterns of 16S rRNA diversity at a depth of millions of sequences per sample. *Proc. Natl. Acad. Sci. U.S.A.* 108, 4516–4522. doi: 10.1073/pnas.1000080107
10. Chen, Y., Cheng, J. J., and Creamer, K. S. (2008). Inhibition of anaerobic digestion process: a review. *Bioresour. Technol.* 99, 4044–4064. doi:

10.1016/j.biortech.2007.01.057

11. Clauwaert, P., Van der Ha, D., Boon, N., Verbeken, K., Verhaege, M., Rabaey, K., et al. (2007). Open air biocathode enables effective electricity generation with microbial fuel cells. *Environ. Sci. Technol.* 41, 7564–7569. doi: 10.1021/es0709831

12. Debuy, S., Pecastaings, S., Bergel, A., and Erable, B. (2015). Oxygen-reducing biocathodes designed with pure cultures of microbial strains isolated from seawater biofilms. *Int. Biodeterior. Biodegradation* 103, 16–22. doi: 10.1016/j.ibiod.2015.03.028

13. Decker, K. L. M., Potterb, C. S., Beboutc, B. M., Des Maraisc, D. J., Carpenterd, S., Discipulo, M., et al. (2005). Mathematical simulation of the diel O, S, and C biogeochemistry of a hypersaline microbial mat. *Fems Microbiol. Ecol.* 52, 377–395. doi: 10.1016/j.femsec.2004.12.005

14. Des Marais, D. J. (1995). The biogeochemistry of hypersaline microbial mats. *Adv. Microb. Ecol.* 14, 251–274. doi: 10.1007/978-1-4684-7724-5_6

15. Des Marais, D. J. (2003). Biogeochemistry of hypersaline microbial mats illustrates the dynamics of modern microbial ecosystems and the early evolution of the biosphere. *Biol. Bull-Us* 204, 160–167. doi: 10.2307/1543552

16. Feehily, C., and Karatzas, K. A. G. (2013). Role of glutamate metabolism in bacterial responses towards acid and other stresses.*J. Appl. Microbiol.* 114, 11–24. doi: 10.1111/j.1365-2672.2012.05434.x

17. Finke, N., and Jorgensen, B. B. (2008). Response of fermentation and sulfate reduction to experimental temperature changes in temperate and Arctic marine sediments. *ISME J.* 2, 815–829. doi: 10.1038/ISMEJ.2008.20

18. Fredrickson, J. K. (2015). ECOLOGY. Ecological communities by design. *Science* 348, 1425–1427. doi: 10.1126/science.aab0946

19. Frund, C., and Cohen, Y. (1992). Diurnal cycles of sulfate reduction under oxic conditions in cyanobacterial mats. *Appl. Environ. Microb.* 58, 70–77.

20. Gregory, K. B., Bond, D. R., and Lovley, D. R. (2004). Graphite electrodes as electron donors for anaerobic respiration.*Environ. Microbiol.* 6, 596–604. doi: 10.1111/j.1462-2920.2004.00593.x

21. Grotzschel, S., Abed, R. M. M., and de Beer, D. (2002). Metabolic shifts in hypersaline microbial mats upon addition of organic substrates. *Environ. Microbiol.* 4, 683–695. doi: 10.1046/j.1462-2920.2002.00356.x

22. Guerrero, R., Piqueras, M., and Berlanga, M. (2002). Microbial mats and the search for minimal ecosystems. *Int. Microbiol.* 5, 177–188. doi: 10.1007/s10123-002-0094-8

23. Jeroschewski, P., Steuckart, C., and Kuhl, M. (1996). An amperometric microsensor for the determination of H2S in aquatic environments. *Anal. Chem.* 68, 4351–4357. doi: 10.1021/ac960091b

24. Kapfhammer, D., Karatan, E., Pflughoeft, K. J., and Watnick, P. I. (2005). Role for glycine betaine transport in *Vibrio cholerae*osmoadaptation and biofilm formation within microbial communities. *Appl. Environ. Microbiol.* 71, 3840–3847. doi: 10.1128/AEM.71.7.3840-3847.2005

25. Karhadkar, P. P., Audic, J. M., Faup, G. M., and Khanna, P. (1987). Sulfide and sulfate inhibition of methanogenesis. *Water Res.*21, 1061–1066. doi: 10.1016/0043-1354(87)90027-3

26. Lewandowski, Z., and Beyenal, H. (2014). *Fundamentals of Biofilm Research.* Abingdon: Taylor & Francis.

27. Lindemann, S. R., Moran, J. J., Stegen, J. C., Renslow, R. S., Hutchison, J. R., Cole, J. K., et al. (2013). The epsomitic phototrophic microbial mat of Hot Lake, Washington: community structural responses to seasonal cycling. *Front. Microbiol.*4:323. doi: 10.3389/fmicb.2013.00323

28. Logan, B. E. (2009). Exoelectrogenic bacteria that power microbial fuel cells. *Nat. Rev. Microbiol.* 7, 375–381. doi: 10.1038/nrmicro2113

29. Lovley, D. R. (2008). The microbe electric: conversion of organic matter to electricity. *Curr. Opin. Biotech* 19, 564–571. doi: 10.1016/j.copbio.2008.10.005

30. Lovley, D. R., and Nevin, K. P. (2011). A shift in the current: new applications and concepts for microbe-electrode electron exchange. *Curr. Opin. Biotech.* 22, 441–448. doi: 10.1016/j.copbio.2011.01.009

31. Marshall, C. W., Ross, D. E., Fichot, E. B., Norman, R. S., and May, H. D. (2012). Electrosynthesis of commodity chemicals by an autotrophic microbial community. *Appl. Environ. Microb.* 78, 8412–8420. doi: 10.1128/AEM.02401-12

32. McNeil, S. D., Nuccio, M. L., and Hanson, A. D. (1999). Betaines and related osmoprotectants. Targets for metabolic engineering of stress resistance. *Plant Physiol.* 120, 945–949. doi: 10.1104/pp.120.4.945

33. Moran, J. J., Doll, C. G., Bernstein, H. C., Renslow, R. S., Cory, A. B., Hutchison, J. R., et al. (2014). Spatially tracking C-13-labelled substrate (bicarbonate) accumulation in microbial communities using laser ablation isotope ratio mass spectrometry.*Environ. Microbiol. Rep.* 6, 786–791. doi: 10.1111/1758-2229.12211

34. Nevin, K. P., Hensley, S. A., Franks, A. E., Summers, Z. M., Ou, J., Woodard, T. L., et al. (2011). Electrosynthesis of organic compounds from carbon dioxide is catalyzed by a diversity of acetogenic microorganisms. *Appl. Environ. Microb.* 77, 2882–2886. doi: 10.1128/AEM.02642-10

35. Paerl, H. W., Pinckney, J. L., and Steppe, T. F. (2000). Cyanobacterial–bacterial mat consortia: examining the functional unit of microbial survival and growth in extreme environments. *Environ. Microbiol.* 2, 11–26. doi: 10.1046/j.1462-2920.2000.00071.x

36. Pages, A., Welshb, D. T., Teasdaleb, P. R., Gricea, K., Vacherc, M., Bennettb, W. W., et al. (2014). Diel fluctuations in solute distributions and biogeochemical cycling in a hypersaline microbial mat from Shark Bay. WA. *Mar. Chem.* 167, 102–112. doi: 10.1016/j.marchem.2014.05.003

37. Purvis, J. E., Yomano, L. P., and Ingram, L. O. (2005). Enhanced trehalose production improves growth of *Escherichia coli*under osmotic stress. *Appl. Environ. Microb.* 71, 3761–3769. doi: 10.1128/AEM.71.7.3761-3769.2005

38. Rabaey, K., Rodríguez, J., Blackall, L. L., Keller, J., Gross, P., Batstone, D., et al. (2007). Microbial ecology meets electrochemistry: electricity-driven and driving communities. *ISME J.* 1, 9–18. doi: 10.1038/ismej.2007.4

39. Rabaey, K., Van de Sompel, K., Maignien, L., Boon, N., Aelterman, P., Clauwaert, P., et al. (2006). Microbial fuel cells for sulfide removal. *Environ. Sci. Technol.* 40, 5218–5224. doi: 10.1021/es060382u

40. Reina-Bueno, M., Argandoña, M., Salvador, M., Rodríguez-Moya, J., Iglesias-Guerra, F., Csonka, L. N., et al. (2012). Role of trehalose in salinity and temperature tolerance in the model halophilic bacterium chromohalobacter salexigens. *PLoS ONE*7:e33587. doi: 10.1371/journal.pone.0033587

41. Renslow, R., Babauta, J., Kuprat, A., Schenk, J., Ivory, C., Fredrickson, J., et al. (2013). Modeling biofilms with dual extracellular electron transfer mechanisms. *Phys. Chem. Chem. Phys.* 15, 19262–19283. doi: 10.1039/c3cp53759e

42. Renslow, R., Donovan, C., Shim, M., Babauta, J., Nannapaneni, S., Schenk, J., et al. (2011). Oxygen reduction kinetics on graphite cathodes in sediment microbial fuel cells. *Phys. Chem. Chem. Phys.* 13, 21573–21584. doi: 10.1039/c1cp23200b

43. Rosenbaum, M., Aulenta, F., Villano, M., and Angenent, L. T. (2011). Cathodes as electron donors for microbial metabolism: which extracellular electron transfer mechanisms are involved? *Bioresour. Technol.* 102, 324–333. doi: 10.1016/j.biortech.2010.07.008

44. Schloss, P. D., Westcott, S. L., Ryabin, T., Hall, J. R., Hartmann, M., Hollister, E. B., et al. (2009). Introducing mothur: open-Source, platform-independent, community-supported software for describing and comparing microbial communities. *Appl. Environ. Microb.* 75, 7537–7541. doi: 10.1128/AEM.01541-09

45. Su, M., Jiang, Y., and Li, D. (2013). Production of acetate from carbon dioxide in bioelectrochemical systems based on autotrophic mixed culture. *J. Microbiol. Biotechnol.* 23, 1140–1146. doi: 10.4014/jmb.1304.04039

46. Tremblay, P. L., and Zhang, T. (2015). Electrifying microbes for the production of chemicals. *Front. Microbiol.* 6:201. doi: 10.3389/fmicb.2015.00201

47. Ueki, T., Nevin, K. P., Woodard, T. L., and Lovley, D. R. (2014). Converting carbon dioxide to butyrate with an engineered strain of *Clostridium ljungdahlii*. *MBio* 5:e1636-14. doi: 10.1128/mBio.01636-14

48. van der Meer, M. T. J., Schouten, S., Bateson, M. M., Nübel, U., Wieland, A., Kühl, M., et al. (2005). Diel variations in carbon metabolism by green nonsulfur-like bacteria in alkaline siliceous hot spring microbial mats from Yellowstone National Park.*Appl. Environ. Microb.* 71, 3978–3986. doi: 10.1128/AEM.71.7.3978-3986.2005

49. Villano, M., Aulenta, F., Ciucci, C., Ferri, T., Giuliano, A., and Majone, M. (2010). Bioelectrochemical reduction of CO2 to CH4 via direct and indirect extracellular electron transfer by a hydrogenophilic methanogenic culture. *Bioresour. Technol.* 101, 3085–3090. doi: 10.1016/j.biortech.2009.12.077

50. Wagner, R. C., Call, D. F., and Logan, B. E. (2010). Optimal set anode potentials vary in bioelectrochemical systems. *Environ. Sci. Technol.* 44, 6036–6041. doi: 10.1021/es101013e

CHAPTER 15

APPLICATION OF IRIDIUM(III) COMPLEX IN LABEL-FREE AND NON-ENZYMATIC ELECTROCHEMICAL DETECTION OF HYDROGEN PEROXIDE BASED ON A NOVEL "ON-OFF-ON" SWITCH PLATFORM

Xiangmin Miao, ChaoYang, Chung-Hang Leung and Dik-Lung Ma

Hydrogen peroxide (H_2O_2) is not only an important compound in food, bioanalysis and environmental analyses[1], but also plays an important role in cellular reactions catalyzed by many oxidases. Moreover, $H_2O_2$is also involved in the function and signal transduction of the cell[2,3]. Thus, the sensitive and simple detection of H_2O_2 is of great interest in chemical, biological, clinical and environmental fields. In the past decades, several kinds of strategies such as electrochemical[4,5,6,7], fluorescent[8,9,10] and colorimetric[11,12,13] methods have been developed for the detection of H_2O_2. Among these methods, electrochemical sensing has attracted great attention due to its merits such as high sensitivity, rapid response, and simple operation.

Enzyme-based detection methods for H_2O_2 have been developed due to number of the advantages of enzymatic reactions, including catalytic activity, high efficiency and good selectivity[14]. Unfortunately, natural enzymes are expensive and easily denatured by environmental changes, thus the enzymes used in these sensing methods are susceptible to potential interfering substances[15,16]. Therefore, non-enzymatic electrochemical sensors have received great attention in the development of H_2O_2 sensors. Imlay, Linn and co-workers reported that the antibacterial property of H_2O_2 is due to DNA damage resulting from Fenton-like reactions in the presence of metal ions such as Fe^{2+}[17], it was reported that H_2O_2 could react with certain transition metal ions in low oxidation states and thus produce HO•, which may cleave single-stranded DNA (ssDNA) to DNA fragments[18]. Based on this principle, many biosensors have been developed for biomolecular detection[19,20,21].

Guanine-rich (G-rich) nucleic acid sequences, existing in the region of gene promoter and chromosome telomeres, can fold into a unique G-quadruplex structure in the presence of metal ion, such as potassium ion (K^+)[22,23]. Based on these characteristics, many G-quadruplex-based sensors have been developed for the detection of metal ions[24,25,26], DNA[27,28], small molecules[29,30] and protein biomarkers[31]. Such K^+-mediated G-quadruplex-based sensing platforms are convenient, low-cost and relatively quick. Previously, our group has synthesized different iridium(III) complexes and discovered that some of them have highly specific binding properties to G-quadruplex,

Citation: Miao X, Yang C, Leung C-H, Ma D-L. Application of iridium(III) complex in label-free and non-enzymatic electrochemical detection of hydrogen peroxide based on a novel "on-off-on" switch platform. Scientific Reports. 2016;6:25774. doi:10.1038/srep25774.

and a number of detection platforms were subsequently constructed based on the specific interaction between iridium(III) complexes and G-quadruplex DNA[32,33,34,35,36,37], through detecting the luminescent enhancement of iridium(III) complexes.

Recently, we discovered that some of the iridium(III) complexes can hinder the electron transfer between the electrode surface and the electrolyte mainly because of the hydrophobicity and the large molecular structure of them. Based on of this unique property, we designed the application of one type of the iridium(III) that specific to G-quadruplex (the structure of it were shown in Fig. 1) in label-free and non-enzymatic electrochemical sensing platform for H_2O_2 by using a novel "on-off-on" switch system. As shown in Fig. 1, the first "switch on" state was achieved based on the immobilize of Nafion (Nf) and methylene blue (Nf@MB) composite membrane as inner-layer, and self-assembled positively charged gold nanoparticles ((+)AuNPs) as outer-layer. Successively, the capture probe (CP, guanine-rich nucleic acid) was modified onto the (+)AuNPs surface through an Au-S bond, and the CP would change to G-quadruplex structure in the presence of K^+, followed by the specific binding with an iridium(III) complex using a simple and label-free method[38,39,40,41,42,43,44]. As a consequence, a hydrophobic and large molecule structural layer wad formed on the electrode surface, and resulted in a "switch off" state. Then, upon incubation of the sensor with $H_2O_2$and Fe^{2+}, Fenton reaction would happen and produced tremendous OH• to cleave CP into DNA fragments, leading to the release of such fragments and the iridium(III) complex from the sensing interface, recovering the electrochemical signal and achieving a "switch on" state. Importantly, the electrochemical signal change was proportional to H_2O_2 concentration. Thus, H_2O_2 could be detected by using such proposed method based on monitoring the signal change of MB.

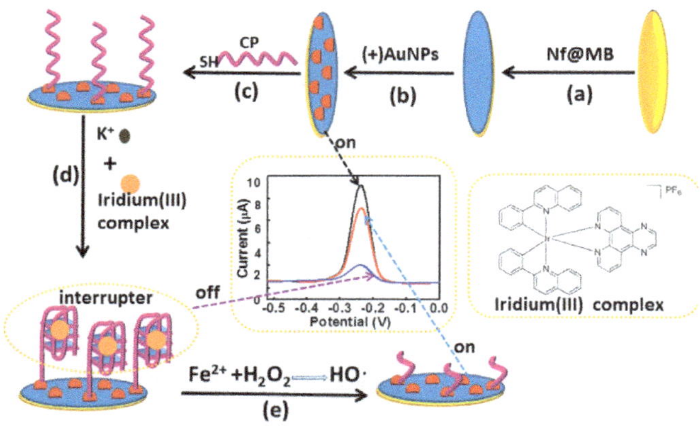

Figure 1: Scheme diagram for the preparation of H_2O_2 **sensor.**

There are several advantages for such a proposed method: Firstly, due to the good stability of negatively charged Nf membrane on the electrode surface, abundant positively charged and highly stable Nf@MB membrane could formed on the electrode, which can greatly improve the stability of the sensor. In addition, MB is highly conductive as a redox indicator, which can great enhance the sensitivity of the sensor. Meantime, about 4 nm of (+)AuNPs, with excellent biocompatibility and good conductivity, was successfully self-assembly onto the Nf@MB membrane, which can greatly enlarge the electrode surface for more CP immobilization. Moreover, to the best of our knowledge, this is the first application of the iridium(III) complex and (+)AuNPs in electrochemical sensor for H_2O_2 analysis. We envision that this "on-off-on" platform could also provide new opportunities for biosensor development.

RESULTS AND DISCUSSION

To investigate the conductivity of the iridium(III) complex, several modified electrodes were characterized by cyclic voltammetry experiments (CVs) in the presence of 5 mM $Fe(CN)_6^{3-/4-}$. As seen from curve a in Fig. 2A, a stable and well-defined redox peak was obtained when the bare gold electrode was scanned in 5 mM of $Fe(CN)_6^{3-/4-}$. After the immobilization of CP on the electrode surface, an obvious decrease of the peak current was observed, which was attributed to the fact that DNA hinders electron transfer (curve b)[45,46]. Then, G-quadruplex structures were formed after the CP-modified electrode was incubated in tris-buffer containing 100 mM of K^+. Subsequently, the interaction of iridium(III) complex with the G-quadruplex led to a further decrease in peak current, because of the high resistance of the electrode interface induced by the adsorption of iridium(III) complex and the formation of a hydrophobic layer on the electrode surface (curve c). Since other organic dyes such as crystal violet (CV) and rhodamine can also specifically bind to G-quadruplex[47,48], we also investigated the conductivity of CV and rhodamine after they bound to the G-quadruplex that immobilized on the electrode. The results demonstrated that CV and rhodamine would accelerate the electron transfer (curve d and e). Therefore, to obtain the step of the "switch off" state, we selected iridium(III) complex as the hydrophobic layer.

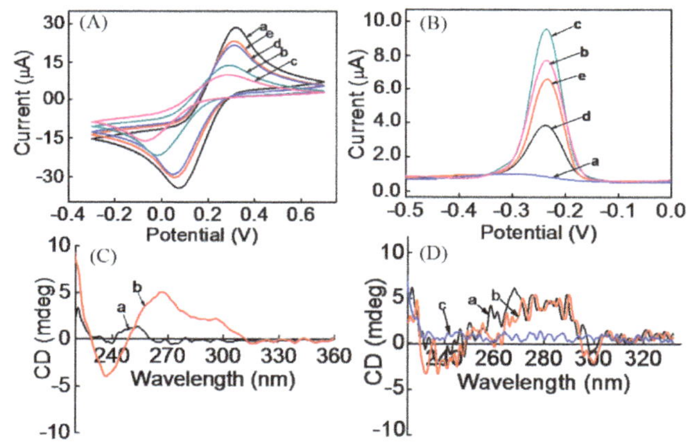

Figure 2: (**A**) Conductivity investigation of the iridium(III) complex and other organic dyes that specific to G-quadruplex in 5 mM of $Fe(CN)^{3-/4-}$: (a) Gold electrode; (b) CP/gold electrode; (c) iridium(III) complex/CP/gold electrode; (d) CV/CP/gold electrode. (e) Rhodamine/CP/gold electrode. (B) DPV of the different modified electrodes in 10mM tris-buffer solution (pH 7.4): (a) Gold electrode; (b) Nf@MB/gold electrode; (c) (+) AuNPs/Nf@MB/gold electrode; (d) iridium(III) complex/CP/(+)AuNPs/Nf@MB/gold electrode; (e) (d) after the incubation with 1.5nM of H_2O_2. (C) Circular dichroism of CP before (a) and after (b) the incubation of it with tris-buffer that contained 100mM of K^+. (D) Circular dichroism of G-quadruplex structure before (a) and after (b) the selective interaction of it with iridium(III) complex, and after the cleavage with H_2O_2 (c).

In order to confirm the successful fabrication of the proposed sensing platform, the fabrication process of the sensor was characterized by DPV experiments. As shown in Fig. 2B, no oxidation peak was observed at the bare gold electrode while an obvious oxidation peak appeared after electro-polymerization of MB on the electrode surface, which is mainly due to the high conductivity of MB as an electron mediator (curve b). Then, the peak current increased after the electro-deposition of AuNPs onto the MB film on account of the excellent conductivity of AuNPs (curve c). However, after the interaction of iridium(III) complex with G-quadruplex form of CP in the presence of K^+, the peak current obviously decreased, since iridium(III) complex would form a hydrophobic and large molecule structural layer on the electrode surface and hinder the electron transfer (curve d). However, upon incubation of the sensor with 1.5 nM of H_2O_2 and Fe^{2+}, a dramatic increase of the peak current was observed (curve e). This increase could be ascribed to the release of iridium(III) complex and CP fragments from the electrode surface after the cleavage of CP by Fenton reaction in the presence of Fe^{2+} and H_2O_2.

To validate the mechanism of the assay, circular dichroism (CD) spectroscopy was performed. As shown in Fig. 2C, there was no obvious peak for CP alone, while a positive Cotton effect peak at around 264 nm and a negative Cotton effect peak at around 236 nm appeared after CP was incubated with tris-buffer containing 100 mM of K^+, indicating the formation of the classic G-quadruplex structure in accordance with literature[49]. No significant change was observed upon the selective interaction of the iridium(III) complex with the G-quadruplex structure of CP (Fig. 2D, curve a and b). However, the intensity at the peak of 264 nm and 236 nm decreased dramatically when the CP was incubated with H_2O_2, these results might be attributed to the cleavage of CP (curve c).

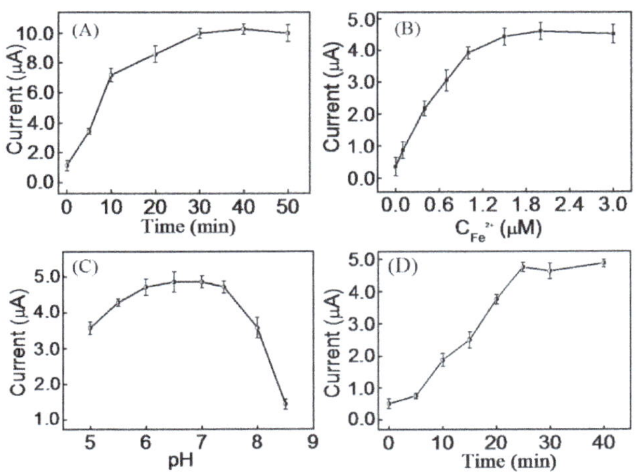

Figure 3: Effect of the self-assembly time of (+)AuNPs on the sensor properties (**A**) Effect of the concentration of Fe^{2+} (**B**), the pH value (**C**) and the incubation time for CP cleavage in the presence of 1.2 nM H_2O_2 (**D**).

Experimental conditions including the self-assembly time of (+)AuNPs, the concentration of Fe^{2+} and the cleavage time of CP by H_2O_2 were optimized. The self-assembly time of (+) AuNPs is an important factor that might affect the properties of the sensor. As shown in Fig. 3A, the current intensity was proportional to the self-assembly time of (+)AuNPs from 0 to 30 min and then reached a plateau. In view of the sensitivity of the sensor, an electro-deposition time of 30 min was chosen for all the experiments.

The role of Fe^{2+} in the Fenton reaction is to catalyze the cleavage of DNA. Thus, Fe^{2+} concentration is an important factor that can affect the degree of CP cleavage. Figure 3B shows that the current signal increased along with the increase of Fe^{2+} concentration up to 1.5 μM in the presence of 1.2 nM H_2O_2, indicating that CP could be cleaved effectively upon addition of 1.5 μM of Fe^{2+}. Thus, we selected 1.5 μM as the optimal concentration of Fe^{2+} in the experiments.

DNA structure depends on the pH environment, so the cleavage activity of H_2O_2 and Fe^{2+} may depend on the pH as well. As shown in Fig. 3C, the current signal increased with the pH over the range from 5.0 to 6.5 in the absence of H_2O_2, and reached a plateau over the range of 6.5–7.4. However, when the pH was higher than 7.4, the current signal decreased dramatically, such result might be due to the fact that an acidic environment can stabilize the Fe^{2+}ions, and higher pH can induce the oxidation of ferrous ions to ferric ions under atmosphere (O_2). Considering the sensitivity of the sensor, we selected 7.4 for further experiments.

Since the second "switch-on" state depends on the cleavage of CP, the cleavage time of CP in the presence of 1.2 nM H_2O_2 was investigated in Fig. 3D. It was found that the current signal increased when the cleavage time increased from 0 to 25 min, and then reached a plateau. To achieve the effective cleavage of CP, 30 min was chosen as the cleavage time of CP.

To evaluate the effect of H_2O_2 concentration on current signal, the proposed sensor was incubated with tris-buffer that contained different concentration of H_2O_2. As shown in Fig. 4A, the DPV signal increased along with the increase of H_2O_2 concentration. According to the calibration plot of Fig. 4B, the linear regression equation was $I = 11.18 + 7.87c$ (c: nM) in a dynamic range from 8.0 pM to 2.0 nM with a correlation coefficient of $R^2 = 0.994$. The detection limit was 3.2 pM calculated by the three-signal method. Here, the electro-deposition of AuNPs onto MB film can greatly enlarge the electrode surface for CP immobilization, and accordingly improve the sensitivity of the sensor. The detection limit of our "on-off-on" detection platform for H_2O_2 detection was lower than other AuNPs[49,50,51,52,53] or MB[54,55,56,57] based electrochemical methods, which can satisfy the demand for H_2O_2 detection in real samples such as food or industrial products. Thus, our proposed sensor could be used to quantify H_2O_2 for food safety and clinical diagnosis.

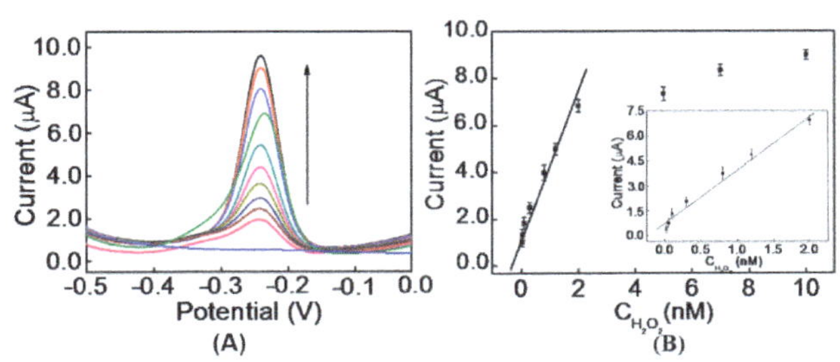

Figure 4: (**A**) DPV experiments for H_2O_2 detection in 10 mM of tris-buffer solution (pH 7.4); (**B**) Calibration curve for the sensor.

The stability of the sensor was investigated by CV. After continuous scanning for 20 cycles, the sensor retained 98.1% of its initial response, indicating acceptable stability. Meantime, the selectivity of such sensor for H_2O_2 was evaluated against other interferences including ascorbic acid, purine trione, glucose and mixed metal ions (contained Pb^{2+}, Cu^{2+}, K^+, Ca^{2+} and Mg^{2+}). From the results in Fig. 5, it could be seen that the current intensity change (ΔI) upon addition of ascorbic acid (150 nM), purine trione (150 nM), glucose (150 nM) and metal ions (150 nM of Pb^{2+}, Cu^{2+}, K^+, Ca^{2+} and Mg^{2+}) were much lower than that of H_2O_2 (1.5 nM). Moreover, the ΔI of the sensor after the incubation of it with a solution containing both H_2O_2 and the interferences was almost the same with that of H_2O_2 (1.5 nM) only. Such high selectivity could be attributed to the highly specific cleavage of DNA based on the Fenton reaction in the presence of H_2O_2 and Fe^{2+}.

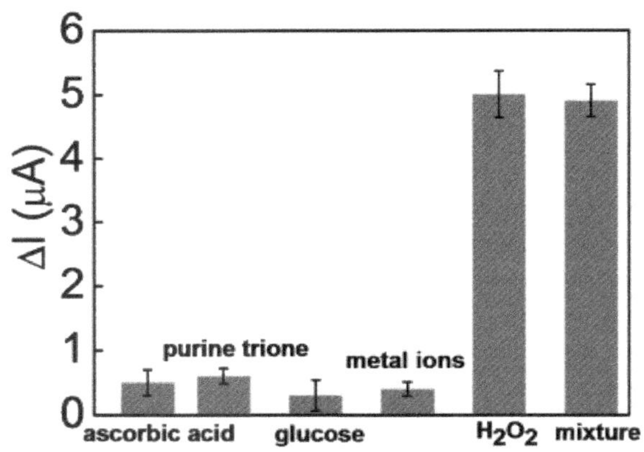

Figure 5: Selectivity of the sensor for H_2O_2 **(1.5** nM) detection against interferences including ascorbic acid (150 nM), purine trione (150 nM), glucose (150 nM) and metal ions (150 nM**).**

H_2O_2 is used as a stabilizer in milk in some European and American countries. However, residual H_2O_2may cause adverse effects to the human body. Thus, the sensitive and rapid detection of H_2O_2 in milk samples is of great importance. To evaluate the robustness of the system for H_2O_2 detection in milk, milk samples were collected and centrifuged at 13600 rpm to remove fat. After corroborating the absence of hydrogen peroxide, such milk samples were diluted ten-fold with tris-buffer (pH 7.4) for recovery experiments. As shown in Table 1, a good recovery ranging from 96.0% to 109% and a relative standard deviation (RSD) between 1.38% and 3.16% were obtained (each sample was tested 3 times). This result indicated that our sensor has potential application for the quantitative determination of H_2O_2 level in real samples.

Table 1: Recovery experiments for H_2O_2 **in milk samples.**

Sample	Added (nM)	Found (nM)	Recovery %	RSD %
	0.10	0.11	110	2.06
Milk	0.30	0.29	96.7	1.38
	0.50	0.48	96.0	3.16
	1.00	1.09	109	2.12

In conclusion, a novel and label-free "on-off-on" switch system was successfully developed for the sensitive detection of H_2O_2. The first "switch on" state was achieved based on the immobilization of Nf@MB and (+)AuNPs. Here, MB was used as the current signal and (+)AuNPs were used to enlarge the surface of the electrode and improve the sensitivity of the sensor. Then, the "switch off" state was obtained based on the formation of a massive hydrophobic layer that contained G-quadruplex structures and iridium(III) complex on the electrode surface. After the incubation of the sensor with H_2O_2, CP would be cleaved into DNA fragments and released from the electrode surface, followed by signal recovering as the second "switch on" state. The novel concept of this "on-off-on" platform was successfully used for H_2O_2 detection with a detection limit down to 3.2 pM. Moreover, experiments proved that such strategy could be applied effectively for H_2O_2 detection in milk samples.

METHODS

Materials and Reagents

Nafion, Methylene blue (MB), rhodamine, thiazole orange (TO), 6-mercapto-1-hexanol (MCH) and gold chloride ($HAuCl_4$) were purchased from Sigma Aldrich (St. Louis, MO). The milk sample (Hokkaido Specially Select 3.6 Milk, UHT Processed whole milk, fat: 4.2 g per 100mL, carbohydrates: 4.8 g per 100 mL, sodium: 41 mg per 100 mL, calcium: 115 mg per 100 mL) was purchased from supermarket in Hong Kong. The iridium(III) complex used in the experiment that specific to G-quadruplex DNA was selected and synthesized according to our previous method[58]. Other chemicals and solvents were of analytical grade and double distilled water was used throughout this study. 10.0 mM of tris-buffer (pH 7.4) was used for H_2O_2detection and 10.0 mM of tris-buffer (pH 7.4, with 100 mM K^+) was used for G-quadruplex formation. The capture probe (CP) was designed according to the literature based on the principle that there are preferential sequences for iron-mediated DNA cleavage. CP was synthesized by Techdragon Inc. (Hong Kong, China) with the sequence of 5′-$GTG_3TAG_3CG_3T_2G_2$-$(CH_2)_6$-SH-3′.

Apparatus

Electrochemical measurements were monitored by a CHI 630C electrochemical workstation (CH Instruments, Inc. U.S.A.). A conventional three-electrode system consisted of a modified working electrode, a platinum wire counter electrode and an Ag/AgCl reference electrode was used in the experiment. CD spectroscopy measurements were constructed by using an Olis 17/UV/VIS/NIR spectropolarimeter at room temperature.

H_2O_2 Sensing Protocol

Gold electrodes were pretreated according to our previous method. Then, 0.5 mg of MB was added into 2 mL of 0.5% Nf ethanol solution and sonicated to obtain a homogeneous suspension contained Nf@MB. Next 5 μL of Nf@MB was cast on the pretreated electrode and dried in the air to obtain an Nf@MB film. After that, 4 nm of (+)AuNPs prepared according to literature[59] was self-assembly onto the Nf@MB film for 30 min, based on the electrostatic adsorption to enlarge the electrode surface for CP immobilization, Subsequently, the modified electrodes were incubated in 10 mM of tris-buffer (pH 7.4) that contained 2.0 μM of CP for 12 h at room temperature (prior to modification, the disulfide bond at the 3′ end of CP was cleaved with tris(2-carboxyethyl)phosphine (TCEP)), followed by reaction with 2.0 μM of MCH for 1 h to block nonspecific sites, and incubated in 10 mM of tris-buffer (pH 7.4) that contained 100 mM of K^+ and 3.0 μM of iridium(III) complex for another 1 h to induce the formation of a hydrophobic layer on the electrode surface. Then, the prepared sensor was incubated with different concentrations of H_2O_2 at 37 °C for 30 min in the presence of Fe^{2+}. Finally, the electrochemical characteristics of the sensor were investigated in tris-buffer by using differential pulse experiments (DPV) from −500 mV to 0.0 mV at room temperature.

ADDITIONAL INFORMATION

Acknowledgments

This work is supported by Hong Kong Baptist University (FRG2/14-15/004 and FRG2/15-16/002), the Health and Medical Research Fund (HMRF/14130522), the Research Grants Council (HKBU/201811, HKBU/204612, and HKBU/201913), French Agence Nationale de la Recherche/Research Grants Council Joint Research Scheme (AHKBU201/12; Oligoswitch ANR-12-IS07-0001), the National Natural Science Foundation of China (21575121), the Guangdong Province Natural Science Foundation (2015A030313816), the Hong Kong Baptist University Century Club Sponsorship Scheme 2015, the Interdisciplinary Research Matching Scheme (RC-IRMS/14-15/06), the Science and Technology Development Fund, Macao SAR (098/2014/A2), and the University of Macau (MYRG091(Y3-L2)-ICMS12-LCH, MYRG2015-00137-ICMS-QRCM, and MRG023/LCH/2013/ICMS and MRG044/LCH/2015/ICMS).

REFERENCES

1. Baghayeri M., Zare E. N. & Lakouraj M. M. A simple hydrogen peroxide biosensor based on a novel electro-magnetic poly(*p*-phenylenediamine)@Fe_3O_4 nanocomposite. Biosens. Bioelectron. 55, 259–265 (2014).
2. Chen X. *et al.* Advances in enzyme-free electrochemical sensors for hydrogen peroxide, glucose, and uric acid. Microchim. Acta 181, 689–705 (2014).
3. Terai T. & Nagano T. Small-molecule fluorophores and fluorescent probes for bioimaging. Arch. Eur. Phys. 465, 347–359 (2013).
4. Liu Q. *et al.* NiO nanoparticles modified with 5,10,15,20-tetrakis(4-carboxyl pheyl)-porphyrin: Promising peroxidase mimetics for H_2O_2 and glucose detection. Biosens. Bioelectron. 64, 147–153 (2014).
5. Zhang Y. *et al.* Fabrication of 2D ordered mesoporous carbon nitride and its use as electrochemical sensing platform for H_2O_2, nitrobenzene, and NADH detection. Biosens. Bioelectron. 53, 250–256 (2014).
6. Lin S. *et al.* Luminescence switch-on detection of protein tyrosine kinase-7 using a G-quadruplex-selective probe. Chem. Sci. 6, 4284–4290 (2015).
7. Wu S. H. *et al.* Amplified electrochemical hydrogen peroxide reduction based on hemin/G-quadruplex DNAzyme as electrocatalyst at gold particles modified heated copper disk electrode. Biosens. Bioelectron. 73, 41–46 (2015).
8. Bortolozzi R., Gradowsk S. & Ihmels H. Selective ratiometric detection of H_2O_2 in water and in living cells with boronobenzo***b***quinolizinium derivatives. Chem. Commun. 50, 8242–8245 (2014).
9. Marks P. *et al.* Highly efficient detection of hydrogen peroxide in solution and in the vapor phase ***via***fluorescence quenching. Chem. Commun. 51, 7061–7064 (2015).
10. Chang H. C. & Ho J. A. Gold nanocluster-assisted fluorescent detection for hydrogen peroxide and cholesterol based on the inner filter effect of gold nanoparticles. Anal. Chem. 87, 10362–10367 (2015).
11. Ge S. G. *et al.* Colorimetric detection of the flux of hydrogen peroxide released from living cells based on the high peroxidase-like catalytic performance of porous PtPd nanorods. Biosens. Bioelectron. 71, 456–462 (2015).
12. Chen S. *et al. In situ* growth of silver nanoparticles on graphene quantum dots for ultrasensitive colorimetric detection of H_2O_2 and glucose. Anal. Chem. 86, 6689–6694 (2014).
13. Wang G. L. *et al.* Ultrasensitive and dual functional colorimetric sensors for mercury(II) ions and hydrogen peroxide based on catalytic reduction property of silver nanoparticles. Biosens. Bioelectron.31, 337–342 (2012).
14. Privett B. J., Shin J. H. & Schoenfisch M. H. Electrochemical sensors. Anal. Chem. 82, 4723–4241 (2010).
15. Zhang B. *et al.* Anodic-stripping voltammetric immunoassay for ultrasensitive detection of low-abundance proteins using quantum dot aggregated hollow microspheres. Chem. Eur. J. 19, 2496–2503 (2013).
16. Gao Z. *et al.* High-resolution colorimetric assay for rapid visual readout of phosphatase activity based on gold/silver core/shell nanorod. ACS Appl. Mater. Interfaces, 6, 18243–18250 (2014).
17. Imlay J. & Linn S. DNA damage and oxygen radical toxicity. Science 240, 1302–1309 (1988).
18. Imiay J., Chin S. & Linn S. Toxic DNA damage by hydrogen peroxide through the Fenton reaction *in vivo* and *in vitro*. Science 240, 640–642 (1988).
19. Li H. & Rothberg L. Colorimetric detection of DNA sequences based on electrostatic interactions with unmodified gold nanoparticles. Proc. Natl. Acad. Sci. USA 101, 14036–14039 (2004).
20. Huang W. T. *et al.* A simple and facile strategy based on Fenton-induced DNA cleavage for fluorescent turn-on detection of hydroxyl radicals and Fe^{2+}. J. Mater. Chem. 22, 1477–1481 (2012).
21. Zhang L. P. *et al.* Fenton reaction-triggered colorimetric detection of phenols in water samples using unmodified gold nanoparticles. Sens. Actuat. B 225, 593–599 (2016).
22. Hu R. *et al.* An efficient fluorescent sensing platform for biomolecules based on fenton reaction triggered molecular beacon cleavage strategy. Biosens.

Bioelectron. 41, 442–445 (2013).

23. Moye A. L. *et al.* Telomeric G-quadruplexes are a substrate and site of localization for human telomerase. Nat. Commun. 6, 7643, 10.1038 /ncomms8643 (2015).

24. Aizen R. *et al.* G-Quadruplex-Stimulated optical and electrocatalytic DNA Switches. Small 11, 3654–3658 (2015).

25. Hao Y. L. *et al.* Amplified colorimetric detection of mercuric ions through autonomous assembly of G-quadruplex DNAzyme nanowires. Biosens. Bioelectron. 52, 261–264 (2014).

26. Xu L. J. *et al.* Turn-on and label-free fluorescence detection of lead ions based on target-induced G-quadruplex formation. Chem. Commun. 51, 8165–8168 (2015).

27. Liu M. *et al.* A SERS/fluorescence dual-mode nanosensor based on the human telomeric G-quadruplex DNA: Application to mercury(II) detection. Biosens. Bioelectron. 69, 142–147 (2015).

28. Li Y. *et al.* A novel electrochemical DNA biosensor based on HRP-mimicking hemin/G-quadruplex wrapped GOx nanocomposites as tag for detection of *Escherichia coli* O157:H7. Biosens. Bioelectron. 63, 1–6 (2015).

29. Shi L. J. *et al.* A label-free hemin/G-quadruplex DNAzyme biosensor developed on electrochemically modified electrodes for detection of a HBV DNA segment. RSC Adv. 5, 11541–11548 (2015).

30. Chen Q. G. *et al.* An enzyme-free and label-free fluorescent biosensor for small molecules by G-quadruplex based hybridization chain reaction. Talanta 138, 15–19 (2015).

31. Xue Q. W., Wang L. & Jiang W. A novel label-free cascade amplification strategy based on dumbbell probe-mediated rolling circle amplification-responsive G-quadruplex formation for highly sensitive and selective detection of NAD^+ or ATP. Chem. Commun. 49, 2640–2642 (2013).

32. Lin Y. P. *et al.* Non-enzymatic sensing of hydrogen peroxide using a glassy carbon electrode modified with a nanocomposite made from carbon nanotubes and molybdenum disulphide. Microchim. Acta182, 1803–1809 (2015).

33. He H. Z. *et al.* G-quadruplexes for luminescent sensing and logic gates. Nucleic Acids Res. 41, 4345–4359 (2013).

34. Ma D. *et al.* Simple DNA-based logic gates responding to biomolecules and metal ions. Chem. Sci. 4, 3366–3380 (2013).

35. Leung C. *et al.* Luminescent detection of DNA-binding proteins. Nucleic Acids Res. 40, doi:10.1093 (2012).

36. Wang M. *et al.* Conjugating a groove-binding motif to Ir(III) complex for the enhancement of G-quadruplex probe behavior. Chem. Sci. 7, 2516–2523 (2016).

37. Ma D. *et al.* Group 9 organometallic compounds for therapeutic and bioanalytical applications. Acc. Chem. Res. 47, 3614–3631 (2014).

38. Leung K. H. *et al.* Label-free luminescence switch-on detection of hepatitis C virus NS3 helicase activity using a G-quadruplex-selective probe. Chem. Sci. 6, 2166–2171 (2015).

39. Lu L. H. *et al.* Detection of nicking endonuclease activity using a G-quadruplex-selective luminescent switch-on probe. Chem. Sci. 5, 4561–4568 (2014).

40. Liu J. *et al.* Iridium(III) complex-coated nanosystem for ratiometric upconversion luminescence bioimaging of cyanide anions. J. Am. Chem. Soc. 133, 15276–15279 (2011).

41. Yu M. *et al.* Cationic iridium(III) complexes for phosphorescence staining in the cytoplasm of living cells. Chem. Commun. 2115–2117 (2008).

42. Ma D. L. *et al.* A luminescent cocaine detection platform using a split G-Quadruplex-selective iridium(III) complex and a three-Way DNA junction architecture. ACS Appl. Mater. Interfaces 7, 19060–19067 (2015).

43. Xiong C. *et al.* Sensitive detection of H_2O_2 and H_2O_2-related reactant with $Ru(bipy)_2(7,8\text{-dimethyl-dipyridophenazine})^{2+}$ and oligodeoxyribonucleotide. Analyst 137, 4428–4434 (2012).

44. Rai P., Wemmer D. E. & Linn S. Preferential binding and structural distortion by Fe^{2+} at RGGG-containing DNA sequences correlates with enhanced oxidative cleavage at such sequences. Nucleic Acids Res. 33, 497–510 (2005).

45. Tuukkanen S. *et al.* Dielectrophoresis of nanoscale double-stranded DNA and humidity effects on its electrical conductivity. Appl. Phys. Lett. 87, 183102, (2005).10.1063/1.2117626

46. Cai L. T., Tabata H. & Kawai T. Self-assembled DNA networks and their electrical conductivity.Appl. Phys. Lett. 77, 3105–3106 (2000).

47. Wang M. *et al.* Label-free luminescent detection of LMP1 gene deletion using an intermolecular G-quadruplex-based switch-on probe. Biosens. Bioelectron. 70, 338–344 (2015).

48. Lu L. H. *et al.* Label-Free luminescent switch-on probe for ochratoxin a detection using a G-quadruplex-selective iridium(III) complex. ACS Appl. Mater. Interfaces 7, 8313–8318 (2015).

49. Zhu J., Zhang L. & Wang E. Measurement of the base number of DNA using a special calliper made of a split G-quadruplex. Chem. Commun. 48, 11990–11992 (2012).

50. Guo S., Xu L. & Xu B. A ternary nanocomposite electrode of polyoxometalate/carbon nanotubes/gold nanoparticles for electrochemical detection of hydrogen peroxide. Analyst 140, 820–826 (2015).

51. Zhang Y., Liu Y. & He J. Electrochemical behavior of graphene/Nafion/Azure I/Au nanoparticles composites modified glass carbon electrode and its application as nonenzymatic hydrogen peroxide sensor. Electrochim. Acta 90, 550–555 (2013).

52. Hong J., Yang W. & Zhao Y. Catalase immobilized on a functionalized multi-walled carbon nanotubes–gold nanocomposite as a highly sensitive bio-sensing system for detection of hydrogen peroxide. Electrochim. Acta 89, 317–325 (2013).

53. Huang K., Niu D. & Liu X. Direct electrochemistry

of catalase at amine-functionalized graphene/gold nanoparticles composite film for hydrogen peroxide sensor. Electrochim. Acta 56, 2947–2953 (2011).

54. Roushani M., Bakyas K. & Dizajdizi B. Development of sensitive amperometric hydrogen peroxide sensor using a CuNPs/MB/MWCNT-C_{60}-Cs-IL nanocomposite modified glassy carbon electrode.Mat. Sci. Eng. 64, 54–60 (2016).

55. Gu T. & Hasebe Y. Peroxidase and methylene blue-incorporated double stranded DNA–polyamine complex membrane for electrochemical sensing of hydrogen peroxide. Anal. Chim. Acta 525, 191–198 (2004).

56. Lin X., Chen J. & Chen Z. Amperometric biosensor for hydrogen peroxide based on immobilization of horseradish peroxidase on methylene blue modified graphite electrode. Electroanal. 12, 306–310 (2000).

57. Lin K., Yin C. & Chen S. An electrochemical biosensor for determination of hydrogen peroxide using nanocomposite of poly(methylene blue) and FAD hybrid film. Sens. Actuat. B 157, 202–210 (2011).

58. Ma D. L. *et al.* A luminescent cocaine detection platform using a split G-Quadruplex-selective iridium(III) complex and a three-way DNA junction architecture. ACS Appl. Mater. Interfaces 7, 19060–19067 (2015).

59. Li Z. B. *et al.* Enhanced electrochemical recognition of double-stranded DNA by using hybridization chain reaction and positively charged gold nanoparticles. Biosens. Bioelectron. 74, 687–690 (2015).

CHAPTER 16

THE EFFECT OF TEMPERATURE ON THE COUPLED SLOW AND FAST DYNAMICS OF AN ELECTROCHEMICAL OSCILLATOR

Alana A. Zülke and Hamilton Varela

Systems with sub-processes evolving on different time scales are ubiquitous in most natural and man-made systems, such as chemical reactions, neuro-biological and ecological systems, to mention just a few. Although there is a clear distance between these many examples of multiple time-scale systems, they find similarities at the mathematical description level[1], being often discussed in terms of linearized models, built without necessarily accounting its peculiarities regarding, for example, experimental conditions or physical-chemical nature[2]. In electrochemistry, oscillating surface reactions represent a category of multiple time-scale systems which are fairly useful to investigate complex behavior at its fundamental level[3,4] and in which experimental constraints impact the evolution of temporal patterns. The intricate nonlinear interaction of reactions with disparate time-scales resulting in complex behavior at the solid/liquid electrified interface has been found to be also useful in shedding light on the mechanisms underlying these surface reactions. In this direction, we have recently tackled some intriguing aspects of the complex dynamics of oscillatory electro-oxidation of small organic molecules on platinum and platinum based surfaces[5,6,7,8,9,10]. We have recently explored[8,9] the effect of the slow and uncontrollable surface changes in the time-series registered during the electro-oxidation of small organic molecules on platinum. The resulting surface deactivation, similarly to an aging process, is revealed by a gradual and slow drift to more positive values of the electrode potential, acting as a bifurcation parameter that ultimately causes the death of oscillations. We proposed a model for the global oscillatory response in which coupled to the (intrinsic) core oscillator responsible for the oscillations itself, is a rather slow and irreversible process, which is responsible for the abovementioned drift. To compensate this drift, we have proposed an empirical experimental strategy consisting in applying a negative galvanodynamic sweep in order to stabilize the potential oscillations by slowing down the natural drift, increasing the time window for the observable oscillations[9]. Results reported in refs [5,8] and [9] strongly support the physical-chemical picture of the slow accumulation of sub-surface oxygenated species as responsible for the observed drift.

Particularly for the formic acid reaction on platinum, the best model system for investigating fundamental aspects of electrocatalysis of small organic molecules, the interest lies in its relevance for energy conversion systems. The complete oxidation to CO_2 yields two electrons and only C-H and O-H bonds cleavages are required. However, despite being the simplest reaction among its class and all the extensive amount of available literature on the matter since the classical works from Parsons *et al.*[11,12], mechanistic issues still trigger some heated discussions on the scientific community[13,14]. Also, the oscillatory oxidation of formic acid has undergone a long history of experimental investigations and numerical simulations[15,16,17,18,19,20,21]. An intriguing aspect of the oscillatory electro-oxidation of formic acid is the occurrence of temperature (over) compensation[7]. The term was used to

Citation: Zülke A A, Varela H. The effect of temperature on the coupled slow and fast dynamics of an electrochemical oscillator. Scientific Reports. 2016;6:24553. doi:10.1038/srep24553.

define the counter intuitive observation of either a temperature independent relation with the oscillatory frequency or even lower frequencies as the temperature is increased for a certain range of operational conditions.

The present contribution aims to add to the comprehension of the coupled time-scales underlying electrochemical oscillators in terms of the dependence on the bulk temperature. We use the electro-oxidation of formic acid on polycrystalline platinum in acidic media as a model system.

RESULTS

The typical responses upon variation of the bulk temperature on the cyclic voltammetry experiments at 50 mVs^{-1} are depicted in Fig. 1a. Overall, the voltammetric profiles are in agreement with the well-known behavior for the system Pt|HCOOH in sulfuric acid media, showing an oxidation wave with an onset potential of ≈800 mV in the forward scan and a reactivation peak in the reverse scan, starting at about 950 mV[11,12]. The contribution of the temperature increase in Fig. 1a is basically seen as an increase in the magnitude of current densities in the whole window. Regarding the hysteresis between positive and negative sweep, no significant effect is observed but a slight decrease of the distance between the direct and reverse peaks follows the temperature increase.

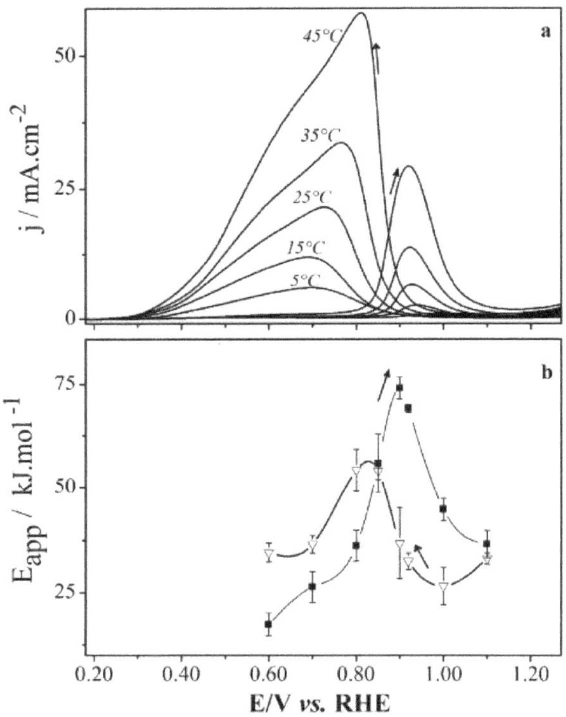

Figure 1: (**a**) Voltammetric profiles (dE/dt = 50 mV.s^{-1}) of the system Ptpoly|0.5 M H_2SO_4 + 1 M HCOOH at different temperatures as indicated. (**b**) Measured apparent activation energies for the electro oxidation of formic acid in the positive sweep (full squares) and in the negative sweep (empty triangles).

Apparent activation energies (E_{app}) were calculated from the slopes obtained by the Arrhenius plots, c.f. Fig. 1b. We estimated E_{app} for the two main oxidative waves as 59 ± 1 kJ. mol^{-1} and 41 ± 2 kJ.mol^{-1}, in the positive scan and in the forward scan respectively, showing good agreement with data available in the literature[22,23,24,25].

Galvanodynamic sweeps at 5 μA.s^{-1} were also performed for five temperatures from 5 to 45 °C, maintaining all other parameters identical. In this case, the obtained time-series were used for mapping the current region in which oscillatory instabilities emerge, allowing us to choose normalized current values (J_n^T, $T = 5, 15, 25, 35, 45$ °C, $0 < n < 1$) for further galvanostatic investigation, using the procedure suggested by Nagao *et al.*[7]. This normalization procedure allows us to compare a set of potential time-series obtained at different temperatures for a given applied current. Complete time-series of the galvanostatic measurements obtained at different temperatures are displayed in Fig. 2.

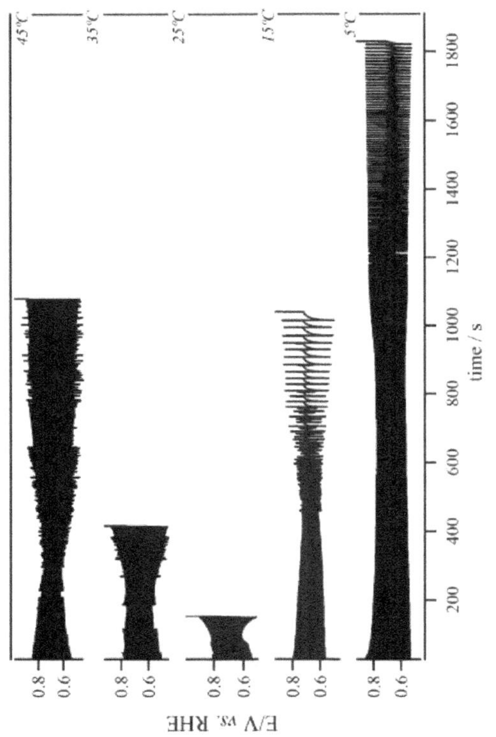

Figure 2: Galvanostatic time-series for a normalized current of 0.5 ($J^T_{0.5}$, see text for details) and at different temperatures.

Typical relaxation-type oscillations with amplitudes ranging from 550 to 850 mV were found for all temperatures[26,27,28]. Not surprisingly, temperature revealed to be an important parameter in the definition of oscillatory frequency, *f*, potential amplitude, ΔE and overall size or stability of the time-series, S_{osc}. For a more reasonable comparison, the complete time-series were divided into three distinguishable families of oscillations (F_1, F_2, and F_3, c.f. Supplementary Fig. S1) that describe the spontaneous evolution of temporal patterns, common to all temperatures investigated here. Potential oscillations starting with high amplitudes of about 300 mV describe well the F_1 region, which gradually evolve to a region where cycles show lower amplitudes and higher frequencies (F_2). The transition from F_2 to F_3 is marked by an increase in amplitude (ΔE) and, again, a decrease in frequency. In F_3, kinetic instabilities of richer complexity such as mixed-mode oscillations (MMO) were found[27,29]. Transition between oscillatory patterns, for instance, from single-period (P^1) to double-period (P^2) and periodic-chaotic sequences[2], were also observed. Those complications led us to opt for oscillatory families F_1 and F_2 for further discussions regarding activation parameters, once the well-behaved dynamics presented a better reproducibility for a wider range of experiments. In Fig. 3 is zoomed a representative region for F_1 highlighting the temperature dependency on the oscillatory dynamics. The overall size of the oscillatory region, S_{osc}, and the oscillatory frequency, f, for all three families depend strongly on the bulk temperature. The average oscillatory frequency ($\bar{f}$) for F_1 and F_2 shows the same tendency as S_{osc}: increasing values from 5 to 25 °C with a subsequent decrease from 25 to 45 °C. We have calculated the distribution of frequency by FFT (Fast Fourier Transform) and confirmed that the dominant frequencies display higher values as temperature is increased from 5 to 25 °C, and lower values as temperature is increased from 25 to 45 °C. A more detailed description of the main qualitative and quantitative information extracted from the potential time-series is presented as Supplementary Material (seeSupplementary Table S1).

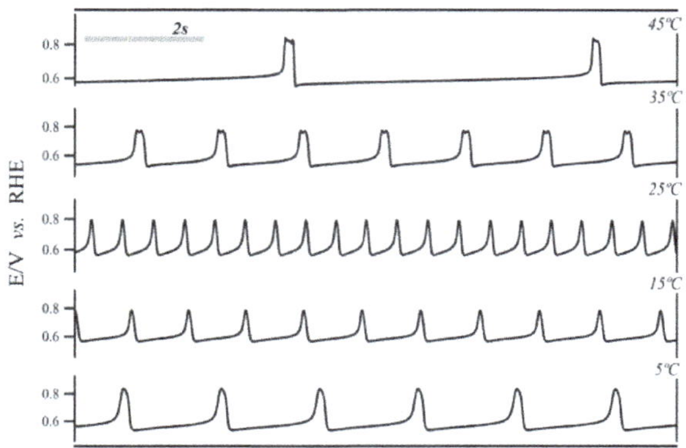

Figure 3: Period length (family F_1) for different temperatures during galvanostatic time-series ($J^T_{0.5}$).

The effect of temperature was also investigated by means of electrochemical impedance spectroscopy (EIS). Plotting the real impedance against the imaginary one, as a function of the modulation frequency , one obtains an impedance spectrum, which can be used to evaluate the stability of the system. Impedance spectra applying DC potentials from 500 to 1200 mV showed the typical qualitative features known for this system[30]. The negative differential resistance (NDR) region was found between 600 and 1000 mV for all temperatures tested.

Nyquist plots showing the general tendency of the temperature dependence on the semi-circles obtained are depicted in Fig. 4. The single semi-circles become smaller with increasing temperature, but there is also seen a break of this tendency for the experiments at 25 °C. Important in the present analysis, EIS can also be used qualitatively to identify Hopf and saddle-node bifurcations in an electrochemical system. Ultimately, a bifurcation is a situation where the impedance spectrum intersects or ends up in the origin of the complex impedance plane. If it cross the negative side of the real axis as $\rightarrow 0$ it corresponds to a saddle-node bifurcation; if however and if the spectrum intersects the negative side of the real axis at a non-zero frequency, the system undergoes a Hopf bifurcation[4]. In this sense, Hopf bifurcation frequencies $_{Hopf}$ were obtained by EIS at different DC potentials, at different temperatures for the formic acid system. We found increasing values of Hopf frequency up to 25 °C, with subsequent decrease for T>25 °C, corroborating with the tendency obtained under galvanostatic conditions.

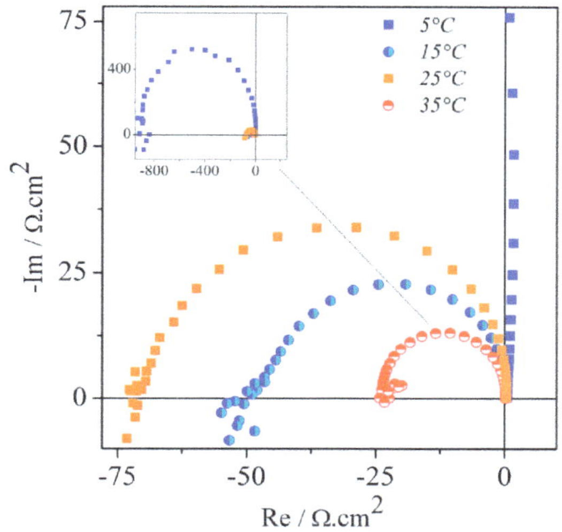

Figure 4: Impedance spectra registered at 800 mV for different temperatures.

Figure 5 summarizes the temperature dependence on most experimental parameters discussed so far. In contrast to the regular dependence observed for the voltammetric current, i.e. higher rates at higher temperatures, the behavior found in all variables studied under oscillatory regime shows a turning point at 25 °C. Activation energies estimated in the range between 5 and 25 °C vary roughly from 40 to 50 kJ mol^{-1}, and, between 25 and 45 °C, from −70 to −95 kJ mol^{-1}. Interestingly, comparable behavior was found in the three oscillation frequencies studied. Implications of these results are discussed in the next section.

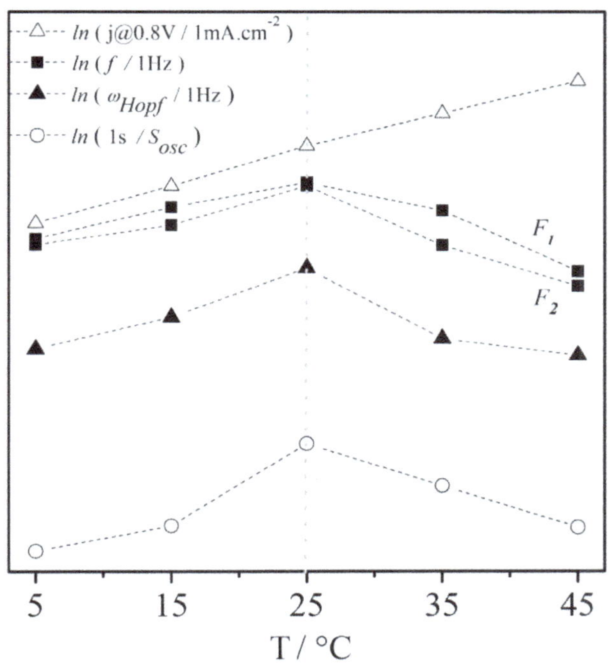

Figure 5: Temperature effect on frequencies f **(full squares)** ω_{Hopf} **(full triangles) and** S_{osc} (empty circles) showing a similar break of tendency at 25 °C.

DISCUSSION

We report the experimental study of the effect of temperature on the fast and slow dynamics observed in a model oscillatory surface reaction, namely the electro-oxidation of formic acid on polycrystalline platinum and in acidic media. Following our previous investigations[7], we explored the nonlinearities intrinsic to this reaction in order to deepener our understanding on the coupling between the two disparate time-scales responsible for the oscillations themselves and for the long-term surface deactivation. The main aspect to be discussed in this seminal approach is the unexpected existence of a turning point in the dynamics at 25 °C, at which the temperature dependence of most variables changes. Now we discuss important dynamic and mechanistic information that result from these empirical observations.

The electro-oxidation of formic acid on platinum and on platinum-based surfaces has been extensively studied as a model surface reaction, with application in energy conversion devices such as low temperature fuel cells[31]. Despite its structural simplicity, formic acid is oxidized in a complex triple parallel pathway, which can be reasonably represented by the scheme given in Fig. 6.

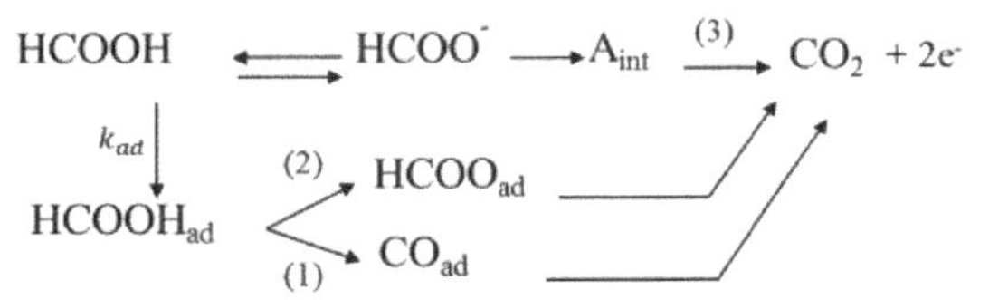

Figure 6: Schematic representation of the multiple parallel routes for the oxidation of formic acid on platinum, in acidic media.

The reaction scheme involves the adsorption of HCOOH followed by a non-faradaic dehydration of $HCOOH_{ad}$ (1) and the formation of chemisorbed carbon monoxide, CO_{ad}. Formic acid also proceeds to adsorbed formate (2) via two oxygen atoms, occupying two surface sites[32]. Due its high stability, formate is unlikely to be a reactive intermediate to CO_2 and might be in equilibrium with its soluble form acting as a mere blocking spectator at lower voltage[33], in contrast with the case of methanol[34]. In practical terms, it is safe to assume that formic acid is mainly oxidized through a non-formate pathway. For the main route, the direct pathway is believed to possess the lower activation energy[35] and the formation of carbon dioxide occurs via a weakly adsorbed intermediate, step (3), formed after the activation of the C-H bond[36], whose nature is still matter of debate. Feliu and co-workers[33] have recently published some new insights on the mechanism of formic acid system in which the authors suggest formate anion interacting with the surface in a C-H down configuration as the active intermediate, A_{int}. At considerably high potentials, the formation of adsorbed oxygenated species occurs, as generically represented by the step (4):

$$Pt + H_2O \rightarrow PtO + 2H^+ + 2e^-$$

and it can assist the oxidation of chemisorbed carbon monoxide via a Langmuir-Hinshelwood (LH) step (5):

$$PtCO + PtO \rightarrow 2Pt + CO_2$$

The effect exerted by the temperature in an electrocatalytic reaction such as the electro-oxidation of small organic molecules discussed here depends on parameters including the surface structure, the applied potential, the dominant current carrier step at the state being investigated, the relative surface coverage of adsorbates, among others[23]. Therefore, even under non-oscillatory conditions, the analysis of the apparent activation energy is a non-trivial task. Results presented in Fig. 1 reveal conventional temperature dependence under voltammetric regime. Simple Arrhenius dependence was found in this extended interval studied, and representative apparent activation energies found in the range between 40 and 60 kJ mol^{-1} are in agreement with previously published data[23,25,33].

During oscillations, the temperature dependence is even more intricate as more than one elementary step is probed in the potential window visited. The potential oscillations obtained at constant applied current reflect the continuous changes in the population of the different adsorbed species, such as CO_{ad}, $HCOO_{ad}$, and oxygenated species;[37] transport of species to/from the electrode surface is unlikely to be responsible for the observed qualitative differences[38]. Comprehensive description and mechanism of the oscillatory electro-oxidation can be found elsewhere[28].

The temperature dependence of steps (1)–(5) globally controls the response under both voltammetric and galvanostatic regimes in different ways. In the first case, the distribution of apparent activation energies, c.f. Fig. 1, reflects the temperature dependence of the rate-determining step at a given applied potential. Under oscillatory regime, different steps are simultaneously probed and the activation energy estimated by global quantities, such as the oscillatory frequency, is a combination of different steps. In this respect, we studied the effect of temperature in terms of the oscillatory frequency for two families of oscillations[7]. In order to infer on the intrinsic dynamics near the Hopf bifurcation we also estimated the Hopf frequency via impedance spectroscopy, c.f. Fig. 4.

As showed in Fig. 5, conventional Arrhenius behavior prevails for temperatures up to 25 °C when the subsequent increase of temperature reveals an opposite, counter-intuitive, tendency: the oscillation frequency decreases for a temperature increase. The latter behavior is referred to as temperature overcompensation and corresponds to a situation of negative apparent activation energy. Temperature (over)compensation has been found in the oscillatory electro-oxidation of formic acid[7,39,40], and ethylene glycol[54]. Important to the present context, in all these cases, this unusual behavior was proved to emerge from the intricate combination of several reaction steps, *vide infra*.

As for the long-term evolution, it has been shown that even though all experimental parameters are carefully controlled, the system slowly evolves and visits distinct dynamic states, c.f. oscillations of type F_1 and F_2, up to the end of oscillations. The slow spontaneous changes that drive the system occurs in a much slower time-scale than that found for the oscillations themselves, referred here to as core oscillator. The time-scales separation ( ), estimated as the ratio between the total oscillation time and typical oscillation period, is about three orders of magnitude, c.f. Supplementary Table S1. The impact exerted by temperature on the long-term evolution was studied by means of the size of the oscillatory region, S_{osc}, obtained for a given normalized applied current, Fig. 2. As seen in this figure, S_{osc} initially decreases with temperature from 5 to 25 °C (note that $1/S_{osc}$ is plotted in Fig. 5), and then it increases in the higher temperature domain. This behavior would apparently imply that the rate of the slow drift is first accelerated with the temperature increase and then slowed down for 25 °C < T ≤ 45 °C. This is however an oversimplified description as the slow drift is chemically coupled to the core oscillator. Before a more appropriate reasoning, we present in the following the current understanding of the mechanism behind this slow changes.

The slow drift observed in oscillatory electrochemical time-series have been long observed and generally attributed to surface processes rather than bulk phenomena such as mass transport or homogeneous reactions[41,42]. We have studied in depth this phenomena in the electro-oxidation of small organic molecules on platinum in acidic media and proposed a mechanism[43,44]. In short, when the formation of surface oxygenated species reaches a certain limit, the place exchange phenomena[45,46] results in the insertion of these species into the platinum lattice, which can be represented generically as,

$$PtO \rightarrow O_{sub}Pt$$

Consequently, the subsurface oxygen, O_{sub}, becomes unavailable to assist the surface oxidation of adsorbed carbonaceous species such as *PtCO*, and also the site $O_{sub}Pt$ becomes less active to catalyze most steps. The accumulation of subsurface oxygen occurs in a much slower time-scale than the potential oscillations and it can be inferred by the increase of the mean potential[43] and also by the accumulation of *PtCO*, in the case of methanol[44].

Bringing together the current understanding of the mechanism of the core oscillator and of the slow surface deactivation, it becomes evident the importance of *PtO* species: it can either assist the oxidation of, say, adsorbed carbon monoxide, step (5), or form $O_{sub}Pt$ and deactivate the surface, step (6). Therefore, *PtO*species chemically couple slow and fast dynamics. The results obtained for the effect of temperature on the size of the oscillatory window, S_{osc}, can be now discussed in terms of Fig. 7, which schematically illustrates the rates of the Langmuir-Hinshelwood (LH) step (5), v_{LH}, and of the slow deactivation process (6), v_d.

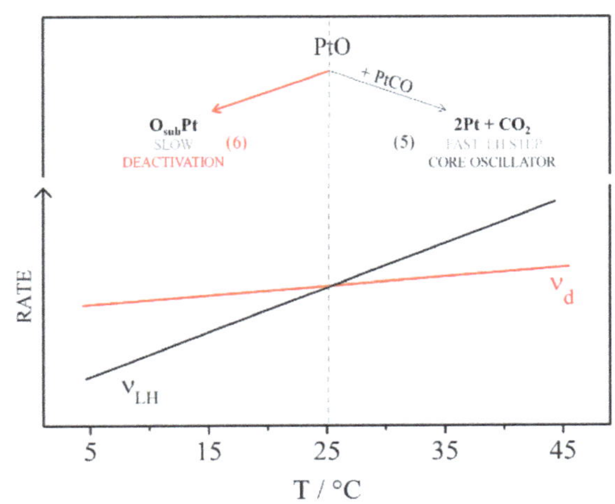

Figure 7: Schematic representation of the coupled slow and fast dynamics in the formic acid galvanostatic oxidation and the temperature dependence observed.

As the main idea introduced in this work, temperature was used to infer on the relative rates of the processes involved in the disparate and coupled time-scales intrinsic to the system. Figure 7 illustrates the two temperature domains observed that can be explained in terms of the *PtO* species. In the low temperature domain, the decrease of the oscillatory window with temperature evidences that the rate v_d is higher than that of the LH step. In the same way, from 25 to 45 °C the LH step is faster and the consumption of oxygenated species *PtO* in the oxidation of adsorbed carbon monoxide, step (5), prevents the place exchange phenomena, step (6), and the time-series becomes longer with the temperature increase, c.f. Fig. 2. Focusing now on the whole temperature range illustrated in Fig. 7, it becomes clear that the activation energy of the LH step is higher than that of the slow surface deactivation, previously attributed to the place exchange phenomena. This observation can also be related to data available in the literature, obtained under non-oscillatory conditions. In fact, activation energies of at least about 100 kJ.mol^{-1} have been found for the electro-oxidation of carbon monoxide on platinum single crystals[47]. In contrast, it seems there is a rather very weak dependence on temperature in the oxide formation step, and presumably on the place exchange process[48]. It has been shown[49,50] that, for a temperature range of 278 ≤ T ≤ 323 K, the increase of oxide formation and reduction is about 26%, under comparable conditions. Moreover, the activation energy for the direct oxidation of formic acid was found to vary from 50 to 60 kJ.mol^{-1} and its dependence on the electrode potential rationalized in terms of the changes in the coverage of anions and oxygenated species[22,51].

An interesting aspect presented in our results is the turning point found for many variables at 25 °C. The formation of adsorbed CO on platinum from formic acid is essentially a chemical step, which has, however, an electrochemical nature since the dehydration of formic acid is potential-dependent. Data available in the literature, obtained by means of CO_{ad} stripping add more important information to the matter, see ref. [47] and references therein. The onset potential for CO_{ad} oxidation significantly shifts towards less positive values as temperature is increased and, in addition to the oxidation charge, the authors indicate a change in the stability of the CO_{ad} adlayer depending on the temperature used. A very interesting result is that the CO_{ad} oxidation charge for Pt(111) shows decreasing values from 270 to 298 K (having its minimum value at 298 K) with a slight increase in the range 298 K < T < 323 K. At 298 K the authors observed the lowest CO coverage($_{CO}$). Once the oscillations found here shows shortest stability window at this very temperature, it seems reasonable to infer that higher oscillatory frequencies are observed for smaller maximum critical $_{CO}$. Taking into account that the time necessary to reach a certain critical $_{CO}$ must increase as $_{CO}$ increases in oscillatory conditions, the observance of *f* increasing for 5° < T < 25° and decreasing for 25° < T < 45 °C can be related to the experimental findings[47] for the temperature dependence on $_{CO}$. In addition, EIS data obtained in the present work reveal smaller values for double layer capacitance, C_{dl}, at 25 °C in comparison with the neighbour values of temperatures tested, strongly suggesting a less blocked surface at this temperature, as shown in Supplementary Fig. S3. Despite the experimental differences (mainly surface structure, species being oxidized, and the reaction regime itself), it is remarkable the turning point found at 25 °C. The way in which the increase in the CO coverage above 25 °C[47] and the higher rates of the LH step observed here during oscillations regime are under numerical investigation at the moment in our Group.

Back to the temperature dependence of the core oscillator, unusual behavior such as temperature compensation and overcompensation can occur in a network of chemical reactions and be reflected in responses such as the constancy of the oscillatory frequency. Overall, these unconventional temperature dependencies can be conveniently rationalized with the concept of antagonistic balance in which period increasing and decreasing sets of elementary reactions and their respective activation energies are combined to result in oscillatory apparent activation energy[52,53]. As the region in which the oscillation frequency decreases with temperature, i.e. T>25 °C, occurs where the LH step, (5), is dominant, we can speculate that this step belongs to the *period-increasing* set. It is important to recall that there is no available method to classify the individual steps as it is difficult to isolate a given elementary reaction and estimate its role when part of a network.

CONCLUSIONS

We report the effect of temperature on the disparate time-scales observed in a model oscillatory surface reaction, namely the electro-oxidation of formic acid on polycrystalline platinum and in acidic media. Already in the short-term dynamics (or of the core oscillator), we found a turning point at 25 °C, where the temperature dependence of most variables changes noticeably. This turning point was also found in the parameter used to study the long-term dynamics. The two regions are characterized by a conventional Arrhenius behavior between 5 °C and 25 °C, and by non-Arrhenius dependence for temperatures between 25 °C and 45 °C, where essentially, the oscillation frequency decreases as the temperature increases. This behavior was related to previously published data on the turning point on the coverage of adsorbed carbon monoxide at 25 °C and also corroborated by the impedance data presented.

The current understanding of the dynamics of the core oscillator and of the slow surface deactivation stresses the dual and pivotal role of *PtO* species, as it assists the oxidation of adsorbed carbon monoxide and forms$O_{sub}Pt$, the long-term surface-blocking species. The competition between these two processes was explained in terms of long-term evolution and the turning point at 25 °C. Based on these findings, we were able to compare the activation energies of both steps and clearly attribute a weaker temperature dependence of the slowly evolving surface deactivation. This observation is also corroborated by independent experimental results and, importantly, obtained under non-oscillatory conditions. The described competition between the two steps also allowed us to speculate on the role of the surface oxidation of carbon monoxide by *PtO*species as being part of the set of reactions that contribute to the increase of the oscillation period with temperature.

The study of temperature effect on both short and long-term dynamics introduced here can be though as a methodology to deepen the current knowledge on some mechanistic aspects. The discrimination of the elementary reactions belonging to the period-in/decreasing sets and the also the determination of activation parameters for individual steps are particularly important to uncover the present and also more complex mechanisms. We are currently exploring such aspects using similar systems and also other experimental parameters, such as the influence of different surface-blocking species.

METHODS

A conventional glass cell equipped with three electrodes was used in our experiments. As working electrode we used smooth polycrystalline platinum with a geometric area of 0.2 cm^2 and roughness factor between 2.0 and 2.5. Current densities (j) were calculated dividing the current applied by the electrochemical active area, which in turn was calculated using the well accepted charge for UPD hydrogen region as 210 μC. cm^{-2}[11,54]. A reversible hydrogen electrode (RHE) was prepared using the same solution employed as supporting electrolyte and was chosen as reference electrode. All potentials were corrected and are referred against RHE at 298 K[55]. A large-area platinum electrode was used as counter electrode. All solutions were prepared using ultra-pure water (Milli-Q system, Millipore, 18.2 MΩ.cm), sulfuric acid (Merck, 98% purity) and formic acid (Sigma-Aldrich, purity ≥98%). The electrochemical measurements were all done by means of a potentiostat/galvanostat PGSTAT 30 (Autolab) equipped with SCANGEN and FRA modules. The temperature of the bulk solution was varied and maintained through a thermostatic bath, monitored using a K-type thermocouple (precision of ±0.5 K). Electrochemical impedance spectroscopy (EIS) was performed in the frequency range from 10 kHz to 0.01 Hz, applying an AC modulation of 10 mV. To assure stationary state, chronoamperometric transients at the identical DC potential to be applied were recorded for at least 500 s after a time-invariant profile had been reached (dj/dt = 0). All experiments described here were performed at [HCOOH] = 1 M dissolved in aqueous [H_2SO_4] = 0.5 M.

Supplementary Material

Table Si: Description of the average features observed for the oscillatory dynamics at different temperatures

T (°C)	$\bar{f}$ / Hz*		%S_{osc}**		$\overline{\Delta E}$ / mV*		Waveform		τ ***	
	F_1	F_2	F_1	F_2	F_1	F_2	F_1	F_2	F_1	F_2
5	0.5	0.8	23.8	19.0	240	110	P^1	P^1	936	1400
15	0.9	1.5	18.0	18.8	250	115	P^1	P^1	900	1500
25	2.5	3.0	28.0	43.7	290	200	P^1,P^2	P^1 , P^2	500	750
35	0.6	1.6	24.3	36.5	280	190	P^1,P^2	P^1 , P^2	250	664
45	0.2	0.4	18.2	9.4	275	150	P^1,P^2	P^1 , P^2	200	400

*Average value for at least 5 cycles. **Average percentage from the overall S_{osc} spent by each family of oscillations with respect to the whole window.***Estimation of time-scale separation as average value of at least 5 cycles ($\tau = \overline{S_{osc}} \times \bar{f}$).

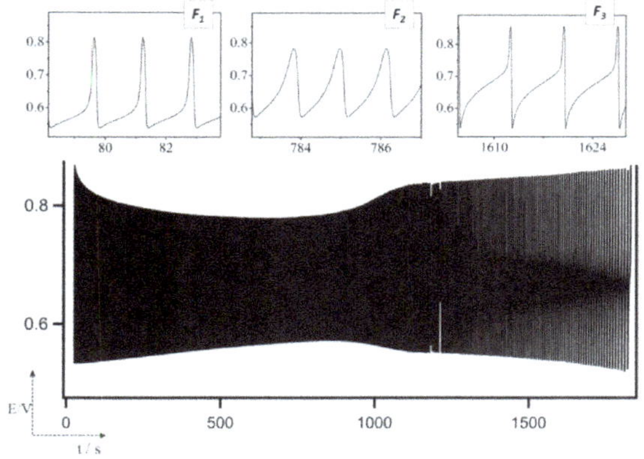

Figure S1: On the top: Three zoomed regions of the potential time series (complete time series shown on the bottom) displaying three distinguishable oscillatory families. Data measured for the system at $J^{5°C}_{0.5}$.

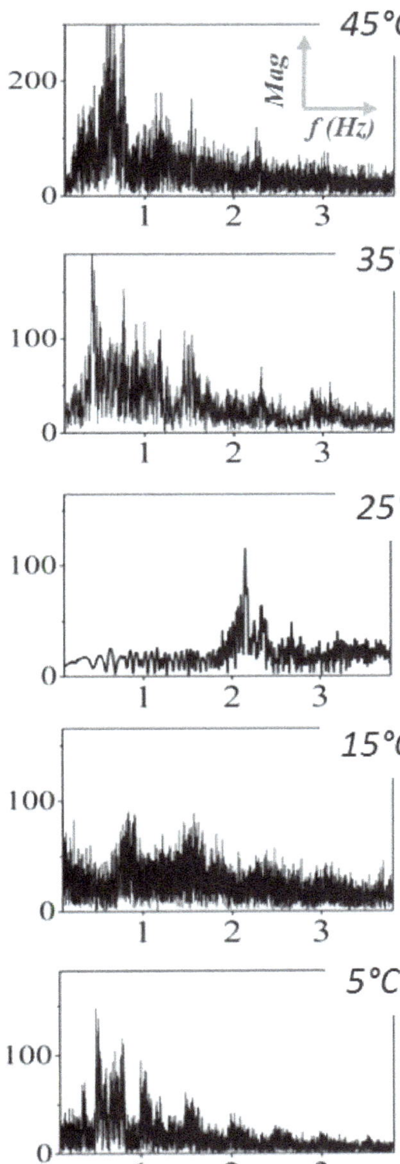

Figure S2: FFT Spectra $J^{T}_{0.5}$ at different temperature as indicated. Magnitude versus frequency (Hz);

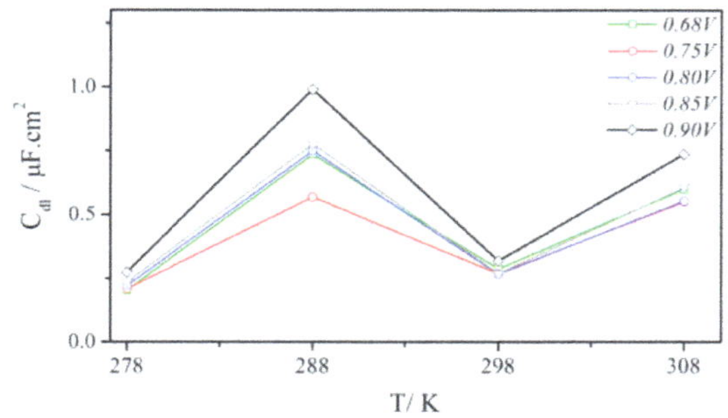

Figure S3: Temperature dependence on the double layer capacitance at different DC potentials as indicated.

Acknowledgments

A.A.Z. and H.V. acknowledge São Paulo Research Foundation (FAPESP) for the scholarship (grant #2013-06205-3) and financial support (grants #2009/07629-6, #2012/21204-0, and #2013/16930-7). HV (#306151/2010-3) acknowledges Conselho Nacional de Desenvolvimento Científico e Tecnológico (CNPq) for financial support. We also express our gratitude to E.G. Machado, L.F. Sallum, and R. Nagao for fruitful discussions and general assistance.

REFERENCES

1. Kuehn C. Multiple Time Scale Dynamics. 191, (Springer International Publishing, 2015).
2. Orlik M. Self-Organization in Electrochemical Systems I: General Principles of Self-organization. Temporal Instabilities. (Springer Science, 2012).
3. Koper M. T. M. Non-linear phenomena in electrochemical systems. J. Chem. Soc. Faraday Trans. 94, 1369–1378 (1998).
4. Koper M. T. M. The theory of electrochemical instabilities. Electrochim. Acta 37, 1771–1778 (1992).
5. Perini N., Batista B. C., Angelo A. C. D., Epstein I. R. & Varela H. Long-Lasting Oscillations in the Electro-Oxidation of Formic Acid on PtSn Intermetallic Surfaces. ChemPhysChem 15, 1753–1760 (2014).
6. Nagao R., Cantane D., Lima F. H. B. & Varela H. The dual pathway in action: decoupling parallel routes for CO_2 production during the oscillatory electro-oxidation of methanol. Phys. Chem. Chem. Phys. 14, 8294 (2012).
7. Nagao R., Epstein I. R., Gonzalez E. R. & Varela H. Temperature (over)compensation in an oscillatory surface reaction. J. Phys. Chem. A. 112, 4617–24 (2008).
8. Nascimento M. A., Nagao R., Eiswirth M. & Varela H. Coupled slow and fast surface dynamics in an electrocatalytic oscillator: Model and simulations. J. Chem. Phys. 141, 234701 (2014).
9. Cabral M. F., Nagao R., Sitta E., Eiswirth M. & Varela H. Mechanistic aspects of the linear stabilization of non-stationary electrochemical oscillations. Phys. Chem. Chem. Phys. 15, 1437–1442 (2013).
10. Sitta E., Nagao R., Kiss I. Z. & Varela H. Impact of the Alkali Cation on the Oscillatory Electro-Oxidation of Ethylene Glycol on Platinum. J. Phys. Chem. C 119,

1464–1472 (2015).

11. Capon A. & Parson R. The oxidation of formic acid at noble metal electrodes. Part I. J. Electroanal. Chem. Interfacial Electrochem. 44, 1–7 (1973).
12. Clavilier J., Parsons R., Durand R., Lamy C. & Leger J. M. Formic acid oxidation on single crystal platinum electrodes. Comparison with polycrystalline platinum. J. Electroanal. Chem. Interfacial Electrochem. 124, 321–326 (1981).
13. Joo J., Uchida T., Cuesta A., Koper M. T. M. & Osawa M. The effect of pH on the electrocatalytic oxidation of formic acid/formate on platinum: A mechanistic study by surface-enhanced infrared spectroscopy coupled with cyclic voltammetry. Electrochim. Acta 129, 127–136 (2014).
14. Brimaud S., Solla-Gullón J., Weber I., Feliu J. M. & Behm R. J. Formic Acid Electrooxidation on Noble-Metal Electrodes: Role and Mechanistic Implications of pH, Surface Structure, and Anion Adsorption. ChemElectroChem 1, 1075–1083 (2014).
15. Raspel F., Nichols R. J. & Kolb D. M. Current oscillations during formic acid oxidation on Pt(100). J. Electroanal. Chem. Interfacial Electrochem. 286, 279–283 (1990).
16. Mukouyama Y., Kikuchi M., Samjeské G., Osawa M. & Okamoto H. Potential oscillations in galvanostatic electrooxidation of formic acid on platinum: A mathematical modeling and simulation. J. Phys. Chem. B 110, 11912–11917 (2006).
17. Koper M. T. M. & Gaspard P. MIxed-Mode and Chaotic Osclllations In a Simple Model of an Electrochemical Oscillator. J. Phys. Chem. 4945–4947 (1991).
18. Kiss I. Z., Munjal N. & Martin R. S. Synchronized current oscillations of formic acid electro-oxidation in a microchip-based dual-electrode flow cell. Electrochim. Acta 55, 395–403 (2009).
19. Raspel F. & Eiswirth M. Current oscillations in the electrochemical oxidation of formic acid at Pt single crystal surfaces. J. Phys. Chem. 98, 7613–7618 (1994).
20. Samjeské G. *et al.* Potential oscillations in galvanostatic electrooxidation of formic acid on platinum: a time-resolved surface-enhanced infrared study. J. Phys. Chem. B 109, 23509–16 (2005).
21. Strasser P., Lübke M., Raspel F., Eiswirth M. & Ertl G. Oscillatory instabilities during formic acid oxidation on Pt(100), Pt(110) and Pt(111) under potentiostatic control. I. Experimental. J. Chem. Phys.107, 979 (1997).
22. Perales-Rondón J. V., Herrero E. & Feliu J. M. On the activation energy of the formic acid oxidation reaction on platinum electrodes. J. Electroanal. Chem. 742, 90–96 (2015).
23. Angelucci C. A., Varela H., Herrero E. & Feliu J. M. Activation Energies of the Electrooxidation of Formic Acid on Pt(100). J. Phys. Chem. C 113, 18835–18841 (2009).
24. Perales-Rondón J. V., Herrero E. & Feliu J. M. Effects of the anion adsorption and pH on the formic acid oxidation reaction on Pt(111) electrodes. Electrochim. Acta 140, 511–517 (2014).
25. Ferre-Vilaplana A., Perales-Rondón J. V., Feliu J. M. & Herrero E. Understanding the Effect of the Adatoms in the Formic Acid Oxidation Mechanism on Pt(111) Electrodes. ACS Catal. 5, 645–654 (2015).
26. Tian M. & Conway B. E. Electrocatalysis in oscillatory kinetics of anodic oxidation of formic acid: At Pt; nanogravimetry and voltammetry studies on the role of reactive surface oxide. J. Electroanal. Chem. 616, 45–56 (2008).
27. Koper M. T. M. In Adv. Chem. Phys. (ed. Prigogine I., Rice S. A.) 82, 161–298 (John Wiley & Sons, Inc., 1992).
28. Krischer K. & Varela H. In Handb. Fuel Cells Fundam. Technol. Appl. (eds. Vielstich W., Gasteiger H. A., Lamm A. & Yokokawa H.) (John Wiley & Sons, Inc., 2010).
29. Nascimento M. A., Gallas J. A. C. & Varela H. Self-organized distribution of periodicity and chaos in an electrochemical oscillator. Phys. Chem. Chem. Phys. 13, 441–446 (2011).
30. Seland F., Tunold R. & Harrington D. Impedance study of formic acid oxidation on platinum electrodes. Electrochim. Acta 53, 6851–6864 (2008).
31. Vielstich W., Yokokawa H. & Gasteiger H. A. Handbook of fuel cells: fundamentals, technology, and applications. Adcances in electrocatalysis, materials, diagnostics and durability. (John Wiley and Sons, 2009).
32. Miki A., Ye S. & Osawa M. Surface-enhanced IR absorption on platinum nanoparticles: An application to real-time monitoring of electrocatalytic reactions. Chem. Commun. 1500–1501 (2002).
33. Perales-Rondón J. V. *et al.* Further Insights into the Formic Acid Oxidation Mechanism on Platinum: pH and Anion Adsorption Effects. Electrochim. Acta 180, 479–485 (2015).
34. Chen Y.-X., Heinen M., Jusys Z. & Behm R. J. Bridge-bonded formate: active intermediate or spectator species in formic acid oxidation on a Pt film electrode? Langmuir 22, 10399–408 (2006).
35. Boronat-González A., Herrero E. & Feliu J. M. Fundamental aspects of HCOOH oxidation at platinum single crystal surfaces with basal orientations and modified by irreversibly adsorbed adatoms.J. Solid State Electrochem. 18, 1181–1193 (2014).
36. Neurock M., Janik M. & Wieckowski A. A first principles comparison of the mechanism and site requirements for the electrocatalytic oxidation of methanol and formic acid over Pt. Faraday Discuss.140, 363–378 (2009).
37. Eiswirth M., Bürger J., Strasser P. & Ertl G. Oscillating Langmuir−Hinshelwood Mechanisms. J. Phys. Chem. 100, 19118–19123 (1996).
38. Chen S. & Schell M. Effects of anion adsorption different from blocking surface sites deduced from instabilities in the oxidation of formic acid. J. Electroanal. Chem. 504, 78–88 (2001).
39. Sitta E., Nascimento M. A. & Varela H. Complex kinetics, high frequency oscillations and temperature compensation in the electro-oxidation of ethylene glycol on platinum. Phys. Chem. Chem. Phys 12, 15195–206

(2010).

40. Schmidt T. J., Grgur B. N., Markovic N. M. & Ross P. N. Oscillatory behavior in the electrochemical oxidation of formic acid on Pt(100): rotation and temperature effects. J. Electroanal. Chem. 500, 36–43 (2001).

41. Neher G., Pohlmann L. & Tributsch H. Mixed-Mode-Oscillations, Self-Similarity, and Time-Transient Chaotic Behavior in the (Photo-) Electrochemical System p-$CuInSe_2/H_2O_2$. J. Phys. Chem. 99, 17763–17771 (1995).

42. Okamoto H. *et al.* Oxidation of formic acid and methanol and their potential oscillations under no or little water conditions. Electrochim. Acta 136, 385–395 (2014).

43. Nagao R., Sitta E. & Varela H. Stabilizing Nonstationary Electrochemical Time Series. J. Phys. Chem. C. 114, 22262–22268 (2010).

44. Boscheto E., Batista B. C., Lima R. B. & Varela H. A surface-enhanced infrared absorption spectroscopic (SEIRAS) study of the oscillatory electro-oxidation of methanol on platinum. J. Electroanal. Chem. 642, 17–21 (2010).

45. Angerstein-Kozlowska H. & Conway B. E. The real condition of electrochemically oxidized platinum surfaces*: Part I. Resolution of component processes. Electroanal. Chem. Interfacial Chem. 43, 9–36 (1973).

46. Conway B. Electrochemical oxide film formation at noble metals as a surface-chemical process. Prog. Surf. Sci. 49, 331–452 (1995).

47. Herrero E. *et al.* Temperature dependence of the COads oxidation process on Pt(111), Pt(100), and Pt(110) electrodes. J. Electroanal. Chem. 567, 139–149 (2004).

48. Schmidt T. J., Grgur B. N., Gasteiger H. a., Behm R. J. & Ross P. N. Effect of Temperature on Surface Processes at the Pt (111)-Liquid Interface : Hydrogen Adsorption, Oxide Formation, and CO Oxidation. J. Phys. Chem. B 103, 8568–8577 (1999).

49. Alsabet M., Grden M. & Jerkiewicz G. Comprehensive study of the growth of thin oxide layers on Pt electrodes under well-defined temperature, potential, and time conditions. J. Electroanal. Chem. 589, 120–127 (2006).

50. Jerkiewicz G., Alsabet M., Grden M., Varela H. & Tremiliosi-Filho G. Erratum to: 'Comprehensive study of the growth of thin oxide layers on Pt electrodes under well-defined temperature, potential, and time conditions' Journal of Electroanalytical Chemistry, 589 (2006) 120–127. J. Electroanal. Chem.625, 172–174 (2009).

51. Capon A. & Parsons R. The oxidation of formic acid on noble metal electrodes. Part II. J. Electroanal. Chem. Interfacial Electrochem. 44, 239–254 (1973).

52. Ruoff P. Antagonistic balance in the oregonator: about the possibility of temperature-compensation in the Belousov-Zhabotinsky. Phys. D Nonlinear Phenom. 84, 204–211 (1995).

53. Ruoff P., Christensen M. K., Wolf J. & Heinrich R. Temperature dependency and temperature compensation in a model of yeast glycolytic oscillations. Biophys. Chem. 106, 179–192 (2003).

54. Capon A. & Parsons R. The oxidation of formic acid at noble metal electrodes Part III. Intermediates and mechanism on platinum electrodes. J. Electroanal. Chem. Interfacial Electrochem. 45, 205–231 (1973).

55. Conway B. E. & Currie J. C. Temperature and pressure effects on surface processes at noble metal electrodes. Part 2.—Volume of adsorbed H and oxygen species at Pt and Au. J. Chem. Soc. Faraday Trans. 1 Phys. Chem. Condens. Phases 74, 1390 (1978).

CHAPTER 17

CURRENT-INDUCED TRANSISTOR SENSORICS WITH ELECTROGENIC CELLS

Peter Fromherz

Keywords: Transistor, extracellular recording, ion channel, neuron, action potential

INTRODUCTION

The electrolyte-oxide-semiconductor field-effect transistor (EOSFET) has been introduced as an ion sensitive sensor (ISFET), where the interaction of ions with the gate oxide induces a changed source-drain current at a constant gate voltage applied to a reference electrode in the bath electrolyte [1]. A more specific interface-induced response can be achieved for biological analytes when the gate oxide is modified by suitable coatings [2]. However, response of an EOSFET may also be induced by ionic currents in the electrolyte, which give rise to a changed electrical potential beyond the electrical double layer of the oxide. The present paper describes the concepts of current-induced transistor sensorics for electroactive cells with a focus on planar cell-substrate junctions. The situation is illustrated in Figure 1 [3]: the adherent cell membrane and the semiconductor chip are separated by a thin cleft of extracellular electrolyte. Ionic and capacitive currents through the membrane give rise to a current flow along the cleft. That ionic current induces a voltage VJ with respect to the bath and modulates the electronic source-drain current of a transistor at constant gate voltage.

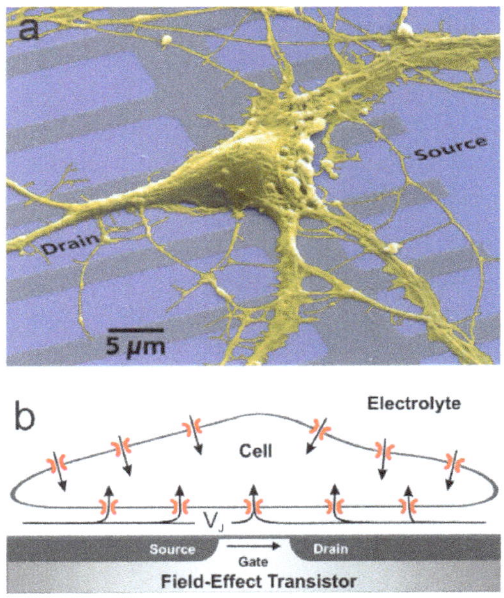

Figure 1. Planar cell-substrate junction with an EOSFET (electrolyte-oxide-semiconductor field-effect transistor), implementing the coupling of ionic and electronic currents: (**a**) Electronmicrograph of a cultured nerve cell from rat hippocampus on the silica surface of a silicon chip with an array of EOSFETs (by permission [3]); (**b**) Schematic cross section (not to scale). A cleft of about 50 *nm* with extracellular electrolyte separates the cell membrane and the silica surface. Ionic and capacitive currents through the cell membrane give rise to a current flow along the cleft. The ionic current induces a voltage V_J with respect to the bath and modulates the electronic source-drain current at a constant gate voltage applied to a reference electrode in the bath.

Citation: Fromherz, P. Current-Induced Transistor Sensorics with Electrogenic Cells. Biosensors 2016, 6, 18.

Three issues are addressed in particular: (i) the cell-substrate junction is described as a planar capacitive core-coat conductor, which allows relating the cellular currents with changes of the electrical potential and changes of ion concentrations in the junction; (ii) in certain situations, the current-induced response is superposed by the interface-induced response of an ion-sensitive gate due to changed ion concentrations; and (iii) the inhomogeneous distribution of ion conductances in the cell membrane is crucial for a current-induced transistor response with intact electrogenic cells.

This paper is a scholarly overview of concepts and does not intend to provide a comprehensive review of the field, though some inconsistencies in the literature are addressed. The examples for various kinds of transistor recordings with electroactive cells are chosen from studies in my own lab.

The structure of the paper is as follows: First, the physics of the EOSFET is considered in order to identify the current-induced and the interface-induced transistor response with electrogenic cells. Then, the theory of a planar capacitive core-coat conductor is presented, which relates cellular currents with changes of the electrical potential and of ion concentrations in a cell-substrate junction. The dynamics of a core-coat conductor is described, and the reduction to a one-compartment model is derived for constant and for changing ion concentrations. On that basis, the transistor recording of voltage-activated ion channels is considered for an intracellular voltage-clamp. In this context, the role of an ion-sensitive gate oxide and the issue of inhomogeneous membrane conductances are pointed out. Current-induced transistor sensorics with intact cells is described for neuronal action potentials recorded at soma and axon, and for a receptor-cell-transistor biosensor.

Several technical issues have to be envisaged for cell-based transistor sensorics: (i) The gate must be smaller than a cell, and far smaller than a traditional transistor sensorics gate. As a consequence, the noise power, which is reciprocal to the gate area, becomes a problem; (ii) Current-induced extracellular voltages are in the range of a few microvolts to a few millivolts, lower than usual interface-induced signals. The signal-to-noise ratio is a critical issue; (iii) The transistor signals vary due to a variability of the cells, to the variable interactions of cells and substrate, and to a variable position of the transistor in the cell-substrate junction. These problems are not addressed here.

TRANSISTOR SENSORICS WITH CELLULAR CURRENTS

First some general aspects of the EOSFET are considered to avoid misunderstandings. On that basis, the interface-induced and current-induced responses of transistor recording are identified for electrogenic cells.

Gate voltage and electrical potential. The source–drain current of an EOSFET is determined by band bending, the drop of the electrical potential in the semiconductor $\Delta\varphi_{Si}^{S} = \varphi_{Si}^{S} - \varphi_{Si}^{\infty}$ from its surface at the gate oxide to the bulk [4]. Band bending is controlled by a gate voltage V_G^{EOS} that is applied between the wires joined to a reference electrode (usually Ag/AgCl) in the bulk electrolyte and to the backside metal contact of the bulk semiconductor. (Practically, a voltage $V_{Si} = -V_G^{EOS}$ is applied to the semiconductor with the electrode on ground potential.) We consider the relation of V_G^{EOS} with the difference $\varphi_E^{\infty} - \varphi_{Si}^{\infty}$ of the electrical potentials in the bulk electrolyte and in the bulk semiconductor, and then the relation of that difference with the band bending $\Delta\varphi_{Si}^{S}$.

The metal wires (identical materials) are in electronical equilibrium with the metal contacts (different materials) of the electrolyte and of the semiconductor, respectively, and these contacts are in electronical equilibrium with the redox solution of the reference electrode and with the bulk semiconductor, respectively. These equilibrations are described as alignments of the electrochemical potential of electrons (Fermi energies) $\tilde{\mu}_k^{el} = \mu_k^{el} - e_0\varphi_k$ (e_0 elementary charge) in the different phases with components of an electronic chemical potential and the electrical potential. (For the quasi-Fermi energy of a redox solution see [5].) As a consequence, the gate voltage controls the difference $\tilde{\mu}_{red/ox}^{el} - \tilde{\mu}_{Si}^{el}$ of the electronic electrochemical potentials in the redox solution and in the semiconductor [6]. When we take into account the electrical equilibration of the redox solution and the bulk electrolyte through a salt-bridge with $\varphi_{red/ox} = \varphi_E^{\infty}$, the relation between the gate voltage and the difference of the electrical potentials in the bulk electrolyte and the bulk semiconductor is expressed by Equation (1).

$$-e_0 V_G^{EOS} = \left(\mu_{red/ox}^{el} - \mu_{Si}^{el}\right) - e_0\left(\varphi_E^{\infty} - \varphi_{Si}^{\infty}\right) \tag{1}$$

The electronic chemical potentials in the redox solution and in the semiconductor are important for the value of the flatband and the threshold voltage of an EOSFET (see Appendix A), but irrelevant for changes of the gate voltage around a working voltage of the field-effect transistor above the threshold of strong inversion with $\delta V_G^{EOS} = \delta\left(\varphi_E^{\infty} - \varphi_{Si}^{\infty}\right)$.

The overall difference $\varphi_E^{\infty} - \varphi_{Si}^{\infty}$ of the electrical potentials can be partitioned into several components according to Equation (2a) with the potential drops $\Delta\varphi_E^S$ and $\Delta\varphi_{Si}^S$ in the electrolyte and in the semiconductor (both from the surface at the oxide to the bulk), with the potential jumps $\Delta\varphi_{OxE}$ and $\Delta\varphi_{OxSi}$ across the interfaces oxide-to-electrolyte and oxide-to-semiconductor, and with the potential drop $\Delta\varphi_{Ox}$ across the oxide (electrolyte-to-semiconductor). The latter is split into contributions of a fixed charge σ_{OxSi} at the oxide/semiconductor interface and of the space charge in the semiconductor according to Equation (2b) for a p-channel transistor in depletion approximation (gate capacitance c_{Ox}, permittivity $\varepsilon_{Si}\varepsilon_0$, donor concentration ND) [4].

$$\varphi_E^{\infty} - \varphi_{Si}^{\infty} = -\Delta\varphi_E^S - \Delta\varphi_{OxE} + \Delta\varphi_{Ox} + \Delta\varphi_{OxSi} + \Delta\varphi_{Si}^S \tag{2a}$$

$$\Delta\varphi_{Ox} = -c_{Ox}^{-1}\left[\sigma_{OxSi} + \sqrt{2\varepsilon_{Si}\varepsilon_0 e_0 N_D\left(-\Delta\varphi_{Si}^S\right)}\right] \tag{2b}$$

The fabrication of a chip determines the potential drop $-\sigma_{OxSi}/c_{Ox}$ as well as the potential jump $\Delta\varphi_{OxSi}$. The conditions of the electrolyte determine the potential jump $\Delta\varphi_{OxE}$ (Si=O and Si-OH groups, oriented H_2O, double layer of $Si\text{-}O^-$ groups and adsorbed ions) [7]. The term quasi-

Helmholtz layer is used in analogy to a metal/electrolyte interface with $\Delta\varphi_{OxE} \equiv \Delta\varphi_{Hh}$. The conditions of the electrolyte also determine the potential drop $\Delta\varphi_E^S$ with a component $\Delta_{\varphi GCh}$ in a diffuse electrical double layer (Gouy-Chapman) and a component $\Delta_{\varphi E}$ in the free electrolyte beyond the double layer, which is induced by a current flow in the electrolyte. $\Delta_{\varphi E}$ vanishes in equilibrium for the electroneutral electrolyte. Considering Equations (1) and (2a), a changed gate voltage δV_G^{EOS} is partitioned into a changed current-induced drop of the electrical potential in the free electrolyte, a changed potential drop across the oxide/electrolyte interface (with Gouy-Chapman and quasi-Helmholtz layer), and a changed band bending (with the changed potential drop $\Delta\varphi_{Ox}^{spch}$ due the space charge) according to Equation (3).

$$\delta V_G^{EOS} = -\delta\Delta\varphi_E - \delta\left(\Delta\varphi_{GCh} + \Delta\varphi_{Hh}\right) + \delta\left(\Delta\varphi_{spch}^{Ox} + \Delta\varphi_S^{Si}\right) \quad (3)$$

Transistor and sensor. Equation (3) can be read in two ways: (i) In a transistor mode, the conditions in the electrolyte are constant. A changed gate voltage δV_G^{EOS} induces a changed band bending with a changed source-drain current above the threshold of strong inversion; (ii) In a sensor mode, the gate voltage is held constant. A changed electrical potential in the free electrolyte or across the Gouy-Chapman and quasi-Helmholtz layers gives rise to a changed band bending. The resulting change of the source-drain current can be expressed as an apparent change of the gate voltage with Equation (4).

$$\delta V_{Gapp}^{EOS} = \delta\Delta\varphi_E + \delta\left(\Delta\varphi_{GCh} + \Delta\varphi_{Hh}\right) \quad (4)$$

The interface-induced response accounts for the classical ISFET sensorics, whereas the current-induced response is a typical issue of the transistor sensorics with electrogenic cells.

Interpretation. The goal of transistor sensorics with electrogenic cells is an interpretation of a signal δV_{Gapp}^{EOS} in terms of the generating cellular currents. According to the core-coat conductor theory of a cell-substrate junction, as considered in the next section, cellular currents induce changes $\delta V_J = \delta\Delta\varphi_E$ of the electrical potential in the junction and changes δc_J^i of the ion concentrations, which give rise to an interface-induced response $\delta\left(\Delta\varphi_{GCh} + \Delta\varphi_{Hh}\right)$.

The interpretation of a signal δV_{Gapp}^{EOS} is involved if the interface-induced response plays a role. In a first step, the interface-induced response must be identified in the total transistor signal. This task is difficult because the peculiar (slow) dynamics of ion concentrations affects the interface-induced response as well as the current-induced response. In a second step, a calibration in terms of concentration changes δc_J^i is required. In principle, such a relation is obtained by titrating a cell-free transistor with the ion concentration c_E^i in the bath electrolyte yielding an ion sensitivity of the gate per decade S_{GE}^i according to Equation (5) for a limited concentration range.

$$\delta\left(\Delta\varphi_{GCh} + \Delta\varphi_{Hh}\right) = S_{GE}^i\,\delta\left(\log c_E^i\right) \quad (5)$$

However, the chemical nature of the gate surface is affected by cell adhesion. With an unknown ion sensitivity $S_{GJ}^i \neq S_{GE}^i$ in the junction, an assignment of a changing concentration δc_J^i is not possible.

An unequivocal interpretation of transistor signals with electroactive cells is achieved if the interface–induced response plays no role with $\delta V_{Gapp}^{EOS} = \delta\Delta\varphi_E$. Two situations can be envisaged: (i) the activation of cellular ion currents is so short that changes of the ion concentrations in the junction can be neglected; and (ii) the gate is chemically passivated such that changing ion concentrations are not able to induce an interface-induced response.

Finally, it may be noted that the sensor concept as sketched here for the EOSFET is also applicable for an electrolyte-oxide-metal-oxide-semiconductor (EOMOS) transistor and – *mutatis mutandis* – for other potentiometric electrodes (seeAppendix B).

PLANAR CAPACITIVE CORE-COAT CONDUCTOR

Ionic and capacitive currents through an adherent cell membrane induce changes of the electrical potential and of the ion concentrations in a cell-substrate junction. These relations are described in terms of a planar capacitive core-coat conductor.

The thin film of extracellular electrolyte in a cell-substrate junction is insulated from the intracellular electrolyte by the plasma membrane and from the semiconductor by an oxide layer (Figure 1b). For cultured mammalian cells, the distance of the two insulating coats is $d_J \approx 50$ nm as measured by fluorescence interference contrast (FLIC) microscopy [8]. It is far larger than the Debye length of the electrolyte with $l_{Debye} \approx 1$ nm and far smaller than the width of the adhesion area with a radius in the range of $a_J \approx 10$ μm. The condition $l_{Debye} << d_J << a_J$ is similar to a relation for neuronal axons, where a thin thread of intracellular electrolyte is insulated from the bath by the cylindrical plasma membrane with $l_{Debye} << a_M << l_{axon}$ for Debye length, radius, and length.

The electrical features of an axon are characterized by three properties: (i) the cytoplasmatic core as well as the bath are electroneutral beyond thin space charge layers at the membrane; (ii) the electrical charge in these space charge layers determines the voltage across the membrane by a capacitor relation; and (iii) the profiles of the ion concentrations and of the electrical potential across the narrow core are neglected as compared with the profiles along the core. As a consequence, axons are described as 1D linear capacitive core-coat conductors with ionic currents through the coat and along the core, usually for constant intracellular ion concentration [9] but occasionally also for changing ion concentrations [10]. With the same rationale, cell-substrate junctions have been described as 2D planar capacitive core-coat conductors for constant [11] and for changing ion concentrations [12,13].

Balance of ion flux. The ion flux through the adherent membrane in a cell-substrate junction gives rise to a changing (in time) ion concentration in the core and to a changing (in space) ion flux along the core. The flux along the core is driven by gradients (in two dimensions) of the electrical potential φ_J (drift mobility u^i_J) and of the concentration c^i_J (diffusion coefficient $D^i_J = k_B T u^i_J$, thermal energy $k_B T$). The flux balance is described by Equation (6) with the ionic charge number z_i and the electrical current i^i_{JM} per unit area of the adherent membrane [14].

$$-\nabla\left[d_J z_i^2 e_0^2 u^i_J\left(c^i_J \nabla\varphi_J + \frac{k_B T}{z_i e_0}\nabla c^i_J\right)\right] + d_J z_i e_0 \frac{\partial c^i_J}{\partial t} = i^i_{JM} \tag{6}$$

Balance of charge. Changing ion concentrations give rise to a changing electrical charge in the junction. A net charge $\sum d_J z_i e_0 c^i_J \neq 0$ in the core electrolyte would violate the electroneutrality condition. That charge is located in thin space charge layers along membrane and semiconductor oxide, respectively, with counter-charges induced across the two coats [15]. A changing charge is related with changing differences $\varphi_J - \varphi^\infty_{cell}$ and $\varphi_J - \varphi^\infty_{Si}$ of the electrical potential in the junction with respect to cell and semiconductor by a differential capacitor relation. With a capacitance c_M of the membrane and an average capacitance c_S of the substrate (differing from the local capacitance c_{Ox} of the gate oxide), we obtain Equation (7). (A different approach has been chosen in [16], where the net charge is assigned to a space charge in the electrolyte film. See Appendix C).

$$\sum d_J z_i e_0 \frac{\partial c^i_J}{\partial t} = c_M \frac{\partial\left(\varphi_J - \varphi^\infty_{cell}\right)}{\partial t} + c_S \frac{\partial\left(\varphi_J - \varphi^\infty_{Si}\right)}{\partial t} \tag{7}$$

Current balance and limit of constant ion concentrations. The balance of electrical charge (Equation (7)) can be combined with the dynamics of the ion concentrations (Equation (6)). We obtain Equation (8) for the balance of the total current, when the electrical potentials in junction, cell, and semiconductor are expressed as voltages with respect to the bulk electrolyte with $V_J = \varphi_J - \varphi^\infty_E$, $V_M = \varphi^\infty_{cell} - \varphi^\infty_E$, and $V_{Si} = \varphi^\infty_{Si} - \varphi^\infty_E$ defining a junction capacitance $c_J = c_M + c_S$ (Distinguish capacitance c_J and ion concentrations c^i_J!).

$$-\sum\nabla\left[d_J z_i^2 e_0^2 u^i_J\left(c^i_J \nabla V_J + \frac{k_B T}{z_i e_0}\nabla c^i_J\right)\right] + c_J \frac{\partial V_J}{\partial t} = c_S \frac{dV_{Si}}{dt} + c_M \frac{dV_M}{dt} + \sum i^i_{JM} \tag{8}$$

For small changes of the ion concentrations, the diffusion term in Equation (8) can be cancelled and the concentrations can be excluded from the spatial derivative of the drift term. With a conductivity $1/\rho_J = \sum z_i^2 e_0^2 u^i_J c^i_J$ of the electrolyte in the junction, we obtain Equation (9).

$$-\nabla\left(\frac{d_J}{\rho_J}\nabla V_J\right) + c_J \frac{\partial V_J}{\partial t} = c_S \frac{dV_{Si}}{dt} + c_M \frac{dV_M}{dt} + \sum i^i_{JM} \tag{9}$$

This "2D cable equation" for a planar capacitive core-coat conductor with constant ion concentrations is homologous to the common 1D cable equation for neuronal axons. Two peculiarities, however, must be noted: (i) the voltage in the junction V_J depends on the voltage changes in the two different adjacent media of cell and substrate; and (ii) at its periphery, the core conductor is open to the bath electrolyte.

Transistor response. The capacitive and ionic currents through the adherent membrane of a cell-substrate junction give rise to a coupled dynamics of the extracellular ion concentrations c^i_J and of the extracellular electrical potential φ_J, as described by Equations (6) and (7). The apparent change of the gate voltage δV^{EOS}_{Gapp} for an EOSFET in the junction (Equation (4)) is caused by a current-induced response due to a changing voltage in the junction and by an interface-induced response due to changing ion concentrations according to Equation (10).

$$\delta\,\Delta\varphi_E = \delta V_J \tag{10a}$$

$$\delta\left(\Delta\varphi_{GCh} + \Delta\varphi_{Hh}\right) = \sum S^i_{GJ}\,\delta\left(\log c^i_J\right) \tag{10b}$$

DYNAMICS OF PLANAR CAPACITIVE CORE-COAT CONDUCTOR

The time constants for changes of the voltage V_J and of the ion concentrations c^i_J in a cell-substrate junction are crucial for the response of a transistor to cellular currents. We consider a circular junction (radius a_J, area $A_J = a_J^2\pi$) with a homogeneous sheet resistance $r_J = \rho_J/d_J$.

Voltage dynamics. The dynamics of the junction voltage V_J for constant ion concentrations is described by Equation (11a) with a radial coordinate a (at a constant bias voltage of the semiconductor). The ionic membrane currents depend on the conductances per unit area g^i_{JM} according to Equation (11b), which are driven by the transmembrane voltage and the reversal voltage V^i_{M0} (determined by the difference of the logarithmic concentrations across the membrane).

$$-\frac{1}{r_J}\left(\frac{\partial^2}{\partial a^2} + \frac{1}{a}\frac{\partial}{\partial a}\right)V_J + c_J \frac{\partial V_J}{\partial t} = c_M \frac{dV_M}{dt} + \sum i^i_{JM} \tag{11a}$$

$$i^i_{JM} = g^i_{JM}\left(V_M - V_J - V^i_{M0}\right) \tag{11b}$$

First, we apply a voltage step ΔV^0_M to a cell without membrane conductances for a boundary condition $V_J(a_J)=0$ and an initial condition $V_J(t=0)=0$. The sharp pulse of capacitive current induces a junction voltage $\Delta V^0_M c_M/c_J$, which decays according to Equation (12) with a relaxation of modes (Bessel functions J_0, J_1, zeros $z_{0,n}$ (n=1,2,..) of J_0) [17] (p. 199).

$$V_J(a,t) = \Delta V^0_M \frac{c_M}{c_J}\sum_{n=1}^{\infty}\frac{2}{z_{0,n}}\frac{J_0\left(z_{0,n}a/a_J\right)}{J_1\left(z_{0,n}\right)}\exp\left(-\frac{z_{0,n}^2}{r_J c_J a_J^2}t\right) \tag{12}$$

To reveal the response to an ion current, we apply a voltage step to a cell with a single ion conductance g^i_{JM} starting from $V_M = V^i_{M0}$. Besides the capacitive response, an increasing junction voltage is induced according to Equation (13) for $|V_J| << |V_M - V^i_{M0}|$ with a stationary parabolic profile [17] (p. 204).

$$V_J(a,t) = \Delta V^0_M g^i_{JM}\frac{r_J A_J}{4\pi}\left[\left(1 - \frac{a^2}{a_J^2}\right) - \sum_{n=1}^{\infty}\frac{8}{z_{0,n}^3}\frac{J_0\left(z_{0,n}a/a_J\right)}{J_1\left(z_{0,n}\right)}\exp\left(-\frac{z_{0,n}^2}{r_J c_J a_J^2}t\right)\right] \tag{13}$$

The complete response is described by a superposition of the Equations (12) and (13). The dynamics is characterized by the time constant $\tau_J^{(1)} = r_J c_J A_J / 5.78\pi$ of the lowest mode.

Typical parameters for cultured mammalian cells (HEK293, hippocampal neuron) are a_J=10 µm, $r_J = 10\,M\Omega$, $c_M = 1\,\mu F/cm^2$, $c_J = 1.3\,\mu F/cm^2$, and $g_{JM}^i \approx 10\ mS/cm^2$. The time-dependent profiles $V_J(a,t)$ upon a voltage step $\Delta V_M^0 = 100\ mV$ are plotted in Figure 2 for the capacitive current and for the ionic current [18]. The capacitive current induces an initial voltage of 77 *mV*. The ionic current gives rise to a stationary parabolic voltage profile with an amplitude of 2.5 *mV*. The dynamics is dominated by the time constant $\tau_J^{(1)}$ =2.25 µs.

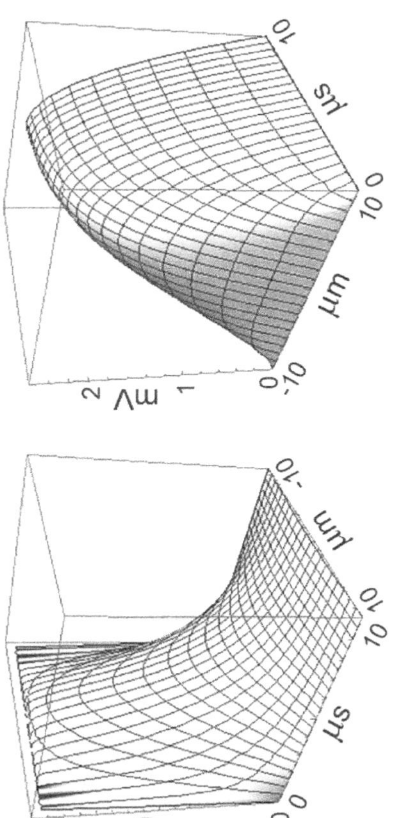

Figure 2. Time-dependent profiles of the extracellular voltage $V_J(a,t)$ in a circular cell-substrate junction upon intracellular stimulation by a voltage step of 100 *mV*: (**a**) response to capacitive current; and (**b**) response to ionic current. The dynamics in the microsecond range is determined by the time constant of the lowest mode of the core–coat conductor. For the parameters, see text. Note the different orientation and different voltage scales of the diagrams.

Quasi-stationary response. The electrical time constant of a cell-substrate junction is in the microsecond range and far shorter than the time scale of the natural dynamics of capacitive and ionic currents of cells in the range of 0.1–1 *ms*. As a consequence, the junction voltage is almost relaxed upon stimulation. The 2D cable equation (Equation (9)) can be reduced to Equation (14) for a homogeneous junction and small voltages $|V_J| << |V_M - V_{M0}^i|$.

$$-\frac{d_J}{\rho_J}\nabla^2 V_J = c_M \frac{dV_M}{dt} + \sum g_{JM}^i \left(V_M - V_{M0}^i\right) \tag{14}$$

In this quasi-stationary situation, the planar capacitive core-coat conductor appears as a sheet conductor that is driven from the cell by a 2D (capacitive and ionic) current-source density. There is a certain analogy to a volume conductor with a 3D current-source density [19,20]. Considering Equation (13), the quasi-stationary voltage profile is parabolic for a circular junction according to Equation (15).

$$V_J(a,t) = \frac{r_J A_J}{4\pi}\left(1 - \frac{a^2}{a_J^2}\right)\left[c_M \frac{dV_M}{dt} + \sum g_{JM}^i \left(V_M - V_{M0}^i\right)\right] \tag{15}$$

Concentration dynamics. The concentration dynamics of ions in a planar core-coat conductor is governed by drift and diffusion according to Equation (6). The drift term is determined by the electrical potential, which depends on the concentration dynamics of all ions (Equation (7)). When we disregard the drift, we obtain Equation (16) for the pure, independent ion diffusion in a homogeneous circular junction.

$$-D_J^i\left(\frac{\partial^2}{\partial a^2} + \frac{1}{a}\frac{\partial}{\partial a}\right)c_J^i + \frac{\partial c_J^i}{\partial t} = \frac{i_{JM}^i}{d_J z_i e_0} \tag{16}$$

A step of ion current Δi_{JM}^i induces a concentration change according to Equation (17) for a boundary condition $c_J^i(a_J) = c_E^i$ and an initial condition $c_J^i(t=0) = c_E^i$, with the concentration c_E^i in the bulk electrolyte. The dynamics is characterized by the time constant $\tau_J^{(1)} = A_J/5.78\pi D_J^i$ of the lowest mode.

$$c_J^i - c_E^i = \frac{\Delta i_{JM}^i}{z_i e_0 d_J}\frac{A_J}{4\pi D_J^i}\left[\left(1 - \frac{a^2}{a_J^2}\right) - \sum_{n=1}^{\infty}\frac{8}{z_{0,n}^3}\frac{J_0(z_{0,n}a/a_J)}{J_1(z_{0,n})}\exp\left(-\frac{z_{0,n}^2 D_J^i}{a_J^2}t\right)\right] \tag{17}$$

For $\Delta i_{JM}^i = 1\,mA/cm^2$ (with $g_{JM}^i = 10\ mS/cm^2$ and $V_M - V_{M0}^i = 100\ mV$), $D_J^i = 2\cdot 10^{-5} cm^2/s$, d_J=70 nm, and a_J=10 µm, a stationary concentration change up to 18.6 *mM* is attained with a time constant $\tau_{J,diff}^{(1)} = 8.7\ ms$. The dynamics of ion concentrations is far slower than the dynamics of the voltage. It is illustrated by Figure 2b, if the time axis is rescaled by a factor $\tau_{J,diff}^{(1)}/\tau_J^{(1)} \approx 4000$. A significant concentration change appears only if the duration of the ion current is comparable with the time constant $\tau_{J,diff}^{(1)}$.

Electrical characterization of cell-substrate junction. The capacitive transient upon stimulation by a voltage-step (Equation (12), Figure 2a) can be used to determine the sheet resistance of a cell-substrate junction. More reliable, however, is the application of an ac voltage with an amplitude $\underline{V}_M(\omega)$ (angular frequency ω). The frequency-dependent voltage profile $\underline{V}_J(a,\omega)$ in a homogeneous circular junction without membrane conductance can be derived from Equation (11) according to Equation (18) with the modified Bessel function I_0. This approach has been used for leech neurons with an array of transistors in the cell–substrate junction [21].

$$\underline{V}_J(a,\omega) = \underline{V}_M \frac{c_M}{c_J}\left[1 - \frac{I_0(\gamma a)}{I_0(\gamma a_J)}\right], \gamma^2 = i\omega r_J c_J \tag{18}$$

A related approach is the application of an ac voltage $\underline{V}_{Si}$ to the capacitor structure of a chip. The response $\underline{V}_M - \underline{V}_J$ Jacross the adherent membrane has been probed with a voltage-sensitive fluorescent dye [22,23]. The relation of stimulus and response is similar to Equation (18) but more involved due to the coupled response of junction and cell. An average sheet resistance of r_J=7.7MΩ has been obtained for HEK293 cells on a silica surface coated with fibronectin and of r_J=12MΩ for hippocampal rat neurons on poly–lysine [24].

A particularly elegant method is the recording of the thermal voltage fluctuations in a cell-substrate junction [25]. The autocorrelation function of $V_J(t)$, averaged over a circular sensor with a radius a_S in the center of a circular junction, is expressed by Equation (19) [26].

$$\Phi^S(\Delta t) = \frac{k_B T}{c_J a_S^2 \pi}\sum_{n=1}^{\infty}\frac{4}{z_{0,n}^2}\left[\frac{J_1(z_{0,n}a_S/a_J)}{J_1(z_{0,n})}\right]^2 \exp\left(-\frac{z_{0,n}^2}{r_J c_J a_J^2}\Delta t\right) \tag{19}$$

Practically, the power spectral density (PSD) of the voltage fluctuations is probed, and the data are fitted with a theoretical PSD, which is obtained by Fourier transformation of the autocorrelation function. For low frequencies, it is given by Equation (20) [26].

$$S_f^S = 4k_B T r_J\left(\frac{1}{8\pi} + \frac{1}{2\pi}\ln\frac{a_J}{a_S}\right) \tag{20}$$

The PSD for rat neurons on poly–lysine has been probed with an EOSFET at low frequency with $S_f^S = 4.7 \cdot 10^{-14} V^2/Hz$ [25]. According to Equation (20), with a_S/a_J=0.25, a sheet resistance of r_J=11MΩ is obtained, which is similar to the resistance evaluated from the ac stimulation.

The PSD for a wide range of frequencies has been probed for neurons from *Lymnaea stagnalis* using an array of EOMOS transistors, as shown in Figure 3 [26]. The data are fitted with a sheet resistance r_J=109MΩ and a junction capacitance (whole cell) of c_J=0.74 μF/cm². The unusually large sheet resistance, which indicates a particularly close adhesion, was due to a special preparation of the dissociated snail neurons. The perfect fit of the data over five orders of magnitude of frequencies confirms that the cell-substrate junction is indeed a capacitive core-coat conductor.

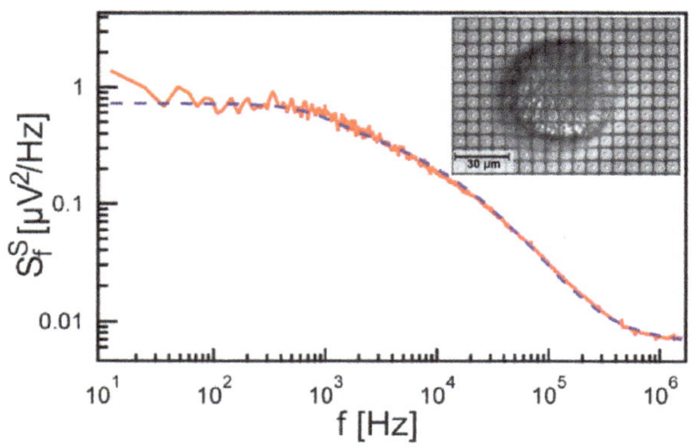

Figure 3. Power spectral density (PSD) of the fluctuating extracellular voltage V_J in a cell–substrate junction (by permission [26]). The PSD of a dissociated neuron from*Lymnaea stagnalis* is probed with an array of EOMOS transistors (insert). The logarithm of the net PSD per frequency interval for the sensor transistor in the center of the adhesion area (after subtracting the PSD without cell) is plotted versus the logarithm of the frequency (red trace). The data are fitted by the PSD of the thermal voltage fluctuations in a planar capacitive core-coat conductor with a sheet resistance $r_J = 109M\Omega$ and a capacitance (whole cell) of $c_J = 0.74\,\mu F/cm^2$ (blue dashed trace).

ONE-COMPARTMENT MODEL

The typical issue of a cell-semiconductor assembly is the electrical coupling of the ionic and the electronic system across the planar junction. The profiles of the electrical potential and of the ion concentrations are not so important. For that reason, the planar core-coat conductor can be drastically reduced: drift and diffusion of ions along the differential area elements of the junction are replaced by global drift and diffusion from a single compartment to the bath electrolyte. On the one hand, this simple approach provides a surprisingly good approximation for the dynamics of the planar core-coat conductor, and, on the other hand, it is adequate if a cell-substrate junction is not sufficiently well defined with respect to the profiles of resistance, capacitance, and membrane conductance and to the position of a transistor. The "one-compartment model" is derived from the planar core-coat conductor for constant ion concentrations as well as for changing concentrations.

One-compartment model for constant ion concentrations. We start with the 2D cable equation (Equation (9)) for a homogeneous circular junction (Equation (11)). We replace the divergence of the electrical current by a global current from a junction compartment to the bath according to Ohm's law (formally by a substitution $-\nabla^2 \Rightarrow \eta_J/A_J$) and obtain Equation (21a).

$$g_J V_J + c_J\frac{dV_J}{dt} = c_M\frac{dV_M}{dt} + \sum g_{JM}^i\left(V_M - V_J - V_{M0}^i\right) \tag{21a}$$

$$g_J = \eta_J\frac{d_J}{A_J\rho_J} \tag{21b}$$

The Ohmic conductance per unit area g_J as defined by Equation (21b) combines three crucial parameters of a cell-substrate junction, the distance d_J, the area A_J, and the resistivity ρ_J. The scaling factor η_J is obtained by comparing the one-compartment model with a circular core-coat conductor as follows.

When a voltage step ΔV_M^0 is applied to a cell without membrane conductance, the capacitive current induces a junction voltage, which decays according to Equation (22a). Starting from $V_M = V_{M0}^i$ for a single ion conductance g_{JM}^i, the voltage step gives rise to an increasing junction voltage according to Equation (22b) for $|V_J| << |V_M - V_{M0}^i|$ with a stationary voltage $\Delta V_M^0 g_{JM}^i/g_J$.

$$V_J = \Delta V_M^0\left(c_M/c_J\right)\exp\left(-t\,g_J/c_J\right) \tag{22a}$$

$$V_J = \Delta V_M^0\left(g_{JM}^i/g_J\right)\left[1 - \exp\left(-t\,g_J/c_J\right)\right] \tag{22b}$$

When the time constant τ_J=c_J/g_J is compared with the time constant $\tau_J^{(1)}$ of the circular core-coat conductor (Equation (12)), we obtain η_J=5.78π. When the stationary voltage is compared with the stationary parabolic voltage profile (Equation (13)), the scaling factor is η_J=4π or η_J=8π, respectively, if the maximum voltage or the average voltage is considered.

The one-compartment model for constant ion concentrations corresponds to a simple equivalent electrical circuit with a single node assigned to the cell-substrate junction and an Ohmic seal conductance [27]. The formal equivalent circuit, however, does not provide a relation of the seal conductance with the structural parameters of a cell-substrate junction as it is obtained when the model is derived from the planar core-coat conductor [28].

One-compartment model with changing ion concentrations. A one-compartment model can be also defined for changing ion concentrations [13]. We start with the drift-diffusion equation (Equation (6)) for a homogeneous junction. The divergence of the ion flux is replaced by the global flux from a junction compartment to the bath with Ohm's law for the drift and with Fick's first law for the diffusion. We obtain Equation (23a), where the ion conductivity $1/\rho_J^i$ and reversal voltage V_{JM0}^i for the adherent membrane depend on the ion concentration according to Equations (23b) [29].

$$d_J z_i e_0 \frac{dc_J^i}{dt} + \frac{\eta_J d_J}{A_J \rho_J^i}\left[\left(\varphi_J - \varphi_E^\infty\right) + \frac{k_B T}{z_i e_0}\frac{1}{c_J^i}\left(c_J^i - c_E^i\right)\right] = g_{JM}^i\left(V_M - V_J - V_{JM0}^i\right) \quad (23a)$$

$$\frac{1}{\rho_J^i} = z_i^2 e_0^2 u_J^i c_J^i,\ V_{JM0}^i = \frac{k_B T}{z_i e_0}\ln\frac{c_J^i}{c_{cell}^i} \quad (23b)$$

The voltage $V_J = \varphi_J - \varphi_E^\infty$ is related with the total electrical charge according to the differential capacitor relation of Equation (24a) (at a constant bias voltage of the semiconductor). For a constant membrane voltage, the integral capacitor relation of Equation (24b) can be used considering an initial condition $c_J^i(t=0) = c_E^i$ and the electroneutrality condition $\sum z_i e_0 c_E^i = 0$ for the bath electrolyte.

$$\sum d_J z_i e_0 \frac{dc_J^i}{dt} = c_J \frac{dV_J}{dt} - c_M \frac{dV_M}{dt} \quad (24a)$$

$$\sum d_J z_i e_0 c_J^i = c_J V_J \quad (24b)$$

TRANSISTOR RECORDING OF ION CONDUCTANCES

Voltage-activated membrane conductances are crucial for the transistor sensorics of electrogenic cells. As examples, we consider the inactivating sodium conductance and the non-inactivating potassium conductance of the classical Hodgkin-Huxley (HH) model [30]. They are described by maximum conductances $g_M^{\infty Na}$ and $g_M^{\infty K}$ and by the open probabilities of independent gates p=m,n for activation and p=h for inactivation according to Equation (25a). The open probabilities obey rate equations (Equation (25b)) with the rate constants α_p and β_p, which depend on the transmembrane voltage.

$$g_M^{Na} = g_M^{\infty Na} m^3 h,\quad g_M^K = g_M^{\infty K} n^4 \quad (25a)$$

$$dp/dt = \alpha_p(1-p) - \beta_p p \quad (25b)$$

Quasi-stationary response. The time constant τ_p for the dynamics of the gates at common transmembrane voltages is in a range of 0.1–1 *ms* for the activation of the Na conductance and in a range of 1–10 *ms* for the inactivation of the Na conductance and the activation of the K conductance. Thus, the dynamics is far slower than the electrical relaxation of the cell-substrate junction but faster than the ion diffusion in the junction. The relation $\tau_J \ll \tau_p < \tau_{J,diff}$ has two consequences: (i) the junction voltage follows the membrane current in a quasi–stationary state; and (ii) the change of the extracellular ion concentrations due to the inactivating Na conductance is small, whereas it may be large for the non-inactivating K conductance if it is held open by a constant intracellular voltage.

To elucidate the transistor response of ion channels, the voltage in an adherent cell is controlled by a whole-cell patch-pipette. With this approach a transistor signal δV_{Gpp}^{EOS} can be compared with the total current through the cell membrane I_M such that a current-induced and an interface-induced responses can be distinguished. As a guide, various numerical simulations of the capacitive core-coat conductor are performed with the HH conductances (rate constants at 18.5 °C) using the one-compartment model with changing ion concentrations according to Equations (23) and (24a) with an area A_J=200 μm², a distance d_J=70 nm, and tabulated diffusion coefficients. A scaling factor $_J$ = 5.78π takes into account the partial spatial averaging by a transistor [31].

Voltage-clamp with constant ion concentrations. Almost constant ion concentrations in the junction are expected for the Na conductance due to the fast inactivation and the high extracellular Na concentration. This assumption is confirmed by a simulation with $g_{JM}^{\infty Na} = 20\,mS/cm^2$ (extracellular electrolyte 50 *mM*·NaCl, 2 *mM*·KCl, intracellular electrolyte 12 *mM* Na ions).

Considering Equation (21), the quasi-stationary junction voltage is described by Equation (26a) for $|V_J| << |V_M - V_{M0}^{Na}|$ at a constant intracellular voltage. The current through the whole membrane with an area AM and an average conductance $\bar{g}_M^{Na}$ is expressed by Equation (26b).

$$V_J(t) = \frac{1}{g_J} g_{JM}^{Na}(t)\cdot\left[V_M - V_{M0}^{Na}\right] \quad (26a)$$

$$I_M(t) = A_M \bar{g}_M^{Na}(t)\cdot\left[V_M - V_{M0}^{Na}\right] \quad (26b)$$

If the dynamics of the conductance is the same in the whole membrane, the voltage V_J and the current I_M exhibit the same waveform and are proportional according to Equation (27).

$$V_J = \frac{1}{g_J A_M}\frac{g_{JM}^{\infty Na}}{\bar{g}_M^{\infty Na}} I_M \quad (27)$$

Na_v1.4 channel. Transistor recordings under voltage–clamp have been reported for the sodium channel Na_v1.4 in HEK293 cells at an extracellular Na concentration of 50 *mM* [32]. The Na conductance was activated by depolarizing steps of the intracellular voltage. The transistor response δV_{Gapp}^{EOS} and the membrane current IM reflected both the typical waveform

of a transient sodium current with an activation phase and an inactivation phase. The apparent change of the gate voltage is plotted in Figure 4 versus the total membrane current. There is a perfect proportionality, which is fitted with a relation $\delta V_{Gapp}^{EOS} = 260\,k\Omega \cdot I_M$. The result indicates that the transistor signal is indeed due to a pure current-induced response with $\delta V_{Gapp}^{EOS} = V_J$. By ac stimulation and recording the response of transistor and current, a parameter $1/g_J A_M = 308\,k\Omega$ has been measured. Considering Equation (27), a depleted conductance in the adherent membrane with $g_{JM}^{\infty\,Na} = 0.85\,\overline{g}_M^{\infty\,Na}$ was estimated.

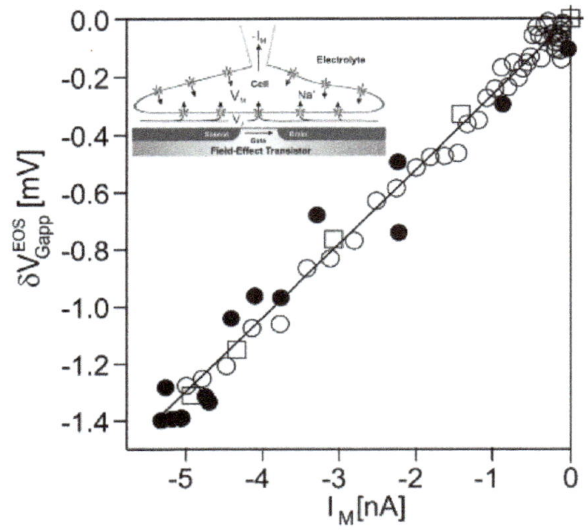

Figure 4. Transistor recording of the sodium channel $Na_v1.4$ in a HEK293 cell at constant intracellular voltage (by permission [32]). The apparent change of the gate voltage δV_{Gapp}^{EOS} is plotted versus the membrane current IM upon a depolarization from –90 mV to –20 mV. (Scheme: cell with whole-cell patch-pipette). The white squares refer to the activation phase and the white circles to the inactivation phase. In addition, the minima of the voltage and of the current transients are plotted for 15 different depolarizations (black dots). The data are fitted by the relation $\delta V_{Gapp}^{EOS} = 260\,k\Omega \cdot I_M$.

Voltage-clamp with changing ion concentration. Changing ion concentrations are expected under voltage-clamp for the non-inactivating K conductance at the low physiological K concentration. The situation is illustrated by a simulation with $g_{JM}^{\infty\,K} = 20\,mS/cm^2$ (extracellular electrolyte 5 mM·KCl, 135 mM·NaCl, intracellular electrolyte 40 mM K ions). The cell is depolarized by 140 mV for 15 ms starting from the reversal voltage of potassium. The results are plotted in Figure 5. We have to distinguish three levels: (i) the current-induced response for constant ion concentrations; (ii) the effect of the ion concentrations on the current-induced response; and (iii) the additional effect of the ion concentrations on the interface-induced response.

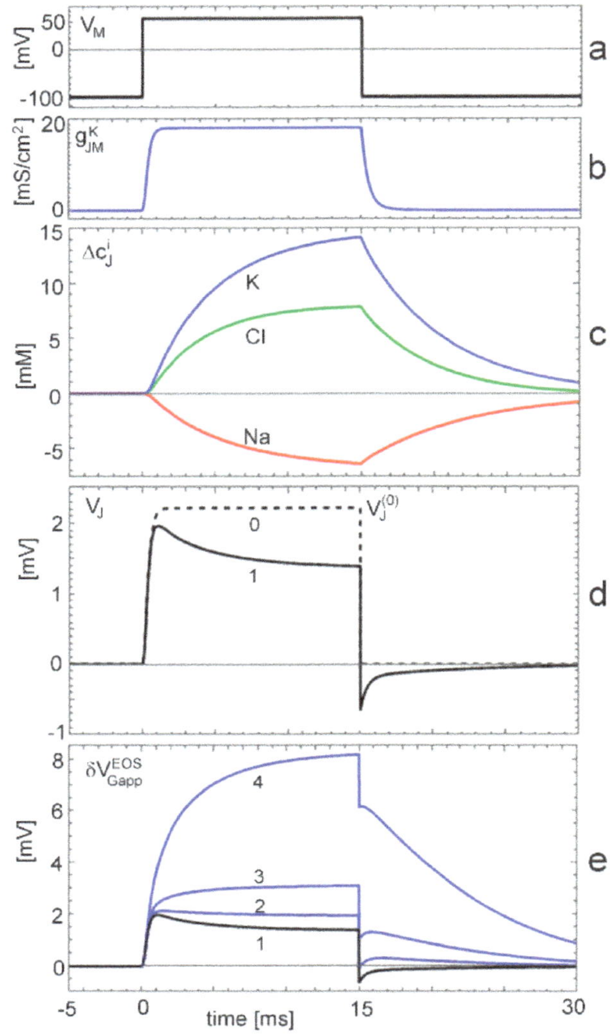

Figure 5. Computed dynamics of a cell-substrate junction for a non-inactivating K conductance under voltage-clamp at a low extracellular K concentration. One-compartment model of the capacitive core-coat conductor with the K conductance of the Hodgkin-Huxley model. For the parameters, see text. **(a)** Intracellular voltage V_M; **(b)** membrane conductance g_{JM}^K; **(c)** changes $\Delta c_J^i = c_J^i - c_E^i$ of extracellular concentrations; (The relative change $\Delta c_J^i / c_E^i$ is high for K with $c_E^K = 5$ mM but low for Na and Cl with $c_E^{Na} = 135\,mM$ and $c_E^{Cl} = 140\,mM$). **(d)** current–induced voltage for constant ion concentrations (trace 0) and for changing ion concentrations (trace 1); and **(e)** transistor response for a gate oxide without ion sensitivity (trace 1), and with ion sensitivities of 1 mV (trace 2), 3 mV (trace 3), and 12 mV (trace 4) (all traces without capacitive transients).

The junction voltage for constant ion concentrations is plotted in Figure 5d as trace 0. Upon depolarization, a stationary level of $V_J^{(0)} = 2.2\,mV$ is reached within a millisecond. The dynamics is determined by the activation of the potassium conductance (Figure 5b). The stationary voltage disappears suddenly upon repolarization to the reversal voltage.

The junction voltage with changing ion concentrations is shown in Figure 5d as trace 1. Within a millisecond, it reaches almost the level of $V_J^{(0)}$. It relaxes within several milliseconds to a stationary voltage of V_J=1.4 mV. This decay corresponds to the dynamics of the ion concentrations (Figure 5c). Upon repolarization, the voltage drops suddenly to a negative level and relaxes in a fast phase due to the dynamics of the ion conductance and in a slow phase due to the dynamics of the concentrations.

The K concentration in the junction increases from 5 *mM* to 15 *mM* (Figure 5c). The logarithmic change $\log c_J^K / c_E^K \approx 0.5$ may give rise to an interface-induced response (Equation (10)). The total transistor signal δV_{Gapp}^{EOS} is plotted in Figure 5e for an insensitive gate (trace 1) with $\delta V_{Gapp}^{EOS} = V_J$ and for three ion sensitivities S_{GJ}^K. For a low sensitivity of 1 *mV* (trace 2), the interface-induced response compensates partially the ion-dependent dynamics of the current-induced response upon depolarization and repolarization. For a sensitivity of 3 *mV* (trace 3), the transistor signal exhibits two positive components upon depolarization, a fast response, which is determined by the dynamics of the conductance, and a slow response, which is determined by the dynamics of the K concentration. Upon repolarization, the transistor signal drops suddenly due to the disappearance of the membrane current. Then, the interface-induced response relaxes with the ion concentrations. For an ion sensitivity of 12 *mV* (trace 4), the current-induced and the interface-induced response are fused to a smooth, large signal up to $\delta V_{Gapp}^{EOS} = 8.5\ mV$. The interface-induced component is six times larger than the current–induced response. Upon repolarization, the signal drops suddenly due to the disappearance of the membrane current. Then, the interface-induced response relaxes with $\log c_J^K$.

Transistor recordings of non–inactivating K conductances under voltage clamp have been reported for hippocampal neurons [33] and for HEK293 cells with the recombinant channels Maxi-K [12,34], Kv1.3 [13,35], and EAG [16].

Kv1.3 channel. The Kv1.3 channel has been studied on chips that were cleaned with an alkaline detergent [13] and with an acidic detergent [35]. Recordings of the "acidic experiment" are plotted in Figure 6a. Upon depolarization, an effective change of the gate voltage up to about 2.5 *mV* appears within 2 *ms*. It decays partially during depolarization. Upon repolarization, the signal drops suddenly and a transient negative signal appears. Recordings of the "alkaline experiment" are displayed in Figure 6b, wherein a smooth increase of the transistor signal up to 7.5 *mV* is observed. Upon repolarization, the signal drops suddenly by 2 *mV*, and the remaining amplitude relaxes slowly. (The conditions of the two experiments differ with respect to channel expression, cell size, and transistor quality; the results can be compared only with respect to the dynamics and the relative amplitudes of the fast and slow components).

For the same ion channel, the waveforms of the transistor signals are rather different. When we compare the highest traces of Figure 6a,b with the simulation of Figure 5e, it is apparent that the waveform of the "acidic experiment" resembles trace 1 for a pure current–induced response with $\delta V_{Gapp}^{EOS} = V_J$, whereas the waveform of the "alkaline experiment" resembles trace 4 with a superposed interface-induced response for an ion sensitivity 12 *mV* of the gate. The different ion sensitivities must be related with the cleaning procedures. In fact, after cleaning with the alkaline detergent, the chip surface was highly hydrophilic, whereas it was extremely hydrophobic after cleaning with the acidic detergent due to the deposition (selfassembly) of a cationic amphiphile, which blocks the negative charges of silica.

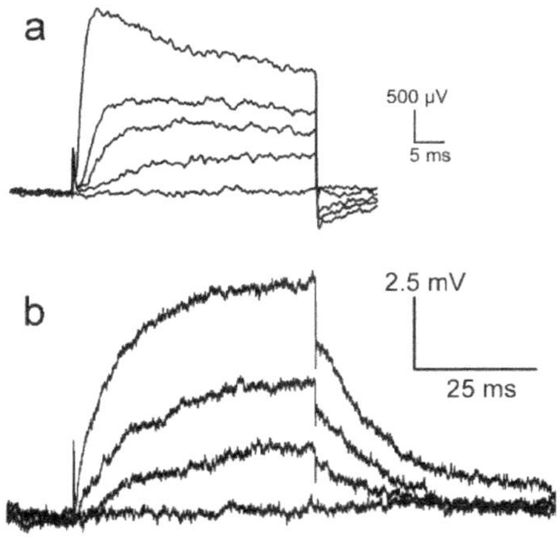

Figure 6. Transistor recordings of the potassium channel Kv1.3 in HEK293 cell at a constant intracellular voltage with an extracellular K concentration of 5 *mM*. The apparent change of the gate voltage δV_{Gapp}^{EOS} is plotted versus time. (**a**) Gate oxide cleaned with an acidic detergent: depolarizations to −25, −15, −5, and +40 *mV* starting near the reversal voltage of potassium, repolarization after 40 *ms* [35]; (**b**) Gate oxide cleaned with an alkaline detergent: depolarizations to −20, 0, and +50 *mV*, repolarization after 50 ms (by permission of Springer [13]). The conditions of the experiments differ with respect to channel expression, cell size, and transistor quality. A comparison can refer only to the dynamics and the relative amplitudes of the fast and slow components.

It appears that rather complex transistor signals can be described in terms of a capacitive core-coat conductor with changing ion concentrations. However, two problems must be considered.

(i) The core-coat conductor theory is not able explain the transistor recordings in terms of the cellular currents with a changing extracellular voltage and changing ion concentrations as long as the ion sensitivity of the gate in the cell-substrate junction is unknown. A calibration of a cell-free transistor is not adequate. For example, the sensitivity $S_{GJ}^K = 12\ mV$ that matches the "alkaline experiment" is distinctly lower than the sensitivity $S_{GE}^K \approx 45\ mV$ that was obtained by titration [13]. The discrepancy must be assigned to a modification of the chip surface by cell adhesion such that negative oxide charges are partially compensated by adsorbed poly-lysine and extracellular matrix protein. (Such a compensation of negative surface charges by a cationic polymer had been observed for a monomolecular lipid film using a fluorescent probe [36]).

(ii) The pattern of the transistor signals (Figure 6) is matched quite nicely by the computations (Figure 5e). However, the interface-induced response is slower than the computed dynamics of the ion concentrations. A low ion mobility in the electrolyte film of the cell-substrate junction can be excluded because the resistivity in the junction is the same as in the bulk electrolyte [24]. The effect may be due to a slow kinetics of ion binding

[13,37]. *EAG channel.* Current-induced and interface-induced signals have been also distinguished in transistor recordings of the EAG potassium channel under voltage-clamp [16]. The current-induced response was low around 0.2 mV due to a low membrane current. Also the interface-induced response was low due to a low ion sensitivity of 0.7 mV of the gate (maybe caused by a cleaning procedure with an incubation in 20% H_2SO_4). The data were explained in terms of the one-compartment model with constant ion concentrations for the current-induced response and by a Poisson-Nernst-Planck approach to compute the changed ion concentrations, which gave rise to the interface-induced response (see Appendix C).

Maxi-K channel with high extracellular K concentration. In a first report on transistor recording of a recombinant ion channel (Maxi-K) under voltage clamp [34], a high extracellular potassium concentration of 100 mM has been applied. The intention was to suppress the interface-induced response, which had been observed when the transistor was cleaned with the alkaline detergent [12]. The results are displayed in Figure 7. Upon depolarization, the noisy signals with low time resolution indicate stationary transistor signaly up to 1 mV, which are proportional to the total membrane current with $\delta V_{Gapp}^{EOS} = 73\,k\Omega \cdot I_M$.

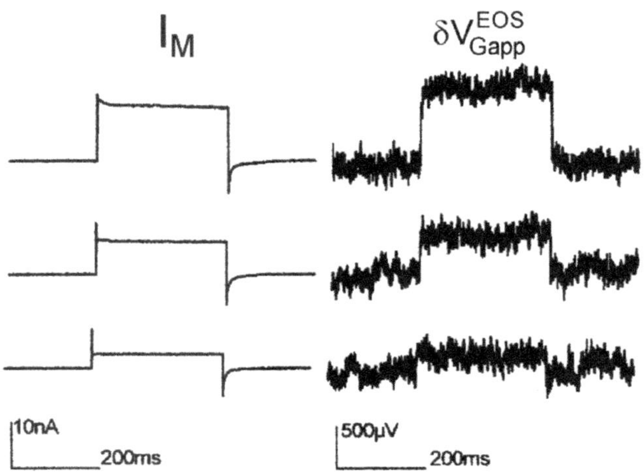

Figure 7. Transistor recording of Maxi-K potassium channel in HEK293 cell at a constant intracellular voltage with a high extracellular K concentration of 100 mM [34]. (**Left**) Total membrane current; (**Right**) Transistor recording. Depolarizations from −100 mV to 20, 45, and 58 mV.

A simulation (extracellular electrolyte 100 mM·KCl, 35 mM·NaCl, intracellular electrolyte 140 mM K ions) with an ion sensitivity of 12 mV for the gate oxide yields a biphasic waveform of the transistor signal that resembles trace 3 in Figure 5e with a fast current-induced response $V_J^{(0)}$ =0.85 mV and a stationary signal δV_{Gapp}^{EOS} =1.05 mV that is reached within several milliseconds. Thus the high extracellular K concentration lowers drastically the interface-induced component (*cf.* Figure 5e), but does not abolish it completely. On the basis of the simulation, a relation $V_J^{(0)}$=60 kΩ·IM is estimated for the current-induced component of the Maxi-K recording. With a weighting factor $1/g_J A_M$=22 kΩ (measured by ac stimulation), an enhancement of the Maxi-K conductance in the adherent membrane of $g_{JM}^{\infty K} = 2.7\,\bar{g}_M^{\infty K}$ is obtained using Equation (27) (in comparison to a value of 3.4 estimated in [34]).

K conductance in hippocampal neurons. A first transistor recording under voltage-clamp with voltage-activated potassium conductances has been reported for cultured hippocampal neurons [33]. The membrane current and the transistor signal are plotted in Figure 8. (The chips were cleaned with a mild dish detergent). Upon depolarization the membrane current exhibits an inactivating A-type potassium conductance and a non-inactivating K-type potassium conductance. The transistor response is quite different. The apparent change of the gate voltage increases rapidly by about 80 μV and further more slowly to a stationary level of 120 μV. Upon deactivation, the voltage drops suddenly by about 80 μV with a subsequent slow relaxation. The fast and slow signal were not distinguished in [33].

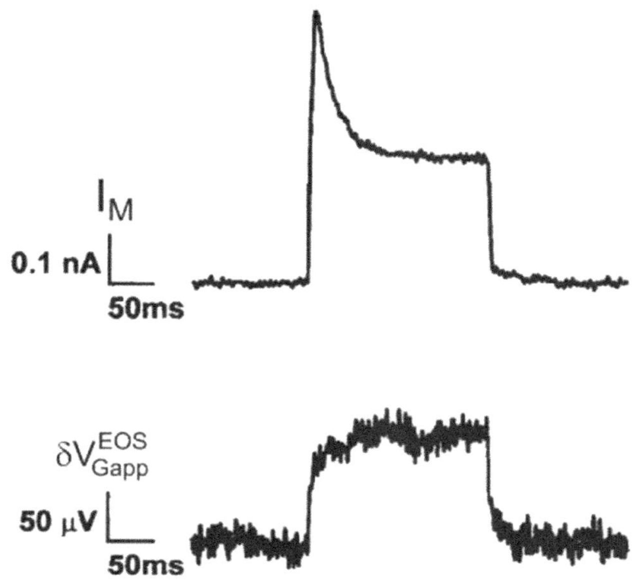

Figure 8. Transistor recording of potassium conductances in cultured hippocampal rat neuron at a constant intracellular voltage [33]. (**Top**) Total membrane current; (**Bottom**) Transistor recording. After a prepolarization for 200 ms to −110 mV, the cell was depolarized to 20 mV and repolarized to −50 mV. The current exhibits an inactivating A-type potassium conductance and a non-inactivating K-type potassium conductance. The transistor signal is assigned to the K-type conductance with a fast current-induced response and a slow component due to the changing potassium concentration. There is no transistor response to the A-type conductance.

It appears that the transistor does not record an A-type conductance in the junction at all, whereas it responds to the non–inactivating K-type conductance with a biphasic waveform that resembles again trace 3 in Figure 5e. Attributing the fast signal to a current-induced response $V_J^{(0)}$=80 µV and considering the stationary membrane current I_M=0.25 nA, we obtain a relation $V_J^{(0)}$=320 kΩ·I_M. With a weighting factor $1/g_J A_M$=90 kΩ (obtained from a capacitive transient), an enhancement of the K-type conductance in the adherent membrane $g_{JM}^{\infty K,K} = 3.5\,\bar{g}_M^{\infty K,K}$ is obtained using Equation (27) (in comparison to a value of 5.7 estimated in [33]), whereas the A-type conductance in the adherent membrane is low with $g_{JM}^{\infty K,A} << \bar{g}_M^{\infty K,A}$.

TRANSISTOR RECORDING OF NEURONAL ACTION POTENTIALS

An action potential (AP) in neurons arises from the dynamics of the membrane voltage and of voltage-activated conductances. In the Hodgkin-Huxley model, the AP is an almost monophasic positive transient $V_M(t)$ of the membrane voltage with a fast rise due to a Na inward current, a slower decay due to a K outward current with an inactivating Na current, and a minor negative tail [30]. Two issues must be considered when transistors are applied for extracellular recording of APs in an intact cell: (i) the dynamics of the voltage in the closed cell must be taken into account; and (ii) a local compensation of the ionic and capacitive currents in the adherent membrane must be avoided.

Current compensation. Let us consider a free neuron with a homogeneous membrane in a homogeneous electrolyte. The global current balance for the whole cell (membrane area A_M) with homogeneous ion conductances g^i_M is described by Equation (28).

$$A_M\left[\sum g^i_M\left(V_M - V^i_0\right) + c_M\frac{dV_M}{dt}\right] = 0 \tag{28}$$

The global current balance implies a local compensation of the ionic and capacitive currents in every area element of the membrane. There is no current flow to the surrounding electrolyte and no extracellular voltage that could be recorded. To obtain an extracellular voltage, the membrane must be inhomogeneous with an enhanced current in one region and a depleted current in another region. For a homogeneous capacitance, that inhomogeneity is created by an inhomogeneous distribution of ion conductances, which is given by the intrinsic structure of a cell or induced by the interaction with the substrate.

Current balance for an adherent cell. We consider a two–domain model of the cell membrane with ion conductances g^i_{FM} in the free membrane (area A_F) and ion conductances g^i_{JM} in the adherent membrane (area A_J). Considering that the conductances are activated for a short time during a neuronal AP, the effect of changing extracellular ion concentrations is disregarded. We use the one-compartment model for constant ion concentrations according to Equation (21) with a quasi-stationary dynamics of the junction voltage. For small voltages $|V_J| << ||V_M - V^i_{M0}||$, we obtain Equation (29a). The current balance of the closed cell is expressed by Equation (29b).

$$\sum g^i_{JM}\left(V_M - V^i_{M0}\right) + c_M\frac{dV_M}{dt} = g_J V_J \tag{29a}$$

$$\sum\left(A_F g^i_{FM} + A_J g^i_{JM}\right)\left(V_M - V^i_{M0}\right) + \left(A_F + A_J\right)c_M\frac{dV_M}{dt} = 0 \tag{29b}$$

Two simple models are considered for inhomogeneous ion conductances with $g^i_{JM} \neq g^i_{FM}$ in the adherent and free membrane.

Leak conductance in soma-substrate junction. In the first model the cell-substrate junction is formed by part of a neuronal soma with low voltage-activated conductances. The interaction with the substrate is assumed to induce a leak conductance g^{leak}_{JM} with $V^{leak}_{M0} = 0$. The junction voltage is determined by the leak current and the capacitive current through the adherent membrane according to Equation (28a). We obtain Equation (30) when we give up the condition $|V_J| << |V_M|$.

$$V_J = \frac{1}{g_J + g^{leak}_{JM}}\left(g^{leak}_{JM}V_M + c_M\frac{dV_M}{dt}\right) \tag{30}$$

For a low leak conductance, the junction voltage is dominated by the capacitive current and proportional to the first derivative of the intracellular voltage. If an almost monophasic AP is created by voltage-activated conductances in the free membrane (*e.g.*, in the axon hillock), a biphasic positive-negative transient $V_J(t)$ is induced (A-type response). If the leak conductance in the junction is large, the voltage transient $V_J(t)$ is proportional to the intracellular voltage, and the response to an AP is monophasic positive (B-type response).

In fact, two kinds of transistor signals had been observed in the original transistor recordings of action potentials for adhering leech neurons [38] with a small response δV^{EOS}_{Gapp} resembling dV_M/dt and a large response δV^{EOS}_{Gapp} resembling $V_M(t)$. The interpretation of these recordings in terms of an A-type and B-type response of the leak model was given in a subsequent paper [39]. It was illustrated by a test experiment.

Defined Gaussian transients $V_M(t)$ with σ=1.25 ms were applied to leech neurons using a whole-cell patch-pipette. A-type and B-type responses of the transistor signal δV^{EOS}_{Gapp} were observed for different cells, as shown in Figure 9 [40]. The two waveforms can be parametrized according to Equation (30) for the A-type response with g^{leak}_{JM}=0.1 mS/cm^2, g_J=25 mS/cm^2 and for the B-type response with an enhanced leak conductance g^{leak}_{JM}=5 mS/cm^2 (responsible for a changed waveform) and a reduced junction conductance g_J=10 mS/cm^2 (responsible for a larger amplitude).

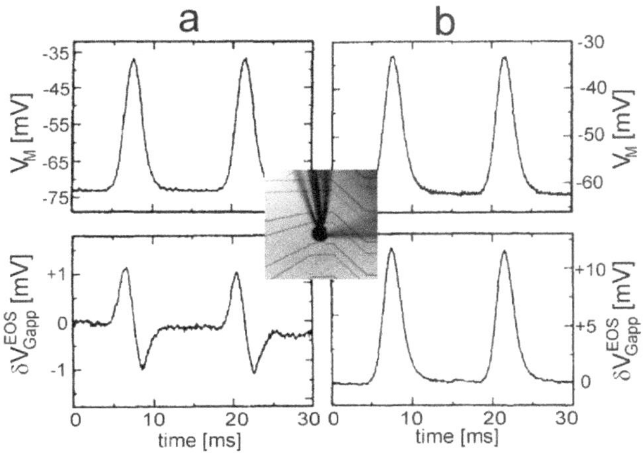

Figure 9. Transistor recording of leech neurons upon stimulation by a Gaussian transient of the intracellular voltage [39]. (**a**) Soma-substrate junction with low leak conductance; (**b**) Soma-substrate junction with an enhanced leak conductance and a reduced junction conductance. The upper graphs display the intracellular voltage $V_M(t)$, and the lower graphs the apparent change of the gate voltage $\delta V^{EOS}_{Gapp}(t)$. The micrograph shows a leech neuron on an EOSFET with a whole-cell patch-pipette used to apply the Gaussian voltage transients and with an impaled micropipette used to record the intracellular voltages.

Reversible transitions between A-type and B-type recordings of action potentials have been observed when neurons were pressed toward the substrate and when the force was released again [41,42]. These transitions were connected with enhanced/reduced leak conductance and a reduced/enhanced junction conductance, respectively. The mechanism is not clear.

Depleted/enhanced voltage-activated conductances. In the second model all voltage-activated conductances in a neuronal membrane are depleted or enhanced in the cell-substrate junction by the same factor μ_J with $g^i_{JM} = \mu_J g^i_{FM}$. When all ion currents are eliminated from Equations (28a) and (28b), considering Equation (21b), the striking Equation (31) is obtained as a relation between the extracellular and the intracellular voltage [43].

$$V_J = \frac{(1-\mu_J)\, A_J A_F}{A_F + \mu_J A_J} \frac{r_J c_M}{\eta_J} \frac{dV_M}{dt} \qquad (31)$$

For complete depletion (μ_J=0), the junction voltage is determined by the capacitive current through the adhesion area A_J. For a homogeneous membrane (μ_J=1), there is no response at all due to local current compensation. For strong enhancement (μ_J>>1), the junction voltage is determined by the ionic current through A_J, which appears in Equation (31) as a negative capacitive current through A_F. The first limit corresponds to a soma–substrate junction without leak conductance as considered above (biphasic positive-negative response to an AP). The second limit corresponds to a junction where the local ion currents dominate with a biphasic negative-positive response to an AP.

Dissociated leech neurons have been attached with their axon stump to an array of EOSFETs [44], and an AP was elicited with an impaled micropipette. A biphasic negative-positive transistor signal was observed, as shown in Figure 10. The current-induced response $\delta V^{EOS}_{Gapp} = V_J$ is induced by the large local ion currents in the axon stump. Due to the global current balance in the cell, the signal reflects the inverted capacitive current through the soma membrane. The asymmetry of the biphasic $V_J(t)$ with a deeper trough corresponds to the asymmetry of an almost monophasic $V_M(t)$.

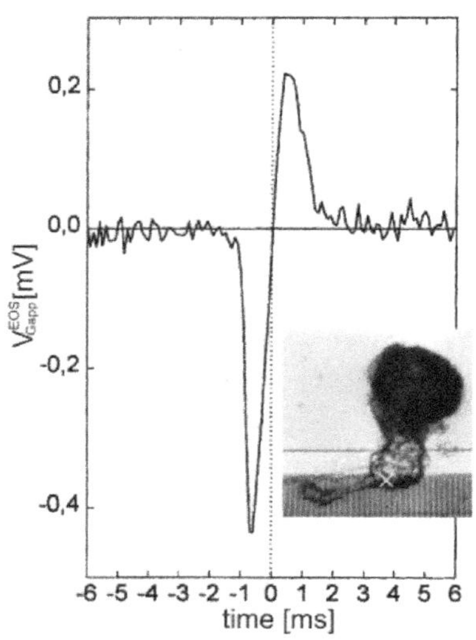

Figure 10. Transistor recording of an action potential beneath the axon stump of a leech neuron (by permission [44]). The apparent change of the gate voltage δV^{EOS}_{Gapp} was observed with the transistor of a linear array as marked in the micrograph. The biphasic negative-positive signal reflects the extracellular voltage $V_J = \delta V^{EOS}_{Gapp}$ as induced by the local ion currents in the cell-substrate junction.

Various waveforms of transistor recordings of APs were reported for cultured hippocampal rat neurons [3,45] as well as for dissociated invertebrate neurons from *Lymnaea* [46] and *Aplysia* [47]. They were assigned to a selective enhancement and depletion of different voltage-activated conductances in the adherent membrane.

TRANSISTOR RECORDING OF PROPAGATING ACTION POTENTIAL

When an action potential propagates along an axon, the different components of the membrane current (capacitive, sodium, potassium) are separated in space. As a consequence, local membrane currents exist, even for a structurally homogeneous axon. An extracellular voltage is induced, which can be probed by transistor recording.

Propagating AP. The membrane current per unit area of an axon is balanced by the changing current along the core according to Equation (32) (cylinder radius a_M, core resistivity ρ_{cell}).

$$g^{Na}_M \left(V_M - V^{Na}_{M0}\right) + g^K_M \left(V_M - V^K_{M0}\right) + c_M \frac{\partial V_M}{\partial t} = \frac{a_M}{2\rho_{cell}} \frac{\partial^2 V_M}{\partial x^2} \qquad (32)$$

The coupling of the membrane voltage with the voltage–activated ion conductances gives rise to a wave $V_M(x \mp v_{AP} t)$ that propagates at a velocity v_{AP} with an invariant waveform. The voltage profile at an arbitrary time is plotted in Figure 11(top) as computed with the original Hodgkin-Huxley parameters $c_M = 1\,\mu F/cm^2$, $\rho_{cell} = 35.4\,\Omega cm$, $g^{\infty Na}_M = 120\,mS/cm^2$, and $g^{\infty K}_M = 36\,mS/cm^2$ (rate constants at 18.5 °C) for a radius a_M=0.5 μm, typical for an unmyelinated mammalian axon. The velocity is v_{AP}=0.86 m/s. The wave is almost monophasic with a steep rise and a smooth drop: The amplitude is 90 *mV* with a full width at half maximum of 260 *μm* in space and 300 *μs* in time. The voltage wave is connected with a current wave $i^{tot}_M(x \mp v_{AP} t)$, which is proportional to the curvature of the intracellular voltage in space as well as in time (because the voltage obeys a wave equation) according to Equation (33).

$$i^{tot}_M(x,t) = \frac{a_M}{2\rho_{cell}} \frac{\partial^2 V_M(x,t)}{\partial x^2} = \frac{a_M}{2\rho_{cell}} \frac{1}{v^2_{AP}} \frac{\partial^2 V_M(x,t)}{\partial t^2} \qquad (33)$$

This current wave induces a drop of the electrical potential in the surrounding. For a cable-substrate junction, the voltage with respect to the bulk electrolyte is partitioned into a voltage drop ΔV_J in the junction and a voltage V_E in the surrounding electrolyte with V_J=ΔV_J+V_E. We consider separately a cable in a homogeneous electrolyte and a cable-substrate junction.

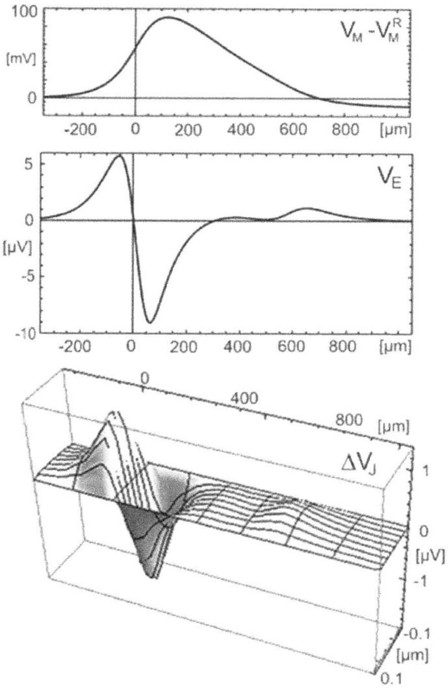

Figure 11. Computed voltage profiles along a cylindrical cable for a propagating action potential at an arbitrary time. (**Top**) Profile of the intracellular voltage with respect to the resting voltage $V_M(x) - V_M^R$ for the Hodgkin-Huxley model with a cable radius a_M=0.5 µm. The action potential propagates in negative x-direction with a velocity of 0.86 *m/s*; (**Center**) Profile of the extracellular voltage $V_E(x,a_M)$ at the surface of the cable in a homogeneous electrolyte with aresistivity ρ_E=100 Ωcm; (**Bottom**) Profile of the extracellular voltage drop $\Delta V_J(x,y)$ in a cable-substrate junction with respect to the surrounding electrolyte for a lane–shaped junction (different scales along and across the junction). For the parameters, see text.

Homogeneous electrolyte. The quasi–stationary response of the electrical potential in a volume conductor (resistivity ρ_E) to a current source density i i_E^{csd} is described by a 3D Poisson equation $-\nabla^2 \varphi_E = \rho_E i_E^{csd}$ [19]. For a propagating AP, the contributions of all currents along the cylinder are superposed. The extracellular voltage $V_E(x,a,t)$ at a radial distance $a \geq a_M$ is given by Equation (34) when the membrane current per unit length $2\pi a_M i_M^{tot}(x,t)$ is placed along the cylinder axis [48]. (A logarithmic divergence of the integral is avoided due to the finite width of the current wave). A reaction of the extracellular voltage is neglected with $|V_E|$<1 mV.

$$V_E(x,a,t) = \frac{\rho_E}{4\pi} \int_{-\infty}^{+\infty} \frac{2\pi a_M i_M^{tot}(x?t)}{\sqrt{a^2 + (x - x')^2}} dx' \tag{34}$$

The current $i_M^{tot}(x,t)$ is obtained from the Hodgkin-Huxley model for a_M=0.5 µm. Equation (34) is evaluated with a resistivity ρ_E=100 Ωcm typical for a culture medium. The profile of the extracellular voltage at the surface of the axon $V_E(x,a_M,t)$ is plotted in Figure 11 (center). The wave is almost biphasic positive-negative and distinctly narrower than the monophasic wave of the intracellular voltage. Peak and trough are separated by 120 *µm* in space and 140 *µs* in time. The amplitudes are 6 *µV* and −9 *µV* [49].

Cable-substrate junction. When a cylinder is attached to a planar substrate, a lane-shaped junction is formed with a distance d_J that increases from a minimum d_J^0 at the midline towards the cylinder radius a_M. We replace that profile by a planar junction with a distance d_J^0 and match its resistance by choosing a half-width $y_J = \sqrt{2a_M d_J^0}$, where the actual distance would be $d_J = 2d_J^0$. This planar junction is driven by the current $i_M^{tot}(x,t)$. The quasi-stationary response of the electrical potential is determined by a 2D Poisson equation $-\nabla^2 \varphi_J = r_J i_{JM}^{tot}$ (Equation (14)). The curvature along the lane is neglected because the membrane current changes smoothly along the axon. The voltage drop $\Delta V_J(x,y,t)$ with a boundary condition $\Delta V_J(x,\pm y_J,t)=0$ is expressed by the stationary parabolic response across the lane-junction according to Equation (35).

$$\Delta V_J(x,y,t) = \frac{\rho_J y_J^2}{2 d_J} \left(1 - \frac{y^2}{y_J^2}\right) i_M^{tot}(x,t) \tag{35}$$

For an adhering axon, we choose d $d_J^0 = 10$ nm as the smallest distance reported for a cell membrane on a silicon chip [50]. The half-width of the junction is y_J=0.1 µm for a_M=0.5 µm. With ρ_J=100 Ωcm, the sheet resistance is r_J=100 MΩ. The profile $\Delta V_J(x,y)$ at an arbitrary time is plotted in Figure 11 (bottom). The response along the lane is almost biphasic positive-negative with peak and trough separated by 120 *µm* in space and 140 *µs* in time. The amplitudes in the center of the junction are 1.3 *µV* and −1.9 *µV*.

The total voltage in the cable-substrate junction is obtained by superposing the voltage drop in the junction and the voltage in the surrounding electrolyte. Considering that the current of an adhering cable flows into one half-space, the amplitudes of the voltage in the electrolyte are 12 *µV* and −18 *µV*. The contribution of the electrolyte is by an order of magnitude larger than the contribution of the cable-substrate junction with 1.3 *µV* and −1.9 *µV*.

Transistor recording of axon in retina. The extracellular voltage of a propagating AP has been studied by transistor recording of axons in the rabbit retina [51]. The retina was mounted with ganglion cells down on an array of EOMOS transistors. Maps of the transistor response δV_{Gapp}^{EOMOS} are displayed in Figure 12. The velocity of the AP is 1.5 *m/s*. The transistor response is biphasic positive-negative with peak and trough separated by 120 *µm* in space and 80 *µs* in time. The amplitudes are about 500 *µV* and −675 *µV*.

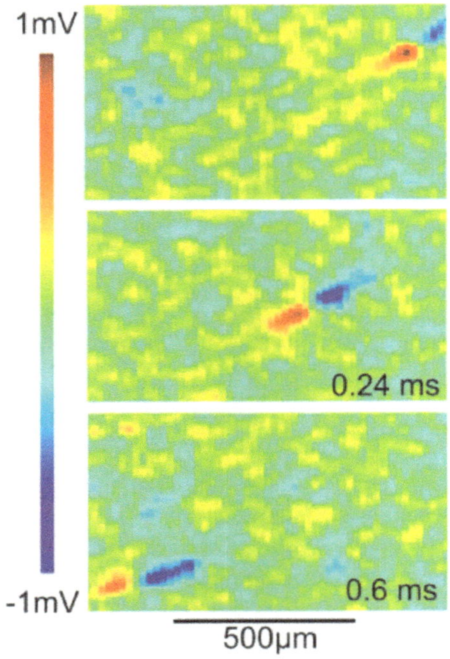

Figure 12. Transistor recording of propagating action potential in a rabbit retina [51]. Color-coded maps of the apparent changes of the gate voltage δV_{Gapp}^{EOMOS} recorded with an array of EOMOS transistors at three different times. The action potential propagates from right to left at a velocity of 1.5 *m/s*. The waveform is biphasic positive-negative and represents the profile of a current-induced extracellular voltage with δV_{Gapp}^{EOMOS} =VE. Peak and trough are separated by 120 μm in space corresponding to a separation of 80 μs in time.

The asymmetric biphasic waveform and the separation of peak and trough correspond to the computation shown in Figure 11. The agreement indicates that the transistor response is due to a current-induced response with $\delta V_{Gapp}^{EOMOS} = V_E$. The velocity may be matched by choosing a larger radius or lower intracellular resistivity according to the relation $v_{AP} = \sqrt{K \cdot a_M / 2\rho_{cell}\, c_M}$ with K=10.47ms^{-1} [30]. Striking is the large amplitude of the recordings. Two issues must be considered: (i) for an axon near the substrate, even without close adhesion, the whole membrane current flows into a half-space with the mounted retina; and (ii) the retina has a resistivity of about 1250 Ωcm [52]. Thus, the result of the computation shown in Figure 11 must be multiplied by a factor 25, yielding amplitudes of 150 μV and −240 μV. The remaining difference with respect to the experiment may due to a larger axon radius or higher membrane conductances.

Transistor recording with adhering axon in culture. The extracellular voltage of a propagating AP has been studied for adhering axons of cultured mammalian neurons using an array of metal electrodes on a CMOS chip (see Appendix B) [53]. The amplitudes of the recordings were in the range of 10 μV. Peak and trough of biphasic signals were separated by about 150 μs. These values are in agreement with the computation shown in Figure 11.

The propagation of an AP along an adhering mammalian axon has also been probed with an array of nanowire transistors on a planar substrate [54]. Almost monophasic positive transients of the nanowire current were reported corresponding to an extracellular voltage in the range of −85 mV [55] or −6.6 mV, respectively [56]. Even the lower estimate is four orders of magnitude larger than computed for the voltage drop in a cable-substrate junction and three orders of magnitude larger than computed for the voltage in the surrounding electrolyte. It is difficult to see how the parameters could be changed to avoid that discrepancy. Moreover, the waveform of the reported transients, which resemble the inverted intracellular voltage of a local AP, are difficult to reconcile with the waveform of the extracellular voltage of a propagating AP [57].

RECEPTOR-CELL-TRANSISTOR SENSOR

Transistor recording with intact cells is an approach to develop biosensors with cellular receptors that bind agonists or pharmaceuticals. The receptor may activate an ion conductance in the plasma membrane, which gives rise to a current-induced response of the transistor. Considering that ligand-activated conductances remain open for a long time due to slow inactivation, a change of the extracellular ion concentrations is expected. An evaluation in terms of a current-induced response requires a gate oxide with a weak ion sensitivity. First, the concept is considered with the one-compartment model for constant ion concentrations (quasi-stationary response, small voltages), and then the prototype with a serotonin receptor is described.

Receptor-activated conductance. We assume that the receptor-activated conductance with a high reversal voltage V_{M0}^{rec} dominates the membrane. According to the two-domain model of Equation (29), the current balances in the junction and in the cell are expressed by Equation (36), with a receptor-activated conductance g_{JM}^{rec} in the adherent membrane and an average conductance $\overline{g}_{M}^{rec}$.

$$g_{JM}^{rec}\left(V_M - V_{M0}^{rec}\right) + c_M \frac{dV_M}{dt} = g_J V_J \qquad (36a)$$

$$\overline{g}_{M}^{rec}\left(V_M - V_{M0}^{rec}\right) + c_M \frac{dV_M}{dt} = 0 \qquad (36b)$$

When the conductance is activated, the intracellular voltage relaxes rapidly towards the reversal voltage according to Equation (36b). According to Equation (36a), there is no stationary junction voltage, only a short transient, which is difficult to detect.

Supplementary K conductance. A stationary driving voltage $V_M - V_{M0}^{rec}$ is established if a voltage-activated potassium conductance g_M^K is added with $V_{M0}^K < V_{M0}^{rec}$. The junction voltage results from a superposition of the ion currents of both conductances according to Equation (37a). According to Equation (37b). The intracellular voltage approaches a resting voltage $V_{M0}^K < V_M^R < V_{M0}^{rec}$.

$$g_{JM}^{rec}\left(V_M^R - V_{M0}^{rec}\right) + g_M^K\left(V_M^R - V_{M0}^K\right) = g_J V_J \qquad (37a)$$

$$\overline{g}_{M}^{rec}\left(V_M^R - V_{M0}^{rec}\right) + g_M^K\left(V_M^R - V_{M0}^K\right) = 0 \qquad (37b)$$

If the potassium current is eliminated, we obtain Equation (38), where the junction voltage is determined by the driving voltage $V_M^R - V_{M0}^{rec}$ and the inhomogeneity of the receptor-activated conductance.

$$\left(g_{JM}^{rec} - \overline{g}_M^{rec}\right)\left(V_M^R - V_{M0}^{rec}\right) = g_J V_J \qquad (38)$$

On this basis, the transistor sensorics is described as follows: a cell stays at its low natural resting voltage with a deactivated receptor conductance and a deactivated K conductance. The agonist is added, and current flows into the cell through the receptor-activated conductance. The intracellular voltage rises such that the K conductance is activated. K current flows out of the cell, and a resting voltage $V_{M0}^K < V_M^R < V_{M0}^{rec}$ is established. The superposition of the two ion currents through the adherent membrane yields a junction voltage $V_J>0$ for a depletion $g_{JM}^{rec} < \overline{g}_M^{rec}$ and a junction voltage $V_J<0$ for an enhancement $g_{JM}^{rec} > \overline{g}_M^{rec}$. Finally, the receptor-activated conductance is inactivated, the intracellular voltage drops, the K conductance is deactivated, and the natural resting voltage is reestablished.

Inhomogeneity induced by diffusion. It is not necessary that the inhomogeneity of a receptor-activated conductance is structurally implemented. When an agonist is added to the bath, it first reaches the free membrane of an adherent cell and, with a delay, the adherent membrane due to diffusion of the agonist along the cell-substrate junction. The first phase with an activated free membrane is equivalent to $g_{JM}^{rec} < \overline{g}_M^{rec}$. The second phase with an activated adherent membrane and an inactivated conductance in the free membrane is equivalent to $g_{JM}^{rec} > \overline{g}_M^{rec}$. Prerequisite of such an effect is a matched dynamics of diffusion and inactivation.

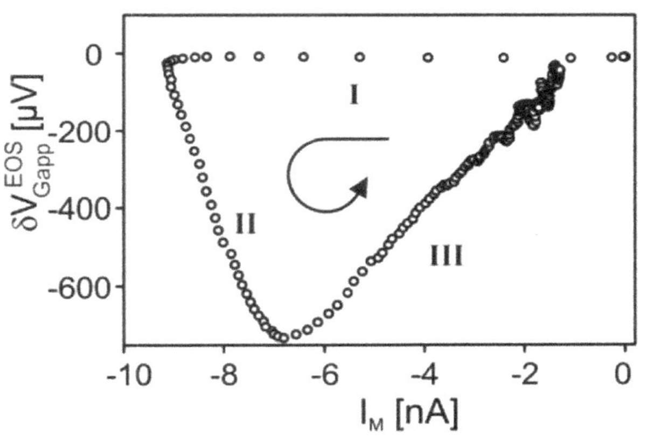

Figure 13. Transistor recording of the serotonin receptor 5-HT3A in a HEK293 cell for constant intracellular voltage (by permission [5]). The apparent change of the gate voltage δV_{Gapp}^{EOS} is plotted versus the whole-cell current IM upon fast addition of $100\mu M$ serotonin at an intracellular voltage of -120 *mV*. The time direction is indicated by the arrow (data interval 2.5 *ms*). The recording is obtained with a low ion sensitivity of the gate oxide. Three phases are distinguished: (I) increasing whole-cell current without transistor signal (about 20 *ms*); (II) decreasing whole-cell current and increasing transistor signal (about 70 *ms*); and (III) decreasing whole-cell current and decreasing transistor signal.

The delayed activation of the adherent membrane by diffusion has been demonstrated for the serotonin receptor 5-HT3A in a HEK293 cell under voltage-clamp [58]. A transistor with a weak ion sensitivity was used as described above. Serotonin (100 μM) was added with a double-barreled pipette. The apparent change of the gate voltage δV_{Gapp}^{EOS} is plotted in Figure 13versus the total membrane current IM. Three phases can be distinguished: (i) the membrane inward current increases for 20*ms* without response of the transistor due to an activation of the free membrane; (ii) the inward current decreases and the apparent change of the gate voltage increases for 70 *ms* due to inactivation of the free membrane and activation of the adherent membrane; and (iii) both the membrane current and the transistor signal disappear due to complete inactivation.

Transistor sensorics with serotonin receptor. A complete biosensor with the 5-HT3A receptor has been implemented with a supplementary Kv1.3 channel in HEK293 cells on a transistor with a low ion sensitivity of the gate [35]. A result is shown inFigure 14. When serotonin is applied, a positive transistor signal of about 600 μV appears. Within about 100 *ms,* it changes to a negative signal of about -450 μV, which relaxes slowly. The whole signal is abolished if the specific antagonist tropisetron is added.

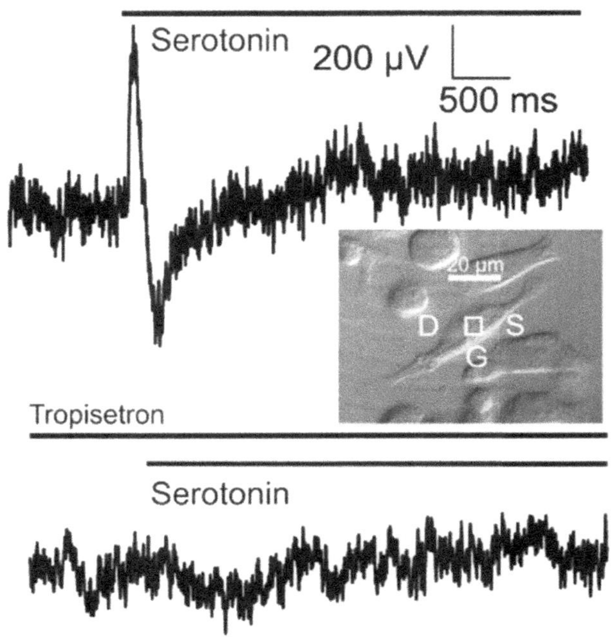

Figure 14. Transistor recording of the serotonin receptor 5-HT3A with the potassium channel Kv1.3 in HEK293 cell [35] for a low ion sensitivity of the gate oxide. The upper trace shows the apparent change of the gate voltage δV_{Gapp}^{EOS} upon application of $100\mu M$ serotonin. The biphasic transient reflects the current-induced extracellular voltage in the cell-substrate junction. The lower trace is a control experiment with the same cell in the presence of 1 μM tropisetron, a selective antagonist of the 5-HT3A receptor. The micrograph shows the position of cell and transistor (source, gate, drain).

The biphasic waveform of the transistor record is expected for a current-induced response with δVEOSGapp=VJ. The two phases correspond to a switched inhomogeneity of the membrane from depletion to enhancement of the receptor-activated conductance in the junction, as caused by diffusion and inactivation. The asymmetry of the positive and negative wave and the small amplitudes are due to the strong overlap of the two phases as the adherent membrane becomes activated at a time when the free membrane is only partially inactivated.

CONCLUSIONS

Current-induced transistor signals with electrogenic cells in a cell-substrate junction are well understood in terms of a planar capacitive core-coat conductor as it is documented by a wide range of experiments. Apparent changes of the gate voltage are caused by the current-induced extracellular voltage in the junction beyond the electrical double layer of the gate oxide. The signals are determined by the structure of the cell-substrate junction (area, cleft width, resistivity) and by the ion conductances of the cell with an inhomogeneous distribution in the adherent and free membrane. Interface-induced signals due to changing ion concentration are important for longer activations of voltage-gated conductances under voltage-clamp. They play a minor role for transient activations of ion conductances as with firing neurons and can be avoided if the ion sensitivity of the gate oxide is suppressed.

Three issues should be considered with respect to a development of a reliable current-induced transistor sensorics: (i) The inhomogeneous distribution of ion conductances is important. A distinct inhomogeneity of ion conductances in the adherent membrane is documented with recombinant ion channels in HEK293 cells, with cultured mammalian neurons, and with dissociated invertebrate neurons. However, these effects were given without control. The cellular mechanisms for the localization of membrane proteins must be taken into account and applied for the development of designed cell-substrate junctions; (ii) The passivation of the gate oxide with respect to ion binding is important. Chemical processes must be developed that block the interaction of the gate with ions, a goal that is opposite to the goal of classical ISFET sensorics; (iii) The position of recording transistors with respect to the adhering cells is important. Accidental positions and manual positioning with a pipette are not adequate. Controlled local adhesion with a chemically modified surfaces must be considered for optimal recordings.

Abbreviations

The following abbreviations are used in this manuscript:

CMOS	complementary metal oxide semiconductor
EOMOS	electrolyte oxide metal oxide semiconductor
EOSFET	electrolyte oxide semiconductor field–effect transistor
FLIC	fluorescence interference contrast
HEK239	human embryonic kidney
ISFET	ion sensitive field-effect transistor
MOSFET	metal oxide semiconductor field-effect transistor

Appendix A. Flatband Voltage of EOSFET

The flatband voltage of an EOSFET can be described in analogy to a MOSFET by introducing workfunctions (for the semiconductor and the redox solution of the reference electrode) and surface dipole voltages [6]. The workfunctions are defined as differences of the electrochemical potential of electrons in the vacuum (just beyond the atomic forces) and in the material with $\Phi_k = \left(\mu^{el,vac} - \mu_k^{el}\right) - e_0\left(\varphi_k^{vac} - \varphi_k^{\infty}\right)$, and the surface dipole potentials as negative differences of the electrical potential across the free surfaces with $-\chi_i = \varphi_i^{vac} - \varphi_i^S$. We eliminate the difference $\mu_{red/ox}^{el} - \mu_{Si}^{el}$ in Equation (1), use the approximations $\chi_{red/ox} \approx \chi_E$ and $\varphi_k^S \approx \varphi_k^{\infty}$, express the difference $\varphi_E^{\infty} - \varphi_{Si}^{\infty}$ according to Equation (2), and introduce the differences of the electrical potential across the quasi-Helmholtz layer, across the Gouy-Chapman layer, and in the free electrolyte. For $\Delta\varphi_{Si}^S = 0$ and $\Delta\varphi_E = 0$ the flatband voltage of a p-channel transistor is given by Equation (A1).

$$V_{EOS}^{FB} = \left(\Phi_{red/ox} - \Phi_{Si}\right)/e_0 - \left[\Delta\phi_{GCh} + \Delta\phi_{Hh} - (\chi_{Ox} - \chi_E)\right] - Q_{OxSi}/C_{Ox} + \left[\Delta\phi_{OxSi} - (\chi_{Ox} - \chi_{Si})\right] \quad \text{(A1)}$$

The two brackets account for changes of the difference of the electrical potential between oxide and electrolyte and between oxide and semiconductor, respectively, when the interfaces are formed from the free surfaces. (Formally they describe a difference of the electrical potential in the vacuum as it can be seen when the transfer of a probe charge across the interface is identified with an indirect transfer through the two free surfaces).

Unfortunately, Equation (A1) cannot be used to estimate the flatband voltage in analogy to a MOSFET. There, effective workfunctions from the bulk gate metal and from the bulk semiconductor, respectively, across the interface to the vacuum of the gate oxide can be measured by a photoelectronic approach. For an EOSFET, the workfunction of the redox solution can be obtained from the redox potential [5,6]. However, the jump of the electrical potential across the oxide/electrolyte interface is unknown. It may be fairly large due to the strong hydration of the polar surface [59]. The flatband voltage of the EOSFET must be measured.

Appendix B. EOSFET and other Sensors

The concept of the current-induced transistor sensorics of the EOSFET with an exposed gate oxide can be applied for other potentiometric sensors, which are used for electrogenic cells to record voltage changes with respect to a reference electrode: (i) With EOMOS transistors on a CMOS chip [60], a thin metal gate is intercalated between the oxide exposed to the electrolyte and the genuine gate oxide. It is used to adjust the working point in calibration phases, but is free floating in recording phases. Every change of the electrical potential in the free electrolyte and at the oxide/electrolyte interface has the same effect as with an EOSFET; (ii) With metal electrodes on a CMOS chip [53] the exposed electrodes are in close contact (and electronically equilibrated) with the gate of a MOSFET. If a polarizable metal is used (*i.e.*, without electronic equilibration with the electrolyte), a changing electrical potential in the free electrolyte or at the metal/electrolyte interface is projected onto the electrical potential at the surface of the gate oxide. Of course, the interface-induced signal differs from an EOSFET due to the different chemistry of the metal surface; (iii) Traditional planar electrodes [61] are electronically equilibrated with a metal wire that leads to the amplifier. For a polarizable electrode, a changing electrical potential in the electrolyte or at the interface is projected onto the wire. Perturbations of amplitude and waveform may occur due to the large separation of electrode and amplifier, and it is advisable to apply suitable calibration voltages to the reference electrode. A certain disadvantage of all arrays with uncoated electrodes is the chemically heterogeneous surface of the substrate, which can affect tissues and cultured cells.

Appendix C. Capacitive Core–Coat Conductor and Poisson–Nernst–Planck Approach

A cell-substrate junction has a dielectrically inhomogeneous structure with a profile of the dielectric constant $\varepsilon(z)$ from semiconductor to oxide to electrolyte to membrane to cytoplasm. With a permittivity $\varepsilon\varepsilon_0$, and an electrical potential φ, the relation between the electrical displacement and the space charge density ρ^{ch} is expressed by Equation (C1).

$$-\nabla(\varepsilon\,\varepsilon_0\,\nabla\varphi) = \rho^{ch} \tag{C1}$$

A short exercise shows that the difference of the electrical potential across an insulator between two electroneutral bulk electrolytes is determined by thin opposite space charge layers as expressed by a capacitor relation. This result is the basis for the physics of biomembranes. It is applicable for a cell-substrate junction with its two insulating coats if the electrolyte film is distinctly wider than a Debye length of the electrolyte: a cell-substrate junction is a planar capacitive core-coat conductor in analogy to the linear capacitive core-coat conductor of a neuronal cable [9,10,11,12,13]. The theory with changing ion concentrations is described by a drift-diffusion equation (Equation (6)) for the balance of the ion flux and by a capacitor relation (Equation (7)) between the electrical charge and the electrical potential.

In a Poisson-Nernst-Planck approach for a cell–substrate junction [16], the drift-diffusion (Nernst-Planck) equation is the same as for a capacitive core–coat conductor. However, the electrical charge is assigned to the electrolyte film with a space charge density ρ_E^{ch}, which determines the curvature of the electrical potential along the junction according to the (Poisson) Equation (C2).

$$-\varepsilon_E\,\varepsilon_0\nabla^2\varphi_E = \rho_E^{ch} \tag{C2}$$

The Poisson-Nernst-Planck approach disregards the capacitive coats of the cell-substrate junction. It describes the slice of an infinite homogeneous volume conductor. So it is certainly incomplete. As a matter of fact, the current-induced voltage in a cell-substrate junction was computed in terms of the capacitive core-coat conductor using the one-compartment (circuit) model for constant ion concentrations (Equation (1) of [16]). In a separate step, the changing ion concentrations were computed with the Nernst-Planck and the Poisson equation (Equations (A1)–(A5) of [16]). The effect of the changing ion concentrations on the current-induced response was not taken into account. This separate computation of the voltage and the concentrations is arbitrary and without any foundation. Moreover, it is also inconsistent: in the circuit model, the charge is proportional to the voltage, and the electrical time constant is in the microsecond range, whereas in the Poisson-Nernst-Planck approach, the charge is proportional to the curvature of the voltage profile, and the electrical time constant is in the nanosecond range.

Despite its deficiencies, the approach yielded results (for the voltage-clamp situation) that are similar (though not identical) to those obtained with the capacitive core-coat conductor for two reasons: (i) the missing concentration dependence of the current-induced voltage (trace 1 in Figure 5e) can be compensated by the interface-induced response with an appropriate ion sensitivity of the gate oxide (trace 3 in Figure 5e); and (ii) the Poisson equation, which substitutes the curvature of the voltage profile by the electrical charge has a similar effect as the formal substitution $-\nabla^2 \Rightarrow \eta_J/A_J$ used to derive the one-compartment model. A relation for the drift component of the concentration dynamics is obtained from Equations (23) and (24b) that is analogous to Equation (A4) of [16] when the permittivity of the core electrolyte is replaced by the capacitance of the coat as $\varepsilon E\varepsilon_0 \Rightarrow c_J A_J/d_J\eta_J$. The time constant is in the range of 1 *µs* instead of 0.1 *ns*, a difference that is irrelevant when the quasi–stationary state is evaluated. Thus the approach of [16] may be considered as a rough approximation of the capacitive core-coat conductor.

REFERENCES AND NOTES

1. Bergveld, P. Thirty years of ISFETOLOGY. What happened in the past 30 years and what may happen in the next 30 years. *Sens. Actuators B Chem.* **2003**, *88*, 1–20.
2. Matsumoto, A.; Miyahara, Y. Current and emerging challenges of field-effect transistor based bio-sensing. *Nanoscale* **2013**, *5*, 10702–10718.
3. Voelker, M.; Fromherz, P. Signal transmission from individual mammalian nerve cell to field-effect transistor. *Small* **2005**, *1*, 206–210.
4. Pierret, R.F. *Semiconductor Device Fundamentals*; Addison-Wesley-Longman: Reading, PA, USA, 1996.
5. Gerischer, H.; Ekardt, W. Fermi levels in electrolytes and the absolute scale of redox potentials. *Appl. Phys. Lett.* **1983**, *43*, 393–395.
6. Fromherz, P. Threshold voltage of the EOSFET: Reference electrode and oxide-electrolyte interface. *Phys. Status Solidi A* **2012**, *209*, 1157–1162.
7. Bousse, L.; De Rooij, N.F.; Bergveld, P. The influence of counter-ion adsorption on the ψ_0/pH characteristics of insulator surfaces. *Surface Sci.* **1983**, *135*, 479–496.
8. Lambacher, A.; Fromherz, P. Luminescence of dye molecules on oxidized silicon and fluorescence interference contrast microscopy of biomembranes. *J. Opt. Soc. Am. B* **2002**, *19*, 1435–1453.
9. Jack, J.J.B.; Noble, D.; Tsien, R.W. *Electrical Current Flow in Excitable Cells*; Clarendon: Oxford, UK, 1975.

10. Quian, N.; Sejnowsky, T.J. An electro-diffusion model for computing membrane potentials and ionic concentrations in branching dendrites, spines and axons. *Biol. Cybern.* **1989**, *62*, 1–15.

11. Weis, R.; Fromherz, P. Frequency dependent signal-transfer in neuron-transistors. *Phys. Rev. E* **1997**, *55*, 877–889.

12. Straub, B. Kopplung von Feldeffekttransistoren mit Rekombinanten Ionenkanälen. Ph.D Thesis, Technical University München, München, Germany, 2001.

13. Brittinger, M.; Fromherz, P. Field-effect transistor with recombinant potassium channels: Fast and slow response by electrical and chemical interaction. *Appl. Phys. A* **2005**, *81*, 439–447.

14. The positive inward current of the junction corresponds to the outward current of the cell, which is positive by convention.

15. The total number of ions and the number of free ions in the junction are not distinguished. With a capacitance of $1 \mu F/cm^2$ and a voltage of $10\ mV$ the charge per unit area in a 50 *nm* thick electrolyte corresponds to a concentration of 20 μM of monovalent ions, which is small compared with the extracellular ion concentrations of 10–100 *mM*.

16. Wrobel, G.; Seifert, R.; Ingebrandt, S.; Enderlein, J.; Ecken, H.; Baumann, A.; Kaupp, U.B.; Offenhäusser, A. Cell-transistor coupling: Investigation of potassium currents recorded with p- and n-channel FETs. *Biophys. J.* **2005**, *89*, 3628–3638.

17. Carslaw, H.S.; Jaeger, J.C. *Conduction of Heat in Solids*, 2nd ed.; Clarendon: Oxford, UK, 1959.

18. A negative curvature of the electrical potential along the core as displayed in Figure 2 must be compensated by a positive curvature across the core for an electroneutral electrolyte without space charge according to the Poisson equation of electrostatics. For a narrow core the corresponding potential profile across the core is very shallow. It is disregarded in a 2D description of the planar core-coat conductor, and profiles of the electrical potential along the core and along the coats are not distinguished.

19. Mitzdorf, U. Current-source density method and application in cat cerebral cortex: Investigation of evoked field–potentials and EEG phenomena. *Physiol. Rev.* **1985**, *65*, 37–100.

20. The physical meaning of the Poisson equation for electrical potential and current-source density is different for a volume conductor and a core-coat conductor. For a volume conductor the Poisson equation describes a quasi-stationary profile of the electrical potential after relaxation of the space charge density (within about 0.1 *ns*). For a core-coat conductor it refers to a quasi-stationary potential profile (built up within about 1 μs) that is compatible with a charging of the coat capacitance and with an Ohmic potential drop along the core.

21. Weis, R.; Müller, B.; Fromherz, P. Neuron adhesion on a silicon chip probed by an array of field-effect transistors. *Phys. Rev. Lett.* **1996**, *76*, 327–330.

22. Braun, D.; Fromherz, P. Imaging neuronal seal resistance on silicon chip using fluorescent voltage-sensitive dye. *Biophys. J.* **2004**, *87*, 1351–1358.

23. Gleixner, R.; Fromherz, P. The extracellular electrical resistivity in cell adhesion. *Biophys. J.* **2006**, *90*, 2600–2611.

24. The average resistance r_J = 7.7 $M\Omega$ of HEK293 cells on fibronectin at an average distance d_J = 74 *nm* corresponds to a resistivity $_J$ = 57 Ωcm. The average sheet resistance r_J = 12 $M\Omega$ of rat neurons on poly-lysine at an average distance d_J= 54 *nm* corresponds to a resistivity $_J$ = 65 Ωcm. Within the variability of the measurements, these resistivities are in agreement with the resistivity $_E$ = 68 Ωcm of the bulk electrolyte.

25. Voelker, M.; Fromherz, P. Nyquist noise of cell adhesion detected in a neuron-silicon transistor. *Phys. Rev. Lett.* **2006**,*96*, 228102.

26. Zeitler, R.; Fromherz, P. The thermal voltage fluctuations in the planar core-coat conductor of a neuron-semiconductor interface. *Langmuir* **2013**, *29*, 6084–6090.

27. Regehr, W.G.; Pine, J.; Cohan, C.S.; Mischke, M.D.; Tank, D.W.J. Sealing cultured invertebrate neurons to embedded dish electrodes facilitates long-term stimulation and recording. *Neurosci. Meth.* **1989**, *30*, 91–106.

28. With regard to the single node of the equivalent electrical circuit, the term "point-contact model" had been introduced 11 for the one-compartment model, a rather unfortunate expression that gave rise to the misinterpretation of a structural point-like cell-substrate junction.

29. The one–compartment model with changing ion concentrations as defined here differs from the model suggested in 13 in the diffusion term, which is now expressed by Fick's first law. A comparison of numerical solutions with the full planar core-coat conductor exhibits a very good agreement with respect to the dynamics and the amplitudes of the voltage and of the ion concentrations.

30. Hodgkin, A.L.; Huxley, A.F. A quantitative description of membrane current and its application to conduction and excitation of nerve. *J. Physiol.* **1952**, *117*, 500–544.

31. The one-compartment model is used here because the profiles of the voltage and of the concentrations along the junction are not relevant for the issues considered.

32. Schmidtner, M.; Fromherz, P. Functional Na^+ channels in cell adhesion probed by transistor recording. *Biophys. J.***2006**, *90*, 183–189.

33. Vassanelli, S.; Fromherz, P. Transistor probes local potassium conductances in the adhesion region of cultured rat hippocampal neurons. *J. Neurosci.* **1999**, *19*, 6767–6773.

34. Straub, B.; Meyer, E.; Fromherz, P. Recombinant Maxi-K channels on transistor, a prototype of iono–electronic interfacing. *Nat. Biotech.* **2001**, *19*, 121–124.

35. Peitz, I.; Fromherz, P. Electrical interfacing of neurotransmitter receptor and field-effect transistor. *Eur. J. Phys. E* **2009**,*30*, 223–231.

36. Fromherz, P. Assembling of proteins at lipid monolayers.

In *Electron Microscopy at Molecular Dimensions*; Baumeister, W., Vogell, W., Eds.; Springer: Berlin, Germany, 1980; pp. 338–349.

37. Woias, P.; Meixner, L.; Amandi, D.; Schönberger, M. Modelling the short-time response of ISFET sensors. *Sens. Actuators B Chem.* **1995**, *24*, 211–217.
38. Fromherz, P.; Offenhäusser, A.; Vetter, T.; Weis, J. A neuron-silicon junction: A Retzius cell of the leech on an insulated-gate field-effect transistor. *Science* **1991**, *252*, 1290–1293.
39. Fromherz, P.; Müller, C.O.; Weis, R. Neuron-transistor: Electrical transfer function measured by the patch-clamp technique. *Phys. Rev. Lett.* **1993**, *71*, 4079–4082.
40. An inverted voltage scale was used in 39 for the transistor signal, which corresponds to an apparent change of the bias voltage of the semiconductor.
41. Jenkner, M.; Fromherz, P. Bistability of membrane conductance in cell adhesion observed in a neuron-transistor. *Phys. Rev. Lett.* **1997**, *79*, 4705–4708.
42. Cohen, A.; Shappir, J.; Yitchaik, S.; Spira, M.E. Reversible transition of extracellular field potential recordings to intracellular recordings of action potentials generated by neurons grown on transistors. *Biosens. Biolelectron.* **2008**, *23*, 811–819.
43. Fromherz, P. Extracellular recording with transistors and the distribution of ionic conductances in a cell membrane.*Eur. Biophys. J.* **1999**, *28*, 254–258.
44. Schätzthauer, R.; Fromherz, P. Neuron-silicon junction with voltage-gated ionic currents. *Eur. J. Neurosci.* **1998**, *10*, 1956–1962.
45. Vassanelli, S.; Fromherz, P. Transistor records of excitable neurons from rat brain. *Appl. Phys. A* **1998**, *66*, 459–464.
46. Zeck, G.; Fromherz, P. Noninvasive neuroelectronic interfacing with synaptically connected snail neurons immobilized on a semiconductor chip. *Proc. Natl. Acad. Sci. USA* **2001**, *98*, 10457–10462.
47. Cohen, A.; Shappir, J.; Yitchaik, S.; Spira, M.E. Experimental and theoretical analysis of neuron–transistor hybrid electrical coupling: the relationships between the electro-anatomy of cultured *Aplysia* neurons and the recorded field potentials. *Biosens. Bioelectron.* **2006**, *22*, 656–663.
48. Plonsey, R.; Barr, R.C. *Bioelectricity, a Quantitative Approach*, 2nd ed.; Springer: Berlin, Germany, 2007.
49. At a radial distance $a = 2\ a_M$ the voltage is still 84% of the maximum at the cable surface due to a logarithmic decay of the voltage near the cable.
50. Braun, D.; Fromherz, P. Fluorescence interference-contrast microscopy of cell adhesion on oxidized silicon. *Appl. Phys. A* **1997**, *65*, 341–348.
51. Zeck, G.; Lambacher, A.; Fromherz, P. Axonal transmission in the retina introduces a small dispersion of relative timing in the ganglion cell population response. *PLoS ONE* **2011**, *6*, e20810.
52. Zeitler, R.; Fromherz, P.; Zeck, G. Extracellular voltage noise probes the interface between retina and silicon chip.*Appl. Phys. Lett.* **2011**, *99*, 263702.
53. Bakkum, D.J.; Frey, U.; Radivojevic, M.; Russell, T.L.; Müller, J.; Fiscella, M.; Takahasi, H.; Hierlemann, A. Tracking axonal action potential propagation on a high–density microelectrode array across hundreds of sites. *Nat. Commun.***2013**, *4*, 2181.
54. Patolsky, F.; Timko, B.P.; Yu, G.; Fang, Y.; Greytak, A.B.; Zheng, H.; Lieber, C.M. Detection, stimulation, and inhibition of neuronal signals with high-density nanowire transistor arrays. *Science* **2006**, *313*, 1100–1104.
55. Fromherz, P.; Voelker, M. Comment on "Detection, stimulation, and inhibition of neuronal signals with high-density nanowire transistor arrays". *Science* **2009**, *323*, 1429b.
56. Timko, B.P.; Patolsky, F.; Lieber, C.M. Response to comment on "Detection, stimulation, and inhibition of neuronal signals with high-density nanowire transistor arrays". *Science* **2009**, *323*, 1429c.
57. It may be argued that the structure of an axon-nanowire-substrate junction is not sufficiently well known to estimate the response of a nanowire. Two extreme situations are a nanowire that is directly attached to the insulating core of the membrane and a nanowire that penetrates the membrane. In the first case the probed voltage would be extracellular as it is described by a cable-substrate junction. In the second case the probed voltage would be intracellular and opposite to the reported signal.
58. Peitz, I.; Völker, M.; Fromherz, P. Recombinant serotonin receptor on transistor, a prototype for cell-based biosensorics. *Angew. Chem. Int. Ed.* **2007**, *46*, 5787–5790.
59. Janacek, J.; Netz, R.R. Interfacial water at hydrophobic and hydrophilic surfaces depletion versus adsorption. *Langmuir* **2007**, *23*, 8417–8429.
60. Lambacher, A.; Jenkner, M.; Merz, M.; Eversmann, B.; Kaul, R.A.; Hofmann, F.; Thewes, R.; Fromherz, P. Electrical imaging of neuronal activity by multi-transistor-array (MTA) recording at 7.8 μm resolution. *Appl. Phys. A* **2004**, *79*, 1607–1611.
61. Gross, G.W.; Wen, Y.; Lin, J.W. Transparent indium-tin oxide electrode patterns for extracellular multisite recording in neuronal cultures. *J. Neurosci. Meth.* **1985**, *15*, 243–252.

INDEX

M

N

O

P

R

S

T

W